Spice Crops

Spice Crops

E.A. Weiss

Agricultural Adviser
Eaglemont, Victoria, Australia

CABI *Publishing*

CABI *Publishing* **is a division of CAB** *International*

CABI Publishing	CABI Publishing
CAB International	10 E 40th Street
Wallingford	Suite 3203
Oxon OX10 8DE	New York, NY 10016
UK	USA
Tel: +44 (0)1491 832111	Tel: +1 212 481 7018
Fax: +44 (0)1491 833508	Fax: +1 212 686 7993
E-mail: cabi@cabi.org	E-mail: cabi-nao@cabi.org
Website: www.cabi-publishing.org	

A catalogue record for this book is available from the British Library, London, UK.

Library of Congress Cataloging-in-Publication Data

Weiss, E.A.
    Spice crops / E.A. Weiss
        p. cm.
    Includes bibliographical references (p.).
    ISBN 0-85199-605-1 (alk. paper)
      1. Spice plants. 2. Spices. 3. Condiments. I. Title.

SB305.W45 2002
633.8'3--dc21

                               2001052651

ISBN 0 85199 605 1

Typeset in Palatino by Columns Design Ltd, Reading.
Printed and bound in the UK by Biddles Ltd, Guildford and King's Lynn

Contents

_____

Acknowledgements

The author wishes to acknowledge with thanks the persons or organizations who supplied the following photographs or drawings: Prof. J.E. Armstrong, 5.1, 5.2; Buderim Ginger Ltd, Queensland, Australia, 11.9, 11.10, 11.13, 11.14; Department of Agriculture, Jamaica, 6.6, 6.8; Department of Agriculture, Kuala Lumpur, Malaysia, 3.2; Department of Primary Industry, Queensland, Australia, 5.3, 6.4, 9.4, 9.5, 9.6, 9.7, 9.8, 9.9, 11.11, 11.12; Mr S.I. Mtumbwa, Harare, Zimbabwe, 9.10; National Herbarium, Dar es Salaam, Tanzania, 6.1, 6.2, 6.3; Mr B. Perera, Colombo, Sri Lanka, 3.3, 11.19; Dr B.E.J. Small, Sydney, Australia, I.4; Spices Board of India, 4.3, 11.1, 11.3, 11.4; Yandilla Pty Ltd, Wallenbeen, NSW, Australia, 2.8. The remainder are from the author's personal collection.

The author also wishes to express his thanks to the spice organizations, individual traders and growers who supplied valuable data or information on crop production, and the many research workers who sent reprints of their papers, quoted in the references. Finally to my son John who provided essential assistance in accessing databases, and once again to my wife, Winnie, who produced the manuscript on disk.

Introduction

The first authentic records of spice and herb usage are on clay tablets from the Sumarian Kingdom about 3000 BC (Bottero, 1985), and many spices were used or imported into Egypt for embalming, as incense, ointments, perfumes, poison antidotes, cosmetics and medicines. In ancient times spices were considered primarily as medicines rather than condiments. It was not until the 1st century AD that spices became widely used in western cooking, following the discovery by a Greek merchant, Hippalus, in AD 40 of the monsoons of the Indian Ocean, until then monopolized by the Arabs to ship spices from the east to the west. This prompted the Romans to increase greatly their seaborne trade with India, with black pepper the most important spice (Miller, 1969). The popularity of spices increased their culinary use to excess, epitomized by the Roman gourmet and epicure, Apicius, who wrote ten books on cooking, later included in the *De Re Coquinaria*. His recipes included bread sprinkled with cumin, boiled ostrich flavoured with coriander and pepper, and roast flamingo accompanied by toasted sesame seeds!

The Romans were responsible for the dissemination of knowledge on spice usage, and also the distribution of Mediterranean spice and herb plants throughout their empire; for example, they introduced some 400 aromatic plants including mustard to Britain. When the Gothic King Alaric reached Rome in September AD 408, he demanded a ransom for not sacking the city, including the equivalent of 1400 kg of pepper in addition to the usual gold and silver. The Roman spice trade was disrupted by the rise of the Arabs and their conquest of the Middle East and North Africa, until the Mediterranean was reopened in the 12th century. During this period the supply of herbs and spices in Europe for both culinary and medicinal uses was satisfied basically from monastery and similar gardens.

During the reign of the English king Henry II, a pepperers' guild of wholesale merchants was established in London in 1180, subsequently incorporated into a Spicers' Guild, succeeded in 1429 by the Grocers' Company; grocer derives from the French *vendre en gros*, a wholesaler. Spicers became apothecaries and finally medical practitioners.

When Columbus landed on a Caribbean island in 1492, he brought back samples of the local spices, and subsequent traders shipped to Europe a variety of capsicum peppers including chillis, vanilla and pimento. Vasco da Gama arrived at Malabar in 1498 and returned with black pepper and cinnamon, and later voyages reached the Moluccas, a source of cloves and nutmegs. The maritime struggles between the Portuguese, Dutch and English were basically to control this most valuable spice trade

(Pearson, 1996), and laid the foundations for the subsequent colonial empires, which persisted into the 20th century. These developments are noted in the introductions to the various chapters. Spices of all kinds are now available worldwide, and for sale in markets everywhere (Fig. I.1).

Since the original introduction of spices to international trade, their uses have multiplied and demand has risen (Rosengarten, 1969; Stobart, 1977). Further increase in consumption, however, is likely to be mainly a function of population growth, assisted by consumer demand for natural ingredients in foods. This latter trend should be encouraged by producing countries, as it will also help maintain demand for the high value essential oils and oleoresins derived from spices, currently under pressure from synthetic alternatives.

The definitions of a spice or herb are numerous, and the *Oxford English Dictionary* defines a spice as 'a strongly flavoured or aromatic substance of vegetable origin obtained from tropical plants, commonly used as condiments' and herbs as 'plants whose stems, leaves or both are used as a food flavouring, in medicines or for their scent'. The author has chosen the following definition, which will be followed in this text: aromatic or fragrant products from tropical plants used to flavour foods or beverages are considered a spice; those from temperate plants can be either a spice, usually the fruit or seed, or a herb when green parts are used. The plants described are those the author considers the most important on an international basis and some, including pepper, are major commodities. While classification of spice crops by plant characteristics, fruit size, essential oil content and constituents is common, it is not wholly satisfactory, as considerable differences exist between cultivars from the same general region. This problem could, as with essential oil crops, probably best be resolved by using chemotaxonomy to define relationships.

Spices are generally labour-intensive, well suited to manual production, and can be integrated into agricultural systems in less-developed countries. The introduction of inexpensive drying and cleaning equipment, and development of sophisticated but affordable processing plants, including supercritical gaseous extraction whose products could replace distilled oils, make it relatively easy to upgrade the basic material to high-priced value-added products. The many developments in agricultural methods noted in the text are directly applicable to smallholder production or to the introduction of large-scale commercial crops and will materially assist spice producers in competing with synthetic alternatives, whose only asset is that they are cheap (Figs I.2–I.4).

Some constraints exist; for example, smallholders who have to purchase fertilizer on a tight budget may well find it more profitable to apply the minimum fertilizer necessary and increase the seed yield by greater attention to weeding and harvesting, as the author has frequently demonstrated. The gross factors affecting seed yield, oil content and composition (controlling organoleptic quality) are basically genetic, but are influenced by environmental and agronomic

Fig. I.1. Spice shop in Isfahan bazaar – Iran.

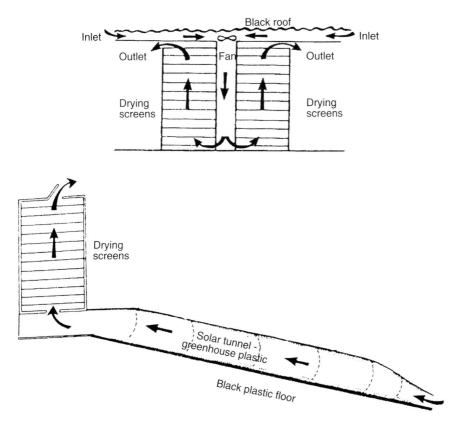

Fig. I.2. Types of easily constructed solar dryers.

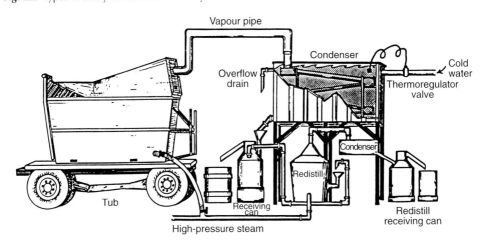

Fig. I.3. Equipment for lifting, chopping and loading herbage into trailer/still body.

methods. However, it is very difficult significantly to increase seed or oil yield if the cultivar grown is genetically unable to respond to the higher standards.

While chemicals will continue to be the main method of pest and disease control, newer and less toxic techniques are being developed, which will also solve the problem

Fig. I.4. Integrated system for obtaining essential oil.

of residues on, or in, spices and herbs. Australia's CSIRO scientists examined the molecular structure of the influenza virus and discovered a way of blocking its effects. The same approach has now produced ecdysone, an insect hormone that prevents moulting and thus kills susceptible insects. Such new agents can often be targeted to specific pests since the reaction varies among insects, thus avoiding beneficial insects. These new control agents will also overcome resistance problems as the target site is fundamental to their life cycle. A new wax phosphine block, developed jointly by CSIRO Entomology and United Phosphorus Limited of India, allows controlled release of the gas and is effective against most storage pests with no significant residues. Complicating the effective control of insect pests is the lack of detailed studies on the degree of infestation requiring control in a particular spice. Adding to the problem is the considerable variation in damage caused by an insect in different countries, or at different stages of plant growth. This information is important, since the data indicate that a narrow margin exists between the insect population a plant can tolerate and that which is economically damaging.

Trap-cropping is of great value where the main infestation period of a particular pest is known. Strips or small plots on the boundary of the cropped area are sown in advance of the main crop, or sown with a crop more attractive to the pest. The trap crop is then sprayed or destroyed thus reducing pressure on the main crop. This technique is particularly useful for small-holders or cooperatives, where the main date of sowing can be mutually agreed and supervised, and where pesticides are little used. Spraying of field edges early in the season also prevents pest build-up and thus reduces subsequent damage. Prophylactic spraying may be necessary, since some insects have a very short adult life, and by the time numbers have become large enough to attract attention, major damage has occurred. Since many spice crops are attractive to honeybees and other insects that have been shown to increase yield, these must be considered in any spraying programme. Where commercial apiarists have hives in the vicinity, due warning should be given of intention to spray, or the operation carried out when bees are least active, usually the late afternoon and evening. Chemical control of fungal diseases is difficult, although some systemic fungicides can reduce the severity of attack, and the breeding of resistant cultivars remains the main long-term solution. An interesting technique to identify new virus diseases has been developed in Australia, where scientists are using DNA to establish the identity of specific members of the tospovirus family, spread by thrips. These viruses are especially damaging in local capsicum crops.

Integrated pest or disease management is an excellent technique for achieving a high degree of relatively low-cost pest and disease control. It is unfortunately seldom effective in smallholder crops where it would be of the greatest benefit, as the cooperation necessary is difficult to achieve. Additionally, in respect of tree spices, the often small number of trees owned by individuals requires a high degree of cooperation to ensure that control is effective, and this is usually lacking. No recommendations for pest or disease control are given in the text as active ingredients, formulations and regulations regarding their use rapidly change.

Water is essential to plant growth and in regions with insufficient rainfall irrigation is necessary. Increasing human populations with their demands for domestic supplies are in conflict with rural producers, and efficient water use is essential to maintain crop production. Methods of determining soil moisture levels to control water applications are vital to ensure optimum usage, and simple equipment such as tensiometers should be installed in all commercial crops as illustrated in Fig. 9.7. One suitable technique is that of partial rootzone drying (PRD), which maintains both wet and dry soil within the rootzone. A significant feature of the technique is that it requires only half of the rootzone to be irrigated at any one time, and can save up to 50% of water without significantly affecting yield.

The future growth in demand for spices will probably be mainly due to the increase in world population as noted, but development of niche markets can be highly profitable. Organically grown or genetically modified (GM)-free crops are examples, and Buderim Ginger of Australia, a world leader in ginger production, successfully entered the organic market in the UK in 2000 via the Tesco chain of stores, which predicted UK sales of organic food products would quadruple by 2005 to £1 billion. Another local example is the production of fresh red ginger shoots, myoga, in Tasmania for the Japanese off-season market. The demand for organically grown foods has increased dramatically in the last decade, and concurrently for organically grown herbs and spices. This market provides great potential for specialized spice growers, as their higher priced produce will, in effect, be equal to value adding. An important use of spices is in medicine and an international body, the Joint Evaluation Committee for Food Additives (JECFA) has been estab-

lished scientifically to evaluate the medicinal value of spices and any long-term toxicological effects on humans; initial research will be on turmeric.

A very important aspect of spice marketing is quality control, especially the reduction in microbial and other contamination, previously a major industry problem. Such contamination can be reduced to very low levels by substituting modern and relatively inexpensive equipment for traditional methods, including forced air dryers, mechanical and steam cleaners, and efficient and hygienic storage to control pest infestation. These methods are being installed in major spice-producing countries such as India, and noted in the text. The ever more stringent regulations governing the international trade in spices and their derivatives for culinary use and in food processing will ensure that spice producers who ignore such specifications will eventually lose their markets in favour of quality producers. Keeping up with the latest national standards has been difficult, but has been greatly simplified by the introduction of national standards websites. While this may be of little direct interest to many producers, it allows exporters and importers to ensure their products comply, and by selecting particular sources of supply will eventually force producers to conform or cease production. Some websites include Standards Australia (www.standards.com.au); British Standards (available on CD-ROM or on www.ihs. com.au) and standards for India (www. indianspices.com/index.htm). For this reason no detailed standards are included in the text, although the European Spice Association Specifications are given as an appendix for information.

The processing of spices, including powders for retail sale, production of essential oils and oleoresins, is usually under strict quality control, and the products sold under reputable brand names and are thus free from contamination. Since processing varies with the spice concerned, this operation has been discussed in each chapter. Spice essential oils and their production are described in detail in the companion volume, *Essential Oil Crops* (Weiss, 1997).

This book should have been written before *Essential Oil Crops* as the author originally began working on tropical spices but was diverted to essential oils, which are frequently derived from tropical spices. Thus *Spice Crops* was the victim of a reverse take-over! Some similarity in the text of both is inevitable as many spices also produce important essential oils. The emphasis, however, is quite different. In the text, all units are metric and values are in US dollars, unless otherwise stated.

1

World Production and Trade

_____

The demand for tropical spices by European nations drove the great voyages of exploration, which culminated in the arrival of the Portuguese on the eastern coast of India in 1498, and on the Spice Islands (Indonesia) shortly after. In the second half of the 16th century, their ships carried some 65,000 Portuguese quintals (about 3300 t) of spices, mainly pepper, to Lisbon annually, and laid the foundations of the modern international spice trade, which in the late 20th century reached 500,000 t, worth about $1.75 billion.

A review of world spice production and trade up to 1980 is included in Purseglove (1981a,b), with further reviews found in the publications of the International Trade Centre (UNCTAD/GATT, 1974, 1982; UNCTAD/WTO, 1998). Spice production in India is covered by Pruthi (1998), and currently on the Indian Spice Board's website (www.indianspices.com/html/s0410trd. htm). Indian production and exports are quoted extensively in this chapter (Indian Spices Board, 1999, 2000), because of the leading position India occupies in international trade and its general influence on world prices, together with statistics from the USA and the European Union as leading spice importers. Details of spice production, imports and exports for selected countries are shown at the end of this chapter, and additional information is covered under each crop in the text.

Regional trade groupings, including the European Union (EU), the North American Free Trade Association (NAFTA), the South American Mercosur, and to a lesser extent the Association of South East Asian Nations (ASEAN), will cause, or have already caused, major changes in the international spice trade and which countries dominate in a specific spice or its derivatives are noted in the text.

The annual international trade in spices is derived from import and export statistics, but either through incompetence, corruption or smuggling of tropical spices in particular, national and regional production and trade is distorted. Some spices are small in volume but large in value, for example saffron, while the avoidance of customs and excise duties on other spices exacerbates this problem. Despite this, the overall total is sufficiently accurate to allow trends to become apparent and prices established. Thus, international trade in spices generally is expected to increase by 3–4% annually, while prospects for individual spices are mentioned in the following sections.

Statistics indicate that the legal international trade in spices nearly tripled in volume in the four decades up to 2000 and is now about 550,000 t, but the value rose six times to nearly $2 billion, a rise caused not only by increased usage but by a world population explosion. However, volume is not an accurate comparison since many producing countries now process the raw spice into value-added products of smaller volume, such as oleoresin.

Five spices – black pepper, capsicums, cinnamon, ginger and turmeric – plus their derivatives, normally account for at least three-quarters of world trade, with pepper accounting for about 33%, capsicums 22%, seed spices 15%, tree spices 14%, turmeric 8%, ginger 6%, cardamom 4% and vanilla 2% (although these ratios can change in a particular year). The foremost spice importer is the EU (with Germany being the leader), but the USA and Japan are the two largest single country importers, respectively.

Anise

The two main anise spices, anise and star anise, are generally not classified separately in international trade, but information may be available from individual countries' export data. Mexico has became one of the major non-European growers and exporters of anise, and will probably become the largest if the recent rate of expansion to satisfy NAFTA demand continues. India exported 1–2 t of seed annually in the early 1900s although the trade later languished; currently India is a leading exporter, and in the decade to 1999, seed exports rose from 5 t to nearly 2000 t. In the 1980s, annual world production of anise oil was estimated at 40–50 t, the main European producers and exporters being the former USSR, Spain and Poland, with France the main importer. Indian exports from 1995 to 1999 ranged between 25 and 30 t, valued at $2.2 million. In the 1950s, the development of synthetic aniseed oil captured a large proportion of the natural oil's market and thus world demand for distilled oil is unlikely to increase significantly.

Anise, Star

China and Vietnam are the major producers and exporters of Chinese star anise fruits and essential oil, with China accounting for the majority of both. In Vietnam, annual production of Chinese star anise fruits is esti-

mated at 2000 t, with 1600 t exported to Cuba, China and the Russian Federation. Annual production of star anise essential oil in China is estimated at 300–500 t, and Vietnam exports 200–250 t, mainly to France, the Czech Republic and Slovakia. Vietnamese oil imported into China is blended with the local product and exported mainly to France. In the late 1990s, the estimated annual value of Chinese star anise essential oil was about $5 million, and the international price between $6.5 and $11.5 per kg. Indian production of star anise is small, with occasional exports (2 t in 1995).

Capers

World production of capers is estimated to be several thousand tonnes and unlikely to expand, although population growth in South-east Asia could increase demand for mariana capers, easily satisfied by increased local production. In this region, India irregularly exports up to 3 t, while Japan has imported between 120 and 140 t annually in recent years. USA imports have varied from 500 to 700 t in the same period, exceeded by Venezuela with 800–1200 t. Caper imports to many developed countries, including Australia, are frequently in retail packs and thus not separately identified, while unprocessed capers may be included in the spice category Not Elsewhere Specified (NES).

Caraway

The major producers of winter-type caraway are The Netherlands, Poland, Hungary and Russia; the spring type is produced mainly by Egypt and western India, and in the decade to 1999, Indian caraway seed exports ranged from 2 to 25 t although they produced virtually no oil. The USA is a major importer of caraway fruit and oil, with average annual imports in the mid-1990s of 3000 t fruit and 4 t oil. In 1997–1998, the US price averaged $1/kg seed and $33/kg oil. Average annual world production of caraway oil is estimated at 30 t, with a total value of about $1 million.

Caraway is internationally traded and as the market is usually oversupplied, caraway prices are generally low. However, in periods of scarcity, prices increase rapidly and dramatically; production then increases again, resulting in overproduction. However, a large proportion of caraway is grown for domestic or very local sale, or collected from the sometimes numerous wild plants. An annual caraway cultivar locally adapted to the cooler mountain climates of South-east Asia could satisfy local medicinal and culinary demand, currently supplied from India. Unless a major change in demand occurs, such as the regular extraction of carvone from caraway seed in the food processing and pharmaceutical industries, continued overproduction and low prices will remain a disincentive to crop improvement.

Cardamom

India was traditionally the world's largest producer of cardamom until overtaken by Guatemala. The Indian total has varied greatly due to drought and outbreaks of mosaic disease, especially in the mid-1960s and again in the 1996/7 season. The main producing States are Kerala, Karnataka and Tamil Nadu. National production totalled 2300 t in 1968, 4600 t in 1978, 8000 t in 1998, and nearly 10,000 t in 2000, although Spice Board reports indicate at least 2000 t remained with growers. Exports continue at the low level of 650–750 t annually, partly due to the high domestic price. Indian cardamoms are considered to be of superior quality by international traders.

Cardamoms were introduced to the province of Alta Verapaz, Guatemala, in about 1920 and subsequently to other south-western provinces, Costa Rica and Honduras. The spice is little regarded in this region and virtually all production is exported. Guatemala is now the world's largest and a very low cost producer at about 12,000 t in 2000, mainly exported, but the supply and quality are erratic. Costa Rica is expanding production and could become a large exporter, and both countries will benefit from closer relations with NAFTA. Sri Lanka is now a minor cardamom producer and a high proportion is exported: 125 t in 1968, rising to nearly 200 t in 2000. In South-east Asia the governments of Laos and Vietnam are supporting cardamom as a smallholder crop and production is increasing, mainly for export.

Average annual world trade in both large and small cardamoms until the mid-1980s was about 6000 t, with India and Guatemala being the major suppliers, but Guatemala has since become the dominant exporter, and its low prices have increased the total market. World consumption is also expanding as the population increases, and thus a small annual increase in total production is sustainable since high-cost producers will lose their market share. The Asian international trade is heavily dependent on the substantial quantities imported by the Middle East and Gulf States, Scandinavia, Germany and Russia. About 25% of Indian exports go to Sweden, with Saudia Arabia as the second largest customer, and supplying Asian peoples now living in the EU, North America and Australia is a growing trade.

World availability of cardamom oil varies annually reaching 50 t in some years; for example, Indian exports ranged between 0.5 and 25 t in the period 1995–1999. Oil is produced in the main cardamom growing countries, particularly Guatemala, India and Sri Lanka, with small amounts distilled in major importing countries. Oleoresin production also varies, with India exporting 0.2–8 t over the same period, and with the growing use of supercritical extraction methods, oleoresin production could expand at the expense of distilled oil.

Cassia

Chinese

The harvested area in China in 1998, mainly in Guangxi and Guandong Provinces, was estimated by the FAO at 35,000 ha producing 28,000 t. Annual exports of dry bark from China during 1966–1976 was 1250–2500 t. In 1987–1997, the USA annually imported about 600 t of cassia bark and 350 t of leaf oil, mainly from China and Vietnam. Since Chinese domestic demand for leaf oil is considerable, production must be approaching 650 t to sustain the quoted level of exports. From 1991 to 1994 leaf oil was quoted on the USA market at $30–35/kg.

Indian

Bark is traded locally as Indian cassia bark or Indian cassia lignea and the leaves as tajpat leaves. Bark exports in the period 1989–1999 have varied from 500 t to 4000 t, mainly influenced by weather conditions during the stripping season and regional prices, which ranged from Rs4/kg to Rs25/kg. Powdered bark exports rose from 800 t to nearly 10,000 t in 1999, while cassia oil exports ranged from 0.50 to 2 t, and cassia oleoresin from zero to nearly 20 t in 2000. Over the 1989–1999 period, exports of tajpat leaves also varied substantially, for the same reasons as affected bark, from 300 t to 1500 t, going mainly to the neighbouring countries of Pakistan and Bangladesh but also to satisfy demand from the large number of expatriate workers in the Gulf States. Tajpat powder exports rose from virtually nil to 20 t, valued at $16,000 in 1998–1999.

Indonesian

Only the bark is economically important and the harvested area in Indonesia in 1998 was estimated by the Food and Agriculture Organization (FAO) at 60,000 ha, with a production of 40,000 t. The main producing region is Padang (West Sumatra). In the years 1991–1994, the USA imported annually about 13,000 t quills from Indonesia, worth $23 million. Since 1997, production and exports have been interrupted by the civil and economic disturbances and statistics are unreliable; unofficial reports indicate production of 25,000–30,000 t with substantial amounts smuggled to Malaysia and Thailand.

Vietnamese

The harvested area in Vietnam in 1999, mainly from the highlands inland from Saigon especially around Binh Dinh, was estimated by the FAO at 6250 ha with a total production of about 500 t. Exports of Vietnamese bark in the 10 years to 1976 varied from 5 t to 300 t. The USA was a major buyer until China re-entered that market.

Chilli and Capsicum

World production and trade in *Capsicum* species is difficult to quantify, since many countries do not differentiate between them, or do when exported but not when imported or vice versa! Since chilli and bell peppers serve different end-users, this lack of data obscures market trends. Chilli peppers and the paprika types of bell pepper, either whole or as powder, are internationally traded as spice, but most bell peppers are consumed as a vegetable, not necessarily in the country of origin.

Since the 1980s, the world market has radically changed, with many producing countries now processing locally and the various free trade associations affecting regional trade. For example, in 1957 the East African Community of Kenya, Uganda and Tanganyika exported 340 t, 249 t and 78 t of dried chillis, respectively, but following independence in the 1960s, the Community was disbanded, and exports became increasingly erratic. Domestic consumption rose as the population in Kenya and Uganda exploded and, combined with civil disturbances, reduced export availability and also quality. Introduction of NAFTA and its recent relationship with Mercosur has significantly changed the regional chilli pepper trade.

Total world production of peppers has been estimated at 14–15 million t annually in the 5 years to 2000. India is probably the world's largest producer, and according to the Indian Spice Board, total chilli pepper production averaged 800,000 t in the 5 years to 2000, with nearly 1 million t in the 1986/87 season; Andra Pradesh State contributes about half of the annual production.

India used to consume domestically about 95% of its chillis, and up to 1990 exports reflected a domestic surplus, but with steadily increasing production, exports reached 65,000 t in 2000. Chilli powder exports in the period 1994–1999 ranged erratically from 4500 t to nearly 15,000 t. Similarly, chilli oleoresin fluctuated between 10 t and 100 t and capsicum oleoresin between 200 t and 300 t. However, paprika oleoresin rose steadily from 225 t to 300 t.

China is also a large producer with a stated 30,000 ha annually planted to peppers of all kinds. Most are consumed domestically and in general only a surplus is exported. In the USA, the state of New Mexico leads in pungent pepper production at 12,000 ha, with California the leading bell pepper producer at 10,000 ha. Hungary is traditionally the home of paprika, with the main production areas in the Szeged and Kalocsa regions.

North America and the EU are the leading importing regions. Chile, China, Mexico and Pakistan are the principal suppliers of pungent peppers to the USA, with Hungary, Morocco and Spain being the main suppliers of paprika peppers. Mexico exports mainly to NAFTA because of favourable terms of trade. US chilli pepper consumption rose from 45 million kg (Mkg) in 1980 to nearly 220 Mkg in 2000 according to the Economic Research Branch of the US Department of Agriculture (USDA), with the supply moving south to Mexico, which is now one of the world's leading producers, and production continues to expand. To arrest any further drain, which is rapidly eroding their multimillion dollar industry, US pepper producers, principally in New Mexico, formed the Chilli Pepper Task Force, which commissioned the New Mexico State University's Chilli Pepper Institute to investigate methods of reducing local pepper production costs, especially improved mechanical harvesting and processing methods.

The EU is virtually self-supporting in paprika production, but imports most of its pungent pepper requirements. Japan has turned from a major to a minor exporter, while production and exports from African and some South-east Asian countries have been disrupted by wars.

Cinnamon

Cinnamon has been introduced to many tropical countries, but significant commercial production is limited to India, Indonesia and Sri Lanka, with lesser amounts from Madagascar, Malaysia and the Seychelles. Reviews of international trade in cinnamon bark and oils can be found in Wijesekera *et al.* (1976) and Purseglove *et al.* (1981a,b), in addition to the general surveys quoted above. Demand for cinnamon (and cassia) as spice has always been strong although some substitution occurs based on price, while competition from synthetic alternatives does not noticeably affect the trade.

Sri Lanka was traditionally the largest producer and exporter, but in recent years has been displaced in volume by Indonesia, although Sri Lankan quills are considered superior. In 1997, Indonesia exported approximately 6000 t annually, 900 t to the EU which was valued at about $17 million. Subsequent production figures from Indonesia are suspect as smuggling to neighbouring countries is rife, but the total harvested area in 1998/1999 and 1999/2000 was estimated by the FAO at 24,000 ha producing 12,000 t. Indonesian bark exports in the period 1989–1994 were, in tonnes, 3293, 2957, 3958, 7857, 3379 and 3890, with the USA, Singapore and The Netherlands being the main markets. Economic problems and civil disorder in Indonesia from 1997 has disrupted but not ended the trade, and the dramatic fall in the value of the rupiah lowered the world price of cassia and cinnamon bark generally. Indian bark exports are small and listed under Other Spices in statistics, while bark powder and cinnamon oil amount to 2–3 t each annually.

Cinnamon bark oil and oleoresin are generally prepared in importing countries, but cinnamon leaf oil originates mainly from producing countries. In the mid-1990s, Sri Lanka exported annually about 115 t of leaf oil and less than 3 t of bark oil, with the main markets being in the USA and EU.

Clove

Reviews of the trade in cloves and clove derivatives to the early 1970s can be found in Willems (1971) and Adamson and Robbins (1975), in addition to the general surveys quoted above. Ignoring domestic usage in producing countries, which in total far exceeds the international market especially in Indonesia, global demand for cloves has increased slowly to around 5500 t, and this trend is expected to continue in line with world population increases.

Zanzibar, now part of Tanzania, was for a long time the world's largest exporter of cloves, but following the revolution of 1964, production slumped. Plantations were split into uneconomical 1.5 ha plots, and growers received from the government only 4% of the export price. This was below the cost of production and there was no incentive to maintain healthy trees or replant; most trees are at least 60 years old. Production fell to less than 1000 t, rising slowly from about 3000 t in 1995 to 6000 t in 1998, well below the pre-1964 average of 9000 t, and far less than the annual 20,000 t of the decades to 1960. However, the very low prices paid to Zanzibari growers over the period 1965–1975 encouraged extensive smuggling to the mainland where prices were higher, and thus official production figures are understated. In 1997, the Zanzibar government announced a support programme for clove producers to rehabilitate the industry, but this was to little effect as no funds were forthcoming. New plantations established on the Tanzanian mainland near Tanga in 1994/5 are now in bearing and small amounts are exported.

Indonesia was initially the world's biggest clove importer, but following disruption of the Zanzibar industry, it became the world's largest producer and consumer of cloves, mainly used in kretek (crackling) cigarettes whose production in 2000 was 225 billion units. Clove production rose slowly but steadily in the period 1952–1962 from 3000 t to 7000 t, but following the 1964 revolution on Zanzibar previously noted, Indonesian clove production rapidly increased. By 1972, it was 15,000 t, reaching 25,000 t in 1982, and continued to rise reaching the massive total of 110,000 t from 512,000 ha in 1997, far above domestic demand of some 30,000 t according to the Statistical Bureau (Biro Pusat Statisik, Djakarta). Manipulation of the market by the Clove Marketing and Buffer Agency (BPPC), headed by one of President Suharto's sons, resulted in clove stocks soaring to 270,000 t, for most of which growers were not paid and suffered a loss of Rs542 billion ($216 million). The subsequent market collapse, low prices and the ongoing internal surplus of cloves resulted in an official programme to replace 200,000 ha of clove trees with coffee, vanilla and food crops. Domestic demand is now steady at around 80,000 t, but since the economic crisis of 1997, statistics for production and exports are unreliable. A substantial illegal trade in cloves to neighbouring countries exists, and their use and re-export distorts regional trade figures.

Madagascar was another beneficiary of the Zanzibar debacle and production rose steadily from 6000 t in 1968 to 15,000 t in 1998, but this trend is not likely to continue in the long term. Relative newcomers to the clove trade are Sri Lanka and Brazil, but with the ongoing surplus of cloves, both are unlikely to expand production much beyond domestic demand.

Coriander

A significant feature of the international trade in coriander seed is the wide annual fluctuation in the total amount available and the relative ranking of supplying countries. This became more obvious in the mid-1990s following the violent change of governments in Russia and Eastern European countries, with the subsequent disruption to their economies.

The total area sown annually to coriander worldwide probably averages 600,000 ha, as Indian statistics state that in the decade 1970–1980 the area sown nationwide averaged 250,000 ha, touched 600,000 ha in 1987/88 and ranged between 400,000–450,000 ha up to 2000. Total world seed production is estimated at 600,000–650,000 t, with India producing between 150,000 and 200,000 t in the decade up to 1999. A large proportion of seed produced in a specific country is used domestically and does not enter international trade, although a high world price can increase a particular year's exports. India, for example, normally exported around 7–10% of its total, rising from about 8000 t in 1991 to 22,000 t in 2000, worth $13 million.

Main seed importers are NAFTA, principally the USA and Canada, the EU, mainly the UK, The Netherlands and Germany, and Japan in the Far East. Of the ASEAN countries Indonesia is the major importer averaging around 6000–7000 t annually, including 1000–2000 t from India; Singapore is a major re-exporter. Most seed is traded as whole seed, but about 2500 t of coriander seed powder is now available annually. Distilled seed oil is produced mainly in Eastern Europe, with Russia traditionally the major supplier, in some EU countries, particularly The Netherlands, and in the USA. Oleoresin is mainly manufactured in industrialized countries to meet domestic demand.

Recent developments are an expanding demand for dried coriander leaves, a market already well established from Mexico to the USA where it is popularly known as cilantro, and for the organically grown herb and spice in Europe (Hoppe, 1996).

Cumin

World production of cumin is impossible to quantify, is complicated by the spice often being included or confused with caraway in national statistics, but has been estimated at 50,000 t in recent years. Iran is a major producer, ranging between 8000 and 15,000 t annually, and until the revolution of 1979 most was exported to Western countries. India was subsequently the main beneficiary of Iran's removal from this market and increased its production and exports. Pakistan and Turkey are minor exporters. Cumin is an important spice on the world market, and annual demand is expected to remain steady or slightly increase. World trade in cumin essential oil is estimated at

12–20 t annually, and oleoresin at about 2–3 t; the latter is produced mainly by manufacturers for their domestic market.

Cumin is mainly produced by smallholders in India with about 60,000 ha sown in recent years. India has exported cumin for over 100 years, and in the first years of this century shipped about 500 t annually to Europe; exports were irregular for decades but increased steadily with the fall in Iranian exports. In the 1975/76 season, the Indian planted area was about 90,000 ha, and with some hiccups steadily rose to nearly 350,000 ha in the 1998/99 season, with production increasing from 30,000 t to 140,000 t. Exports over the same period also rose from 2500 t to 6000 t in 1996/97, but jumped to 16,000 t and 10,000 t in the two subsequent seasons. Exports of cumin powder reached nearly 400 t in the 1998/99 season, with seed oil exports of around 1.5 t annually.

The main importers are EU countries, the USA, which imported about 6000 t of seed annually in the mid-1990s, the Gulf States with Yemen, and Japan in the Pacific region.

Dill

World dill seed production is increasing very slowly, as a surplus on the relatively small international market quickly depresses even further its generally low price. Since dill can easily be grown in home gardens, a very few plants easily satisfy domestic requirements.

Total world production of dill seed is unquantified, but the largest commercial producers are India and Pakistan (although this may include sowa), China, western Russia and neighbouring countries, Hungary and Egypt. India had 15,000 ha under dill in the 1992/93 season yielding about 11,000 t of seed falling steadily to around 5000 ha and 2500 t in the 1997/98 season. Exports over the decade 1989–1999 averaged 700 t. Major users are the USA (about 600 t annually), Japan (50 t) and Germany (30 t). Trade in dill seed powder is between 40 and 45 t annually, and dill seed oil about 40 t, with India supplying about half in recent years.

Dill herb production is mainly by smallholders to supply a local market, with large-scale producers serving international markets in Scandinavia, and Germany where nearly 7000 ha were under herb and spice production in 1996–1998. Worldwide production of dill herb oil is estimated at 100–150 t with a value of about $1 million; the largest producers are the USA with about 1000 ha under dill and a large but unknown amount in China. Only small amounts of herb oil are traded internationally.

Fennel

Few trade statistics are available for fennel alone, but India, China and Egypt are the main international suppliers. India averaged 15,000 ha under fennel in the period 1979–1995, rising sharply to 20,000 ha in 1997–1999; total annual yield was erratic but averaged 1 t ha^{-1}. Seed exports in the period 1975–1990 varied between 500 t and 2000 t, the fluctuation due mainly to the ruling international price. From 1991, exports rose steadily to nearly 6000 t in 1998/99, with a high of 12,400 t in 1997/98. Fennel powder exports also rose over the period from less than 10 t to 60 t in 1998/99. In 1988, the USA imported 1750 t of seed valued at $1.8 million, and in the 1990s, the annual value averaged $3.4 million. World trade in fennel essential oil is estimated at 28 t, valued at $700,000 in the mid-1990s. Most oil and oleoresin is produced in a specific country to satisfy domestic demand.

Fenugreek

Statistics on world fenugreek production and international trade are generally lacking, but are dominated by India as the major exporter. International demand is generally considered to be static or rising slowly, but the surge in India's exports would appear to indicate either new uses or growing demand.

In India, fenugreek is grown commercially mainly in the states of Gujarat, the most important, Rajasthan, Uttar Pradesh and Tamil Nadu. The total area sown to fenugreek nationally has varied between 35,000 and 45,000 ha in the 20 years to 2000, producing about 1.25 t ha^{-1}. In the period 1975–1990, Indian exports ranged between 1000 and 3000 t year^{-1}, rose sharply to 6000 t in 1991–92 and 7000 t in 1993–94, reached a record 15,000 t in 1995/96 worth $5.6 million, and subsequently fell to between 6000 t and 10,000 t year^{-1}. Major importers of Indian fenugreek are, in order of importance, Japan, the United Arab Emirates (UAE) and Yemen, and South Africa. As a single unit the EU is the largest destination, with the UK, France, Germany and The Netherlands being the largest consumers.

Since 1991, Indian powdered fenugreek seed exports ranged from 45 t to 215 t. Annual exports of seed oil have varied from zero to 4 t in the last decade, and oleoresin exports from 0.2 to 4 t. Oleoresin was traded in the USA in 1999 at $25–30/kg.

Ginger

World production of ginger is unquantifiable as a substantial proportion is consumed by producers or sold on local markets; recent estimates vary from 350,000 t to 400,000 t annually, with India contributing 250,000 t in 1998/99. The total trade in ginger products is also unknown, but as only the dried rhizome and its derivatives can be regarded as a spice, only export figures covering this sector are of value. International trade figures are distorted by the substantial illegal trade in dried rhizomes, especially in the South-east Asian region; for example, in mid-2000 the Australian Customs Service seized 24,000 kg of fresh rhizomes from Thailand valued at A$250,000.

Prospects for ginger and its derivatives on the world market indicate a steady, if small, annual increase, but in individual countries substantial expansion is possible to satisfy growing local demand at reasonable cost as in the Philippines, or by low cost exports as from Guatemala. A general and major constraint on increasing local ginger production is the difficulty in obtaining sufficient high-quality planting material.

India is the largest producer and exporter of ginger, with other major producers and exporters (in alphabetical order) being Australia, China, Indonesia, Jamaica, Nigeria and Taiwan, with the considerable production in Malaysia, the Philippines and Thailand being mainly consumed domestically. Major importing countries are the EU, North America and the Middle East, each taking just under one-third of world exports.

Exports of dried rhizomes have fluctuated, as several producing countries including India and China have begun producing oil and oleoresin, and civil wars in African countries, including Nigeria and Sierra Leone, have disrupted their production and trade, while disease reduced Fiji's exports to virtually zero from 1996 to 1998. India's exports of dried rhizomes have steadily risen over the last 40 years from 2000 t in 1958/59 to 28,000 t in 1998/99, although with substantial annual variation. Exports from China and Indo-China may also include rhizomes of other ginger species or unrelated plants, as noted in Chapter 11.

World trade in ginger derivatives is steadily increasing as more countries add value to their crop by producing ginger powder, oil and oleoresin, and food processors in developed countries adapt processes to utilize the advantages of these products. Ginger powder traded is now approaching 1500 t annually with India supplying up to 1200 t; ginger oil fluctuates between 100 and 200 t with India supplying about half. Ginger oleoresin traded internationally is only a proportion of world production as most is manufactured and utilized nationally especially in the EU and USA, with the exception of India, whose production is mainly exported (between 50 t and 100 t annually).

Laurel

Wild trees have been harvested for centuries in many Mediterranean countries to produce dried leaves, particularly Turkey, to which laurel is indigenous. The international trade in dried laurel leaves (bay leaves) is approximately 6000 t, Turkey normally supplying two-thirds, with Albania, Israel and Morocco supplying lesser amounts (Baser, 1997), while Indian exported 25–75 t in recent years. The EU imported between 800 and 900 t dried leaf annually in the 1990s.

Laurel leaf oil is of comparatively recent origin and is produced commercially in Georgia, before and since independence from the former USSR, with small amounts in Italy, Spain and Turkey from local plants, and in other European countries from imported dried leaves. Laurel oil is currently of little importance, but has a number of advantages over dried leaves for food manufacturers. Should demand for oil increase, production could be rapidly expanded, as laurel is an easily established and managed plantation crop, which can be fully mechanized. There were, for example, about 7000 ha of such plantations in the Republic of Georgia in 1997.

Mustard

The combined mustards are a leading spice by volume in world trade, and second only to black pepper in value. Seeds of *Brassica juncea* and *Sinapis alba* account for about 40% and 60% of the total while the international trade in *Brassica nigra* is minimal. World consumption of mustards, with the exception of *B. nigra*, is expanding slowly and the current producing countries will be able gradually to increase supplies. However, any major increase in world availability will inevitably reduce international prices of seed and derivatives.

Canada, India and the USA are the world's largest producers of mustard seed with annually about 200,000 ha, 150,000 ha and 7000 ha sown to *B. juncea* and *S. alba*, averaging just over 1 t ha^{-1}. Major markets for the North American seed are in the EU, with most oil and oleoresin consumed domestically. Indian exports of seed in the decade to 1999 varied between 600 t and 1300 t, and mustard seed powder between 4 t and 24 t, with major markets in neighbouring countries, the Middle East and the EU. Exports of mustard seed oil ranged from 1 t to 6 t over the period, while mustard oleoresin, a more recent product, varied widely from 0.5 t to 22 t in the 5 years to 1999.

The major importers of mustard seed and its derivatives are in the EU where production of condiment mustard in recent years has accounted for about 135,000 t of the world total estimated at 160,000 t annually. Main consumers are in France, Germany, the UK and the USA, with consumption per head ranging from about 0.3 kg in the UK to 1 kg in France.

Nigella

Although several tonnes of nigella seed are traded annually, few reliable data on production and trade are available, as it is usually included under Other Spices in statistics.

Nutmeg and Mace

The most important current nutmeg producers are Indonesia and the Caribbean island of Grenada, which dominate international trade, while the considerable production in Sri Lanka, India and Malaysia is mainly consumed domestically and their exports do not affect world trade. The area under nutmeg in India is about 4500 ha with yield varying from 700 to 2000 kg ha^{-1}. Half of Indonesia's annual production is usually consumed domestically but this is true for only a small proportion of Grenada's. Demand for nutmeg and mace is virtually static in terms of world per capita consumption, and thus growth will match the population increase. Advances in agronomy, plantation management, marketing and distribution would probably allow current world demand of about 20,000 t of nutmeg and mace to be produced on about 15% of the present planted area, and thus the first country to apply these techniques could dominate future world trade.

Reviews of world trade in nutmeg and mace are contained in Guenther (1952), in addition to the general reviews quoted above. Annual world production is currently estimated at 17,000 t of nutmeg and 3000 t of mace, with 60% originating in Indonesia and 30% in Grenada, although this proportion was disrupted from 1997 by the economic collapse of Indonesia. In 1987, the Indonesian Nutmeg Association and Grenada Co-operative Nutmeg Association

agreed to allocate annual sales to Indonesia of 6000 t nutmegs and 1250 t mace, and to Grenada 2000 t nutmegs and 350 t mace, with any shortfall by one being allocated to the other. However, similar to other such commodity agreements, it was seldom honoured by participants and is now virtually moribund.

The area planted to nutmeg in Indonesia was 67,000 ha in 1994 producing 14,000 t of nutmegs and exports of nutmeg and mace for the years 1989–1994 in tonnes (mace in parentheses) of 2470 (465), 6392 (1050), 7335 (1549), 4658 (1180), 7500 (1150) and 7900 (1400). Subsequent figures are unreliable as the state-owned exporting companies have collapsed, and little control over, or record of, exports is maintained. However, on the island of Banda a nutmeg growers cooperative was formed in 1999 under the patronage of the ruler, Sultan Alwi, which is rejuvenating nutmeg production and replanting abandoned plantations. The estimated 500,000 trees on Banda during the 18th-century Dutch occupation had fallen to around 30,000.

The international trade in nutmeg and mace oils and oleoresins is about 150 t annually, but all are also produced in importing countries for domestic usage in varying amounts and for special purposes. This market has been significantly affected by the introduction of supercritical gaseous extraction methods and equipment.

Major nutmeg and mace importers are the USA (around 2000 t nutmeg and 200 t mace annually) and Japan (550 t and 50 t). Imports of the various oils vary annually and are generally related to price.

Pepper

Black pepper is the world's most important spice, both in volume and value, and a major world commodity, but total production varies substantially mainly due to adverse weather conditions in producing countries. In the decade to 2000, production ranged between 170,000 t and 240,000 t and exports between 125,000 t and 180,000 t. The International Pepper Community (IPC) was formed to protect the interests of pepper producers with India, Indonesia, Sarawak and Brazil as founder members, but later expanded to include other countries. IPC members account for about 85% of world production and nearly 90% of world exports.

World pepper demand is increasing at 2.5–3% annually, mainly due to the increase in world population rather than an expansion in usage. Ironically, the lowest per capita consumption is in South-east Asian producing countries whose peoples prefer chillis. The steady demand has encouraged smallholders in the traditional producing countries to maintain production, but the advent of relatively new low-cost producers in China and Vietnam has introduced an element of instability into the world market. The growing relationship of Mercosur with NAFTA gives Brazil a major advantage. This accounts for the massive expansion in Brazil's planted area and production, which, if sustained, will inevitably lead to a change in the traditional ranking of world pepper producers and radically influence the direction of world trade.

Average pepper production and export in tonnes by the main producers (in alphabetical order) over the period 1989–1999, plus exports in 1999 and 2000, is shown in Table 1.1. The main pepper producers in the period were India and Indonesia, with Sarawak (Malaysia) third, but in the mid-1990s China overtook Malaysia; Thailand and Vietnam became significant exporters; and Sri Lanka languished, while Brazil's production rapidly increased to 40,000 t in 2000, mainly exported. Countries currently producing small quantities for export are Costa Rica,

Kampuchea, Madagascar and Tonga, while others, including the Philippines, consume their production domestically. To illustrate the effect of inclement weather on pepper-growing regions, the total offered on the world market in the crop years 1992/93, 1993/94 and 1994/95 was 163,000 t, 109,000 t and 117,000 t, respectively, while Indonesia lost two-thirds of total production in one year due to drought. Thus, pepper prices tend to fluctuate widely from year to year, depending mainly on availability and to a lesser extent on speculators.

The area planted to pepper in 1997/98 was estimated by the FAO at 365,000 ha: about 200,000 in India, 75,000 in Indonesia, 26,000 in Sri Lanka, between 10,000 and 12,000 ha each in China and Malaysia, 7500 in Vietnam and 3500 in Thailand, with Brazil at 12,000 and expanding rapidly.

The main importers of pepper, in ranking order (excluding Singapore and Hong Kong which are basically re-exporters), with their averages over the period 1993–1997 are shown in Table 1.2.

Pimento

In 1755, about 440,000 pounds of berries were exported to Europe from Jamaica and exports rose steadily to around 10 million pounds (4.5 Mkg) by 1900, reaching a record of 25.6 million pounds (11.6 Mkg) in 1908. Leaf oil was first produced in 1920. World demand for pimento is static, but for pimento oil is slightly increasing, and this trend is likely to continue in the long term.

Table 1.1. Average pepper production and export in tonnes by the main producing countries over the period 1989–1999, plus exports in 1999 and 2000.

Country	Production	Exports	1999	2000
Brazil	25,400	23,300	16,600	20,400
India	58,600	31,800	47,400	21,000
Indonesia	47,700	41,600	36,000	52,000
Malaysia	21,500	21,600	21,600	23,500
Sri Lanka	4,400	3,700	3,700	4,800
Thailand	8,600	2,300	550	650
Other countries	30,600	19,200	4,000	7,000
World total	196,800	143,500	161,000	166,000

Table 1.2. Average pepper imports and re-exports in tonnes by the main importers over the period 1993–1997.

Country	Imports	Re-exports
USA	46,616	1,748
Germany	17,518	2,584
The Netherlands	12,772	9,598
France	8,847	1,104
Japan	7,464	–
United Kingdom	6,458	561
Russia	5,820	–
Spain	4,315	152
Canada	4,036	61
Poland	3,527	155
Egypt	3,517	166
Italy	3,283	136
South Korea	3,033	24
Belgium	2,900	970
Saudi Arabia	2,827	–
Other countries	46,163	3,706
World total	217,185	

Total world trade in pimento is currently between 3000 and 4000 t, valued at some $5–7 million. Jamaica is the world's largest producer of pimento and accounts for about 70% of world exports, with the remainder shared by Central American countries and other Caribbean islands. Major importers are the USA, the EU (mainly the UK and Germany), Sweden, Finland and Canada. Leaf oil production reportedly varies between 30 and 60 t per annum, virtually all distilled on Jamaica, and destined mainly for the USA and UK.

Saffron

Production is unofficially estimated at around 65–75 t, much of which is consumed where produced or in neighbouring countries. Considering that 100,000–140,000 flowers yield only 1 kg of high-quality saffron, this is a huge amount. World trade in saffron is unquantifiable as the spice's small volume and high value encourages smuggling. The largest producer is probably Iran, followed by Spain, which together account for 90% of the total. Saffron is seldom mentioned in international trade statistics, but is unlikely to exceed 45–50 t with high annual variation.

Most countries' official imports are usually less than 5 t. Major importers are Saudi Arabia and the Gulf States for domestic use, the USA and the EU, which also re-export saffron in retail packs and mixtures. Demand for saffron as a spice is slowly increasing as the world's population rises and developed world consumers enhance their diets with a wider range of food flavours. A major constraint on increased saffron production is the chronic shortage of good-quality corms.

Turmeric

The world trade in dried turmeric rhizomes is based mainly on export statistics from major producing countries at about 20,000 t, plus a large and unquantifiable movement between many countries of Asia and the Pacific region. Future trade prospects are for a steady demand as the population in major markets increases.

The major producer is India and its exports dominate world trade. The area under turmeric rose steadily from 80,000 ha in 1970 to 120,000 ha in 1990, and to nearly 145,000 ha in 2000, and production from 150,000 t to 600,000 t in 2000. Exports of dried rhizomes over the same period rose from 11,000 t to an estimated 40,000 t, plus a small amount of fresh rhizomes. Powdered turmeric exports in the years 1990–1999 rose from 6000 t to 12,000 t, turmeric oil from 0.5 t to 4.0 t and oleoresin from 150 t to 250 t, and these three derivatives are expected to continue this upward trend. Minor producers include Bangladesh, China, Indonesia, Myanmar, Pakistan and Taiwan. A number of Central and South American countries produce and export small amounts of turmeric, the most important being Jamaica, Brazil, Haiti and Peru.

In general there are two major regional markets; one comprises the North African countries, the Middle East, Iran, Japan and Sri Lanka, accounting for about 75% of international trade and supplied by Asian producers; the other comprises North America and the EU, accounting for about 15% and supplied mainly by India and the South American producers. In the two decades to 1995, the USA imported about 1800 t of turmeric annually worth some $2 million, and in the years 1997–2000 imported 2045 t, 2300 t, 2600 t and 2800 t mainly from India, with an average value of nearly $4 million. Iran was a major consumer at 4500 t annually prior to the revolution when all imports were greatly reduced, but has recently increased its imports from India from 1000 t in 1994 to nearly 5000 t in 1999.

Vanilla

World production of vanilla (cured pods) has varied widely due mainly to inaccurate recording, but production probably reached nearly 3000 t in 1999 and is expected to remain around that level for the foreseeable future. A major factor influencing natural vanilla usage is price, and if it rises too steeply, natural vanilla could be replaced by synthetic substitutes. Should this occur then vanilla markets could be permanently lost. Supporting usage, and thus the price, is the worldwide trend by consumers for natural ingredients in foods and regulations in major consuming countries requiring product labels to state if natural vanilla or artificial vanillin is included.

Annual vanilla production figures have to be carefully interpreted, as some national volume statistics may combine green and cured vanilla under the one heading, yet the latter is 70% lighter than the former. In addition, the time taken in processing may make accurate identification of the year of production uncertain, while unreported changes in stock levels also produce distortions. For example, in 1985 Madagascar had carryover stocks of 1500 t of processed beans, with a similar amount expected from that season's harvest, an overhang that reduced the world price from $70/kg to $60/kg. Thus, figures for total world trade are an unreliable guide to average production, as annual export and import statistics may also vary due to different national accounting methods and periods.

The international trade in vanilla was for many years controlled by a French-administered cartel, Univanille, composed of Madagascar, Réunion and Comoro, now in abeyance due to the entry of several other producers who refused to join and the virtual abandonment of vanilla growing on Réunion. A major adverse effect on world vanilla sales and prices was the 1984 decision by the Coca-Cola Company to stop using vanilla as a flavouring in its cola drinks. Domestic consumption of vanilla beans by many producing countries is negligible, Indonesia being an exception, thus virtually all production enters world trade. Although vanilla's value is insignificant in the total imports of consuming countries, it constitutes an important source of income for Indian Ocean producers.

In the period 1991–1997, Indonesia had between 12,000 and 15,000 ha under vanilla (replacing cloves as noted), with annual production reaching almost 2000 t in 1999; Madagascar produced between 500 and 1000 t and Comoros averaged around 200 t. Total exports of the major producers in the decade to 2000 rose from about 1500 t to 2600 t, with the market share for Indonesia being 50%, Madagascar 35% and Comoro 12%. Other very minor producers and exporters are India, Malaysia, the Philippines, Tonga, Thailand and Uganda, with Singapore a re-exporter.

Mexico, which was once a world leader, virtually abandoned vanilla production, and currently exports between 10 and 15 t annually to the USA. However, introduction of NAFTA has apparently reawakened interest in vanilla growing.

The leading importing country in the decade to 2000 was the USA, and in the years 1997–1999 imports were 2000 t, 2000 t and 1400 t, valued at $43,000, $38,000 and $9,500, with the downward trend in volume expected to continue. Other significant importers in 1998–1999 were Germany at 330 t, France at 300 t, Canada at 180 t, Japan at 70 t, Switzerland at 45 t and Australia at 15 t.

A price differential exists based on origin, with best-quality Bourbon vanilla (produced on the Indian Ocean islands) selling on the US market for $70,000/t in the years 1988–1992, while Indonesian vanilla sold for $26,000/t. Prices for what is termed extraction vanilla were considerably lower. In the last decade prices have tended to fall although generally retaining the origin differential; according to official trade statistics, ,the average US import price in 1997–1998 from all sources was about $20/kg free on board (FOB). While major consumers of natural vanilla are the USA, France, Germany and Japan, these countries are also important producers of synthetic vanillin.

International Spice Statistics

Spice imports into selected countries in the period 1997–1999 from official import statistics are shown in Table 1.3 for the EU, Table 1.4 for Japan and Table 1.5 for the USA.

Indian exports of selected spices are shown in Table 1.6. India exports to about 150 countries, but 20 make up 85% of the total sales value. The top seven, in ranking order, are the USA, UK, Sri Lanka, the UAE, Japan, Germany and Singapore, which account for half of sales to the top 20 countries.

International prices for selected spices in mid-2001 were as shown in Table 1.7.

Table 1.3. Main spice imports into the UK in the period 1997–1999.

Spice	1997		1998		1999	
	Q	V	Q	V	Q	V
Anise (seed)	146	399	208	403	167	417
Capers	325	689	281	590	270	443
Caraway (seed)	266	378	111	68	68	79
Cardamoms (whole)	617	1,744	387	1,831	387	2,832
Chilli	4,894	15,255	5,241	15,741	4,678	11,571
Cinnamon (ground)	234	514	292	658	251	406
Cinnamon (whole)	1,016	1,628	1,098	1,947	887	1,556
Clove	366	414	293	511	346	639
Coriander (seed)	4,908	4,377	3,347	3,137	3,655	2,347
Cumin (seed)	2,550	4,157	3,149	5,132	2,688	4,041
Fennel and juniper	819	1,731	714	1,585	902	1,938
Ginger	8,592	13,803	10,086	11,483	9,261	12,343
Mace	126	535	53	357	69	594
Nutmeg	351	989	2,939	3,565	216	701
Pepper (ground)	1,247	4,856	1,182	4,559	1,309	5,482
Peppers (whole)	6,372	25,920	4,563	26,100	4,173	22,198
Poppy (seeds)	694	747	410	638	355	384
Thyme and bay	452	1,622	416	1,493	391	1,484
Turmeric	2,509	2,694	2,203	2,633	2,354	2,513
Vanilla	225	3,510	290	4,089	385	4,644

Q = quantity in tonnes; V = value in thousand US$.

Table 1.4. Main spice imports into Japan in the period 1997–1999.

	1997		1998		1999	
	Q	V	Q	V	Q	V
Capers	133	783	119	699	119	545
Chilli	11,299	33,648	9,881	23,177	10,336	28,404
Cinnamon (whole)	1,462	3,286	1,310	3,001	971	2,188
Cinnamon (ground)	347	671	489	908	509	967
Clove	340	517	338	550	301	1,068
Curry (blended)	93	491	71	360	83	447
Ginger	91,168	113,855	91,034	74,602	91,685	74,011
Mace	37	281	46	527	50	587
Mustard (seed)	10,030	5,041	9,325	4,651	10,223	5,455
Nutmeg	496	1,663	370	1,713	634	4,264
Pepper (ground)	1,497	8,428	1,306	9,637	1,049	7,996
Pepper (whole)	6,726	31,986	5,881	36,515	6,976	38,840
Poppy (seed)	430	454	296	347	345	530
Spices NES	2,458	17,125	1,819	9,357	2,268	11,980
Thyme and bay	303	1,154	435	1,375	354	1,105
Turmeric	3,642	3,805	4,404	5,813	3,946	4,847
Vanilla	95	4,263	69	2,962	91	4,033

Q = quantity in tonnes; V = value in thousand US$; NES = not elsewhere specified.

Table 1.5. Main spice imports into the USA in the period 1997–1999.

	1997		1998		1999	
	Q	V	Q	V	Q	V
Anise (seed)	1,212	2,842	1,449	3,739	1,354	3,156
Capers	732	2,744	509	1,697	584	1,723
Caraway (seed)	3,140	2,926	3,238	2,650	3,405	2,585
Cardamoms (whole)	256	976	353	1,326	224	1,902
Chilli	37,401	74,371	54,253	89,232	57,143	89,866
Cinnamon (whole)	16,431	31,228	18,134	27,204	16,346	22,059
Cinnamon (ground)	1,065	2,045	1,325	2,053	1,220	1,968
Clove	1,431	2,215	1,158	1,726	1,356	2,711
Coriander (seed)	3,102	2,844	3,269	2,888	3,591	2,285
Cumin (seed)	6,561	11,165	7,165	10,942	7,121	9,533
Fennel and juniper	3,419	4,414	3,753	4,138	3,472	4,112
Ginger	13,837	17,796	14,036	16,448	15,583	17,487
Mace	147	796	198	1,393	230	1,926
Nutmeg	1,924	4,414	1,502	5,185	1,746	8,598
Pepper (ground)	973	3,583	1,405	5,701	2,084	9,469
Pepper (whole)	51,072	189,960	41,900	217,893	54,381	265,996
Saffron	10	4,762	12	5,488	9	5,287
Thyme and bay	2,234	6,275	1,900	4,858	2,314	6,159
Turmeric	2,045	3,021	2,284	4,140	2,643	4,024
Vanilla	2,198	43,238	1,941	38,059	1,361	28,214

Q = quantity in tonnes; V = value in thousand US$.

Table 1.6. Indian exports of selected spices in the period 1995–2000. (Data from the Spice Board of India.)

	1995–96		1996–97		1997–98		1998–99		1999–2000(E)	
	Q	V	Q	V	Q	V	Q	V	Q	V
Cardamom (large)	1,677	1,224	1,628	1,210	1,648	1,265	1,424	1,191	1,000	1,551
Cardamom (small)	527	1,297	226	870	370	1,267	475	2,521	550	2,760
Celery	2,678	625	3,780	802	3,317	799	3,991	969	2,550	681
Chilli	56,165	19,546	50,051	20,145	51,779	15,890	61,253	21,661	58,700	23,394
Coriander	11,541	2,243	12,574	3,137	23,734	6,435	20,685	4,589	12,250	2,585
Cumin	3,871	1,739	6,375	3,438	16,281	8,136	10,723	6,011	4,250	2,945
Fennel	2,594	752	4,850	1,789	12,368	3,582	5,279	1,538	3,400	1,148
Fenugreek	15,138	1,867	8,891	1,205	6,006	987	10,082	1,915	8,750	1,771
Ginger	18,483	3,892	29,737	5,924	28,268	7,263	8,778	4,065	7,800	2,912
Pepper	26,244	19,630	47,893	41,232	35,907	49,636	34,864	63,811	42,100	86,498
Turmeric	27,050	4,620	23,019	5,845	28,875	8,307	36,522	12,455	32,250	10,460
Spice oil and oleoresins	1,912	11,502	2,358	15,901	2,419	23,153	2,750	30,077	2,825	28,546
Total	169,779	70,836	193,281	103,397	212,871	128,619	198,725	152,702	176,424	165,250

Q = quantity in tonnes; V = value in million US$.
NB, aniseed, cinnamon, cassia, saffron, ajwan seed, dill seed, poppy seed and mustard are not specified separately.

Table 1.7. International prices for selected spices in mid-2001.

Spice	Origin/grade	Market	Price $ t^{-1} CIF
Black pepper	Malabar (Garbled MG-1)	New York	3,085
	Lampong	New York	3,305
	Sarawak	New York	3,305
	Brazil	New York	3,305
Cardamom	Guatemala Bold green	Saudi Arabia	17,000
	Indian Asta (6–7 mm)		Not quoted
Cassia	China	New York	1,325
Cinnamon	Ceylon H2 Cinnamon (soft bark)	New York	5,730
Clove	Madagascar/Zanzibar	New York	6,945
	Sri Lanka (hand-picked)	New York	8,705
Coriander	Canada	New York	615
Chillis	Indian Sannam-4 (stemless)	New York	1,255
Cumin	Turkey/Pakistan	New York	2,535
	India	New York	2,755
Fennel	Indian ASTA	New York	1,325
	Egyptian	New York	1,325
Fenugreek	India/Turkey	New York	795
Ginger	India Cochin	New York	1,655
	China (whole)	New York	1,390
Turmeric	India Madras Finger	New York	1,300
	India Alleppey Finger	New York	1,455

CIF = cost, insurance and freight.

2

Cruciferae

The *Cruciferae* contains about 300 genera and 3000 species, mainly herbs native to the temperate regions, but some are also widely cultivated in the tropics. The genus *Brassica* L. contains about 160 species including the very important *Brassica napus* L. and *Brassica campestris* L. (oilseed rape), source of the edible oil canola and a major competitor to soybean, plus a range of cultivated vegetables including cabbage and cauliflower (*Brassica oleracea* L. species), swede (*Brassica rutabaga* L.), turnip (*Brassica rapa* L.) and Japanese mustard or water mustard in China (*Brassica japonica* (Thb.) Sieb.). The most important spice producers are *B. juncea* (L.) Czern. (Indian mustard), *B. nigra* (L.) Koch. (black mustard), and *Sinapis alba* L. (white mustard), included in this chapter for convenience. *Brassica* and *Sinapis* genera are closely related, but some easily recognizable differences are that *Sinapis* has pale green leaves, petals with short claws and fruits with bristles, whereas *Brassica* often has grey-green leaves, petals with larger claws and smooth fruits. The generally accepted name of mustard is considered to be derived from the Latin *mustum ardens*, burning must, since the seed was sometimes ground with grape must.

Mustards

Among the first accurate references to mustard seed are those found on Sumarian clay tablets of around 2000 BC , one quoting a contemporary proverb:

> When a poor man has died, do not try to
> revive him!
> For when he had bread he had no salt
> And when he had salt he had no bread.
> When he had meat he had no mustard,
> And when he had mustard he had no meat.

A prescription to relieve pain was a poultice composed of fennel, seeds of *Vitex*, *Juniperus excelsa*, tamarisk, tragacanth, *asafoetida*, *Lolium* and mint, mixed with flour, boiled in mustard water and bandaged on the patient. In general, salt and mustard were the major Sumerian condiments. The earliest references to mustard in Chinese literature apparently relate to *B. japonica*, translated as water mustard, whose leaves were eaten (Chang, 1976), and there is no reference to true mustard seed until much later, apparently introduced from India along the Spice Road. India was a major producer of mustard and remains so today.

Mustard seed gathered from wild plants has been used as a spice in Europe for thousands of years, and was popular with the Greeks and Romans, who also ate the young green plants as a vegetable. The Greek philosopher Pythagoras, 503 BC, recommended a mustard poultice to treat scorpion stings, while 100 years later another Greek, Hippocrates, listed a number of external and internal medical uses for the seeds.

So important was the spice in Roman times that the Emperor Diocletian fixed its price in AD 301. Initially seed was ground and sprinkled on food, but over time the seed became the basis for more sophisticated products. For example, seed was ground with honey and vinegar, or mixed with grape must. A very interesting use of mustard seed is contained in the reported exchange between King Darius 3rd of Persia and Alexander the Great. Darius sent a bag of sesame seed to Alexander symbolizing his vast army, who replied with a bag of mustard seed to signify not only the equal number of his soldiers, but also their powerful energy.

Mustard is mentioned in the Christian Bible, and by the 10th century the monks of St Germain des Pres near Paris, France, were famous for growing mustard and it became increasingly popular as a flavouring. At a fete given by the French Duke of Burgundy in 1336, guests consumed 70 gallons of mixed mustard in one sitting! A Burgundian named Boornibus discovered a method of pressing mustard into dry tablets, which continued to be manufactured in Dijon up to the 17th century. In 1634, the vinegar and mustard makers of this town were granted the exclusive right to make mustard and in return 'were required to wear clean and sober clothes, keep only one shop in the town so that it would be obvious where any bad mustard originated, and put their names on casks and stone jars'. From 1937, Dijon mustard became an appellation controlled by French law, and the city remains a market leader in the preparation of specialized wet mustards exported worldwide.

Mustard as a condiment was introduced to England by the Romans together with some 400 aromatic plants, and mustard seeds were found when excavating a Roman site at Silchester, Hampshire. Seeds were apparently ground in special small querns, and one appeared on an inventory of a house in London's Cornhill, dated 1356, owned by a gentleman named Stephen le Northenes. Subsequently, seed was ground by millers and supplied as a dry powder, to be made into a paste and sold in parchment-covered earthenware pots. Tewkesbury was the famous mustard centre of England, referred

to in Shakespeare's *Henry IV Part 2*, commenting on a gentleman thus: 'His wit's as thick as Tewkesbury mustard!' A set of Dame Alice de Bryene's household accounts for the year 1418–1419 shows she 'used 84 pounds of mustard seed bought for one farthing per pound from Stourbridge Fair'.

A description of contemporary mustard preparation can be found in Sir Hugh Plat's *Delights for Ladies*, published in the late 17th century.

> It is usual in Venice to sell the meal of mustard in their markets, as we do flour and meal in England: this meal by the addition of vinegar in two or three days becometh exceeding good mustard, but it would be much stronger and finer if the husks were first divided by sieving, which may easily be done, if you dry your seeds against the fire before you grind them. The Dutch iron handmills, or an ordinary pepper mill, may serve for this purpose. I thought it very necessary to publish this manner of making your sauce, because our mustard which we buy from the chandlers at this day, is many times made up with vile and filthy vinegar such as our stomach would abhor if we should see it before the mixing therof with the seeds.

English mustard was commonly supplied as dry powder to be made up as required, and at the beginning of the 18th century a Mrs Clements of Durham became famous for a finely ground mustard powder, known as Durham Mustard, which was given the Royal seal of approval. Production expanded and the centre of powder production moved north from Tewkesbury (Stobart, 1997). In the middle of the 19th century a Norwich miller, Mr Jeremiah Colman, greatly refined mustard powder and supplied a very fine, pure (i.e. without husks) powder with a standard flavour. The Colman name became virtually synonymous with mustard in England, and made the family's fortune. It also produced the famous saying, 'It's not the mustard that people eat that made Colman rich, but that left on the plate!' Colman's mustard was produced in Norwich with seed specially grown by local farmers until quite recently, but *B. nigra* seed has now been virtually replaced as the principal ingredient by *B. juncea* seed. A result is that the standard powder is less pungent. Colman's later introduced a range of flavoured mustards, including Savora, which are more popular outside Britain. From the 18th century onwards the common types of mustard used today began to be defined, as described in the section Products and specifications.

Botany

An important review of the cytotaxonomy of the *Brassica* and several closely related genera is that of Harberd (1972), although some confusion remains, especially in the close relationships between species in *Brassica* and *Sinapis*. Harberd examined 85 *Brassica* species, and divided the genus into cytodemes, composed of species with the same chromosome number which were cross-fertile, and those whose chromosome number might differ and which were cross-infertile. Four of the most important agricultural species are diploid, three are allopolyploid, and each belongs to a separate cytodeme. The four diploid species are *B. nigra* ($2n = 16$), *B. oleracea* ($2n = 18$), *B. campestris* and *B. rapa* ($2n = 20$), plus *Brassica tournefortii*, and the three allopolyploids *B. napus* and *B. juncea* ($2n = 36$) and *Brassica carinata* ($2n = 34$). *Sinapis alba* has $2n = 24$.

Crosses between the allopolyploid *B. juncea*, *B. napus* and other cultivated rapes have been recorded. Studies on *Brassica* spp., including mustard cultivars, have shown that hereditability for yield components is generally low, and important characters contributing to yield are number and weight of siliqua/plant, number of seeds/siliqua and 1000-seed weight (Arthamwar *et al.*, 1995; Rai, 1995). *B. juncea* cultivars have also been classified on glucoside content.

The major objective of plant breeders generally was to increase seed yield ha^{-1} and seed oil content. Other important aims were to increase low-temperature tolerance, improve stem strength, reduce shattering and introduce resistance to specific pests and diseases. More recently, increased drought and salt tolerance have become important, with the acceptance by growers and breeders of capability of mustard to extend its range. The major problems in India are low yield,

late maturity, shattering and lack of pest and disease resistance, and pure line selection was initially the main method used for yield improvement. The most recent advances have been made by developing different plant types, with more branches, pods and heavier seeds, and by using a growing period adapted to a specific region, since it is considered that this approach is more suited to existing cultural systems.

Recent studies have shown that selection for oil content can quickly increase this attribute. It is apparently easier to raise seed oil content than greatly improve seed yield, as the first is less influenced by environment and its genetic relationships are simpler. Plant size is highly correlated with seed yield per plant, with a smaller correlation between the number of pods on the main raceme and yield. The number of secondary and tertiary racemes that reach maturity is a major contributor to increased yields. Of importance and great assistance to plant breeders concerned with oil yield and oil constituents has been the development of techniques for determining the oil content of single seeds without impairing viability, including nuclear magnetic resonance spectrometry (NMR).

A very recent development has been the breeding in Canada of *B. juncea* cultivars whose seeds produce an oil almost identical to canola oil and meal. This was basically achieved by lowering the glucosinolate and erucic acid levels, and raising that of oleic acid. Field trials have indicated that the oil has all the essential characteristics of canola oil, and animal feeding studies have shown that the meal is equal, or superior, to canola meal. An asset of these new cultivars is that they may be better adapted to lower rainfall areas than oilseed rape. In India three other brassicas are cultivated in addition to *B. juncea*, known as rai: *B. campestris* var. toria, *B. campestris* var. brown sarson and *B. campestris* var. yellow sarson, the last two being named for their seed colour.

Brassica juncea. *B. juncea* (L.) Czern (synonym *Brassica rugosa* Prain.) is known in English as brown or less correctly Indian mustard, in India, generally as *rai* but also as *raya* or *laha*, in Sri Lankan as *kaduga*, in German as *rutensenf* or *sareptasenf*, in Italian as *senape indiana* and in Spanish as *mostaza de Indias.*

There are no botanically acceptable subspecies of *B. juncea*, although a number of Indian cultivars have local names. Germplasm collections are maintained in India, the USA and The Netherlands. *B. juncea* is now the most commercially important of the three commonly available mustard seeds and, although probably originating in Africa, is widely cultivated from Eastern Europe to China and Japan. In Asia, *B. juncea* is a very important vegetable and oilseed crop.

B. juncea is an erect annual herb growing up to 1.5 m, with yellow flowers and pungent seeds (Fig. 2.1). Plants form a rosette

Fig. 2.1. *B. juncae* shoot showing leaves, flowers and fruit.

after emergence and before stem elongation. The sturdy central stem is circular in section, up to 1.5 cm, and is much branched, glaucous and slightly hairy. Leaves are variable below; above they are alternate, lobed, exstipulate and margin notched. Lower leaves are not clasping the stem and are on long petioles up to 20 cm long, with a very large ovate terminal segment; in colour they are a bright, mid- to grass-green. Growth regulators applied to young plants can affect general growth, seed characteristics and constituents. Gibberellic acid increased the number of pods and seed weight (Khan, 1996a, b); paclobutrazol increased siliqua size and affected seed constituents (Setia *et al.*, 1996), while Ethrel plus nitrogen increased seed yield and seed oil content (Khan, 1996a,b).

Leaves may be eaten as a pot herb, but must be boiled twice as they contain the toxic glucoside, sinigrin. An analysis of leaves gave the following percentages: water 90.6, protein 2.6, fat 0.4, carbohydrates 4.8, fibre 1.0 and ash 1.6, plus vitamin A at 610 international units (IU), and ascorbic acid at 62 mg per 100g.

Floral initiation becomes apparent when the stem apex enlarges and small domes appear around the edge, which quickly develop into small, stalked spheres, while developing leaves curl over to cover the developing buds. Flowers are small, 3–5 mm long on bractless terminal racemes 7–10 cm long, longer on some cultivars, and generally a pale yellow although they may be brighter (Fig. 2.2). Flowers are bisexual, with four narrowly elliptical sepals, 3–4 × 1.5 mm, spreading horizontally, six stamens in two whorls, ovary superior, and sessile. Flowering may last for several weeks and is an impediment to mechanical harvesting.

The fruit is a four-sided siliqua up to 3–5 cm, with a slender beak, not appressed to rachis, composed of two carpels separated by a replum forming two chambers and dehiscing lengthwise from bottom to top. Pods contain 4–10 small, round seeds up to 1 mm diameter, with approximately 6250 seeds per 100 g; the testa is wrinkled, minutely pitted, dark reddish to very dark brown or almost black, and pungent but not equal to *B. nigra* seed. Seeds usually have no

Fig. 2.2. *B. juncae* flowers.

dormancy period, similar to oilseed rape. Seeds normally contain mucilage, sinigrin, sinapic acid and sinapine (its choline ester), and 25–35% fixed oil consisting mainly of glycerides of erucic, eicosenoic, arachidic, nonadecanoic, behenic, oleic and palmitic acids. Composition is one method of distinguishing *B. juncea* seeds from close relatives (Velisek *et al.*, 1995).

The glucoside sinigrin (potassium myronate) reacts with the enzyme myrosin (myrosinase) in the presence of water to produce a volatile oil containing mainly allyl isothiocyanate (0.25–1.5%, usually around 0.9%), glucose and potassium bisulphate, responsible for the pungent aroma and flavour of all mustard seeds. The operation of the myrosinase–glucosinolate system and the hydrolysis products involved has been reviewed (Bones and Rossiter, 1996). The seeds of some Eastern cultivars also contain an allergen designated *Bra jIE*, the cause of allergenic reactions to certain proteins, and similar to an allergen obtained from *Sinapis alba* (Monsalve *et al.*, 1993).

About 25 minor components have been identified. The volatile oil content varies from 1% to 2% and is affected by the extraction method. Seed of *B. juncea* from widely separated geographical regions, i.e. Europe, India and China, varies in its pungency and non-volatile oil content, and thus is not interchangeable in mustard formulations. Further characteristics of mustard oils are described in the Products and specifications section.

Brassica nigra. *B. nigra* (L.) Koch. (synonyms *Sinapis nigra* L., *Brassica sinapioides* Roth, *Sisymbrium nigrum* (L.) Prantl.) is known in English as black (sometimes brown) mustard, in French as *moutarde noir*, in German as *schwartzen senf*, in Italian as *senape negra*, in Spanish as *mostaza negra* and in India generally as *Banarsi rai*. There are no botanically acceptable subspecies but a number of local cultivars have been developed. Germplasm collections are maintained in the national gene banks of Canada, India and the USA.

B. nigra is adapted to a range of temperate and subtropical climates, and most probably originated in the region from West Asia to Iran, but currently occurs wild in the Mediterranean region, and throughout central Europe, the Middle East and the Ethiopian highlands. It is a naturalized escape in parts of Britain and the USA, and natural selection for earliness amongst volunteer plants may cause black mustard to become a troublesome weed.

Black mustard is a tall, annual, branched herb, with small yellow flowers and small blackish seeds. The taproot is well developed, and it has a central sturdy stem up to 3.0 m in height (generally 2.5 m), which is green or slightly glaucous, glabrous to bristly and much branched, with secondary branching common. The leaves are midgreen, alternate and on long petioles; the lower leaves are large, up to 16 × 5 cm, pinnatifid or pinnatilobed, usually with two lower lobes and a much larger terminal lobe; the central leaves are moderately lobed. The lower and central leaves are irregularly dentate and are often partly bristly hairy; the upper leaves become gradually smaller up the stem, and are narrow, lanceolate, entire and glabrous. The young leaves are pungent when crushed and can be used sparingly in salads.

The inflorescence is axillary or terminal bractless racemes, together arranged paniculately. The flowers are bright yellow, bisexual, up to 8 mm in size and on short pedicels; the four sepals are narrowly elliptical, 3–4 × 1.5 mm, spreading horizontally; the four petals are clawed, obovate and 6–8 × 2–2.5 mm; the six stamens consist of an outer whorl of two shorter and an inner whorl of four longer, the pistil being slightly shorter than the longest stamens. The ovary is sessile, superior and elongated, consisting of two united chambers and a style ending in a semi-globose stigma. The main flowering period in the northern hemisphere is mid-summer to early autumn. Pollination is by insects, mainly bees and pollen beetles such as *Meligethes* spp.

The fruit is a smooth four-sided, pod-like capsule (siliqua) up to 2.5 cm in length, with a short beak at apex. It is erect, closely appressed to the rachis, and dehisces longitudinally when mature. Pods contain up to 12 seeds, and are globose, up to 1 mm in diameter. The testa is black to dark reddish-brown, minutely pitted and yellow within (Fig. 2.3). In experiments to test the longevity of seeds, those of *B. nigra* buried in glass bottles were still viable after 40 but not 50 years (Quick, 1961). *B. nigra* is a copious nectar producer and yields a mild-flavoured, light-coloured honey.

Seed composition is generally similar to *B. juncea* and contains a glycoside sinigrin, an enzyme, protein, mucilage, and 25–30% of a non-volatile oil, and between 1–2% volatile oil. Seed from different regions and the various cultivars vary in volatile and non-volatile oil content and oil pungency. The seed has generally been replaced in mustard formulations by that of *B. juncea*. The glucosinolate sinigrin on hydrolysis releases the volatile allyl isothiocyanate responsible for the pungent taste of black mustard; the glucosinolate content ranges from 110 to 140 μmol g^{-1}. Average seed content is 8% water, 29% protein, 28% fat, 19% carbohydrates, 11% fibre and ash 5 g per 100 g seed containing 0.4 g Ca, 0.6 g P and 21 mg Fe. In addition, seeds contain 0.6 g beta-carotene equivalent, 0.4 mg thianine, 0.31 mg riboflavin and 7.3 mg niacin per 100 g seed. 1000-seed weight usually between 2 and 4 g, but varies with cultivar.

Black mustard germinates almost immediately after sowing and the first leaves are usually visible within 48 h. Early growth is very rapid and flower initiation may start as early as 2 weeks after germination, but usually occurs after 4–6 weeks. Fruit is mature in a further 4–8 weeks. Fruits mature in sequence and shatter easily when ripe; black mustard is thus unsuitable for commercial mechanized production.

Sinapis alba. *S. alba* L. (syn. *B. alba* (L.) Rabenh., *B. hirta* Moench.) is known in English as white mustard, in the USA as yellow mustard, in French as *moutarde blanc*, in German as *weiber senf* or *senfsaat*, in Dutch as *mosterd*, in Italian as *senape bianca*, in Spanish as *mostaza silvestre* and in Arabic as *khardal*. Germplasm collections are maintained in the USA, Canada and The Netherlands. *S. alba* is a naturalized escape in England, the USA and elsewhere, and a close relative of *S. arvensis* L. (charlock).

Fig. 2.3. Mustard seed.

S. alba is native to the Mediterranean region and West Asia, now widely distributed in cool temperate regions, and grown commercially in eastern and northern Europe, and especially on the Canadian prairies where conditions favour large-scale cultivation.

White mustard is an annual herb growing up to 120 cm in height, although it is generally half this, with yellow flowers and pungent brown seeds. The taproot is long and thin, and the stem erect, stout, ribbed, bristly, with hairs inclined down, and branching on the upper part. The leaves are usually alternate, oval, deeply three-lobed, with margins crenate to dentate, and up to 15 cm in length, although this is variable. The petiole is longest on larger leaves, which are mid- to dull green in colour. This mustard is also grown for green or dried fodder, and as a green manure.

The flowers are bright yellow, in terminal racemes, with four obovate petals 10 × 4 mm, and with the base narrowly clawed. The four sepals are 4–5 mm long and self-sterile. The six stamens comprise an outer whorl of two shorter and an inner whirl of four longer. The pistil is similar in length to the inner stamens; the ovary is elongated and sessile, with the style ending in a semi-globose stigma. The main flowering period in the northern hemisphere is mid-summer to early autumn. The fruit is a hairy, three-veined siliqua, 2.5–4.0 cm × 3–7 mm, with the beak curved, flat and usually the same length or longer than the pod; it is generally non-shattering. It contains 4–6 small seeds 1.5–2.0 mm diameter, and the testa is minutely pitted, yellowish and white within; there are about 6000 seeds per 100 g (Fig. 2.3). The seed contains up to 30% fixed oil, proteins, mucilage, sinapine, the glucosinolate sinalbin, which on hydrolysis with the enzymes myrosin and glucosinolases yields p-hydroxybenzyl isothiocyanate, and p-hydroxybenzylamine and associated compounds. The seed is robust and germinates well in almost any conditions and type of seedbed, and together with the non-shattering fruits allows fully mechanized cropping.

The seeds are often ground with seeds of B. nigra to produce table mustard, but are much less pungent. Cracked seeds prevent mould and bacterial growth and are used as a preservative in pickles and similar products. White mustard is grown commercially mainly to supply seed for the 'mustard and cress' market, and is only a minor salad vegetable in India. However, rape is increasingly replacing S. alba as the mustard component, with Lepidium sativum L. (garden cress), and both are consumed as seedlings and usually sold in small cartons.

All three mustards flourish under virtually identical conditions when cultivated, but B. nigra and S. alba are now seldom grown as large-scale commercial crops. All further references to mustard in the text refer to B. juncea, its seed and derivatives, with specific reference if necessary.

Ecology

The three species used to produce mustard are each native to a specific region, but so popular are their pungent seeds that all three have been widely introduced into other countries of the Old and New Worlds, and escapes have become naturalized and have flourished. The ease with which they readily naturalize is the reason why they have become significant weeds of other brassica crops, and their cultivation is often discouraged where oilseed rape is a major crop, for instance in Australia. A warm temperate climate is most suitable for mustard, but regional cultivars have greatly extended its range, especially in the Indian region. In general it is not climate that determines where mustard is grown commercially, but the ability to fully mechanize mustard seed production. Thus, the large B. nigra plants unsuitable for mechanization are now grown only where manual labour is available, in particular in Ethiopia, and this is why the shorter B. juncea is the most commercially important.

Mustard has a very wide photoperiodic range since it grows naturally or is cultivated from northern Russia, with its prolonged periods of daylight, to tropical India, with a 12 h day. Thus, local cultivars are well adapted to their environment. Mustard is usually sown in spring in Europe and North America and harvested in autumn; in India it is generally sown in early October and harvested in February. In all regions, however, sowing dates vary to suit local cultivars.

Annual temperature in major mustard growing regions averages 6–27°C, but mustard can be grown in the tropics during the cool season with a temperature around 25°C. Temperatures above 30°C generally retard growth and if prolonged in the absence of adequate soil moisture may desiccate plants. With adequate soil moisture, plants can survive short periods at 40°C during the vegetative phase. Data indicate that within the optimum range, a higher average temperature during the reproductive phase and lower during flowering and seed filling gives the highest yields. High temperatures at flowering can desiccate flowers and severely reduce seed yield. Thus, sowing to ensure flowering does not coincide with such periods is important.

Mustard is resistant to low temperatures and a light frost when established, but seedlings are frequently badly damaged. However, if a late frost kills up to 50% of plants, it better to leave the remainder to mature rather than resow, since mustard has a considerable ability to produce greater individual growth to compensate for a lower population. Laboratory trials in India have shown that freezing plants for 2 h at the vegetative stage did not affect seed lipid and fatty acid content, but at the reproductive stage, lipid content and erucic acid were reduced while linoleic acid increased (Gupta, 1996). Use of liquid nitrogen for seed preservation depended on the seed moisture content to ensure seed viability (Neeta, 1996).

A rainfall of 750–850 mm is desirable, with 450–500 mm falling mainly prior to flowering, since moisture stress at this time is the most damaging. Seedlings are very susceptible to waterlogging for other than short periods, and prior to flowering, excess soil moisture generally reduces yield although often having little visual effect on mature plants. On well-drained soils, a rainfall above 1000 mm is tolerated. High wind has little effect on immature plants, but can be damaging to plants nearing harvest by flattening the crop and delaying harvesting.

Hot winds may cause shattering of pods, but the degree is a cultivar characteristic. In dry areas of India, mustard is replaced by *Eruca sativa* Lam., which produces an edible oil.

Soils

In its natural habitat, wild mustard will grow on a wide range of soils, but light to sandy loams are preferred for commercial crops. In general, those soils that produce good crops of oilseed rape, *B. campestris* var. *oleifera* (canola), are suitable for mustard. A soil with a neutral pH reaction is preferred, but mustard will produce good crops on soils ranging from pH 6.0 to 8.0 if the fertilizers chosen are related to soil reaction. A major requirement is that soils must be free-draining or drained, and under such conditions mustard has been grown successfully on soils with a substantial clay content from Europe to India and China. Unlike vegetable brassica crops, mustard is generally unsuitable for planting into harvested paddy fields. Peat soils, which are usually too acid for mustard, were improved by the addition of wood ash and lime in Indonesia (Suryanto and Lambert, 1994).

Salt tolerance is a very valuable characteristic in crop plants and likely to become more important as salt contamination of arable land continues to grow. Mustard has a degree of salt tolerance, but it is presently little exploited for this valuable characteristic. In The Netherlands, for example, rape is accepted as one of the first crops that can be grown on land reclaimed from the sea.

Fertilizers

Mustard responds well to added nutrients, either organic manures or chemical fertilizers, which generally raise yields by increasing the number of pods and seed weight. However, overall plant growth also increases, and while this may be of little consequence to smallholders, for commercial crops the optimum ratio between the greater flow of vegetative material through combine harvesters and increased seed yield must be determined to ensure profitable fertilizer use. Where little information is available on which fertilizer should be applied to mustard, the types and levels used on local wheat or canola crops are suitable until optimum rates are established.

Farmyard manure and other bulky organic residues, including castor poonac and tank silt in India, are usually applied only by smallholders, and increase yield by supplying nutrients, conserving moisture and improving soil texture. Green manures must be ploughed in well before planting, as their decomposition can induce a temporary soil moisture and nitrogen shortage. Organic wastes suitable for applying by fertilizer spreaders are used on large-scale commercial crops, including piggery wastes spread in the autumn and composted sewage sludge in Canada, eastern European countries and Russia.

Chemical fertilizers appear to have only a minor effect on seed constituents at the levels normally applied, but as with other brassica crops, nitrogen may modify seed-oil and protein content. However, these reactions can normally be disregarded since the gross effects of agricultural systems and the environment are far more important, while higher fertilizer rates or water applications do not compensate for sowing outside the optimum period. Where large areas of mustard are to be planted, it is essential to determine the most profitable combination of applied water and nutrients, since there can be substantial variation in cultivar response.

Mustard is relatively leafy in its early stages, and a nitrogen (N) shortage will retard leaf development and reduce yield. Additionally, the retention of the largest possible leaf area for as long as possible has been shown to increase yields by maintaining the flow of assimilates to flowers and young pods. This yield increase is basically due to the greater number of seeds and pods reaching maturity, rather than a general increase in pod or seed weight. An adequate N supply not only encourages leaf development, it can materially assist in retaining leaves in active photosynthesis over this period. Although more N than phosphorus (P) or potassium (K) is absorbed during early growth, maximum N uptake occurs at main flowering period.

Where no trials have been carried out to determine the N levels necessary, the local wheat or canola mixture can be used as noted, for example, a compound mixture of NPK at a ratio of 10:25:25 at 200 kg ha^{-1} placed in the seedbed. However, 30–40 kg N ha^{-1} in the seedbed is widely considered the maximum, with a top dressing of 60–80 kg N ha^{-1} applied when plants are some 20 cm in height, but not later than bud formation. As a guide for areas where mustard has not previously been grown, N should be applied at twice the P level, with the major portion as a top dressing.

Where large amounts of N are required, split applications give higher seed yield, but the additional work necessary may not be profitable. In many areas with high-intensity rainstorms, split application can be essential to overcome the perennial problem of nutrients being leached or washed away before becoming available to plants. An N topdressing should not be applied when frost remains on leaves, and preferably not when frost can reasonably be expected.

The European Weed Research Council's assessment of the phototoxicity of certain N compounds as overall dressings apply to rape, but are relevant to mustard and are calcium nitrate 1, solid urea 3 and urea spray 4, of a maximum of 10. When liquid N is to be used, a dribble-bar type of applicator is recommended, otherwise crop damage can occur.

The interaction between N and seed characteristics in brassica crops has been widely studied, and this element undoubtedly affects seed constituents. Most often reported is a decrease in seed-oil content directly related to a rise in protein content, with increasing N levels. Nitrogen application may also change the seed-oil constituents altering the various acid ratios. The interaction between nitrogen and water use is discussed in the section on irrigation, and with disease in the relevant section.

Once the basic requirement has been supplied, additional amounts of P appear to have little effect on yield. Mustard is considered to be an efficient extractor of soil P in the presence of adequate soil moisture, and can thus make profitable use of residual P following applications to preceding crops. A minimum seedbed application of 20–35 kg P ha^{-1} is recommended, the variation depending on soil phosphate status. Adequate P generally reduces the time to flowering and maturity, but a deficiency increases this period.

The type of phosphate fertilizer used is apparently of little importance at the same level of P, provided that its availability is similar, since P uptake is high in young plants. When fertilizers are mixed on-farm, diammonium phosphate (DAP) would appear to be superior to other ammoniated phosphates, and single superphosphate to other types when there is a slight sulphur deficiency.

Potassium (K) is considered necessary to ensure adequate uptake of N and P but its effect on mustard is not well documented, although low K levels in India resulted in a reduction in seed-oil content. At normal nutritional levels, K concentration is usually lowest in young leaves and highest in old leaves, the reverse being true when K is deficient. Where K is supplied to other crops in a rotation, it will seldom be necessary to apply large amounts to mustard. As a general guide, the amount of K applied should be half the P level unless there is a known deficiency, as often occurs in India. When K is required, it is best applied in the seedbed, usually as part of a compound mixture, since little information is available on the effect of various types of potassic fertilizers or their relative efficiency.

Other nutrients may also be added. Liming is seldom necessary and usually unprofitable except on very acid soils, and often produces only small yield increases in the presence of adequate N, P and K. On acid soil below pH 5.8 a manganese toxicity can develop, together with reduced resistance to mildew. Sulphur is not normally required by commercial mustard crops especially those receiving single superphosphate with its higher S content, but the nutrient gave significant yield increases in areas where there is a known deficiency, especially in India. The yield increase is often highest in the presence of adequate N, and with winter-sown crops. There is no evidence of any significant effect on seed composition. Boron deficiency can impair germination, retard growth and may also have a minor effect on seed constituents. If boron deficiency symptoms have occurred in previous brassica crops, then mustard will also be affected. Rates of 1–2 kg boron ha^{-1} are normally sufficient, and this can be mixed with the seed or be part of a compound fertilizer. Laboratory trials have shown reactions to copper, iron, magnesium, molybdenum and zinc, but their application to mustard crops is seldom necessary. Mustard may also take up heavy metals including cadmium, but the amount in seeds is well below published limits for human toxicity (see caraway and coriander, Chapter 10).

Cultivation

Mustard seed is very small and the seedbed must therefore be well prepared and level. The amount of cultivation necessary is a measure of the grower's skill, since on some heavy clays that produce high yields, too fine a seedbed will result in compaction or crusting. On more sandy soils, over-cultivation can dry out the seedbed to such an extent that there is no moisture available for germination. Minimum tillage techniques, or direct drilling, are successful on light soils and those liable to blow. Mustard fits easily into any farming system growing small grains with little additional capital expenditure on plant or machinery.

The small mustard seed has limited energy resources and without a suitable tilth much of this energy can be dissipated before the seedling is established. Depth of sowing is also critical for this reason, but on some tropical soils where the surface can quickly dry out, deeper placement is acceptable provided there is no danger of crusting, the soil remains loose, and higher seed rates are used. Seedbed preparation and rainfall should be complementary, and although this is frequently of no importance in the more temperate areas, it can be of vital importance in regions with definite rainy seasons.

Seed can be planted up to 5 cm deep depending on the soil type, but as the top 3–5 cm of topsoil can quickly dry out, deeper sowing may be necessary. In these circumstances the drill furrow should not be completely covered, but allowed to fill naturally. Soil moisture is the controlling factor, for seed must be sown in moist soil if emergence is to be even. Seedbed temperature does not appear to be critical or as important as adequate soil moisture and even depth of planting to ensure a good, regular stand.

Seed from the previous season should be used although mustard seed apparently has no significant dormancy period. Storage of commercial seed in good quality sacks or paper bags is adequate, but breeding or similar stocks may require more careful handling. Seed stored in polyethylene, polyvinyl-lined jute sacks, cloth or ordinary gunny bags gave germination of 92%, 89%, 78% and 78%, respectively, after 30 months in India (Verma et al., 1995). Grading of seed for commercial crops is of no practical value, since it apparently has little effect on the final yield under field conditions.

Since mustard seed is very small, specialized seeding equipment is desirable, but many standard grain drills can be used if they are suitably modified; fitting a grass-seed box is often sufficient. Mixing seed and fertilizer to obtain a more even flow of seed down drill spouts is not to be recommended, since placing the two in contact adversely affects germination. Precision seeder units are available, some electronically controlled, and these offer considerable advantages to growers who may have a variety of small-seeded crops in their rotation.

Rolling after sowing can be beneficial in certain circumstances; in the USA, spool-type rollers produced a soil surface that assisted mechanized harvesting (Parish and Bracy, 1995). Smallholder crops are often broadcast and these plants have more branches and pods but the number and weight of seeds per pod is little different. Plants are thinned to the required spacing about 15 days after emergence.

Seed rates can vary widely with little effect on final yield, in the range 3.0–12.0 kg ha^{-1} mainly depending on sowing technique. For fully mechanized production and combine harvesting, dense stands are preferred to thin, since this promotes more even growth, flowering and ripening. Row widths of 18–24 cm and 36–50 cm are common and are often determined by the equipment available. Since mustard plants are capable of expanding to fill available space, differences in row or inter-row spacing is not critical. Where vigorous weed growth is expected when seedlings are young, wider rows allow greater access to cultivating machinery with less danger of plant damage.

Seed should be dressed with a fungicide and an insecticide where *Psylliodes* beetles or similar species are abundant. Using insecticidal granules instead of dusting gives longer protection, and the mixture of granules and seed may also produce a more even stand by allowing greater control of seed flow. In less developed areas, an increased seed rate is often a more practical alternative. Germination is epigeal, with the cotyledons carried clear of the soil surface. Some cultivars form a rosette before stem elongation.

A significant and increasing reduction in seed yield occurs from sowing after the optimum planting date, and this must be accurately determined for a particular locality. For example, delaying sowing from 1 October to 30 November in Maharashtra, India, decreased seed yield by one-third (Gare et al., 1996). Yield reduction is due to a marked decrease in the number of branches per plant and pods per plant, since racemes are borne terminally on stems and branches. Flowering usually begins when the stem reaches half its mature height and can continue for several weeks. The extent of early flowering has been correlated with high yield, as data indicate that seed from the earliest formed pods is heavier than in later pods; seed-oil content in India also showed a similar ratio (Nanda et al., 1996).

Mustard produces many more flowers than fruit set, and normally only two-thirds of flowers produce fruit. While mustard is basically insect pollinated, th ese may be scarce in many cultivated areas, and trials with rape have demonstrated the benefits of locating hives of domesticated bees close to fields. In some regions, seed yields were doubled, in addition to producing a substantial amount of valuable honey.

Weed control

Mustard seedlings are very susceptible to weed competition and a weed-free seedbed should be the aim. Once stems begin to elongate the crop suppresses all but the most persistent weeds. Two mechanical or hand weedings are usually sufficient and if more are necessary attention should be paid to weed control in the preceding crop, or rotations may have to be modified. Weeds not only compete with a crop for available moisture and nutrients, they also cause harvesting problems and increase its cost. Similar to pest and disease control, weed control should be an integrated operation; for instance, many cereal weeds are more easily controlled in brassicas and vice versa.

Pre-emergence herbicides are applied to commercial mustard crops, but it is emphasized that before their use on large-scale plantings, local trials are necessary. Soil and climate can radically alter crop sensitivity to both compound and application rate. Herbicides used include alachlor, benazolin, butam, carbetamide, dalapon, di-allate, fluchloralin, glyphosate, nitralin, pendimethalin, propyzamide and trifluralin, used singly and in combinations; oxyfluorfen was toxic to mustard in India.

Irrigation

Knowledge of the relationship between mustard's growth and available water is important to ensure that a shortage does not affect growth at any critical period, which can vary between cultivars and country. Generally a water stress at main flowering and pod filling will reduce yield or seed-oil content of most oilseed crops and mustard is no exception. Where irrigation has increased yield, it is frequently by increasing the number of pods per plant and the number and size of seeds/pods. Also affecting yield is the degree to which mustard plants retain their leaves and the extent to which other plant parts can substitute for their assimilate production. This is important in terms of water use, since it is of little value applying water to crops that are basically defoliated or incapable of using soil moisture to increase seed yield. It appears mustard can compensate to a considerable degree for leaf loss and other plant parts are able to produce assimilates that contribute to seed yield. Plants therefore benefit from adequate soil moisture levels during seed maturation. Irrigation also assists in retaining leaves, which continue to produce assimilates for a longer period and this also increases seed yield.

Four irrigations are considered the minimum for fully irrigated crops; one should be pre-planting, and one at main flowering, since a water stress when pods are filling will reduce yields. In the dry Rajasthan area of India, one pre-planting irrigation and one at main flowering/pod filling gave the greatest return in terms of water used and fertilizer applied. In this region, a crop that will produce a locally acceptable yield level with the minimum of water is very valuable, since in many years there is insufficient water available to grow a cereal crop.

Where irrigation is used to supplement rainfall, it is essential to determine accurately the optimum amount of water and number of applications. Trial results indicate that irrigation alone is not sufficient for maximum yield, and optimum water use invariably requires adequate N availability. The correct combination of water and N is well documented, and the two together can double the yield increase over that obtained from either applied separately. Sprinkler irrigation is successful, but is profitable only where the system already exists and capital cost is spread over a range of crops. Row spacing can greatly influence plant water use, and since climatic conditions vary, the importance of local trials to establish the interaction between the plant population and water use cannot be overemphasized. The ideal, however, may well have to be modified to use existing machinery and thus avoid unnecessary capital investment.

High winds increase evapotranspiration, and close spacing frequently reduces water loss by creating suitable microclimates within a crop. Shading decreases transit loss, and closely spaced rows on raised beds with irrigation furrows at wider intervals are most suitable. Watering alternate furrows, or large applications at longer intervals, can assist in increasing yield per unit of water used, or watering at night when daily wind velocity is lowest. The effect of saline soils or irrigation water on mustard indicates a general susceptibility, but as oilseed rape is often the first crop grown on land reclaimed from the sea in The Netherlands, there appears to be considerable potential for more research into this most valuable attribute.

Intercropping and rotations

Smallholder mustard in India is usually strip-cropped with other dry season (*rabi*) crops including oilseed rape, niger and pulses, the proportions depending on prices ruling that season, and is sometimes undersown in wheat or barley. Mustard is also grown as a pure crop in the rotation small grains–legumes–maize–mustard, and may occasionally be grown as a catch crop between a green manure and sugarcane. The economics of various rotations and their profitability have been reviewed by Sushi Kumar *et al.* (2001). An interesting combination was that of mustard with the multi-purpose tree *Prosopis cineraria* (L.) Druce (*khejri*), in semiarid areas of India. Results showed that shade adversely affected mustard seed yield within the canopy area, but the trees yielded products including fuel, which were socially valuable and thus the combination was profitable (Yadav and Blyth, 1996). Soil populations of root knot nematode were significantly reduced by growing mustard annually between rows of vines and ploughing in the plants, in New South Wales, Australia.

Commercial mechanized mustard crops are generally sown following a cereal crop using minimum tillage or direct sowing techniques, but spring-sown crops require a more conventional seedbed. Mustard should preferably not be grown more than 2 years in succession to avoid a build-up of pests and diseases, neither should it precede or follow a cruciferous crop. A preceding cereal crop allows a high degree of weed control of cruciferous and brassica species and reduces subsequent infestation in mustard.

Pests

Mustard, like many other crucifers, is attacked by a wide spectrum of insects from the seedling stage to maturity, but the damage caused by a particular pest complex varies in different countries. Field evidence indicates that the longer a specific cruciferous crop is grown in a particular area, the greater the damage, even where the crop is not normally a component of local agriculture. Some pests noted in this section are at present minor, but in the author's opinion they could become increasingly important. This is especially true where mustard is grown in a rotation including more favoured food plants that can support a high insect population, or that allow a carry-over from one season to another.

The degree of damage caused by insect pests in mustard is better documented than the relationship between insect damage and subsequent seed yield, since visually severe infestations can not always be correlated with subsequent yield (E.A. Weiss, personal observation). This was also noted on rape infested with seed weevil in Britain, and other studies have indicated that rape has a considerable ability to compensate for leaf and bud loss caused by insects. Thus, in India 40–50% defoliation of lower leaves on mustard plants up to 80 days after sowing had little effect on final seed yield (Chhabra *et al.*, 1996). Such findings are of considerable practical value in assessing the degree of loss that mustard can sustain before protective measures are necessary. This applies particularly to insecticidal sprays toxic to pollinating insects. It is unlikely that a single application of insecticide will effectively control several pests, for these usually attack crucifers in succession. However, correct timing of sprays to control *Meligethes* spp. in British rape crops gives acceptable control of *Ceutorhynchus* spp. Combining other aspects of crop husbandry with pest control can be very effective; for example, integrating fertilizer and water applications with pesticides gave the highest yield and net returns in India (Tomar and Gautam, 1995).

The most damaging pests are pollen beetles, seed-pod weevils, flea beetles and aphids, which basically attack the reproductive parts, while insects attacking the vegetative parts are less damaging and often more easily controlled. The majority of pests attacking seedlings above and below ground are non-specific, i.e. cutworms, wireworms, and grasshoppers, especially *Chrotogonas trachypterus robertsi* from West Asia to India, and are usually controlled by seed dressings or soil insecticides.

Pollen beetles, *Meligethes* spp., are generally a serious pest of brassica crops in more temperate climates, in particular *Meligethes aeneus* (Fig. 2.4). This is a greeny-black beetle, 2.5–4 mm in length. In Europe it overwinters in hedges and woodlands, and migrates to crops when the temperature exceeds 15°C. Eggs are laid on buds and larvae hatch to feed on developing flowers, and it is in this period that the most lasting damage occurs. Damaged buds fail to produce pods, leaving a blind stalk on the raceme which can be clearly seen after the normal pods have developed. Once flowers have opened, they are the preferred food, and since mustard plants flower profusely and continuously, the time and degree of the initial attack determines the level of protection necessary; it has been estimated that up to 50% loss of flowers may cause no significant yield reduction.

The widespread seed-pod weevil, *Ceutorhynchus assimilis* (Fig. 2.5), is damaging on a range of vegetable crops, with the related species, *Ceutorhynchus quadridens*, in the former USSR and Hungary being locally important. Published estimates of infestation by seed-pod weevils are consistently high, with up to 75% recorded in Canada in a particular crop. Infestation usually becomes apparent in late spring, and crops are continually invaded throughout the flowering period. *C. assimilis* is a typical weevil, grey in colour and 2–4 mm long. Eggs are laid in young pods, usually singly, and each grub eats several, though seldom all, seeds before boring an exit hole in the pod wall. Similar to damage caused by pollen beetles, early infestation causes less yield reduction than later. The danger level for these weevils is low, and 1–2 adults per plant at main flowering requires control measures. Adult weevils may be harvested when the crop is combined and can be found in the sample. However, at this stage they cause no damage, since they do not feed on mature seed and normally leave the seed as opportunity occurs.

The damage caused by the weevil can be exacerbated by another common cruciferous pest, the brassica pod midge, *Dasyneura brassicae*, also known as the bladder pod midge. This small black midge uses the exit holes of pod weevil grubs to lay its eggs in the pods. Its larvae are small, white or yellow, and there can several in a pod. Attacked pods ripen prematurely and shed their seed, and autumn-sown crops are often the worst affected. To be effective, control measures should be carried out at the yellow bud stage. Since the midge is greatly dependent on holes made by *C. assimilis* to deposit its eggs, control of the latter will considerably reduce damage caused by the midge. Larvae

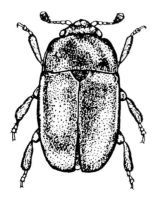

Fig. 2.4. Pollen beetle, *Meligethes aeneus.*

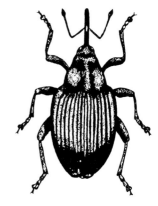

Fig. 2.5. Seed weevil, *Ceutorhynchus assimilis.*

of *Lixus* spp. bore stems and mine leaves, but are of local importance and often confined to one area of a particular crop.

Flea beetles are common pests and are often most damaging on young plants. *Phyllotreta* spp. and *Psylliodes* spp. cause varying degrees of damage almost everywhere cruciferous crops are grown. The beetles are small, 1–2 mm long, shiny, active and dark-coloured, and can occur in very large numbers. Adults eat cotyledons, stems and leaves, while larvae of some species mine stems. The characteristic symptom of flea beetle attack is small, round holes in cotyledons. Fields are often invaded progressively from the edges to the centre, and if infestation is uncontrolled, a crop can be almost completely destroyed. The red turnip beetle, *Entomoscelis americana*, is particularly troublesome in North America, as is *Phyllotreta cheiranthi* in North and East Africa, *Phyllotreta mashonana* in Ethiopia and *Phyllotreta undulata* in Russia, Iran and neighbouring countries.

The armyworm, *Spodoptera exigua*, also known as the beet armyworm and beet webworm, is potentially a most damaging pest where it occurs, together with the related *Spodoptera exempta*. The active caterpillars are up to 25 mm long, and are a darkish green, with two parallel black stripes down their back. Predators, parasites, various fungi and viruses frequently exercise considerable control over larvae numbers, preventing major outbreaks. Another noctuid of importance is *Mamestra configurata*, the bertha armyworm of North America, while *Mamestra brassicae* L., the cabbage moth, is damaging on a wide range of crops from Europe to Japan, but may be only locally important on mustard.

The widely distributed diamond-back moth, *Plutella maculipennis*, feeds on many crucifers. The pale, yellowish-green larvae, up to 125 mm long, feed mainly on the underside of leaves, flowers or young pods, and on the outer surface of older pods that are still green and succulent. The larvae have the habit of wriggling quickly backwards when disturbed, or dropping from plants at the end of a silken thread. *Plutella xylostella* (syn. with *P. maculipennis*) is common in East Africa and Ethiopia. Several white butterflies, *Pieris* spp., can become locally damaging; *Pieris brassicae*, *Pieris napi* and *Pieris rapae* are most often recorded. The last-named is often considered to be a pest only in temperate regions, but in fact its range extends through a wide area of the semi-tropics and elevated tropics. The large white *P. brassicae* is a serious pest in Ethiopia.

Aphids commonly occur in large numbers, the degree of damage being related to the time of infestation rather than its severity. Most frequently recorded is *Brevicoryne brassicae*, the cabbage aphid, in temperate to semi-tropical regions from Europe to India, China and Australia, North and East Africa, Ethiopia and South America. Aphids appear to be most damaging at the rosette stage with severe stunting and distortion of the leaves and developing stem, and at flowering when the developing pods are seriously affected and can fall. In addition to the mechanical damage caused by feeding, *B. brassicae* is also capable of transmitting mosaic virus. Also of wide distribution, and particularly important in Africa, Asia and the Far East, is *Rhopalosiphum erysimi pseudobrassicae* (commonly known as *Hyadaphis erysimi*) the turnip aphid, which also attacks *E. sativa*. In India it is often recorded as *Lipaphis erysimi*, the mustard aphid, and is a serious pest whose depredations may be increased by application of N alone, but reduced when N, P and K are applied in combination (Singh, R.P. *et al.*, 1995). The ubiquitous *Myzus persicae* is generally important for the various diseases it is known to transmit.

Sawflies, mainly *Athalia* spp., can become serious pests in Africa and Asia; *Athalia proxima*, the mustard sawfly, whose larvae feed on leaves and young seedlings in India, Myanmar and Indo-China, can be very damaging. The blackish caterpillars curl and fall to the ground when touched. Several species commonly occur in Africa, especially *Athelia himantopus* and *Athelia sjostedti* in East Africa, while *Athalia rosae*, the turnip sawfly, can be found on crops from Europe through to Asia.

Various sucking bugs have been recorded; in Australia, Rutherglen bugs, *Nysius vinitor*, a major pest of oilseeds, attack pods and continue to do so even after a crop has been windrowed. Another lygaeid, *Nysius cymoides* is common on many crucifers in North Africa and West Asia, and *Lygus rugulipennis* can transmit a virus in Europe. *Bagrada hilaris* f. *cruciferum*, a black and orange bug, occurs from Africa to China and can cause severe damage; in the absence of its main food plants, it will feed on many other crops and indigenous plants, and is thus difficult to control. Another stink bug of wide distribution is the cabbage bug, *Eurydema oleraceum* L.; other species including *Helopeltis* spp., *Campylomma* spp. and *Lygus* spp. have been reported from Mediterranean countries, West Asia and eastern Africa as being persistent minor pests, sometimes locally destructive. The polyphagous *Nezara viridula*, the green stink bug, occurs on plants in North and East Africa and, in the absence of alternative host crops, can be damaging, as can the related *Carpocoris* spp.

Mites are commonly found but seldom reach dangerous levels. The red-legged earth mite, *Halotydeus destructor*, can cause damage to seedlings in Australia and require control. Other species have been reported from Africa, and West Asia to India, but classification of these and allied genera is confused, and many identifications are suspect.

A number of nematodes are recorded from mustard including root-knot nematodes (*Meloidogyne* spp.), root-lesion nematodes (*Pratylenchus* spp.) and the stem nematode, *Ditylenchus dipsaci*. A typical symptom of severe root nematode infestation in a mustard field is a roughly circular patch of dead or stunted plants. Since many of these nematodes are polyphagous, damage is likely to occur where large populations are endemic, and field control is impracticable. Birds, especially pigeons, rabbits, buck and similar small animals can cause considerable crop damage. Drilling autumn-sown mustard into long cereal stubble can reduce pigeon damage; shooting is more permanent and allows some release of growers' feelings!

Diseases

Diseases cause significant yield reduction in mustard, and the most important are described below. The total loss caused by disease is unquantified, and that due to a specific pathogen is generally lacking. Blackleg is one of the most damaging on a world basis, and alternaria black spot can reduce yield by 20%. Seasonal conditions are a major factor influencing the degree or extent of an infestation, together with varietal susceptibility and farming practices. Thus, mustard following mustard or another cruciferous crop invariably suffers greater damage, as does a crop grown from seed harvested from diseased plants. Where diseases are known to be mainly seed-borne, fungicide or warm-water treatment is effective.

A most important disease is blackleg, *Leptosphaeria maculans*, and even in those areas where it was not previously considered to be of major importance (Canada, for example), it is now much more widespread and likely to become more damaging. Symptoms are grey spots dotted with small, black, fruiting bodies on the cotyledons and leaves. Stem lesions may be grey or black and form elongated patches, usually near the stem base. Affected stems often rot or are so weakened that they fracture, commonly at flowering, and a severe infestation can cause major damage. Succeeding mustard crops, or mustard grown close to fields planted to rape, are worst affected, since the spores persist on volunteer and wild brassicas. Spores can also survive on crop residues, mainly roots and stems, for up to 5 years. Crop rotation, clean weeding, burning or ploughing under stubble, and the use of disease-free seed can assist in reducing damage caused by blackleg, but resistant varieties are the long-term solution. The disease is seed- and wind-borne, the latter being most common, and spores can be carried many kilometres to infect areas considered clear.

Clubroot, caused by *Plasmodiophora* spp., varies in importance regionally. An early symptom is plants with yellow leaves wilting at midday, becoming progressively stunted and frequently dying. Roots show swellings and eventually decay (Fig. 2.6). A temperature around 23°C, acid soils and high soil moisture encourage infection. Spores can persist for many years in the soil, and mustard should not be grown following or in close proximity to an infected rape crop.

Fig. 2.6. Club root.

The common damping-off or foot rots are frequently caused by *Pythium* spp., *Phytophthora* spp. and *Thanatephorus* spp. The degree of damage due to a particular pathogen varies regionally, and there may be specific local varietal resistance. The main symptom is a dark-brown or black rot on the stem base of seedlings, sometimes causing a constriction of the stem. Those seedlings that survive a severe attack often produce stunted plants with little seed yield. Another rot caused by *Xanthomonas campestris*, black rot, also attacks many other crucifers and is of economic importance in India, where there are a number of local races. A disease attacking stems, sclerotinia stem rot, is caused by *Sclerotinia* spp. Symptoms are bleached areas several centimetres in length on stems, which later tend to shred. Control is difficult as the disease attacks a wide range of cultivated and wild plants. Canker due to infection by *Phoma lingam* is common and damaging. Symptoms are light-brown or purple lesions on stems and later on pods, and necrotic areas on leaves. Young plants are stunted. The disease is seed-borne and is also spread by infected plant debris (Fig. 2.7).

Symptoms of the widespread and damaging disease, *Alternaria* leaf spot, are black or greyish spots on leaves, stems and pods. On stems, the spots coalesce to form long, irregu-lar blotches, and badly infected pods may split. In a severe infection, almost every plant in a crop may be affected. The disease is seed-borne, and also causes damping-off of young seedlings. Less damaging is ring spot due to *Mycosphaerella brassicicola*. Symptoms are white spots on leaves and large purple to grey speckled lesions on stems.

Mildews caused by *Erysiphe* spp. and *Peronospora* spp. commonly occur and cause varying degrees of damage, but are generally not considered serious enough to warrant control. Black mould, caused by *Clado-sporium* spp., can be very damaging in some years in Canada. The disease not only reduces seed yield and quality but causes respiratory trouble for workers on farms and in stores.

Aster yellows virus has been reported, but appears to be of little importance. Symptoms are oval, watery bladders in place of normal pods. Seed from virus-free sources is the only practical control measure. The virus is reportedly more common in dry regions.

Harvesting

Mustard is mature when the stems and pods become yellow, and when the seeds become very dark, rattle in the pod when shaken and have a moisture content of around 15%. This

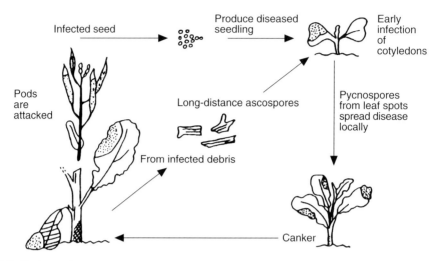

Fig. 2.7. Canker cycle.

is usually some 160–240 days after sowing autumn crops and 80–120 days for spring-sown crops, depending mainly on region and local weather conditions. Once mature, mustard ripens very rapidly and the optimum harvesting period can be very short. Ripening commences at the base and proceeds up the stem, and it is unwise to wait for the crop to become completely ripe before harvesting.

A high proportion of smallholder crops are manually harvested, with plants stooked in the field to dry. If the field is required for a second crop, then plants are transported to the homestead for threshing. This is often by beating plants with sticks, or using one of the now common pedal-threshers. Whichever method is used it should preferably take place on a tarpaulin to ensure no seed is lost. Seed is then winnowed, bagged and placed in storage.

Large commercial crops are usually mechanically harvested, and a range of equipment is available. Many winter-sown crops are windrowed and later threshed with a pick-up combine. It is important a crop be windrowed at a seed moisture content not exceeding 35% to ensure good-quality seed. Windrowing at 45% or above not only reduces the final yield, it also reduces seed-oil and protein content. Windrowing below a seed moisture content of 20% allows plants to become too dry and windrows become

light and fluffy and easily blown about by the wind. This increases shattering loss, and makes pick-up threshing difficult; however, in very windy regions windrowing reduces shattering loss. When late windrowing is unavoidable, a light roller can be used behind the windrower to anchor plants lightly to the stubble. Leaving a 20–25 cm high stubble keeps plants off the ground and allows air to circulate more freely. When the moisture content of the seed has fallen below 14% (some 7–10 days is usually sufficient), windrows can be threshed. Special pick-up attachments on the combine reduce seed loss, but a conventional cutter-bar fitted with long lifting tines can be adequate if used with care.

Spring-sown crops can be combined directly when most pods in the middle of the stalks are yellow and the seeds are brown and have a moisture content of less than 11%. A vertical knife attachment is frequently necessary to divide a heavy standing crop to reduce shattering loss when windrowing or direct combining, but in normal crops a torpedo divider is usually adequate. Windrower/swathers can usually handle some 1.5–2.5 ha h^{-1} in a normal crop. Harvesting in the early morning and late afternoon will reduce shattering loss and night combining has been successful. Very dry seed with 8–9% moisture content often suffers considerable damage during threshing.

A well-threshed, dirty sample that can be recleaned may be more profitable than losing seed out of the back of the combine by endeavouring to do this in the field. Direct combining of standing crops usually produces a dirtier sample than combining from windrows. Current combines contain modifications to suit mustard/rapeseed harvesting, and these produce samples that require little extra attention prior to sale or storage. Desiccation to accelerate harvesting has been successful on canola but no similar data are available for mustard. Yields vary considerably with most smallholders achieving 300–500 kg ha^{-1}, but this can be doubled if selected seed is sown with good management; for commercial crops in more developed countries up to 3000 kg ha^{-1} can be achieved.

Processing

Mustard seed has an a fixed oil content up to 30%, thus quick and efficient drying is essential to prevent heating and rancidity. Seed apparently dries relatively easily, and thus a substantial reduction in moisture content is possible with a single pass through a dryer. Hot or cold air can be used for continuous drying, although smaller amounts of seed can be dried by ducted bin ventilation if the outside air is of low humidity; many cereal driers are suitable for mustard. For bulk drying, 1 m depth of undried seed is recommended, and crust which forms on the surface during drying should be broken. With continuous-flow hot air driers the temperature should not exceed the maker's stated maximum, which can vary from 60 to 80°C depending on the rate of seed flow, and seed should be cooled with dry air. Fixed and volatile oil production is also discussed in the section Products and specifications.

Storage and drying

The high oil content, small seed size and danger of heating require mustard to be handled quickly and efficiently during drying and storage to prevent deterioration and loss of quality. Most combined seed received at storage has a moisture content of between 10 and 25% from standing crops and between 10 and 15% from windrowed. Recleaning is often necessary to prevent contamination and heating in bulk from wet weed seeds and debris.

Clean, dry seed stores well, and is easily moved in sacks or bulk; however, because it is small, round and heavy, it runs freely, and precautions are necessary against transit loss. Seed for export is usually graded to international standards (ISO) or national specifications such as the Indian Agmark. Dry, unbroken seed is generally more free of storage pests than other oilseeds, the hard seedcoat offering considerable protection. However, in India the meal snout moth, *Pyralis farinalis*, rapidly infests improperly stored or damaged seed, and has been recorded as a minor pest elsewhere.

Products and specifications

Mustard produces seed, the most important product, a fixed oil and a relatively unimportant essential oil.

Seed. Mustard seed is the basic material for a wide range of flavourings commonly known as mustard, many with a famous brand name, such as Colman's, or a regional association, such as Dijon. Some major types are later described. The uncrushed seeds can be used as a diaphoretic, diuretic, emetic, expectorant, irritant or stimulant. Tea prepared from the seed is used to cure sore throats and to relieve bronchitis and rheumatism. Hot water poured on crushed seed makes a stimulating foot bath. In the USA, black mustard seed and its derivatives have the status generally regarded as safe, GRAS 2760, and white mustard as GRAS 2761.

ENGLISH MUSTARD. The traditional type consists of finely powdered decorticated black mustard seed, blended with some yellow seed and a little wheat flour, which improves the powder's characteristics by absorbing the fixed oil. This mustard is known as Double Superfine, and is sometimes coloured by the addition of a little turmeric. Warranted Pure is a similar fine mustard in which yellow seed replaces the flour, and can be used by persons allergic to wheat flour and in America and other countries where food laws prohibit wheat flour as an ingredient. Older people who complain that English mustard is not as hot as it used to be are correct, because black mustard seed has virtually been replaced by the more easily available and cheaper but less pungent white mustard seed.

English mustard made correctly with cold water and allowed to develop for 10 minutes is clean and pungent. It is a most useful kitchen spice because it is free from salt, vinegar and strong herbs or other spices. In mayonnaise and salad dressings, mustard helps to stabilize emulsions as well as adding bite. Stale mustard loses a great deal of its essential oil and therefore its pungency. Dry powder in a well-sealed jar will retain its flavour and aroma for a long period. An interesting English mixed mustard named Savora, which is rather mild but aromatic and quite unlike other mustard types, was developed at the turn of the century. It is currently especially popular in South America and is manufactured both in Norwich and Dijon.

Powder mustard was made in Dijon up to the end of the 17th century, and mixed mustard was made in England in Shakespeare's time. National preferences subsequently decreed that most mustard used in Britain is dry, but most mustard consumed in France, virtually the whole of the EU and the USA is wet mixed. The latter is also now gaining a larger market share in Britain.

FRENCH MUSTARD. The two main types of French mustard are pale Dijon or white mustard and the darker Bordeaux mustard. Although the Dijon type represents some 85% of mustard consumed in France, in England French mustard is generally the dark Bordeaux, because Bordeaux was the main French wine-shipping port to England, and the wine was often accompanied by consignments of Bordeaux mustard. This mustard is dark because the seed is not decorticated, and is of low pungency, sour–sweet with vinegar and sugar and contains considerable amounts of tarragon, herbs and spices. Dijon mustard is light in colour, sharp and salty, pungent and hot, but less so than English mustard. Based on black and yellow seed, it is ground wet to a very fine paste and contains no tarragon, although imitations made outside France often do. It is light-coloured as the seeds are decorticated.

The best-known brands are Amora, the market leader, Grey-Poupon and Dijon (founded in 1777) and Bornibus Maille (founded in 1747). Olida from Yvetot in Normandy and Louti from Bordeaux are also popular. Besides the classic mustards there are other important types: Moutarde Florida based on the wine of Champagne is light, clean and very mild; Louit's pimento mustard is flavoured with red pepper; and Maille's Moutarde des trois fruits rouges, which is so red in colour and so mild as scarcely to warrant the name mustard, though it is sour and aromatic, almost with a taste of anise.

OTHER MUSTARDS. German mustard is similar to Bordeaux, being dark, sweet-sour and flavoured with herbs and spices. Dusseldorf is the great German mustard centre. Finally, there is the type of mixed mustard most popular in the USA, which is mild, based usually on yellow seed and rather like a thick piccalilli sauce. It is the type used with the sausage known as 'hot dogs'.

Mustard meal. Mustard meal is made by grinding dry, whole seeds, and should be used fresh or kept in sealed, opaque containers in a cool store. It seldom used domestically as a spice or flavouring. In traditional medicine, mustard meal mixed with water was used extensively as a plaster preparation and to prepare mustard poultices or baths to treat skin ailments, arthritis and rheumatism.

Mustard oil. Mustard oil (the fixed oil) is obtained by expressing whole seeds with a yield of 25–35% and does not contribute to the pungency or aroma of prepared mustard. It is used mainly as a cooking oil, especially in Asia. Commercial oils generally have the following characteristics: straw-coloured to brownish or greenish brown mobile liquid; specific gravity (25°C) 0.914–0.916; refractive index (25°C) 1.4655–1.4670; solidification point −8° to −16°C; saponification number 170–174; iodine number 92–97; insoluble in water, slightly soluble in alcohol, miscible with chloroform and petroleum ether. Significant differences occur in the characteristics of regional oils, and between expressed and expelled oils. The edible oils recently developed in Canada have a high oleic and α-linoleic acid content, and are classed as nutraceutical oils since their high omega-3 fatty acids are considered to reduce the danger of heart attacks and inflammatory diseases. A range of retail products is now available in North America and Australia (Fig. 2.8).

Expressed oil is widely used as a flavour ingredient in processed meat products of all kinds, pickles, condiments, gravies and sauces, frozen dairy desserts, confectionery, baked goods, gelatines, prepared puddings and non-alcoholic beverages. Highest average maximum use level in the USA is 0.02% (about 200 p.p.m.) in gravies. Expressed white mustard oil has a mild odour but is more pungent than black or yellow oils, and is added to prepared mustards to give a stronger taste. Expressed cake is used as stockfeed.

Essential oil. The essential oil is a yellowish, mobile liquid, with an acrid taste and a very pungent, unpleasant odour. The oil is placed in the sulphuraceous, lachrymatory group by Arctander (1960). The oil is obtained from the expressed residues of brown or black mustard seeds. These are macerated in warm water to allow the hydrolysis of sinigrin by the enzyme myrosin as noted previously, and then steam distilled. Yield of oil is 0.5–2.2%, but is generally around 1%, and depends mainly on the particle size of the expressed material. It must be kept in sealed opaque containers in cool storage, and most countries have regulations covering its storage and use, some listing it as a toxic or hazardous chemical. White mustard essential oil is produced by the same method, and has strong pungency but little odour.

Mustard essential oil consists almost wholly of allyl isothiocyanate (above 93%) and is known as allyl mustard oil, virtually duplicated by the easily produced synthetic product, which has all the properties of the distilled oil and is much cheaper; thus, distilled oil has lost much of its former importance. However, the USA and other countries generally require food labels to state when natural oil or the synthetic is an ingredient.

The essential oil is an extremely powerful irritant. It is one of the most toxic essential oils and was raw material for the infamous mustard gas of the First World War. The oil blisters the skin, is lachrymatory and when undiluted it should not be tasted or inhaled. The compound *p*-hydroxbenzyl isothiocyanate from white mustard does not have lachrymatory properties, but has a very pungent taste. Isothiocyanates such as those present in mustard have been implicated in

Fig. 2.8. Mustard oil products.

endemic goiter (hypothyroidism with thyroid enlargement). Volatile oil has strong antimicrobial and fungicidal properties, while the constituent sinigrin is reportedly toxic to certain insect larvae. For this reason it is sometimes used to protect stored grain in India.

Other Brassica mustards

B. campestris L. (rape or field mustard) is grown extensively in India, where it is known as sarson or toria, to produce an edible and cooking oil. Its seeds are not a component of table mustard although it is sometimes used as an adulterant.

B. carinata A. Braun. (Ethiopian mustard) is native to eastern Africa, but it has not been identified as a truly wild plant. It is cultivated in Ethiopia and occasionally in northern Kenya, where it is used as a vegetable, its seed as a condiment, and it is sometimes expressed for the oil.

Unrelated species

These include hedge mustard *Sisymbrium officinale* (L.) Scop., the *erysimon* of the Greek physician Dioscorides; clown's mustard (*Iberis amara* L.); mustard bush (*Salvadora persica* Garc.); garlic mustard (*Alliaria petiolata* (Bieb.) C & G).

3

Lauraceae

This great family of tropical and subtropical plants includes some 40 genera and over 2000 species of trees and shrubs, many aromatic and commercially important. The genus *Cinnamomum* Schaeffer is not well defined, and may contain up to 250 species with many synonyms in common use; a complete revision would be welcome. Members are evergreen trees and shrubs mainly occurring from Asia eastward through Melanesia, the Pacific islands and Australia, with a small number in Central and South America. The most important spice producers are *Cinnamomum cassia* (cassia) and *Cinnamomum verum* (cinnamon), and both also produce important essential oils (Weiss, 1997). The basic chromosome number of the genus is $x = 12$, and *Cinnamomum burmannii*, *C. cassia*, *Cinnamomum loureirii* and *C. verum* are diploids $2n = 24$. The genus *Laurus* L. contains two species, *Laurus nobilis*, the source of bay leaves and laurel oil, and *L. canariensis* Webb & Berth. Laurel is also incorrectly used to prefix the common name of numerous unrelated plants.

The very ancient use of cassia and cinnamon in their natural habitat to flavour food, in social and religious ceremonies and in herbal remedies, and their dissemination along traditional eastern trade routes are well documented. As this trade gradually extended from Asia to Europe, so knowledge and use of the spices grew, creating greater demand. This demand finally sent Europeans sailing in search of their source and thus broke Arab domination of the oriental spice trade, a monopoly the Arabs had maintained for centuries by inventing fantastic stories of cinnamon's origin and the dangers involved in its collection, many retold in the *Thousand and One Nights*. It was to search for the spice islands that Columbus sailed west in 1492 to make landfall in the Americas and to discover gold not spices, while da Gama sailed east, rounded South Africa in 1498 and sailed on to India's pepper coast.

In this chapter, cinnamon will refer to *C. verum* and its products, and cassia similarly to *C. cassia*. Although the designation cassia is well established in world trade and unlikely to be easily changed, it may be misleading as there is in the *Caesalpinioideae* a genus *Cassia*, which includes *Cassia angustifolia* Vahl., whose fruit is the well-known senna pods, and *Cassia siamea* Lam., the common tropical timber, fuel and shade tree.

The botanic origin of commercial cinnamon and cassia products is as follows: Ceylon, Seychelles and true cinnamon from *C. verum*; Canton, China Junk, Chinese, Honan, Kwantung, Kwangsi and Yunnan cassia from *C. cassia*; Annan, Danang, Saigon, Tonkin and Vietnam cassia from *C. loureirii*; Indonesian (or cassia vera), Batavia, Java, Korintji, Koringtoji, Macassar and Padang cassia from *C. burmannii*; Indian cassia may be derived from *Cinnamomum obtusifolium*, *Cinnamomum tamala* or *Cinnamomum sintok*; Philippine cinnamon is obtained from *Cinnamomum iners*, *Cinnamomum mindanaense* and *Cinnamomum mercadoi*, which produces the highest-quality bark. Many of these designations are now of little commercial importance; for example, Chinese cassia bark from the main producing areas in Guangzi (Kwangsi) and Guandong (Kwangtung) provinces is now officially sold as Tung Sing and Si Chang.

Most *Cinnamomum* plantations are valued for spice and essential oil production, and little attention is given to their wood. However, their timber is generally suitable for speciality products including the well-known camphor-wood chests from *Cinnamomum camphora* L., and there may be scope for multipurpose plantations. *Cinnamomum* timber is traded in Malaysia in the medang group, which includes other *Lauraceae* genera.

Identifying the various *Cinnamomum* barks and oils was formerly extremely difficult and time-consuming, exacerbated by the almost universal adulteration. Distinguishing cinnamon from cassia was the most important, since the various cassias can be substituted in end-products. Current analytical methods have virtually solved this problem, provided accurate comparison data or specifications are available, and are also capable of identifying adulterants.

Since cassia products are currently derived almost wholly from cultivated trees, pressure on wild species has been considerably reduced, but germplasm collections are few. The most important, containing germplasm from local wild and cultivated cinnamon species, is at the Indian Institute of Spices Research (IISR), Calicut, India, although a central collection covering the Indian subcontinent and South-east Asia has been proposed (Ravindran and Peter, 1995). Interesting accounts of the cultivation, production and distillation of cinnamon and cassia are contained in Guenther (1950a,b), Brown (1955, 1956), Gildemeister and Hoffmann (1959), and Purseglove (1981a).

Cassia

Commercial cassia bark is obtained from various *Cinnamomum* spp. but only three are important internationally: *C. cassia* (Chinese cassia), *C. burmannii* (Indonesian cassia) and *C. loureirii* (Vietnamese cassia), while bark from *C. tamala* is of importance only in India. Cassia bark was originally obtained from wild trees but growing demand encouraged plantation establishment, with China becoming the major producer and lesser but increasing amounts from Indonesia and Vietnam.

The cassia of European antiquity probably originated in Indo-China and reached the Middle East initially via the sea trade of early Mesopotamian civilizations and later overland via the Silk Road, and subsequently to Greece and Rome. Cassia and its uses were described by the Greek physician Dioscorides in his *Materia Medica* and by Pliny in his *Historia Naturalis*, both of the 1st century AD, and Theophrastus (372–287 BC) in *Enquiry into Plants*, who accurately distinguished between cassia and cinnamon. Both spices are mentioned in the Christian Bible and Jewish Torah, but it is unclear which of the two was involved. In 15th-century England it was considered that 'synamome was for lordes, but canelle (cassia) was for commyn people'. Today in the USA, cassia is preferred over cinnamon, although the powdered product may consist of either a single species or a blend of several.

The three cassia species discussed have a number of characteristics in common, described here to avoid duplication. Deviations will be mentioned in the following sections for each species. They are shrubs or trees, the lower trunk is usually branchless and the sapwood whitish to pale yellow. The leaves with glandular dots are aromatic when crushed; they are usually opposite, alternate or arranged spirally, simple and entire, three-veined more or less prominent and rarely pinnately veined; stipules are absent. The inflorescence is an axillary or terminal cymose panicle of clusters or umbellules of generally whitish or whitish-yellow flowers. The flowers are usually bisexual in threes. The six sepals are subequal, united into a tube at the base and are usually hairy. The fertile stamens are usually nine in number arranged in three whorls, the stamens in the outer two whorls being introrse, those in the inner whorl being extrorse and with a pair of stalked or sessile glands; the anthers usually four-celled. The fourth and innermost whorl consists of three staminodes with slender filaments and empty anthers. The ovary is superior, sessile and one-celled, with a single, pendulous, anatropous ovule; the style is slender, with a discoid or obscurely three-lobed stigma. The fruit is a one-seeded berry, globose or ovoid to cylindrical, with the base surrounded by the enlarged and indurated perianth, the tube often carrying persistent perianth lobes. The seed is without albumen, the testa thin and the cotyledons large, flat, convex and pressed together; the embryo is minute. The essential oils of cassia and cinnamon are all placed in the cinnamom group by Arctander (1960). Pests and diseases attacking cassia species are generally as for *C. verum* and are starred in those sections, and any specifically important will be mentioned.

Chinese cassia

Cassia is regarded in China as a spice of great antiquity, and the earliest herbals of the 3rd and 2nd centuries BC, lost but quoted in later writings, indicate that it was widely known, and its uses were described by Bretschneider (1895), who also stated that the name cassia is derived from the Chinese *kwei-shi*. Cassia bark grew in importance, and when the first Chin emperor, Shih Huang Ti, conquered Nan Yueh in 216 BC, he renamed it Kwei Lin, cassia grove, a name preserved in Kweilin, now Guilin and capital of Guangxi province. The bark was accurately described together with its medicinal uses by T'ao Hun-kin, AD 451–536 (Laufer, 1919). Cassia was mentioned in the *Periplus of the Erythraean Sea*, written about AD 200 (Huntingford, 1976). Kasia bark was listed as shipped from present day Somalia, most likely a re-export of Indian or Indonesian bark, as the author found no *Cinnamomum* spp. native to eastern Africa. Arabs and

Persians later knew cassia bark as *darchini*, from *dar* meaning wood or bark and *chin* meaning China. However, in some countries, including India, *darchini* came to mean true cinnamon! In this section, cassia without qualification refers to Chinese cassia, its products and derivatives.

Botany

C. cassia J.S.Presl. ex Blume. (syn. *Laurus cassia* L.; *Cinnamomum aromaticum* C. Nees) was originally considered a variety of *C. obtusifolium* (Roxb.) C. Nees, which it closely resembles. Chinese cassia is known generally as *kwei* or *kui* in China, and more widely as cassia lignea in international commerce. In English, it is known as Chinese cassia, in French as *cannellier de Chine*, in German as *zimtkassie*, in Swedish, Dutch and Italian as *kassia*, in Japanese as *kashi keihi*, in Laotian as *sa*, in Thai as *kaeng*, in Vietnamese as *qu thanh*, and in Indonesian as *kayu manis cina*.

C. cassia is an evergreen tree growing up to 20 m in height, naturally conical in shape with small white flowers and coarse greyish-brown bark. The tree is aromatic in all its parts, but only the bark, bark oils and leaf oils are economically important. Unlike *C. verum*, all its oils are similar in composition.

The roots are deeply penetrating and extensive when cultivated, with normally a single taproot and several substantial laterals. Cuttings root very slowly and it is often 6–8 months before sufficient roots develop to allow transplantation. The slow rate of above-ground growth is probably due to this tardy root growth, since at 5–7 years, trees may reach only 1.5–2.5 m.

Normally the central trunk reaches 25 cm diameter at breast height (DBH) at 10–15 years, growing up to 70 cm if left uncut. The trunk is normally well branched from low down, and the progressively acute angle of branches gives uncut trees their conical shape. Cultivated trees grown on fertile soil for bark production are usually cut back almost to ground level at 5–7 years old; subsequent regular cutting keeps trees to 3–3.5 m high; prunings, twigs and leaves can be distilled to produce leaf oil. Properly maintained plantation trees may have a useful life of 40–60 years. Trees used to produce cassia buds or fruit are left uncut.

The bark on young shoots and stems is smooth and greyish, and on mature trees is rough and greyish brown. Bark from trees growing at higher altitudes in Guangxi Province is of better quality than trees from lower levels, whose coarser bark is the cassia lignea of commerce. Bark from young shoots is removed, scraped to removed the outer bitter bark, then dried to become the cassia bark of commerce. Bark from mature wild trees is locally considered especially valuable for medicinal use.

The aromatic inner bark is brown to brownish red, the phloem containing irregularly shaped oil cells, and unlike cinnamon may contain 10–12% mucilage. A detailed description of the morphology and histology of various cassia barks is contained in Parry (1962). Inner bark contains 1–4% essential oil and Guangxi bark generally has a higher average oil content than Guandong; the main constituent is cinnamaldehyde at 60–98%. Few data on bark constituents are available, and published analyses of bark oil vary widely, apparently due to similar factors affecting these characters in *C. verum* (q.v.). Main characteristics of commercial oil in the USA are: specific gravity (25°C) 1.045–1.063; optical rotation (25°C) −1° to +1°; refractive index (20°C) 1.6020–1.6060; slightly soluble in water, soluble in an equal volume of alcohol and glacial acetic acid.

The leaves are alternate or nearly opposite, the laminae are oblong-elliptic to lanceolate, acuminate, basically cunate, petioles 1.0–1.5 cm long. Young leaves are light green, flaccid and soft; mature leaves are stiff, thick, leathery, 8–20 × 4–8 cm, shiny dark green above and finely hairy below. Twigs are brown and hairy. Fresh leaves may contain up to 1.6% essential oil containing 80–95% cinnamaldehyde. Leaf callus tissue and cell suspension cultures show production of (−)-epicatechin, procyanidin B2, procyanidin B4 and procyanidin Cl, precursors of condensed tannins, which are assumed to be the main components responsible for the plant's medicinal effects; cinnamaldehyde is not present.

The inflorescence is a lax, spreading, terminal or axillary silky-tomentose panicle, 7.5–15 cm long; the flowers are small, pubescent, white or whitish yellow and in cymose arrangement of threes on short 2–3 mm pedicels; the perianth is pale yellow and deeply divided into six oblong, blunt lobes, abscissing transversely at the base after anthesis. The main flowering period in China is June–August. The cassia buds of commerce are immature fruits.

The fruit is an oval/elliptic, black to blackish-purple pulpy drupe, 1.0–1.5 cm long, containing one seed borne in the shallow cup of calyx lobes. Fruits mature in February–March and are avidly eaten by birds, which are the main seed-dispersal agents. Immature fruits are dried and become the misnamed cassia buds, *Flores cassia*, of commerce. These dried fruit, known as *kui tsz* in China, are a smooth greyish or reddish brown, 6–10 mm long; the calyx is hard and wrinkled, and contains up to 2% essential oils, whose main constituents are aldehydes up to 80%. The morphology and histology of cassia buds has been described in detail (Parry, 1962). The seeds are oval, 1 cm long and dark brown with lighter stripes; there are usually 2500–2700 kg^{-1}.

Ecology

Chinese cassia is a tree of subtropical forests and occurs naturally in south-eastern China, northern Assam and Myanmar, and within 30° N and 30° S latitudes has adapted to a range of environments. Chinese cassia is cultivated mainly in the southern Chinese provinces of Guandong and Guangxi and in Vietnam, and has been introduced to Hawaii, Indonesia, South America, Sri Lanka and the southern USA, where trees generally have flourished but are not commercially cultivated.

Cassia prefers hot, sunny and humid conditions, with temperatures of 15–30°C, but is adaptable to a very wide seasonal range, from the high 30s to −7°C, and also to very large differences in day–night temperatures. The main areas of cultivation in China have an average summer maximum of 38°C and minimum 28°C, and a winter maximum of 15°C and minimum −0.5°C. Mature trees are resistant to sporadic light frost, with the lowest recorded temperature tolerated in China being −2.5°C, in Florida −6.5°C, and below freezing in Paraguay (Anon, 1976; Plucknett, 1978).

Cassia's natural habitat is in the monsoon region where rainfall occurs as very heavy seasonal storms interspersed with dry periods of irregular length. Trees can therefore withstand short periods of waterlogging or dry weather, but prolonged periods of either adversely affect growth. Thus, Chinese plantations are situated on the middle slopes of high hills or low mountains facing north-east. A minimum annual rainfall of 1500 mm is necessary but there is apparently no upper limit where soils drain freely. Chinese plantations are usually below 500 m, and frequently below 100 m, but wild trees are found up to 1500 m. Bark from trees growing at higher elevations is generally thicker and less aromatic. Cassia is a forest tree and thus partially shade-tolerant, especially when young, but mature cassia grows well in full sunshine.

Soils and fertilizers

Cassia grows naturally on well-drained, severely leached hill or mountain soils of low fertility, which are often strongly acid (pH 4.5–5.5); thus cassia is most suitable for areas with few alternative cash crops. When cultivated on hillsides, special terraces are constructed or tree rows follow the contour to prevent erosion, and they are irrigated where possible. Generally the only fertilizers applied to plantations in China and Vietnam are organic wastes or mulches. Fertilizer recommendations for cinnamon are equally applicable to cassia under the same standard of management.

Cultivation

Cassia is usually grown from seed, but also from cuttings or layering. Ripe fruit is obtained from trees left for the purpose and young fruit bagged to protect from birds and insects. Ripe fruit is picked and treated as for cinnamon. Viability falls rapidly and seed should be sown within 7–10 days; germination is normally about 90% after 20–40 days. Seed can be sown in pots or prepared beds, 1.5 cm deep with 3 cm between seeds. Most seedbeds are located under trees and thus intermittent shade is provided naturally until seedlings are strong enough to be planted out, after 1–3 years.

Cuttings are taken from ends of new growth, with one or more leaves, depending on local preference, and allowed to wilt before planting in nursery beds where they remain for up to 2 years. The optimum time for taking cuttings is when flaccid young leaves of new growth become firm and horizontal. Single leaf cuttings, with 5 cm stems and the top axillary bud, must be exposed to full light to promote development. Rooting is slow, taking 2–18 months, and the strike rate is usually about 50%. Preparation of planting holes is generally as for cinnamon, and seedlings are planted out at 15–18 cm high. A spacing of 1 m in 1–1.5 m rows is common in plantations, but spacing is often regulated by the size of individual fields and terraces.

Saplings are cut almost to ground level at 5–6 years; stumps are trimmed and left to regenerate. Subsequently the strongest three to four shoots are allowed to develop and cut when 2–4 cm in diameter. Leaves and rejected bark can be distilled. Some trees are left uncut to produce seed and cassia buds, usually 15 × 30 m apart, and these provide intermittent shade. Plantation life is 40–60 years, although dead or diseased trees are replaced. Weeding and plantation maintenance is generally as for cinnamon in Sri Lanka. Irrigation, especially of young tress, is common where water and topography allow. Water is normally applied to more mature trees only during periods of drought in the growing season. Underplanting with other crops is uncommon.

To obtain maximum leaf and twig production in Hawaii, trees are pruned to provide a vertical framework of stems and branches, at a minimum population of 10,000 trees ha^{-1} (Plucknett, 1978). The Hawaiian climate allows leaves to be harvested three or four times annually.

Harvesting

Bark collection, stripping and drying is basically as for cinnamon in Sri Lanka. Immature fruit (cassia buds) are picked from trees that have been especially left in plantations, or collected from wild trees in October–November. Fruit is dried either in thin layers on shelves in sheds or in the shade outdoors, and is regularly turned and the diseased fruit removed. Artificial drying is virtually unknown.

Distillation

Leaves are collected for distillation when bark is harvested, but prunings and excess foliage can be distilled at any time. Unwanted bark, leaves, twigs, small stems and immature fruit can be distilled together since their oil characteristics are similar, and although quality is lowered, this oil finds a ready market locally. Distillation procedures are generally as for cinnamon (q.v.) with an oil yield of 0.25–0.75%.

Products and specifications

Chinese cassia produces cassia buds and bark used as spices, and leaf and bark oils.

Buds. Chinese cassia buds correctly dried have a slight odour of cinnamon and a warm, sweet, pungent but never bitter flavour. Buds generally resemble small cloves, are usually added whole to foods and are a common ingredient of sweet pickles. While their use is widespread in Asia and the Pacific region, both as a spice and in traditional remedies, buds are less popular in Western cooking where references to cassia in recipes usually mean the bark. Major constituents are as in the oil.

Bark. Chinese cassia bark of superior quality is a uniform light reddish brown, thin and aromatic, with an essential oil content of 1.5–2.0%; constituents are as in the oil. The bark resembles cinnamon quills, but is usually coarser, often thicker and less well prepared. Grading by types is basically similar to cinnamon bark. Many brands were previously available, the best being Honan and Yunnan, but as noted these are now sold as Tung Sing from Guangxi Province and Si Chang from Guandong Province in various grades. The Chinese government has supported attempts to establish a standard product, but as much cassia bark is reprocessed and repacked in Hong Kong, the end product remains variable. Bark is also available in broken grades and pieces. Bark is usually sold retail as quills or powder (Fig. 3.1), and the powder is an ingredient of the well-known Chinese Five Spices.

Fig. 3.1. Dried Chinese cassia bark.

The various bark products are widely used in Chinese herbal medicines to treat a range of ailments, including diarrhoea, gripe, malaria, coughs and chest complaints. Powdered cassia bark is listed in the British *Herbal Pharmacopoeia* as a specific remedy for flatulent dyspepsia or colic with nausea. In the USA, cassia bark whole or powdered has the status generally regarded as safe, GRAS 2289.

Leaf oil. Cassia leaf oil is obtained by distilling fresh or wilted foliage and small twigs. Crude oil is a brownish-yellow to dark-brown viscous liquid, which can deposit a resinous sediment during storage. Its odour is similar to rectified oil, but more tenacious, and is thus preferred by food processors and cola beverage makers to whom the dark colour is no disadvantage. Rectified oil is clear, pale to dark yellow and less viscous, with a persistent, sweet, spicy warm note. The flavour is warm, spicy, balsamic and very sweet. Rectified oil is also used in the flavouring industry, but its clarity enables it to be used in perfumes, toiletries, soaps and similar products. Some 40 constituents have been identified in Chinese leaf oils, not all present in one sample; the major constituents are cinnamaldehyde (70–95%) and coumarin (12–18%) (Lawrence, 1994a). Characteristics of Chinese leaf oil are shown in Table 3.1. Properties of rectified oils reflect those of the parent crude.

Genuine leaf oil differs markedly from many commercially available oils sold as Chinese leaf oil, as adulteration was previously common, but detection is now much more simplified and adulteration is less practised. Chinese export quality oil is officially available in two grades containing 85% or 90% cinnamaldehyde and is lead-free. Although oil quality, characteristics and cinnamaldehyde content are governed by official standards, these are not very efficiently enforced. In European phytomedicine, cassia leaf oil (0.05–0.2 g daily intake) is used in herbal drinks, for loss of appetite and dyspeptic disturbances, and has antibacterial and fungistatic properties.

Bark oil. Bark oil is similar to leaf oil but is of little commercial importance as leaf oil is easily available and cheaper. Bark oil contains cinnamaldehyde as its major constituent up to 95%; other constituents are benzaldehyde, coumarin, methyl salicylate and salicylaldehyde but no eugenol (Vernin *et al.*, 1990). A monograph on the physiological properties of bark oil has been published by the Research Institute for Fragrance Materials. An oleoresin can be prepared by solvent-extracting bark, usually by the end-user. In the USA, cassia bark oil is in the category generally regarded as safe, GRAS 2258/2290.

The leaf and bark oils from other Chinese *Cinnamomum* species have been described (Zhu *et al.*, 1994), and from Chinese cassia accessions grown in Kerala, India (Krishnamoorthy *et al.*, 2000).

Table 3.1. Physical characteristics of Chinese cassia leaf oil.

Characteristic	Crude oil Commercial average	Rectified oil US Pharmacopoeia Standard	Rectified oil British Standard Specification
Specific gravity (25°C)	1.055–1.070	1.045–1.063	1.049–1.067
Refractive index (20°C)	1.600–1.606	1.602–1.615	1.600–1.614
Optical rotation (25°C)	−1°10′−+6°0	−1°−+1°	ng
Acid number	6–15 (up to 20)	ng	< 7.0
Cinnamaldehyde (%)	75–90	> 80	ng
Carbonyl value	ng	ng	> 340
Solubility (v/v 70% alcohol)	1:2–3	1:2	1:3

Figures in columns show the range.
ng, not given.

Other Cinnamomum *species*

A related species is *Cinnamomum deschampsii* Gamble; it is of unknown origin although it has been collected in Peninsular Malaysia and Singapore but was probably introduced. It occurs in the lowlands, in fields and along roads. *C. deschampsii*, in Malaysia known as *kayu manis*, is a bushy shrub up to 1.5 m tall and with a trunk diameter up to 30 cm. The bark is thick, very aromatic and with a very pleasant flavour resembling Chinese cassia, for which it is a substitute and an occasional adulterant. The leaves are coriaceous, the petiole 0.5 cm and the blade oblong to elliptical-ovate, 7.5–15 × 5–7 cm. The inflorescence is a lax spreading panicle with silky flowers. The fruit is an ellipsoidal one-seeded berry, about 1 cm in size.

Indian cassia

Indian cassia products are obtained from *C. tamala* (Buch.-Ham.) Nees & Eberm., native to the subtropical Himalayas from Kashmir to Assam and Bangladesh. In Sanskrit it is known as *tejpatra*, from which most modern names are derived, in Hindi and Bengali as *tejpat* or *tajpat* and in Gujarati as *tamalpatra*.

Two types of *C. tamala* are known, one whose leaf oil is high in cinnamaldehyde, the other in eugenol. Their epidermal characteristics, and those of other north Indian *Cinnamomum* species have been described and differentiated (Akhil and Nath, 1997). The two *C. tamala* types differ in appearance and range, though the latter overlaps, and grow naturally in forests between 900 and 2000 m on the Himalayan foothills. Both grow easily from seed dispersed mainly by birds, and almost pure stands can be found in some areas, for example Sylhet, Bangladesh. Bark from both is used to adulterate *C. verum* bark.

C. tamala (eugenol type) is an evergreen primarily forest tree generally below 8 m, with the main trunk up to 45–50 cm DBH and the bark mucilaginous. The leaves are large, dark green, generally lanceolate, acute and on short, stout petioles. The flowers are on loose terminal panicles. The fruit is a small pulpy, purplish drupe containing one small brown seed. The tree is very long-lived, and in northern India when only leaves are harvested, many trees over 100 years old exist. The eugenol type occurs generally over the natural range of the species in India and Bangladesh, and wild trees make a substantial contribution to the total of harvested leaves and bark. The tree is also widely cultivated in large and small plantations to provide dried leaves as spice.

C. tamala (cinnamic type) is generally slightly smaller, up to 7.5 m tall, with a central trunk of 30–40 cm DBH when mature. The leaves are opposite, glabrous, green to dark green, ovate-oblong, acute and on short petioles; the young leaves are lanceolate, acute and initially pink. The flowers and fruit are as for the eugenol type. The main flowering period is in May and fruits ripen in June–July. This type of *C. tamala* is more restricted in its range, although common in a specific area, and especially plentiful in the hills around Nainital, India.

Cultivation and harvesting

Wild trees supply a significant proportion of the bark and leaves produced, and where wild trees are numerous they virtually become plantations owned by individual families who manage and harvest them. Both types of *C. tamala* are also cultivated, and preparation of seedbeds, care of seedlings or cuttings and transplanting are similar to *C. cassia*. Seedlings are transplanted into the field at 4–5 years old, at a spacing of 3 × 2 m or 3 × 3 m. Shade trees are planted or trees retained in new clearings for the first 8–10 years.

Trees are not usually harvested for bark, and the first leaf harvest is around 10 years of age and continues annually until the tree dies. Plantations receive little regular attention once established, mainly slashing or removal of vegetation and protection from fire. Leaves are harvested after the main monsoon rains have ceased, in October–December or up to March during dry periods, as rain reduces the aroma of the leaves and depresses the oil content and thus their value as a spice. Annual yield is 10–20 kg leaves per tree.

Leaves are dried and sold to traders, or fresh and dried leaves are sold as a spice in local markets. Leaves for distilling are normally partially dried and then sold to a local distillery as few growers own stills. Although cultivation of wild plantations appears perfunctory, analyses of leaves harvested from these trees show an oil content of 45% compared with an average of 33% from wild trees (Bradu and Sobti, 1988). Trees harvested for bark are treated similarly to cinnamon in Sri Lanka.

Products and specifications

Indian cassia supplies bark and bark oil, leaves and leaf oil, but dried leaves are the most important.

Leaves. Leaves are dried and sold as spice, and fresh leaves are added to food and used as a vegetable, and also used to flavour vinegar. The large number of Indian citizens now living overseas has created a minor export demand. Leaves are also designated *Folia malabathri* in medical literature.

Bark. Bark is stripped and processed similarly to cinnamon in Sri Lanka, and although the final product is of lower quality and less uniform, it is cheap and finds a ready market locally and in adjoining countries. The bark is coarser than *C. verum* bark, although possessing the true cinnamon odour, and is a substitute and adulterant for true cinnamon.

Leaf oil. Leaf oil is obtained by steam distilling fresh, wilted or dried leaves and twigs with a yield of between 0.15 and 0.25% from the generally inefficient local stills. It is sold locally. Leaf oil characteristics from the two types of *C. tamala* are shown in Table 3.2. and the main components in Table 3.3. A sample of the cinnamic-type oil distilled from dried leaves bought at a local market had a cinnamaldehyde content of 61%. The oil formulated as an ointment is effective in the treatment of ringworm (Pratibha *et al.*, 1999).

Bark oil. Bark oil is obtained by steam distilling bark, twigs and bark pieces, but is seldom produced and not available commercially. A pale yellow oil obtained from *C. tamala* cinnamic type had 70–85% cinnamaldehyde, with the odour of true cinnamon, but fainter and coarser.

Indonesian cassia

C. burmannii is native to South-east Asia but cultivated only on Java and Sumatra (Indonesia) and in the Philippines. In Indonesia, trees are also frequently planted along roadsides and for shade.

Botany

C. burmannii (C. & T. Nees) Blume. (syn. *Laurus burmannii* C. & T. Nees; *C. mindanaense* Elmer) is known in the trade as Indonesian or Padang cassia. In English it is known as Indonesian cassia, in Indonesia and Malaysia as *kayu manis* and in the Philippines as *kanela* or *kalingag*. In Indonesia, two main forms of *C. burmannii* are recognized, one with red young leaves occuring at higher altitudes, the other with green young leaves at lower altitudes. The first, *korintji*, produces the highest quality bark.

  C. burmannii is an evergreen shrub or small tree, up to 6–8 m high and sometimes higher. The leaves are sub-opposite, the petiole 0.5–1 cm and the laminae are oblong-elliptical to lanceolate 4–14 × 1.5–6 cm, with three ribs often not continued to the apex. The young leaves are red and finely hairy, the mature leaves glabrous and a dull glossy green above and glaucous reddish below (Fig. 3.2). The leaves contain up to 0.5% essential oil whose main constituent is cinnamaldehyde. The inflorescence is a short axillary raceme; the pedicel is 4–12 mm in length; the perianth is 4–5 mm, and the lobes absciss transversely after anthesis; the stamens are about 4 mm and the staminodes 2 mm. The fruit is an ovoid berry up to 1 cm in size (Fig. 3.2).

Ecology and soils

Cassia occurs in Indonesia from sea level to 2000 m but in the important production area of Padang, Sumatra, with an evenly

Table 3.2. Physical characteristics of two types of Indian *Cinnamomum tamala* leaf oil.

Characteristic	Eugenol type	Cinnamic type
Specific gravity	1.025 (20°C)	0.9563–0.9843 (30°C)
Refractive index	1.526 (20°C)	1.4942–1.5038 (30°C)
Optical rotation	+ 16°37′	+0°12′ to −0°6′
Aldehyde content (%)	ng	24–37
Eugenol content (%)	60–80	ng
Solubility (v/v 70% alcohol)	1:1.2	ng

ng, not given.

Table 3.3. Main components of Indian *Cinnamomum tamala* leaf oil (%).

Component	Cinnamic type	Eugenol type
Camphor	0.90	ng
Linalool	15.28	0.2
Pinene	ng	1.5
β-Caryophyliene	7.26	ng
p-Cymene	ng	3.2
α-Terpineol	1.54	0.3
Benzyl cinnamate	1.87	ng
Benzaldehyde	2.00	0.2
Cinnamaldehyde	55.19	ng
Eugenol	4.23	7.8
Eugenol acetate	2.06	1.2

ng, not given.

Fig. 3.2. Fruiting twig of *C. burmannii*.

distributed annual rainfall of 2000–2500 mm, trees flourish between 500 and 1500 m. Fertile sandy loams yield the best quality bark. The main fertilizers applied are crop and animal wastes spread around trees in the dry season, with a small amount of any locally available phosphate placed in planting pits. The Department of Agriculture recommends a similar fertilizer programme to that used on cloves.

Cultivation

Indonesian cassia is usually grown from seed but also from cuttings which are treated similarly to *C. cassia*. Fresh seed, treated as for cinnamon, is essential and germinates in 5–15 days, and seedlings remain in the nursery for 8–12 months. Care is necessary when planting out as damage to taproots increases the incidence of stripe canker. Recommended field spacing is 2–4 × 2–4 m. Weeding, plantation management and harvesting are as for cinnamon in Sri Lanka. An average-sized tree yields about 3 kg of stem bark and 1.5 kg of branch bark. In a crop cycle of 10 years, the total bark yield is about 2 t ha^{-1}.

Products and specifications

Indonesian cassia produces bark, bark oils and leaf oils, and dried buds but only the dried bark is commercially important.

Bark. Bark is available in several grades according to length, colour and quality; good quality bark is a uniform reddish-brown and has a pleasant cinnamon odour with no trace of mustiness. Although the Indonesian government introduced standards for the various grades, adulteration with bark of other *Cinnamomum* spp. or with aromatic bark of unrelated trees is common. Thus, data said to be derived from Indonesian cassia bark or bark oil should be treated with caution.

Bark oil. Bark oil is obtained by steam distilling bark, bark pieces and chips, with yield of 1.5–4.0% for *korintji* and 1.0–3.0% for *padang* (Batavian). The oil is colourless to brownish-yellow, with an agreeable aromatic odour, and a sweet, pungent but not astringent taste. The major constituent is cinnamaldehyde up to 80% plus up to 12% phenols. Minor constituents include α-terpineol, coumarin and benzaldehyde, but apparently no eugenol (Lawrence, 1969); however, another analysis gave 1,8-cineole and α-terpineol as major constituents (Ji Xiao-duo *et al.*, 1991). Trials in Italy indicated that the oil successfully treated honeybee colonies infected with foul brood (Carpana *et al.*, 1996).

Leaf oil. Leaf oil is obtained by distilling fresh or wilted leaves with a yield reportedly between 0.3% and 0.5%. The major constituent is reportedly cinnamaldehyde up to 65%, but a Chinese oil had as its main constituents 1,8-cineole and borneol (Ji Xiao-duo *et al.*, 1991). The oil is similar to cassia leaf oil, but is not available commercially, although it could easily be produced in quantity if the market expanded.

Vietnamese cassia

C. loureirii is the source of Vietnamese cassia, and only the bark is commercially important. Wild trees are harvested or cultivated to a minor extent generally in Indo-China, but Vietnam is the main producer.

C. loureirii C. Nees (syn. *C. obtusifolium* (Roxb.) C. Nees, *C. burmannii* var. *loureirii* C. Nees ex Watt.) is frequently referred to in the literature by its first synonym. A common name for *C. loureirii* in arboreta or botanic gardens is the cassia flower tree. *C. loureirii* is also confused with *Cinnamomum pedunculatum* Presl. (syn. *Cinnamomum japonicum*) native to China, Korea, the Ryukyu Islands and Japan, and with *Cinnamomum sieboldii* Meissner, also native to the Ryukyu Islands, introduced to and widely planted in Japan. Vietnamese cassia occurs naturally from the eastern Himalayas, through Myanmar, Thailand and Malaysia to Indo-China, especially Vietnam.

Vietnamese bark is known in the trade as Saigon, Annam or Tonkin cassia; in English it is known as Saigon cassia or cassia lignea, in French as *canelle de Saigon*, in China as

Annamese cinnamon, in Vietnam as *qu thanh ho* and in Laotian as *khe*. Bark from selected trees in the Thanh-ha province of Vietnam was known as *qu thanh* at the royal Hue court, capital of the Nguyen dynasty. The main production areas in Vietnam have an annual rainfall of 2500–3000 mm, with well-weathered lateritic soils.

C. *loureirii* is a medium sized tree up to 8–10 m in height, similar in shape to C. *cassia*. The leaves are opposite/alternate, green to dull green, three-ribbed, rigid, elliptic or oblong, alternate-acuminate and 7.5–12.5 × 3–5 cm in size; the petiole is 12–15 mm in length. The flowers are very small and yellow/white. The fruit and seed are similar to C. *cassia* but smaller. The wood is reddish-grey, moderately hard and glossy, and is used for furniture and interior house fittings.

C. *loureirii* is usually raised from seed, but also by cuttings, layers or suckers similarly to C. *verum*. Seedlings remain in seedbeds for about 1 year, and are planted out when about 1 m high at a locally favoured spacing – close for thin bark, wider for thick bark. Young trees are pruned of side branches to allow growth of a tall, straight, unbranched trunk. Cultivation is similar to that of Indonesian cassia. First harvesting is at 10–12 years, when the bark is removed from standing trees and larger branches, sometimes with the help of bamboo scaffolding. Horizontal cuts are made 40 cm apart, and vertical cuts 25–35 cm apart, yielding bark slabs. Trees are then felled and inferior bark cut from smaller branches. Stumps are left to regenerate, and later a straight, sturdy shoot selected to become the next tree. Preparation of bark is similar to cinnamon in Sri Lanka, but sometimes bark slabs are retained for processing by buyers.

Products and specifications

Vietnamese cassia produces bark and bark oil, dried buds and leaf oils. Bark is the only product of commercial importance and was sold under a number of regional names with Annam and Tonkin origins considered superior, but most bark has been available under the general name Saigon cassia. The Vietnamese government introduced the name Vietnamese Cassia as a standard without reference to regional origin and established four official grades, thin, medium, thick and broken, corresponding to the four original Saigon grades. However, the grading system is virtually ignored and bark from several *Cinnamomum* spp. or different grades can be found in most consignments. The major bark constituent is cinnamaldehyde, but information on other components and their relative abundance is lacking. Food processors do not usually differentiate between Vietnamese and other cassias in the USA, and spice processors generally sell quills and powdered bark retail as cassia/cinnamon, although consumers reportedly prefer the Vietnamese product.

Bark oil. Vietnamese bark oil is locally produced in very small quantities by hydro or steam distillation, with a reported yield ranging widely between 1.5 and 7.5%, but with averages around 4%. The oil is dark yellow to yellow-brown, the scent aromatic and the taste sweet, pungent and slightly astringent. The major constituent is cinnamaldehyde up to 80–95%, and of the minor components only six are above 0.3% (Lawrence, 1995a). The proportion of minor components is unreported. Caution is necessary when assessing data or analyses purporting to be from Vietnamese bark and oil, similar to Indonesian products. A bark extract was reportedly effective against *Alternaria alternata* in Vietnam (Kim *et al.*, 1996). In Japan a light-brown root bark oil was found to contain 80% cinnamaldehyde, plus camphene, linalool and cineole, but is of technical interest only.

Leaf oil. Leaf oil is obtained by distilling leaves and twigs, fresh or wilted, with a yield variously reported as 0.2–0.4%. Little information is available on Vietnamese distilled oil, but a leaf oil from C. *loureirii* distilled in Japan was a light brown, major components being linalool up to 40%, aldehydes (mainly citral) up to 27% and cineole. In the USA the regulatory status generally recognized as safe has been accorded to Vietnamese cassia, GRAS 2289, its bark oil, GRAS 2290, and its leaf oil, GRAS 2292.

Cinnamon

It was as spices that cinnamon and cassia first became commercially important and cinnamon is still considered one of the finest sweet spices, with cassia a substitute. Reports of cinnamon usage in Dynastic Egypt or ancient China are considered suspect, and probably referred to cassia. Both cassia and cinnamon were familiar spices in Greece and Rome and often used to excess; the Roman Emperor Nero is stated to have burned 1 year's supply of Rome's cinnamon at his wife's funeral. Ceylon cinnamon was mentioned by the Arab writer Kasawini about AD 1275, and by a Minorite friar, John of Montecorvino, in 1293. The Arab traveller Ibn Batutah in 1340 described the huge bark storehouses, *bangasalus,* which then existed in what is now Colombo, and in the 18th century the Dutchman Francois Valentijn described bark harvesting in detail (Arasaratnam, 1978).

The Portuguese captain Vasco da Gama reached Calicut, on the Malabar coast of India, at the end of the 15th century and found the spices he was seeking. He loaded 400 quintals of cinnamon, other spices, and a letter from the Zamorin, the ruler, to the King of Portugal, which da Gama wrote on a palm leaf with an iron pen, stated: 'Vasco da Gama a gentleman of your household came to my country, of whose coming I was glad/My country is rich in cinnamon, ginger, pepper and precious stones/That which I ask you in exchange is gold, silver, corals and scarlet cloth.' On his second voyage da Gama returned to Lisbon with 30,000 quintals of the spices mentioned by the Zamorin (Jones, 1978).

When the Portuguese arrived in Ceylon in 1505 they found that, 'Bark was the king's monopoly and obtained from wild trees which grew on the west coast in a strip 20–50 miles wide and 200 miles long between Chilaw and Walawe', and later the historian Ribeiro described various methods used to collect and prepare bark in the Kingdom of Kotte (*Fatalida historica da Ilha Ceilao,* Lisbon, 1836, trans.). The Portuguese were fearful of having only one source of cinnamon and, faced with growing competition from the Dutch, made several unsuccessful attempts to introduce cinnamon and other spices to their Brazilian colonies.

Systematic cinnamon cultivation in Sri Lanka began between 1767 and 1770 during the Dutch occupation, by a colonist named de Kok. The Portuguese and Dutch maintained cinnamon production as a state monopoly, which was abolished by the British in 1833, who also removed the very high export duty in 1843; subsequently large shipments were made to Europe. Records indicate that the Portuguese extracted an annual tax of 110,000 kg bark while the Dutch produced about 180,000 kg annually; under the British this rose to 500,000 kg. The area under cultivated cinnamon in Sri Lanka was 16,000 ha in 1850, falling to 14,000 ha in 1970 (Wijesekera *et al.,* 1976). The Dutch also introduced cinnamon into their East Indian colonies, particularly Java, and Indonesia is now a major producer and exporter.

Cinnamon was brought to the Seychelles in 1771 by Pierre Poivre (Fock-Heng, 1965), and by 1880 trees had colonized the deforested slopes of Mahe Island, where they became the dominant species of secondary vegetation on the middle and higher slopes of the mountains. Cinnamon also successfully invaded coconut groves and the calcareous coastal flats. The tree was initially ignored in favour of vanilla, but when vanilla prices collapsed, locals turned to cinnamon bark production. In 1908, bark exports reached 1202 t but unrestricted felling of mature trees without replanting reduced exports to 200 t in 1915. Attention then concentrated on leaf oil production, which rose from 15 t in 1916 to 100 t in 1950. Although tourism is now the country's major industry, bark and leaf oil production has steadily increased since 1995 to reach an estimated 600 t in 1998.

Where and when cinnamon oil was first produced is unknown, but it was steam distilled by the Dutch in Sri Lanka during their occupation. A European mention of cinnamon oil is in German documents of the 16th century, when an ordinance was published controlling its price.

Botany

C. *verum* J.S. Presl. (syn. *C. zeylanicum* Blume; *Laurus cinnamomum* L.) is frequently referred to in the literature by its main synonym, *C. zeylanicum*. The generic name is derived from the Arabic or Persian *mama* via the Greek *amomum* meaning spice, and the prefix *chini* to its believed origin. It is known commercially as true cinnamon to distinguish it and its products from those derived from other *Cinnamomum* spp. In English it is known as cinnamon, in Dutch as *caneel*, in Spanish and Portuguese as *canela*, in French as *cannellier de Ceylon*, in Sri Lankan Singhalese as *kurundu*, in Indian as *dalchini* and *ilayangam*, in Sanskrit as *tamalpatra* and in Indonesia and Malaysia as *kayu manis*. The Arabs initially referred to cinnamon as *qirfat-ed-darsini*, corrupted later to *qurfah*, which survives as *qelfar* used to describe Malabar bark. In this section, cinnamon refers to *C. verum* and its products.

There are no germplasm collections of *C. verum*, although one has been proposed for India (Ravindran and Peter, 1995), and little information on breeding programmes, although selection of superior strains is carried out in India (where the improved cultivars Nava Sree and Nithya Sree have been released), Sri Lanka and the Seychelles. Since cinnamon is open-pollinated, vegetative propagation of clonal material is essential to ensure production of trees with a required characteristic. Considerable physical variation exists within the species and distinct local strains have been identified. In Sri Lanka, for example, six named types are differentiated by the taste and aroma of their leaves. Chemical classification would be a more accurate method of differentiation; however, some differences are probably due to local environment, since soil and climate have a substantial effect on tree growth, type of bark and oils, as discussed later.

C. verum is a bushy, evergreen tree up to 15–20 m in height with numerous branches from low down, long, leathery, bright green leaves, small yellow flowers and ovoid blackish fruits. The bark and leaves are strongly aromatic. That three quite distinct essential oils – bark, leaf and root – are obtained from one tree is of particular phytochemical interest, and the method of biosynthesis that can generate or store eugenol in leaves, cinnamaldehyde in stem bark and camphor in root bark is remarkable (Senanayake and Wijeseker, 1990).

Cinnamon commonly produces moderately deep and extensive roots, whose bark is of no value but contains about 3% of a colourless to pale yellowish-brown essential oil, the main ingredient being camphor up to 60%. Seedling root growth is initially rapid, developing a penetrating taproot with numerous spreading laterals. Trees normally have a single central stem up to 20 m in height, although usually 12–15 m, but under cultivation trees are coppiced to 3 m at 3–4 years. The trunk on uncut trees is up to 60 cm DBH and up to 90 cm in the Seychelles, and the bole is buttressed to 60 cm. The branches are numerous and often drooping, beginning low on the trunk. The wood of mature trees is faintly scented, lustrous and varies from light brownish-grey to grey or yellowish-brown, without markings. The timber is light to moderately heavy, usually straight-grained, even-textured and weak, and seasons early but warps, splits, cracks and stains.

The bark on young shoots is smooth and pale brown, and on mature branches and stems is rough, dark brown or brownish-grey. The bark on young shoots is removed, scraped and dried, and becomes the cinnamon quills of commerce. Five bark types are recognized in Sri Lanka, but the major differences are probably more environmental than genetic, since visual differences do not consistently correlate with oil yield or type and thus its value as a spice. A detailed description of the morphology and histology of cinnamon bark has been published (Parry, 1962). Bark is further discussed at the end of this section.

The leaves are opposite, leathery, generally ovate or elliptic, and usually 5–20 × 3–10 cm in size, with the base rounded and the tip acuminate. They are exstipulate, and the petiole is 1–2 cm in length, with the upper surface grooved (Fig. 3.3). The leaves are highly aromatic, shiny and green to bright green above, dull grey-green below,

Fig. 3.3. Leaves and flowers of *C. verum.*

but young leaves may initially have a reddish tint. Leaves have three to five conspicuous longitudinal light-coloured veins running from base to tip. Green leaves contain 0.7–1.2% essential oil, cinnamon leaf oil, whose main constituent is eugenol at 70–95%. Considerable local variation exists in leaf shape and size, and four types are recognized in Sri Lanka based on taste: sweet, neutral, pungent and bitter, but only the last two are distilled. The two leaf oils formerly available in Sri Lanka were regional types, one high and one low in eugenol. Trees with a purple flush of new growth in India had nearly 30% more bark oil than trees that flushed green, and thus flush colour could be used to select seedlings with a high bark oil content (Krishnamoorthy *et al.*, 1988).

The inflorescence is a lax terminal or axillary panicle up to 10 cm long, and the peduncle creamy white and softly hairy, up to 7 cm (Fig. 3.3). The flowers are small, up to 3 mm diameter, and are pale yellow or cream, with an unpleasant fetid smell. The calyx is campanulate and softly hairy, with six acutely pointed segments up to 3 mm, and the corolla is absent. There are nine fertile stamens in three whorls, with two small glands at the base of the stamens of the third whorl; a fourth innermost whorl consists of three staminodes. The filaments are hairy and stout and the anthers are four- or two-celled. The superior unilocular ovary tapers to a short style and contains a single ovule. The main flowering period is April–May in Sri Lanka, and January–February in India.

The fruit is a fleshy ovoid drupe, 1.5–2.0 cm long, black or bluish when ripe, with the enlarged calyx persisting at its base. Fruits mature in 3–5 months, and are edible and attractive to birds, which are the main seed dispersal agent. Ripe fruits have a distinct odour when opened and a flavour resembling juniper. Crushed and boiled ripe fruits yield a fat known as cinnamon suet in India and used in candle-making. The globular brown seeds contain approximately 33% fixed oil and about 1% volatile oil.

Steam-distilled oils obtained in India consisted of 20–33% hydrocarbons and 63–74%
oxygenated compounds; 34 compounds
accounted for 94% of the oil, with *trans*-
cinnamyl acetate and β-caryophyllene
being the major compounds (Jayaprakasha
et al., 1997).

The dried bark contains an essential oil, a
fixed oil, tannin, resin, proteins, cellulose,
pentosans, mucilage, starch and minerals,
while the presence of calcium oxalate crystals indicates low-quality bark. Analysis of
four samples of Indian bark gave the following average proportions of trace elements, in
p.p.m.: K 1590, Ca 1091, Ti < 40, Cr < 35, Mn
179, Fe 89, Zn < 25, Br < 15, Rb 30, Sr 57
(Joseph *et al.*, 1999). Analyses in Sri Lanka
determined bark mucilage content as
1.9–3.7%, and ash of mucilage as 17.2–24.6%.
In general, the major bark constituents and
their abundance are very similar to those of
the essential oil.

Essential oil content of bark ranges
between 0.5 and 2.0%, the main constituent
being cinnamaldehyde at 60–70%. Oil distilled
from outer and inner bark differs, and only
inner bark (quills) is normally distilled for oil
to be used as a spice (see also sections on
Distillation and Products and specifications).

Ecology

Cinnamon is a tree of the wet tropics and has
a somewhat restricted natural range in Sri
Lanka, India and South-east Asia. When
introduced elsewhere, it thrives and is cultivated only where the climate is very similar
to its natural habitat. The unrestricted export
of cinnamon seedlings and related species
from Sri Lanka to Japan in 2000 caused a
major row between local growers, concerned
over the industry's future, and the government. Birds dispersed seeds so widely on the
Seychelles that cinnamon became dominant
in secondary forest on the main islands of
Mahe and Silhouette a century after its introduction. Additionally, cinnamon is not an
attractive browse plant to cattle and once
past the seedling stage is normally ignored.

Cinnamon produces its finest bark in
sunny regions with an average temperature
of 27–30°C, but will grow well in a wide

range of environments, as specimens in
botanic gardens in many countries demonstrate. Bark from these trees is usually of
indifferent quality; for example, bark collected from trees growing in the Botanic
Gardens, Entebbe, where the climate is
cooler and drier than Sri Lanka, was thick,
brittle and of poor quality, and similarly in
Melbourne, Australia (E.A. Weiss, personal
observation).

A rainfall of 2000–2500 mm with no pronounced dry season is the optimum.
Productive regions near Colombo, Sri Lanka,
and Victoria, in the Seychelles, have a similar
annual average of 2365 and 2354 mm,
respectively. Cinnamon grows naturally
from sea level to 1500 m, but 500 m is considered the maximum in Sri Lanka for good-
quality bark, and somewhat higher in India.
Although basically a forest tree, cinnamon
grows well in a more open or low tree environment, and most plantations in Sri Lanka
are unshaded.

Soils

Cinnamon grows naturally on a very wide
range of soils, but soil type directly influences rate of growth. Sandy soil produces
the highest quality bark in Sri Lanka and
India, but a more loamy soil is preferred in
the Seychelles. In Sri Lanka, highest quality
bark is produced on light sandy soils of the
western region around Negombo, while lateritic soils of the southern region produce a
heavier bark of lower quality, although
yield per tree is significantly higher.
Coastal loams are favoured in Madagascar,
but cinnamon also grows on lateritic soil
further inland. Waterlogged soils are
unsuitable and low-lying areas subject to
sporadic flooding should also be avoided,
as trees growing on such soils generally
produce a bitter, low-oil-content bark. Most
growers consider saline soils are unsuitable
for cinnamon, but there are no data to support this opinion. In the Seychelles and on
Zanzibar, healthy trees grow in coastal
coconut plantations within reach of sea
spray, but no comparison of bark obtained
from such trees and those growing elsewhere is available.

Fertilizers

The most common fertilizers applied to cinnamon are animal, plant or still residues, in addition to the mulch resulting from weeding or surplus leaves and twigs at harvest. Where rock phosphate is easily and cheaply available, it is spread around harvested plants before the rains and hoed in. Cinnamon produces a relatively large root system, even when regularly coppiced, and is thus able to scavenge a fairly large soil volume for nutrients. In India and Sri Lanka, a mixture of dung, decomposing leaves, coconut or castor poonac is commonly placed in holes when planting out seedlings. Adding phosphate is becoming more common, and single is preferable to double or triple superphosphate as it contains a small amount of sulphur often beneficial in tropical soils.

The nutrients removed by cinnamon yielding 460 kg bark ha^{-1} are shown in Table 3.4, which also includes other Indian spice crops, and whilst chemical fertilizers are generally recommended, little is used either by large growers or smallholders. An Agricultural Department recommendation for Sri Lanka, also followed in south India, is a 2:1.5:1.5 mixture of urea, rock phosphate and potassium chloride, applied annually to trees 1–3 years old at 40–60 kg ha^{-1}, and to mature trees at 100 kg ha^{-1}, in two equal applications between April and August and between October and January, and combined with weeding which hoes in the fertilizer.

Cultivation

Cinnamon can be grown from seed, but as it is open-pollinated, vegetative propagation from selected parents is recommended, and modern micropropagation methods can be used to produce large numbers of cloned seedlings. Division of rootstocks is probably the quickest method of upgrading a plantation, since high-yielding stools can be selected. Plants are cut back to about 15 cm, sections of rootstock removed with adhering soil and planted out immediately into prepared pits. Stems are ready for cutting in 12–18 months compared with 2.5–3 years from seedlings.

Young two-node cuttings planted in polythene bags filled with a mixture of red soil and coconut coir dust and then placed under a polythene tent had the highest strike rate in the Seychelles and were ready for planting out in 12–18 months. Apical cuttings with three leaves were best in Sri Lanka, and treatment of more mature hardwood cuttings with the growth regulator IBA at 2500 p.p.m. encouraged rooting. Seedlings are planted out just prior to the rainy season with supplementary watering, or just after rains commence; this is normally June–July in India, May–June in Sri Lanka and

Table 3.4. Nutrient removal by Indian spice crops.

Spice	Average yield (kg ha^{-1})	N	P	K	Farmyard manure[b]
		\multicolumn{4}{c}{Nutrient removal (kg ha^{-1})}
Black pepper	2,000	46	10	63	10,000
Cardamom	4,508	2	19		10,000
Ginger	2,800	59	24	11	20,000
Turmeric	4,700	86	31	194	20,000
Clove	90	10	3	8	2,500
Nutmeg	750	8	2	4	2,500
Cinnamon	460	3	1	41	0
Coriander	476	20	3	9	1,500
Cumin	578	19	5	3	2,000
Fennel	1,300	32	13	18	2,000
Fenugreek	1,000	35	18	14	2,500

b, calculated as containing 6 kg, 4 kg and 4 kg of N, P and K, respectively per tonne of farmyard manure.

October–November in the Seychelles. A field spacing of 120 × 120 cm or 90 × 90 cm is recommended in Sri Lanka for commercial plantations, and blanks are infilled to ensure this population is maintained.

When seeds are required it is usually necessary to bag fruits for protection as they are avidly eaten by birds. Ripe fruits are picked and heaped until the pulp rots; seeds are then washed, shade-dried and planted as soon as possible since viability quickly declines. In India, the highest germination rate is obtained by sowing 3–7 days after seed has been harvested and falls quickly to zero after 40 days. Seeds sown within 7 days germinate in 20–25 days, but later sowing extends this to 32–42 days. Seeds should be sown in shaded beds, closely spaced in 20 cm rows and covered with 2–3 cm of soil. Clumps of seedlings may be transplanted into baskets at 4 months and planted out 4–5 months later, or left in the nursery until required. For direct sowing, a 1.5 × 1.5 m or 2.0 × 2.0 m spacing is usual and up to 20 seeds are placed into prepared holes and later thinned. In India, a common spacing is 3 × 3 m with several seeds sown per hole. In general, close spacing is preferred to wide, as the former produces erect stems easier to prune and harvest, compared with horizontal stems at the wider spacing.

Plants are coppiced to 10–15 cm after 2 years and stools covered with soil to encourage shoot formation; 4–6 new shoots are allowed to grow and kept straight by pruning side shoots and excess foliage. Stems are cut at 1.2–5.0 cm in diameter and 2–3 m long, and stools pruned to remove unwanted material and earthed up. Stools normally produce an increasing number of shoots up to 8 years of age, and decline after 10–12 years.

Weed control and irrigation

Manual weeding three or four times annually is necessary during the first 2 years; thereafter twice per year should be sufficient. Herbicides are seldom applied to cinnamon plantations, and it is unlikely that their use would be profitable under present production systems. Few plantations are irrigated on a regular basis for the same reason.

Pests

A variety of insects have been recorded from cinnamon, but most are of little significance; those damaging young stems and bark are the most important, others attacking leaves becoming important only where leaf oil is produced or defoliation greatly reduces stem growth. Insects marked with an asterisk(*) have been reported as pests of other *Cinnamomum* species.

In India and Sri Lanka, caterpillars of the cinnamon butterfly *Chilasa clytia** are the most destructive pest, and normally appear in December when trees produce new growth. Caterpillars are voracious feeders and if uncontrolled quickly defoliate trees, especially in new plantations where leaf loss adversely affects growth, and can also cause death of newly planted seedlings. Shoot borers cause significant damage to individual trees and reduce bark quality; in India and Sri Lanka shothole borers, *Xylosandrus* spp., are frequently recorded.

Leaf miners, especially *Phyllocnistis crysophthalma* and to a lesser extent *Acrocercops* spp.*, become important when they reduce the rate of shoot growth by defoliation. Gall and leaf mites, especially *Eriophyes bois* and *Typhlodromus* spp., caterpillars of the leaf webber, *Sorolopha archimedias**, and arboreal ants, *Oecophylla smaragdina*, can cause similar problems. Young seedlings in the nursery may be eaten off at ground level by *Agrotid* larvae* or mole crickets, *Gryllotalpa* spp.*, and in the field after transplanting. Larvae of various *Popillia* spp., which attack roots, are particularly damaging in nurseries and on young seedlings. Most often recorded are *Popillia complanata* and *Popillia discalis*.

Diseases

The numerous pathogens recorded from cinnamon are mainly of little importance. They are those normally found at a low level of infection on other local trees and seldom warrant control measures. Diseases of *C. verum* also attack other *Cinnamomum* spp. and are marked with an asterisk (*).

A most important disease in India, Sri Lanka and Indonesia is caused by *Corticium salmonicolor** (syn. *C. javanicum*). The initial symptom is the formation of a pale, pinkish-white brittle layer (crust) on young stems or branches, and unless controlled, the infection spreads, destroying bark and finally killing shoots. The pathogen also attacks mango, jackfruit, custard apple and other fruit trees often growing in the vicinity of cinnamon plantations. Affected prunings and other plant parts should be burnt. Also attacking shoots and young stems, but also more mature trees, in the same regions is *Phytophthora cinnamomi**; the most obvious symptom is vertical strips of dead bark beginning at or below ground level. The disease is prevalent on badly drained soils. In Indonesia it has also been recorded on other *Cinnamomum* spp. Although seedlings can be severely damaged and killed, the effect on more mature trees is disfiguring but not fatal.

A stem disease caused by *Exobasidium cinnamomi* can become serious in some years in Sri Lanka, and may also spread to leaves. Symptoms on leaves are small yellowish concave spots whose underside bears greyish-white spore bodies. A stem blight due to *Diplodia* spp. attacks young seedlings in the nursery and when first planted out. Symptoms are small light-brown patches on stems, which spread and cause eventual death. Several root rots are damaging to some extent, including brown rot caused by *Phellinus lamaensis*, white rot by *Fomes lignosus** and two black root rots by *Rosellinia* spp. Symptoms are roots covered with a network of black mycelium and small white starlike spots under the bark. *Rosellinia* spp. attack a range of cultivated trees and shrubs.

Leaf blight caused by *Pestalotia cinnamomi*, known in Sri Lanka as grey leaf spot or blight, can cause severe damage and defoliation. Symptoms on leaves are small yellow spots, becoming grey with central black dots. A closely related species attacks coconut, and cross-infection is suspected. *Glomerella cingulata** can become locally or seasonally severe in India with similar effects on trees. Three other leaf diseases have been recorded, caused by *Aecidium cinnamomi*, *Leptosphaeria* spp. and *Gloeosporium* spp., but the extent of damage was not noted.

Harvesting

Smallholders harvest their own trees, but contract labour is employed on larger plantations. In Sri Lanka, stems are cut during the rains to facilitate peeling, when the red flush of young leaves is beginning to turn green and sap is flowing freely. The finest quality bark is obtained from shoots with uniform brown thin bark, 1.0–1.25 m by 1.25 cm diameter, from central shoots and middle portion of shoots. Stems harvested at 6-monthly intervals produce fine bark, at 8-month intervals a coarser bark, and older stems have prominent nodes making peeling difficult.

Cut shoots are bundled and carried to homesteads or a central store for peeling and drying, and this operation has basically remained unaltered for centuries. Generally only sufficient is cut to provide a day's peeling. A surplus may be left overnight to slightly ferment, as this eases removal of the outer bark by scraping. Bark to be distilled for oil should not be left in wet bundles or become damp, as this encourages mould or fermentation, which adversely affects oil composition. Harvesting is mainly during May–June and October–November.

Bark yield per hectare in Sri Lanka depends mainly on the standard of management. Smallholders average 100–125 kg quills ha^{-1} and 25 kg chips ha^{-1} annually; on well-managed larger plantations, the first crop after 3–4 years yields 56–67 kg quills ha^{-1} increasing to 168–224 kg ha^{-1} in subsequent years. About 63 kg cinnamon chips ha^{-1} and approximately 2.5 t undried leaves ha^{-1} are also obtained, but only 75% of leaves available are usually distilled, the remainder being left as mulch. On well-established plantations in India the first harvest yields 100–120 kg quills ha^{-1} rising to 200–250 kg ha^{-1} in subsequent years. Indonesian bark yields are very variable, reflecting the generally low standard of management, which also results in lower overall bark quality, and smallholders average about 100 kg ha^{-1}. The high level of Indonesian bark production is due to the large number of trees, and improved management would quickly raise both the total and quality.

The Seychelle's bark production depended mainly on felling self-sown trees, and bark oil was formerly the most important cinnamon product exported, but this was later replaced by unselected rough-grade cinnamon bark. Yield varied from a few kg per bush to 45 kg dried bark from a large mature tree. Average annual yield in the 1950s to 1960s was 0.6 t ha^{-1}, but declined rapidly in the 1970s (Lawrence, 1994a). The yield of leaves was usually around 2 t ha^{-1}. Rough bark is also produced in Madagascar using techniques similar to those in the Seychelles. Plantations have received little attention for decades, although the number of wild trees has increased as birds have dispersed seed over large areas. Trees are usually harvested only when the local bark or oil price is high.

An Indian or Sri Lankan plantation can remain profitable for 15–45 years depending mainly on the standard of management, and the effective life can also be extended by selecting at replanting only high-yielding, long-lived stools, or cloned material. Little information on individual tree yield is available, data which would appear to be a basic requirement for improved profitability.

Processing

In Sri Lanka, smallholders may peel and prepare their own bark, but on larger plantations this is undertaken by specialist workers, whose skill and speed are remarkable, and who use only brass or stainless steel tools to avoid staining the bark. The outer bark is first removed using a special curved knife and the cleaned stem then rubbed with a brass bar to loosen the inner bark. Two cuts 30 cm apart are made around the stem, and two longitudinal slits on opposite sides. The inner bark is then carefully separated from the stem. The inner bark curls naturally into the well-known quills, which are joined to increase their length, and filled with smaller quills and pieces to make a near-solid cylinder. These are consolidated by hand-rolling, trimmed to about 1 m, then dried in the shade. Quills are inspected and rolled daily to ensure they remain firm. First-quality quills are a smooth, uniform yellow-ish brown. Smaller quills, bark pieces, etc., are sold separately, and a proportion chopped or ground for local sale or distillation. Seychelles and Malagasy bark, outer and inner together, is peeled from stems or small branches, sun- or oven-dried, cleaned and packed in bales or sacks for export. Very little is distilled or prepared and graded as in Sri Lanka.

Distillation

Cinnamon bark produces two oils, a superior type derived from the inner bark and a lower quality from broken quills, chips and bark. In Sri Lanka, a third oil, *katta thel*, is produced from bark and twigs for local consumption. Oils produced in the same still from quillings, featherings or chips vary in their constituents; chip oil has a very good odour and flavour although containing 20% less cinnamaldehyde and twice the eugenol of bark oil. Root bark oil is produced only when a plantation is uprooted for replanting.

Local stills in Sri Lanka have cone-shaped bodies made of vertical planks of wood, tightly bound by horizontal metal bands, and so well made and fitted that there is little leakage. A charge is about 200 kg leaves, with steam generated in a separate boiler, and distilled for 8–9 h at the height of the bark-peeling season, but up to 24 h in the off-season. Oil yield varies from 0.5–1.0% (about 30–40 kg ha^{-1}), depending on distilling method, time of year leaf is harvested, and whether leaves are fresh or wilted. Many small stills are made from copper built into a stone hearth and heated by a fire beneath. About 23–27 kg chips and 180–225 l water is the usual charge, distilled for 5 h; residual water from one distillation is collected and used in the next. Oil yield is approximately 0.2%. Stills developed by the Ceylon Institute of Scientific and Industrial Research (CISIR) give yields of 0.5–2.0% from chips, although oil quality frequently improves before an increase in yield. Very similar methods are generally used in India, although large private plantations in Kerala State have upgraded their stills and obtained yields of 1.5–2.5%.

Leaf oil is usually complementary to bark production; leaves stripped from shoots, together with small leafy twigs and stems, are left in the field for 3–4 days, then transported to the distillery. Stills in India, Sri Lanka and Indonesia are usually located at a convenient site, since leaves are bulky and difficult to transport, and constructed of materials locally available. Thus, regional leaf oils can be very variable, but where more modern equipment has been installed, oil of excellent quality and type is produced. Seychelles oil is produced by distilling leaves collected during January–September, with an average yield of 1.8–2.0 t ha^{-1}. A charge is 300 kg fresh leaves, distilled for about 6 h. Under the government's 1982–1986 Development Plan, two modern distilleries were constructed, but neither achieved full commercial operation as the local people abandoned cinnamon production for less onerous jobs in the booming tourist industry.

Detailed information on leaf distillation is lacking, with little basic data on the effects of the various methods employed, such as oil yield and any change in oil composition resulting from distilling fresh leaves or leaves wilted for up to 7 days, or charge composition and distillation time. This and similar information is essential to ensure that a distillery operates efficiently and profitably, and greatly assists the integration of leaf oil production with other plantation operations.

Storage and transport

Quills are usually sold to traders for grading on colour and quality, and quills may also be sulphur-bleached to lighten the colour. Cinnamon quills should be stored in sacks, and highest quality bark is wrapped in new sacking, and, more recently, packed in corrugated cartons. Finally, quills are pressed into cylindrical bales of 100–107 cm (the traditional 42 inches), weighing 45–50 kg, for shipment. Bark (quill) trading is becoming more consolidated and larger traders have established their own recognized standards in addition to meeting international standards where necessary (see also next section). The major insect pests of stored

cinnamon and cassia bark and quills include *Lasioderma serricorne, P. farinalis* and *Sitodrepa panicea.* Rodents not only eat quills, but their faeces are a banned contaminant in many importing countries.

Cinnamon oils are frequently sold as unrefined crude in 200 l drums, or refined in 50 and 200 l drums. All oils should be stored and transported in full, sealed drums in cool conditions. Refined oil should be kept sealed until required for use or repacking. When less than a full drum of refined oil is required, the remainder must be resealed as soon as possible in a smaller, full container to maintain its quality.

Products and specifications

Cinnamon trees produce the commercially important bark used as spice, bark and leaf oils and oleoresins.

Bark. Commercial cinnamon bark, traded as quills, is a dull pale brown, up to 5.5 mm thick, the inner surface somewhat darker, and finely striated longitudinally. The finest quality quills destined for the retail market have a uniform colour, a delicate fragrance and taste warm, sweet and pleasant (Fig. 3.4). Sri Lankan quills are considered superior. The organoleptic properties of the bark are basically determined by the composition of the steam-volatile oil, plus trace amounts of coumarin, the only important non-steam-volatile constituent (see Bark oil below).

Sri Lankan bark is locally divided into many grades, but is traded commercially for export in the following types and qualities:

- Quills: yellowish-brown, cylindrical, approximately 1 m × 10 mm diameter; scraped, smooth, thin, rolled with occasional scars, and fine, light-coloured wavy lines running lengthways and clearly visible. A pleasing, fragrant odour and a warm, sweet, aromatic taste. Quills are further separated into compound (filled) quills and simple quills.
- Quillings: broken pieces of quills of various grades, length and diameter; similar but not equal to the aroma and taste of sound quills.

Fig. 3.4. Cinnamon quills.

- Featherings: shavings and small pieces of bark remaining after processing inner bark into quills; similar in taste but not aroma to quillings.
- Chips: small pieces of mature, unpeelable bark, occasionally fragments of inner bark, greyish-brown within, and deficient in both aroma and taste.

Quills are also graded by quality and size from H, M, C to Alba, the finest. H quills are the largest and of uneven colour, while Alba quills are the smallest, pencil-sized, smooth and uniform in colour with no blemishes, and sold at a premium. The large H-grade quills are the most frequently adulterated by inclusion of poorer bark. Discoloured quills may be bleached using a sulphur treatment. Unscraped and scraped bark is also locally traded, and together with the last three grades is most frequently distilled to produce bark oil. Bark from other countries is also graded but not to the same extent. However, grading is complicated by the use of the designation cinnamon to include cassia in some countries, and to cover either indiscriminately in others.

Bark grading to International Standards Organisation (ISO) requirements will eventually become universal. ISO 6539–1983 already covers cinnamon from Madagascar, the Seychelles and Sri Lanka, and products that do not meet these requirements for the spice will be downgraded in major international markets. The American Spice Trade Association (ASTA) specification requires cinnamon to contain not more than: two dead insects, 1 mg lb^{-1} mammalian excreta, 2.0 mg lb^{-1} other excreta, 1% mould by weight, 1% insect defiled by weight and 0.5% extraneous foreign matter by weight.

The main use for bark is as spice in quills or powder. Small quills are added to meat and similar dishes and removed after cooking, are an ingredient in sweet pickled products, and are added to hot punches and similar drinks. Ground bark or powder is commonly added to curries, pilaffs and stews in the Middle East and Asia, where its antibacterial properties due to the phenol content may help counteract the effects of bad meats. Powder is preferred in confectionery, pies, baked products and a variety of sweet dishes. In Mexico, cinnamon mixed with chocolate is a favourite beverage, and cinnamon powder is often sprinkled on chocolate drinks in fast-food outlets worldwide. Both bark and powder are frequently adulterated in their countries of origin, but not the retail products sold by major food processors in western countries. However, in the USA cassia is often labelled cinnamon, not to deceive but because consumers and the US Food Drug and Cosmetic Act do not discriminate between them.

Oils. A striking feature of the three cinnamon oils is the difference in the major constituent: cinnamaldehyde up to 60% in bark oil, eugenol up to 80% in leaf oil and camphor up to 60% in root bark oil. The finest quality oils are obtained by solvent extracting distillate, or supercritical carbon dioxide extraction. The main constituents of Sri Lankan cinnamon oils are shown in Table 3.5, and the main characteristics of bark oils from selected origins in Table 3.6.

Table 3.5. Main constituents of Sri Lankan cinnamon oils (%).

Constituent	Leaf oil	Bark oil	Root-bark oil
1,8-Cineole	0.15	1.65	15.2
Camphor	ng	trace	60.0
Linalool	1.5	2.3	1.2
α-Terpineol	0.15	0.4	3.8
Cinnamaldehyde	1.3	74.0	3.9
Cinnamyl acetate	0.8	5.0	0.3
Eugenol	87.0	8.8	5.0

ng, not given.

Table 3.6. Characteristics of cinnamon bark oils from selected origins.

Characteristic	Sri Lanka	Seychelles	Madagascar	British Pharmacopoeia	Essential Oil Association, USA
Specific gravity	1.023–1.040[a]	0.943–0.976[b]	1.016[b]	1.010–1.030[a]	1.010–1.030[c]
Optical rotation	0° to −1°8′	−2°30′ to −5°10′	−2°34′	0° to −2°	0° to −2°
Refractive index (20°C)	1.581–1.591	1.52843–1.53271	1.5746	1.573–1.595	1.573–1.591
Solubility (v/v 70% alcohol)	1:2–3	1:10	1:2.5	1:3	1:3[c]
Aldehyde as cinnamaldehyde (%)	65–76	22–84	61.4	60–70	55–78
Eugenol content (%)	4–10	6–15	10	ng	ng

[a] 20°C; [b] 15°C; [c] 25°C, ng, not given.
Figures in columns show the range.

Bark oil. Bark oil is obtained by steam or hydro-distillation with cohobation. Oil distilled by different methods is basically similar in characteristics and composition but quite different in odour and taste, properties of paramount importance to end-users. Cinnamon bark contains water-soluble volatile aromatic components, recovered by extracting distillation water and adding the extract to water-distilled oil. This oil is known as complete oil and considered superior to distilled bark oil generally available.

Cinnamon bark oil is a pale to dark yellow oily liquid; its odour is strong, warm, sweet, spicy and tenacious and its taste sweet and pungent but not bitter. Lower-grade commercial oils are usually darker and lack the powerful odour and tenacity. Oil from first-grade Sri Lanka quills, often known as Ceylon oil, is considered the best quality (Jayatilaka *et al.*, 1995). Seychelles bark oil is less favoured than Sri Lankan as it has a harsher odour, probably due to the presence of camphene and camphor.

The major oil constituent is cinnamaldehyde (Lawrence, 1998a), but other components impart the characteristic odour and flavour distinguishing this oil from other *Cinnamomum* bark oils. Bark oil is placed in the cinnamon group by Arctander (1960). The powerful characteristic note of high-quality oil is believed to be due to a combination of methyl-*n*-amylketone and other aldehydes and ketones, but the complicated relationship between oil composition and organoleptic properties is still unclear. Analyses of leaf and bark oils are shown in Table 3.5, but the analytical method used influences the results. The proportion of main components in good quality Sri Lanka bark oil varies; aldehydes can range from 51 to 76% (average 55%) and eugenol from 5 to 18%. A high proportion of chips in distilled material can increase the eugenol content, and quillings and featherings the aldehyde content.

Main oil use is to flavour foods, beverages, pharmaceutical and dental preparations, cosmetics and perfumes. Bark oil is considered to have a strong germicidal and fungicidal activity; cinnamaldehyde is anaesthetic, antipyretic, hypotensive, hypothermic and sedative. However, persons using fresh cinnamon oil can develop cinnamon sensitivity, expressed as allergic dermatitis and/or urticarial lesions. Symptoms usually disappear on cessation of cinnamon oil use (Woolf, 1999). The antifungal activity of oils from a number of *Cinnamomum* spp. has been described (Mastura *et al.*, 1999).

Adulteration of bark oil is common, most frequently with cinnamon leaf oil, cassia oil, clove leaf oil, eugenol and cinnamaldehyde, which is mass-produced from coal-tar bases, especially toluene. Cinnamon bark oil is costly, $385 kg^{-1} in 1993, but leaf oil is much cheaper at $8.25 kg^{-1} in 1994, but still more expensive than clove leaf oil at $2.70 kg^{-1} in 1994, an alternative source of eugenol.

Root-bark oil is colourless to pale yellowish-brown, similar in odour to stem-bark oil but weaker, camphoraceous and lacking in fragrance. The major component is camphor up to 60%, which crystallizes out on standing, plus 1,8-cineole up to 16% and eugenol up to 5%. Oil yield from root bark at 1–2.8% is higher than leaf and stem bark. Root-bark oil is produced only in Sri Lanka when an exhausted plantation is replanted and is usually added to other cinnamon oils.

Leaf oil. The oil is a yellow to brownish-yellow liquid, with a warm, spicy and somewhat harsh odour, lacking the richness of bark oil; the taste is slightly bitter, burning, very spicy and powerful. The oil is placed in the dry-woody, pronounced spicy group by Arctander (1960). Eugenol content of Sri Lankan leaf oil can be 60–65% or 70–85% depending on the district of origin. Seychelles leaf oil is a valued source of eugenol, usually above 90%, with phenols 78–95% and aldehydes 5%. Madagascar oil has a eugenol content of 70–90%, and a distinguishing feature is the relatively high benzyl benzoate content; neither is generally available.

Major users of leaf oils are food processors and pharmaceutical companies. Leaf oil as a source of eugenol has lost ground to the cheaper clove leaf oil, except when eugenol is needed for conversion to iso-eugenol used in confectionery products. The only common adulterant is clove leaf oil, used when cinnamon leaf oil is higher in price; the reverse also occurs!

Monographs on the physiological properties of cinnamon bark oil and cinnamon leaf oil have been published by the Research Institute for Fragrance Materials (RIFM). In the USA, the regulatory status generally recognized as safe has been accorded to cinnamon, GRAS 2289, cinnamon bark oil, GRAS 2290/2291, and cinnamon leaf oil, GRAS 2292.

Oleoresin. This is usually prepared by extracting cinnamon bark with organic solvents; the yield using ethanol is 10–12%, using benzene 2.5–4.3%, and more recently 1,1,2-trichloro-1,2,2-trifluoroethane has been used (Bernard *et al.*, 1989). Oleoresin is a deep reddish- or greenish-brown rather viscous liquid, with a volatile-oil content of 16–65%. Its major use is in flavourings, cake and similar mixes, pickles, prepared meats, convenience foods and related products where flavour stability at high temperature is important. Oleoresins are normally commercially available dispersed on various bases and usually sold on a volatile-oil content of 23–27% or 65%.

Other Cinnamomum *species*

Cinnamomum culitlawan (L.) Kosterm. (syn. *Cinnamomum culilaban* (L.) J.S. Presl.; *Cinnamomum culilawan* (Roxb.) J.S. Presl.; *Laurus culitlawan* L.) is known in Indonesia as *kulitlawang, kayu teja*. It is native only to Indonesia, specifically the Moluccas, Ambon and adjacent islands, and apparently very uncommon. It was introduced into Malaysia and India at the end of the 19th century, and was initially cultivated in both, but later abandoned. The bark smells of cloves (*kulitlawang*) and is used both as a spice and medicinally. The bark and its oil (lawang oil) are used as a constipating agent and as a medicine against cholera. The oil is rich in eugenol and was used as raw material for the synthesis of vanillin.

South-east Asian *Cinnamomum* species used as substitutes or adulterants for cinnamon and the commercially important cassias include: *Cinnamomum bejolghota* (Buch.-Ham.) Sweet (syn. *C. obtusifolium* (Roxb.) C. Nees) occurring from northern India to Indo-China, southern China and the Andaman Islands; *Cinnamomum impressinervium* Meissner in Sikkim; *Cinnamomum glaucescens* in Nepal; and *C. deschampsii* Gamble. in Malaysia, where a number of other *Cinnamomum* species and their oils have been described (Ibrahim and Goh, 1992).

Bark or oils obtained from other *Cinnamomum* spp. have been described; from China, Japan and Taiwan (Fujita *et al.*, 1974; Lin and Hua, 1987); India (Khana *et al.*, 1988); Indonesia (de Guzman and Siemonsma, 1999); Nepal (Sthapit and Tuladha, 1993); Sri Lanka (Sritharan *et al.*, 1994); and a general survey of bark and oils from a number of cinnamon species (Jirovetz *et al.*, 2000).

Unrelated species

The genus *Litsea* Lam. contains about 200 species widely distributed in tropical Asia, Australia and the Pacific, with many members producing fruit or seeds containing a fixed or volatile oil (Weiss, 1997). Bark from *Litsea diversifolia* Hk.f. produces a poor-quality cinnamon extract. A South American tree, *Aniba canelilla* (H.B.K) Mez, has a cinnamon-scented bark used locally as a cinnamon substitute, and is probably the explorer Humbolt's 'Cinnamon of the Amazon', described in his monumental 23-volume *Voyage de Humbolt et Bonpland aux Regions Equinoxiales* (1805–1834).

Laurel

The genus *Laurus* L. has two members, *Laurus nobilis* L., commonly known as bay laurel, and *Laurus azorica* (Seub.) Franco (syn. *L. canariensis* Webb & Berth.), native to the Canary Islands, hence the common name Canary Island laurel, although a number of other species were formerly included in the genus. A chemically distinct form has been reported from China, with leaves high in eugenol. Linnaeus in 1737 choose the Latin *Laurus* for this genus and the Greek *Daphne* for the quite different spurge-laurel and related species.

In classical Greece, *L. nobilis* was sacred to the god Apollo, and according to legend, when the nymph daughter of the earth goddess Gaia, Daphne, was pursued by Apollo, the slayer of her bridegroom, she entreated the Gods to give her assistance, so they changed her into a laurel tree. Apollo then crowned himself with a circlet of laurel leaves, and declared the tree sacred to his divinity. A garland of woven laurel leaves was awarded as a symbol of honour or victory in Rome, but it is said that Julius Caesar preferred a crown of Alexandrian laurel (*Ruscus racemosus*), as its broader leaves covered more of his bald head! In the Middle Ages, distinguished men were crowned with a wreath of berried laurel, hence the English title of Poet Laureate. University graduates were known as Bachelors from the Latin *baccalaureus* (*bacco*, a berry and *laureus*, of laurel) and they were forbidden to marry as this would distract them from their studies; hence the general designation in Europe of unmarried men as bachelors. Laurel was also believed to provide protection against lightening strike, and was also said to confer the gift of prophecy; thus, a dying laurel tree in a garden predicted a disaster. In Shakespeare's play *Richard II*, an actor says, 'Tis thought the King is dead; we will not stay / The bay trees in our country are all withered.'

Laurel has long been used in herbal medicine; an infusion of fruit was used to suppress profuse menstruation and to hasten childbirth, and an extract of bark and leaves was frequently prescribed to alleviate kidney disorders and respiratory problems. In his *Grete Herbal* of 1526, Peter Treveris recommended, 'A paste of powdered bay berries mixed with honey applied to the face, will treat against all manner of red things that come in young folks faces.' Berry oil is also believed to alleviate stomach disorders, and was also used in veterinary medicine. Laurel is commonly planted in gardens and in tubs to produce fresh leaves for culinary use. In this text, the designation laurel applies to *L. nobilis*, its products and derivatives.

Botany

L. nobilis L. (syn. *L. undulata* Miller) is known in English as bay or sweet laurel, in French and Dutch as *laurier*, in German as *lorbeer*, in Italian as *allauro* or *lauro*, in Spanish as *laural*, in Swedish as *lager*, in Arabic as *ghar* and in Turkish as *defne* (tree) or *defne yapragi* (leaf). The chromosome number of laurel is variable; a polyploid sequence from 36 to 72 has been demonstrated, and laurel has been allocated the most common number, $2n = 48$. Apomixis (agamospermy) has been observed in tetraploid laurel. No germplasm collections are recorded.

Laurel is a leafy evergreen tree, generally 5–10 m, but when cultivated is usually pruned to below 3 m. Bark on mature trees is greyish brown, and on younger stems is smooth and shiny, often with a reddish tint. Small branches and twigs are terete. An essential oil can be obtained from young shoots and bark; the main constituent in bark oil is 1,8-cineole up to 40%, but α-terpinyl acetate up to 45% is found in wood oil (Kekelidze, 1987). The trunks can be used as fence posts, and the flexible twigs woven into rough baskets.

The leaves are borne on short petioles, 0.5–1.5 cm long, and are alternate, somewhat leathery, variable in size being 5–15 × 1.5–3.0 cm, entire, with margins slightly undulate, lanceolate to oblong-lanceolate, with the base cunate and the apex acute; there are 10–12 pairs of lateral veins prominent on both sides. Young leaves may be violet red; mature leaves are dark shiny green above, pale yellowish green below and glandular. Oil is contained in cells that occur below the adaxial and abaxial epidermis of leaves, and can be distinguished early in growth. A detailed description of leaf structure and development has been published (Maron and Fahn, 1979). Leaves from trees growing in the heavily polluted streets of Athens were thicker, accumulated condensed tannins, and palisade cells had a larger number of chloroplasts. Leaves have a faintly anise-like odour, taste aromatic, and are slightly bitter when fresh, which disappears during drying. Turkish laurel leaves have a very delicate flavour favoured by international traders and thus command a premium. Further details of leaf composition and characteristics are given at the end of this section.

The flowers are inconspicuous, yellowish-green, and there are four to five in small axillary umbels, with sexes usually on separate trees (Fig. 3.5). The peduncle is 2–12 mm long and softly hairy. There are four bracts, the inner rounded, 0.7–1 cm, the outer smaller and glabrous outside and silky haired inside. The pedicel is 2–5 mm,

Fig. 3.5. Laurel flowers.

with four ovate-oblong sepals 4.4–6 mm, which are obtuse with both sides hairy. The female flower with four staminoids is connected at the apex, and the ovary is one-celled; the style is short, and the stigma somewhat enlarged. The male flowers have 8–12 fertile stamens, an outer whorl without glands, and two and three whorls with two sessile, reniform glands in the centre of the filaments; the anthers are ellipsoidal, two-celled, introse and pistil-liod. Flowering in Europe is normally in late spring or early summer. An essential oil distilled from flowers in France had a high β-caryophyllene content at 10% compared to leaf oil, and also higher contents of viridiflorene (12.2%), germacradienol (10.1%), β-elemene (9.7%), and E-ocimene (8.0%) (Fiorini et al., 1997).

The fruit is a small, hard, ellipsoid berry, about 15 mm diameter × 1–2 cm long, shining dark purple to black when ripe, and surrounded by the persistent perianth. The fruit (seed) contains an aromatic fatty oil up to 25%, containing some 20 fatty acids including: lauric 11–52%, palmitic 5–17%, oleic 15–40%, and linoleic 17–23%. Petroleum ether extract of fruit contains the characteristic compounds 10-hydroxyoctacosanyl tetradecanoate, 1-docosanol tetradecanoate and 11-hydroxytriacontan-9-one (Hafizoglu and Reunanen, 1993).

Dried leaves contain water 4.0–10%, protein 7.0–11%, fat 4.0–9.0%, and carbohydrates 65%. Dried leaves also contain (per 100 g) ash about 4g (including Ca 1 g, P 110 mg, K 530 mg, Na 20 mg, Zn 40 mg, Fe 530 mg) plus thiamine 0.01 mg, riboflavin 0.42 mg, niacin 2 mg, ascorbic acid 47 mg and vitamin A 6185 IU; 1725 kJ per 100 g. Fresh leaves also contain catechin, which is preserved by drying; purified catechin occurs as white needles with a melting point of 93–95°C. Cuticular waxes present on leaf surfaces are readily extracted during distillation, and affect oil composition so must be removed by additional refining. Young buds and leaves contain 3-octulose, a rare compound in nature.

Detailed leaf analyses are lacking, and analyses of leaf oil are used to compare leaf quality, aroma, etc. The oil's monoterpene content is mainly responsible for the leaf aroma, with 1,8-cineole up to 50% being the most important, plus α-terpinylacetate up to 8%, sabinene to 6% and α-pinene up to 5%. Some 140 compounds have been identified in the oil, 75 in amounts of at least 0.1%, but which make any significantly contribution to leaf organoleptic properties remains unclear. Time of picking may have some influence on taste and odour of laurel leaves and leaf oil; in Israel 1,8-cineole content was highest in spring (60%) and lowest in summer (40%) (Putievsky et al., 1984).

Regional variation in oil content is well documented; from fresh leaves being 0.5–3.5%, and from dried leaves 1.25–2.50%. However, essential oil content and major constituents from fresh leaves vary substantially even between adjacent trees; in Georgia (former USSR), for example, 1,8-cineole content ranged from 30 to 40%, α-terpinyl acetate from 14 to 21%, and sabinene from 7 to 11%. The oil content of young leaves in Turkey is substantially higher than in mature shoots (Anac, 1986), but in Russia highest leaf-oil content was in mature leaves (Bagaturiya, 1985). The main characteristics of average commercial oils are within the following range: specific gravity (25°C) 0.905–0.929; optical rotation (20°C) +19° to +26°; refractive index (20°C) 1.465–1.470; saponification number 15–45; soluble in 1:1 volume 80% alcohol. Regional oils differ substantially in their main characteristics (Weiss, 1997).

Ecology

Laurel probably originated in Asia Minor where large natural stands are common, but is now indigenous to many Mediterranean and European countries and has been widely introduced elsewhere. Laurel dislikes shade and favours regions with long sunny periods and an average temperature range of 8–28°C. Data indicate that day/night temperatures of 18–25°C and 4–12°C respectively are the optimum. Although tolerant of low temperatures when mature, and despite the fact that plants will regenerate if frost kills off foliage, laurel is not generally frost resistant and young trees and seedlings usually succumb to a severe frost unless well mulched.

Wild plants will grow on an annual rainfall of 300–500 mm, but laurel flourishes between 1000–1500 mm; above this, soil must be free draining. Laurel is intolerant of waterlogging and should this coincide with very cold periods, mortality is high. Laurel generally grows below 700 m, the upper limit being determined by the degree and occurrence of frost, but grows well as a garden plant in frost-free areas of the Kenya highlands between 1500 and 2000 m. Low temperature due to altitude affects the general growth of trees, can reduce leaf size and may also reduce oil content of leaves. In Turkey and Greece, oil content is generally higher in leaves from trees growing at or near sea level than from those growing on inland hills (O. Anac, Istanbul University, personal communication).

Considerable regional and even district variation in leaf and leaf-oil composition exists, especially in relation to phenol and phenol-ether content, partly due to ecological factors but mainly to a dominant regional strain or local type. Seasonal and agronomic factors, however, especially time of picking, are far more important in determining the type of oil produced.

Soils and fertilizers

Laurel flourishes on fertile, well-drained soils, with a neutral to slightly acid reaction, but will grow in soils from pH 4.5–8.5, although at the lower and higher limits growth, leaf and leaf oil may be reduced and the ratio or abundance of constituents altered. Plants are very adaptable under good management, and soil *per se* is less important than other agronomic factors. Information on the effect of fertilizers on general plant growth, leaf-oil content, composition and characteristics is lacking. In Russia, spent leaves were returned to plantations as mulch and as protection against frost damage to roots.

Cultivation

Laurel is generally uncultivated but can easily be grown in commercial plantations, or existing natural stands managed on a more systematic basis when this is profitable. Plants may be grown from seed but germination is often irregular. Only fresh seed should be used as viability rapidly falls with storage. Soaking seed for 12–24 h in warm water can increase percentage germination and hasten emergence, and when sown in polythene bags containing a potting mixture resulted in 98% germination in Georgia. Shading and mulching seedbeds can increase germination and early seedling growth, but the time and expense involved are too great except for special crops.

Propagation is usually by 10–12 cm cuttings, preferably with an apical bud, taken from mature sections of the current season's shoots in late summer, for planting out the following spring. Cuttings root readily if correctly planted, tended and treated with a fungicide, and a higher rooting rate is achieved using growth hormones. Significant variation exists in leaf-oil content and oil components between individual trees; thus, selecting a high-yielding parent from which to obtain cuttings should quickly increase average oil yield per hectare, and micropropagation rapidly produces large numbers of plantlets.

Laurel grown as a rainfed tree crop to provide cash income for smallholders requires a final triangular spacing of about 6 m between mature trees. Thus, an initial 3 × 3 m spacing will allow sufficient bushes to provide an income while trees are growing. The most profitable Russian plantations are lines of bushes maintained as hedges by mechanical harvesting, at a spacing of 0.5 × 2 m (Chkhaidze and Vadachkoriya, 1989; Chkhaidze and Kechakmadze, 1991; Ebanoidze, 1997; Gabuniya and Ebanoidze, 1998). Weeding is generally manual and there are no records of herbicides being used.

It is unlikely that laurel could be profitably grown as a fully irrigated crop, but where water is available, supplementary irrigation can maintain growth and thus leaf supply. Irrigated plantings in Israel averaged 5.5 kg green leaves from trees 7–9 years old (oil yield 91 l ha^{-1}). Interplanting laurel with low-growing annual crops is possible in the first 2–4 years to reduce establishment cost. In less-developed or drier regions, annual intercropping of more widely spaced laurel rows with short-season rain-grown crops could be the most profitable.

Pests and diseases

Little information is available on the economic damage caused by insects, and most records are from garden plants. The abundance of wild trees ensures a sufficiency of leaves, even when individuals or small areas may suffer severe damage. For example, very heavy but local infestations of armoured scales, *Aonidia lauri*, can occur on laurel in Turkey, Greece and Morocco.

Diseases are far more important and often widespread in any specific population. The two most damaging diseases of wild laurel are a root rot caused by *Phytophthora* spp., which also attacks other *Lauraceae*, and leaf spots caused by *Colletotrichum* spp.; *P. cinnamomi* can damage plants at any age and growth stage. Symptoms in young plants are a general yellowing and wilting of leaves and failure of stems, and a severe attack is fatal. Such plants in plantations should be uprooted, burnt and replaced by healthy stock. On more mature plants, leaves turn yellow and fall and twigs may die back, while a severe infection may cause general defoliation and often death; in a less severe infection only part of the tree dies. Symptoms of infection by *Colletotrichum nobile* are brown spots becoming progressively larger and coalescing, causing leaves to desiccate and fall; unless controlled, leaf loss can be severe. Commonly, a general but low level of infection prevails with only minor damage to leaves, but it is accompanied by a substantial fall in leaf-oil content. Routine spraying may be necessary in areas where the disease is endemic. Chemicals can leave residues on leaves or in oil, and analyses should always be carried out to determine if this is so and the spraying programme adjusted accordingly. Residues of DDT and BHC in dry leaves have been detected.

Harvesting

Well-rooted spring-planted cuttings should be ready for first harvest in autumn of the same year when irrigated, or in the following year if rain-grown. Three leaf harvests per year are possible in commercial plantations, but two cuts generally give the highest yield of green material. One annual or three cuts every 2 years were the most profitable on mechanically harvested hedge-grown laurel in Russia. Cutting twice to 40 cm above ground in Georgia gave twice the yield of dry leaves at 4900 kg ha^{-1} over cutting once to 8–10 cm (Chkaidze and Vadachkoriya, 1985), but irrigated plants cut once annually to 60 cm in Israel gave a calculated yield of 9000 kg fresh leaves ha^{-1}.

Leaves grow vigorously for 30–40 days, more slowly for the next 15–20 days, generally cease growing after approximately 60 days and may remain on trees for 1–3 years. The yield of green leaves per tree will vary with the system of management, since this influences leaf number and growth rate. The variation in leaf yield per tree reported from various countries indicates that selection could quickly increase average yield per hectare. Harvesting is currently mainly manual but hand-held mechanical cutters similar to those used in tea plantations could be easily employed on regularly harvested bushes. Fully mechanized hedge-grown laurel plantations in Georgia are cut using high-clearance harvesters and attached binders to produce sheaves of cuttings for transport to stripping sheds (Ebanoidze, 1997).

Time of harvesting can significantly affect leaf constituents, oil yield and composition, and thus leaf quality and taste. In Turkey, harvesting in September usually gives oils with the highest 1,8-cineole content (Fig. 3.6). In general, leaf-oil content tends to rise as the season progresses and the proportion of newly mature leaves rises. The ratio between major oil constituents also varies over the same period, especially 1,8-cineole, from 25 to 55%, although seldom exceeding 40% in distilled oil. Oil from young leaves in the Russian spring is high in terpinen-4-ol, α-pineol and eugenol, but in summer is high in pinene, sabinene and cineole; the highest 1,8-cineole content is in oil from November-harvested leaves. Berries can be also be harvested, but the demand for dried berries or berry oil is insignificant.

Fig. 3.6. Harvested laurel leaves – Turkey.

Processing

Leaves collected from wild plants are carefully dried in the shade for up to 14 days to retain their green colour and prevent fermentation, sometimes on trays in special sheds, but never in the sun, as this turns leaves brown and makes them unsaleable. Leaves lose about 40% of their weight during drying. Dry leaves are bagged and transported to traders, who grade and pack leaves for export. High-quality leaves destined for international retail packaging may be pressed between boards to prevent curling during drying. When correctly dried, no significant difference exists in the constituents and characteristics between fresh and dried leaves.

Harvested leaves from plantations in Georgia are usually distilled for oil, and the operation is fully mechanized. Bundles of shoots with leaves received from a mechanical harvester are cut and spread for accelerated drying in a tunnel dryer. Leaves are then stripped mechanically from stems for later distilling. The remaining stems, which contain up to 1% essential oil, may also be distilled (Ebanoidze, 1997).

Distillation

Oil yield from fresh laurel leaves is normally 0.5–3.5% and from dried leaves 1.25–2.5%.

The oil content of leaves dried at temperatures between 40 and 70°C under controlled conditions differs little, with the recommended temperature being 60–70°C. Oil from fresh or dried leaves is similar, but the method of distilling has a direct effect on oil yield and composition; in general, water and steam distillation or a combination produce oils of similar composition but varying yield. Oil produced by hydro-diffusion differs significantly in composition, having much less 1,8-cineole and monoterpenes but more monoterpenoid alcohols, acetates and benzenoids (Boelens and Sindreu, 1986). Supercritical carbon dioxide extraction of Turkish laurel leaves produced an oil containing mainly oxygenated monoterpene compounds, around 76%, the main component being 1,8-cineole up to 43%, α-terpinylacetate up to 13%, and α-terpineol to 2.5%. The decrease in monoterpene hydrocarbons over conventionally distilled oils was an advantage, as it improved odour quality (Table 3.7; Ozek *et al.*, 1998).

Products and specifications

The most important products are the dried leaves, incorrectly but generally known as bay leaves, and leaf oil, both used as spice, and berry oil used as a flavouring.

Table 3.7. Comparison of distilled and supercritical CO_2-extracted laurel oils.

	Hydro-distillation	Steam distillation	Supercritical CO_2 extracts	
			40°C/80 bar	50°C/80 bar
Yield %	2.6	1.9	1.3	1.1
Monoterpene hydrocarbons	19.5	16.5	6.8	5.1
Oxygenated monoterpenes	73.6	70.7	75.6	76.4
Sesquiterpene hydrocarbons	–	–	0.8	1.4
Oxygenated sesquiterpenes	–	–	1.1	0.8
1,8-Cineol	49.7	54.2	43.0	40.2
α-Terpinylacetate	5.3	7.6	10.8	13.8

Leaves. Dried laurel leaves produce the spice of commerce, and the growing use of spices in domestic cooking has raised the quality of retail products in international supermarkets. Leaves should be whole, flat, of a uniform light green with a brownish tinge but not brown, with a pleasant sweetly aromatic, camphoraceous, cineolic aroma and sweet spicy taste (Fig. 3.7). Broken leaves of the same quality are acceptable but fetch lower prices. Powdered leaves are generally lacking in both taste and aroma. Laurel is placed in the fresh, spicy woody group by Arctander (1960).

Advertising claims the spice to be free from chemical residues and organically grown; as most leaves are from untended wild plants this last requirement is no problem! As a spice, laurel is considered a vital ingredient of the genuine *bouquet garni*, and its leaves are one of the few spices added to dishes early in cooking as the taste gradually intensifies, but should be removed before serving as their sharp edges can cause injury to the mouth or internal injuries if swallowed. Interestingly, laurel leaves were used to package Turkish liquorice sold in Scandinavia; the leaves were then used as a spice and thus a combination of the two flavours came to be locally accepted as the flavour of laurel leaves.

Fig. 3.7. Dried laurel (bay) leaves.

Crushed or powdered leaves are an essential ingredient of pickling spices and spiced vinegar, and are extensively utilized in the food processing industry, especially for meats. An extract of leaves is considered astringent, stomachic, a stimulant, narcotic, a powerful emmenagogue, and is used in the treatment of infections of the urinary tract and for dropsy. Its traditional use as a treatment for diarrhoea is probably based on the catechin content.

Leaf oil. The oil, which is usually obtained by steam distilling freshly dried leaves, is normally colourless or very pale yellow; its odour is strong, sweet, aromatic and slightly camphoraceous, and its taste is sweet-warm and spicy. The oil is placed in the fresh, spicy-woody group by Arctander (1960). However, as noted, supercritical carbon dioxide extraction produces a finer flavoured oil. The major components are 1,8-cineole up to 40%, linalool and α-terpinyl acetate, up to 10% each; over 100 other constituents have been isolated, and both the abundance and proportion of main constituents can vary considerably depending on geographical origin (Boelens and Sindreu, 1986; Pino and Borges, 1999a). The oil replaces dried leaves in flavouring meats, pickles, processed foods, confectionery, etc., since strict control of the amount used is possible and thus the desired flavour level can be determined in finished products. Oil is a minor ingredient in perfumery, toiletries and soaps. One well-known use in the Middle East is the manufacture of Aleppo's famous *sabun bi ghar*, containing olive oil and 5–20% laurel oil. This golden soap has been produced and exported from Aleppo, Syria, for at least 500 years.

Laurel oil has antibacterial, antifungal and antioxidant activity (Baratta *et al.*, 1998). A monograph on the physiological properties of laurel leaf oil has been published by the RIFM. Laurel leaf oleoresin (commercially bay leaf oleoresin) is produced in very small quantities almost wholly in the USA for a very limited market in the processed food industry.

Berry oil. The oil obtained by steam distilling fresh or semi-dried berries is a pale, greenish or olive yellow mobile liquid; its odour is spicy and camphoraceous, and its taste warm spicy. The oil is placed in the fresh, spicy woody group by Arctander (1960). The major constituent is terpinen-4-ol, and although the oil's composition is similar to leaf oil, relative proportions of components are quite different. Berry oil is produced only on demand, and is occasionally used in perfumery. Expressed berry oil contains mainly glycerol laurate plus a volatile element high in cineole. It is used to treat rheumatic conditions, as a minor condiment and in perfumery material.

Oleoresin. Oleoresin is obtained by solvent-extracting leaves, and is dark green, semi-fluid, with a strong aroma of the spice. It is usually available in the following volatile oil contents (ml per 100 g): 15–20, 25–30, 32–36, 72–80. Its uses are as for the oil. Laurel leaves, leaf oil and oleoresins, and expressed berry oil all have the USA regulatory status generally recognized as safe, GRAS 2124, 2125, 2613 and 2612, respectively.

Unrelated species

Laurel is widely applied as a prefix to the common names of many plants; ironically, dried laurel leaf used as a spice is called sweet bay or bay leaf, although true bay oil and leaves are obtained from *Pimenta racemosa* (Mill.), a member of the *Myrtaceae* (Chapter 6). True laurel leaves can usually be distinguished from non-laurel species by holding a leaf against the light, when the translucent leaf margin of laurel is plainly visible. Cherry laurel, *Prunus laurocerasus* L., has serrated leaves, which contain prussic acid and taste of bitter almond. The mountain laurel, *Kalmia latifolia* L., and the related sheep laurel *Kalmia angustifolia* L., also have toxic leaves.

4

Leguminosae

The *Leguminosae* is one of the three largest families of flowering plants, with about 700 genera and some 18,000 species ranging from herbs to trees. The distinctive structure of the flowers easily identifies members. A most important characteristic of legumes is their ability to maintain a symbiotic relationship with specific bacteria via root nodules, which fix atmospheric nitrogen and make it available to the host plant. The *Leguminosae* is composed of the *Caesalpinioideae*, *Mimosoideae* and the *Papilionoideae*, whose members are characterized by their highly aromatic foliage and long straight pods with an elongated beak. The *Papilionoideae* provides many important crop plants including groundnut (*Arachis hypogaea* L.), soybean (*Glycine max* (L.) Merrill) and the spice fenugreek (*Trigonella foenum-graecum* L.).

Fenugreek

Fenugreek is probably indigenous to the eastern Mediterranean, but its natural distribution is obscure as it has been cultivated since antiquity, the first written record dating to 4000 BC. Its cultivation spread throughout the Arab world, to Europe, Ethiopia, the southern parts of the former Soviet Union, India and China. In many of these countries it is now an acclimatized escape and numer-

ous enough to be harvested. Introduced to South-east Asia by Indian and Arab traders, it has been successfully grown as a small-holder crop on Java, Indonesia, where its seed oil is used as a hair dressing.

In ancient Egypt and Greece, fenugreek was a fodder crop prior to its medicinal properties becoming known. In Dynastic Egypt, seeds were roasted then boiled to produce a hot drink, and fresh leaves or sprouted seeds were used as a vegetable; it was also an ingredient of the important incense, *kuphi*, and embalming herbs. The Greek Pedanius Dioscorides, who served as an army doctor under emperor Nero, recommended in his *De Materia Medica* that fenugreek powder dissolved in wine should be taken to alleviate the pain of gout, and when taken as snuff cured headaches. Pliny, in his *Historia Naturalis* of AD 77, prescribed the powder/wine mixture as a treatment for deafness. The growing of fenugreek, together with many umbelliferous spices, was promoted by the emperor Charlemagne (747–814) and it was grown by Benedictine monks in central Europe in the 9th century. In the Middle Ages in Europe, a paste applied to the head was recommended as a cure for baldness, but so offensive was the smell that it became known as 'Greek excrement'! An extract of the mucilaginous seeds was considered an effective treatment for mouth and stomach ailments. Introduced to Britain in the 16th century, it never became popular as a spice and is little used today.

The Arabs highly prized fenugreek, and its medicinal uses were studied at the famous School of Salerno by Arab physicians. The plant is still highly regarded and grown in the Mediterranean region, West Asia and India for culinary and medicinal purposes, but in northern Europe and North America it is now a minor seed crop and more important as forage. Introduced to Australasia by European settlers, it is grown only by specialist herb producers. Its original introduction to China is unrecorded, but during the 11th-century Sung dynasty was recommended for the treatment of various ailments.

Seeds and leaves of fenugreek are widely used as a culinary spice, to enhance the taste of meat, poultry and marinated vegetables. In North Africa, it is mixed with breads and in the traditional dish halva, while boiled or parched seeds mixed with honey are a snack. In Egypt and Asia, seedlings of 5–8 cm are eaten as a vegetable, and a drink is prepared from dried and ground sprouted seeds. Young shoots and pods are commonly used as a vegetable in India although somewhat bitter, and the dried leaves of cultivar Kasuri Mehti are used to flavour sauces and gravies. Also in India, the seeds are the source of a yellow dye and an industrial mucilage.

Fenugreek is not only an important spice but also has potential in modern medicine. The seed contains diosgenin, an important precursor in the production of several drugs, and as the plant has a shorter growing period than the *Dioscorea* species, which the author grew on a commercial scale in the west Kenyan highlands, fenugreek could become an alternative source. In developed countries, fenugreek is utilized in veterinary rather than human medicines.

Botany

The name *Trigonella foenum-graecum* L. comes from the generic name *Trigonella*, stated to be derived from the Latin for little triangle, referring to the flower shape, and *foenum-graecum*, which is Latin for Greek hay or grass. In English it is known as fenugreek, in

French as *fenugrec senegree*, in German as *bockshornklee*, in Italian as *fieno greco*, in Portuguese as *alforva*, in Spanish as *alholva*, in Dutch as *fenegriek*, in Russian as *pazhitnik*, in Arabic as *helba*, in Ethiopia as *Amarinia*, *abish*, in India as *methe*, in Indonesian as *kelabet* and in Chinese as *hu-lu-pa* or *k'u-tou*.

The genus *Trigonella* L. is not well defined, but contains about 100 species, many native to the Mediterranean and West Asian regions, some also to southern Africa and Australia, but now widely distributed. Fenugreek is probably unknown as a truly wild plant, and naturalized plants are remnants or escapes from cultivation. Long-continued selection resulted in the emergence of regional forms in the Mediterranean region, India and Ethiopia, inaccurately grouped into subspecies; cultivars is more accurate. Fenugreek is a diploid with $2n = 16$. Genepools are maintained at the All Russia Research Institute for Plant Breeding, St Petersburg, and at various Institute of Spices Research sites in India, including Coimbatore, Jagudan and Jobner, where the variability in vegetative, reproductive and other characteristics has been determined (Yadav, 1999; Das *et al.*, 2000).

The major variations reported in the literature concerning herbage and seed composition, in various plant part ratios such as pods/plant, seeds/pod, pod length, seed weight and constituents, indicate the wealth of material available to plant breeders and the potential to substantially increase herbage or seed yield, or a desired derivative. Breeding programmes in India have produced the superior cultivars 'Rajendra Kanti' and 'Lam Selection-1'. An important non-spice use for fenugreek, as noted, is as a potential source of diosgenin and this is discussed at the end of this section.

Fenugreek plants have a well-developed taproot and a spreading, fibrous root system, carrying numerous small, flattened, much-lobed nodules containing *Rhizobium* spp., usually *Rhizobium meliloti*. Spraying growing fenugreek plants with a solution containing the N-fixing bacteria *Klebsiella* (KUPO) gave significant herbage and seed increases in India (Mukhopadhyay and Sen, 1997). The bitter roots are seldom eaten.

Plants normally have a single stem (Fig. 4.1.) but cultivars may have three to six basal branches giving a multi-stemmed appearance. The stem is green to purple, smooth, erect, stiff, up to 140 cm tall but usually 60–90 cm; it is terete, slightly pubescent, longitudinally striate, with internodes hollow with age and profusely branched at all heights. The leaves are alternate, pinnate, with primary pinnae odd-numbered (3–19), trifoliolate and 2–15 cm long; the stipules are triangular, small and adnate to the petiole; the rachis is short and the leaflets obovate or oblong, 1.5–4 cm × 0.5–2 cm, with the upper part of the margin denticulate; the petiole is grooved above, 1–4 cm long. Analyses of fresh leaves show considerable variation in composition, especially in protein content, between 5 and 50%. One analysis gave the following approximate composition (per 100 g leaves): water 87 g, protein 5 g, fat 0.2 g, carbohydrates 5 g, fibre 1.4 g, and ash 1.4 g containing Ca 150 mg and P 50 mg.

Young shoots and thinnings are used as a vegetable although somewhat bitter, and fresh leaves are used to treat flatulence. The glycoside responsible for the bitterness is considered harmless in the amounts normally ingested (Abdel-Barry and Al-Hakim, 2000). Fenugreek for fodder should be cut when pods are forming, and fresh forage is comparable to lucerne hay in digestibility. It is used as a feed supplement to promote animal health, or to improve palatability of poor hay. In Canada, it is grown as a high-quality hay crop. The residue (straw) after threshing is considered a medium-protein roughage suitable for dairy cattle maintenance.

The inflorescence is a terminal, compound umbel, up to 20 cm diameter but usually smaller. The peduncle is variable from 5 to 16 cm; the primary rays (5–30 per umbel) are 0.5–12 cm long, being unequal in length with the shortest in the centre; the secondary rays (pedicels) are variable (10–30 per umbellet) up to 1 cm and unequal in length. The calyx is campanulate and finely pubescent, the tube 4.5 mm, and the lobes 5 × 2 mm; it is standard hood-shaped, obovate, 12–15 × 6–7 mm, clawed, pale yellow, with wings obovate 12 × 2.5 mm; the keel is ladle-shaped, 7 × 2 mm, and bi-auriculate. There are nine stamens; the pistil has a sessile ovary, glabrous style, and a capitulate stigma.

The flowers are white to whitish-yellow, and usually open from 9 a.m. to 6 p.m. with a peak just before midday. Anthesis occurs from about 10.30 a.m. to 5.30 p.m., again peaking at noon. The stigma becomes receptive 12 h prior to flower opening and remains receptive for another 10 h. Flowering occurs 4–5 weeks after sowing and continues for up to 3 weeks. The main flowering period is mid to late summer in Europe.

The fruit is a light green to yellow-brown, ovoid-cylindrical, usually slightly curved schizocarp, 5–19 × 2–4 mm, terminating in a fine beak often equal in length to the pod; it

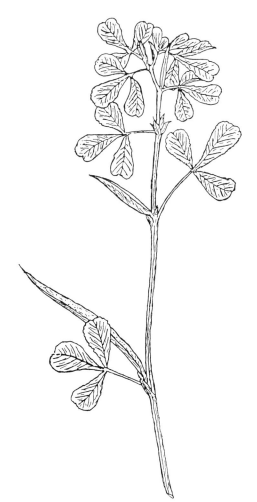

Fig. 4.1. Fenugreek shoot.

is glabrous, with fine longitudinal veins, and indehiscent but splitting at maturity into two mericarps each with five prominent ridges and oil-vittae between the ridges. The pod shape also gives the name 'goat's horn' to the plant. The fruit contains 10–20 small, smooth, brownish seeds, which are oblong-rhomboidal, 3–5 × 2–3 mm, and divided by a deep furrow into two unequal lobes with rounded corners (Fig. 4.2). Seed testa adnate to the pericarp. Weight per 1000 seeds is 10–20 g, depending on cultivar.

Fruit/seed composition and the amount and ratios of the various components vary widely, and are apparently strongly affected by environment and location. Alkaloids present were identified as trigonelline up to 0.13%, choline 0.05%, gentianne and carpaine. The composition of regional cultivars differs considerably, and the following data indicate this range. Fruit may contain (per 100 g): water 8–10 g, protein 15–28 g, fat 6–12 g, carbohydrates 35–45 g, fibre 8–16 g and ash 4–8 g containing Ca 120–220 mg, P 350–450 mg, K 100–120 mg, Na 75–100 mg, Fe 20–24 mg and Zn 2–5 mg.

Seed also contains approximately (per 100 g) vitamin A 130–140 IU, niacin 2–6 mg, thiamine 0.2–0.4 mg, riboflavin 0.25–0.35 mg, tryptophan 250–280 mg, plus flavinoids including vitex and vitexin glucoside; energy value of 1540 kJ. Fruit protein generally has an *in vitro* digestibility of 95%, with 6.5% lysine, is low in sulphur-containing amino acids, and reportedly contains trypsin and chymotrypsin inhibitors. Seed contains a yellow dye occasionally used to colour foods, and mucilage up to 30%, which in India is the source of an industrial galactomannan mucilage. A fatty oil can be extracted from seeds with a 5–8% yield.

The fruit-seed oil is rich in phosphates, lecithin and nucleoalbumin, contains readily absorbed Fe in an organic form, and substances similar to the alkaloids in cod-liver oil that stimulate appetite. A seed extract has wound-healing properties (Taranalli and Kuppast, 1996). However, some persons can be allergic to topical application of extracts or pastes, or ingestion of fenugreek seeds (Patil *et al.*, 1997) and the sotolone content may cause urinary trouble in babies fed artificial maple syrup (Sewell *et al.*, 1999). The possible use of seed extracts in the treatment of diabetes is being investigated (Al-Habori and Raman, 1998; Dhananiay and Gupta, 1999).

Diosgenin. The seed contains saponins, which yield on hydrolysis up to 2.5% steroid sapogenins, mainly diosgenin and its isomer yamogenin, and enzyme treatment of seeds before acid hydrolysis reportedly doubled the diosgenin/yamogenin (d/y) yield. The

Fig. 4.2. Fenugreek seeds.

main factor controlling the d/y level in seeds is genetic, and thus cultivars vary widely in the level of these compounds, as was determined in Slovenia (Zupancic *et al.*, 2001).

Diosgenin content of Indian cultivars ranged from 0.7 to 1.0% on a moisture-free basis, and for Polish cultivars up to 2.0%. In addition to diosgenin, seeds contain small amounts of similar steroids used in the synthesis of hormones for oral contraceptives and corticosteroids, plus the alkaloids trigonelline (0.1 to 0.4%) and choline. Some related saponins have been identified (Brenac and Sauvaire, 1996; Yoshikawa *et al.*, 1997). Distribution of diosgenin in plant parts during growth varies, and in plants treated with benzylaminopurine the highest level (20 mg g^{-1} dry weight) was in young leaves 30 days after germination, while seed diosgenin content increased by nearly 30% (Ortuno *et al.*, 1998). The effect of growth regulators GA, IAA and ethephon on diosgenin synthesis showed that GA gave the greater increase of 77% in 15-day-old plants falling to 68% in plants 30 days old (Ortuno *et al.*, 1999). Diosgenin production from various plant cultures has also been investigated. Root-tip cultures grown *in vitro* with *Agrobacterium rhizogenes* showed the sapogenin content was only about 0.03 g per 100 g dry matter (Merkli *et al.*, 1997).

Ecology

Fenugreek is basically a warm temperate crop with cultivars adapted by long domestication to warmer and colder environments. In warm climates it is usually grown as a cool-season crop, in temperate areas as a summer crop. Plants are killed by frost at any growth stage, and this limits its northern expansion except under protected cropping. It grows well in the Ethiopian highlands where it is widely cultivated, but has not been encountered in the highlands of eastern and central Africa (E.A. Weiss, personal observation). However, it was occasionally found in Indian or Arab gardens at the coast, and known as *uwata*. Fenugreek is common on smallholder plots in Egypt, the northern Sudan and on the Gezira irrigated area, but is never more than a minor crop. In India, it is reportedly naturalized in Kashmir and cultivated extensively in other Indian states, as noted.

Plants will grow well with an annual rainfall of 500–750 mm, and can survive a moderate soil moisture shortage, but above 1250 mm are susceptible to waterlogging on more heavy soils, especially the seedlings. Little is known regarding fenugreek's reactions to light periods or light intensity, although plants generally prefer full sun to shade.

Recommended times for sowing fenugreek are: April–May in Europe and North America; in India, June–July when raingrown, October–November under irrigation; in upper Egypt, October–November on a falling Nile; in the Sudan, July–August; in Ethiopia, where the rainy season varies, it is generally sown at the end of the rains; in south-east Australia, October–November.

Soils

Fenugreek flourishes on well-drained loams or sandy loams, and grows fairly well on gravelly and sandy soils, but clay soils are best avoided. The acceptable pH range is 6.0–8.0. Moderately saline soils or irrigation water are tolerated, with some reduction in herbage or seed yield. Selected strains in India showed a higher tolerance of salinity, and thus could supply genetic material to improve this valuable characteristic in commercial cultivars.

Fertilizers

The nutrient requirements of fenugreek are low if grown on fertile soils, but commercial crops may require added nutrients, as determined by soil analyses and fertilizer trials. If fenugreek is grown in a rotation with crops to which phosphate has been applied, further application is seldom necessary.

An indication of the nutrient requirement of fenugreek in India is shown in Table. 3.4, where an average fruit yield of 1000 kg ha^{-1} removed 35 kg N ha^{-1}, 18 kg P ha^{-1} and 14 kg K ha^{-1} (Sadanandan *et al.*, 1998). In India generally, relatively low levels of N up to 90 kg N ha^{-1} increase herbage and seed yield but response to P is variable, while K has a local level usually unaffected by the other two nutrients (Randhawa *et al.*, 1996a; Halesh *et al.*, 1998; Sheoran *et al.*, 1999). The local value of the crop determines the amount of fertilizer applied and smallholder crops usually receive plant and animal wastes. In India up to 30 t ha^{-1} is recommended by the Agricultural Department, ploughed in well before planting, but few farmers have such amounts and generally apply whatever is available.

Root nodules on fenugreek roots supply most of the plants N requirement when grown for seed, with an additional application applied as a top dressing when high herbage yield is required. In general, about 25 kg N ha^{-1} in the seedbed is recommended and this can be combined with P as diammonium phosphate. Seed must be dressed with a *R. meliloti* strain, preferably M1–1 where the crop has not previously been grown, but commercial producers routinely dress all crops. Plants should preferably not be top-dressed with N when grown for seed as this tends to prolong the vegetative phase and lowers seed yields. Limited data indicate that high levels of N and P depress seed diosgenin levels. Phosphate requirement is moderate and seldom exceeds 12 kg P ha^{-1}, and if the previous crop has been well supplied with this nutrient, half this amount applied to the seedbed is usually sufficient.

The K requirement is frequently higher than that of P, but varies widely, and should be determined by soil analyses and trials to establish the optimum amount. Data indicate that K up to twice the P required is commonly applied to the seedbed as part of an NPK compound fertilizer; in India a 25:25:50 NPK mixture is recommended.

Cultivation

Land preparation as described in the introduction to the *Umbelliferae* (Chapter 10) is suitable for fenugreek. The most important requirement for commercial crops is as level and weed-free a seedbed as possible. Fenugreek is grown from seed, and small-holder crops are usually broadcast at 25–35 kg seed ha^{-1} and harrowed in. A common spacing for commercial crops is 7–10 cm between plants in 20–45 cm rows, but the row width and optimum plant population should suit existing machinery. Cultivars differ in plant size, and this must also be considered. In India, for example, different cultivars are sown to provide either forage or seed.

Seed should be sown not more than 2–3 cm deep, and provided good quality fresh seed is used, should germinate within a week. Germination is epigeal, the first leaf simple and the second usually trifoliolate. Plants flower about 5–6 weeks after emergence and are ready to harvest 8–12 weeks later, the variation being due to cultivar differences. In India, highest yields were obtained in the Punjab by sowing between the last week of October and the first week of November; generally seed is sown from early to mid-April in Europe.

Weed control

Fenugreek seedlings are poor competitors against the often rampant weed growth in many warmer countries, and this emphasizes the necessity for a weed-free seedbed, since early cultivation in general is dangerous until the taproot is well grown. Small-holder crops sown in rows will require two to three hand-weedings, the first depending on the weed growth, but should preferably be delayed until plants are about 10 cm in height. Commercial crops are weeded mechanically as necessary, and this operation must be carried out with care as seedlings are easily uprooted. Depth of operating is important in more mature plants to avoid damage to the wide-spreading and often superficial roots. Herbicides are seldom used.

Irrigation

Fenugreek is usually rain-grown and only where water and labour are plentiful can it be a fully irrigated crop, as in India, upper Egypt and parts of the Sudan. Small, frequent waterings are favoured, and once watered the soil must be kept moist. Some cultivars in India are reported to be tolerant of slightly saline water, which is a valuable characteristic, but this is not general and most cultivars are adversely affected (Yadav *et al.*, 1996).

Intercropping and rotations

Fenugreek is frequently interplanted in smallholder crops with coriander, sesame, beans and chickpea, or sown with lucerne and clover as a mixed fodder crop in India and Egypt. It can be underplanted in orchards, palm and citrus groves. Fenugreek can be included in any arable rotation, for hay, as a cover crop, or as green manure; in India, it may be sown after irrigated cotton. It should preferably not be grown for more than two successive crops, as the pest and disease build-up is too costly to control.

Pests

Information on specific pests damaging fenugreek is limited, but in general those polyphagous insects noted on legume crops in the vicinity also attack fenugreek. For example, *Aphis craccivora* and *M. persicae* have both been noted as causing damage generally, from West Asia to India, while various *Thysanoptera* (thrips) occur on almost all fenugreek crops, with *Scirtothrips dorsalis* an important pest from the Mediterranean to India. The mite *Tetranychus cucurbitae* attacks fenugreek in India, and a severe infestation greatly reduces fruit set. A major pest in the Sudan is *Pachymerus pallidus*, a seed beetle, which attacks a wide range of crops and is difficult to control. Caterpillars of a number of *Lepidoptera* are recorded on fenugreek and many are also polyphagous, including *Diacrisia obliqua*, *Diacrisia orichalcea*, *Prodenia litura*, and especially the mung moth, *Maruca testulalis*, in India (Fig. 4.3). Various nematodes, often unidentified, are reported from fenugreek roots.

Diseases

The diseases attacking fenugreek require more detailed investigation and identification, as many are generally described as root and other rots and wilts. In India, 27 species of fungi were isolated from fenugreek seeds (Prabha and Bohra, 1999). Time of sowing can influence the damage caused by disease; for example, sowing in mid-October compared with the end of November reduced the damage caused by powdery mildew *Eyrsiphe polygoni* and *Leveillula taurica* in Haryana, India, by 30% (Shushi Sharma, 1999). Another mildew, *Peronospora trigonella* (downey mildew) has recently become more common, but the degree of damage is usually minor.

Several leaf diseases cause varying degrees of damage generally or in specific seasons, including rust due to *Uromyces anthyllidis* in India, and leaf spot due to *Cercospora traversiana*, which is seed borne and now occurs in many countries from India west to eastern Europe, eastern Africa including Ethiopia and several countries in South America, and is becoming of greater economic importance. Root and collar rots cause more or less damage in individual

Fig. 4.3. Mung moth, *Maruca testulalis*.

crops, with *Rhizoctonia* spp., usually *Rhizoctonia solani* (Fig. 4.4), and *Alternaria* spp., often *A. alternata*, most frequently reported. A virus similar to bean yellow virus has been noted from India.

Harvesting

Plants are ready for harvesting 3–4 months after sowing when the majority of pods have turned a uniform yellowish brown. Smallholder crops are pulled by hand, and plants are hung upside down to dry in the shade. Since pods are normally indehiscent there is little loss of seed. Plants are later threshed manually as convenient, and the residue is suitable for stockfeed. In Gujarat, India, fenugreek grown as a leafy vegetable is harvested by clipping leaves and young shoots, and plants allowed to regrow until mature. This produces seed yields of about 500 kg ha^{-1}. Commercial crops in Europe and elsewhere are combine harvested with specially adapted machines with front-end attachments, but are frequently cut, windrowed and threshed by a pick-up combine. Alternatively, plants can be cut and bound into sheaves, stooked to dry and later transported to a stationary thresher.

A smallholder seed yield of 600–900 kg ha^{-1} is common; in India, crops grown for seed yielded 1675 kg ha^{-1} and in Egypt, partially irrigated crops yielded 1400 kg ha^{-1}.

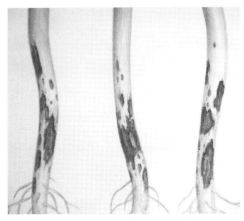

Fig. 4.4. Collar rot due to *Rhizoctonia solani*.

Commercial crops in Europe and Canada grown for seed yielded to 4500 kg ha^{-1}, and for forage up to 10–12 t dry herbage ha^{-1}.

Processing and storage

Commercial seed is cleaned, dried if necessary and stored in moisture- and vermin-proof bins or silos. Once dried, the small seed can be easily handled by modern equipment, bagged or transported in bulk. Export grading in India is to Agmark standards and PFA regulations, and must also conform to ASTA specifications governing fungal, insect and mammalian contamination.

Distilling is as for dill or coriander, and no special requirements or equipment are necessary other than extra cooling to ensure that the distillate is not overheated. Oil should be stored in full, opaque containers in the dark.

Products and specifications

Fenugreek produces fresh herbage, seed used as spice, oil and oleoresins.

Herb. The herb is used fresh as a vegetable or chopped as a flavouring in Asia; the taste is rather bitter. It is not a culinary spice but is often a constituent of freshly prepared curries although not curry powders. Its major use, fresh or dried as hay, is as fodder for livestock.

Seed. Good-quality seed is a smooth, uniform light brown, with a bitter taste and a characteristic but faint smell of burnt sugar, which becomes more pronounced on heating. The seed, either ground or whole, is the well-known culinary spice of southern Europe and west Asia, and a common ingredient of Asian cuisine, especially curries. However, many people dislike both the odour and flavour. Seed is often roasted before being ground and added to dishes as this increases its flavour, but care is necessary as overheated seed is extremely bitter. The characteristic flavour of roasted fenugreek seed is due to degradation of much of the alkaloid trigonelline to nicotenic acid and associated pyridines and pyrroles.

Seed extracts. Extracts are generally prepared by extracting ground seed with alcohol or aqueous solutions. A number of proprietary extracts, liquid and powder, are available, which differ in aroma and flavour mainly due to the time for which seed was roasted prior to extraction. A major use for solid extracts in the USA is in the formulation of imitation maple syrups, when the degree of roasting produces various flavours. Other food uses include frozen dairy desserts, confectionery, baked products, gelatines, meat extracts and processed meats. A fixed oil up to 8% can be obtained on solvent extraction with a strong flavour described as nutty or fishy depending mainly on its age. However, it is seldom produced.

Essential oil. The essential oil obtained by steam distilling seed is a greenish-yellow mobile liquid containing at least 50 components including alkanes, sesquiterpenes and oxygenated compounds, and the proportions vary with cultivar and season (Lawrence, 1987). Major constituents are dihydroactinidiolide, 2,3-dihydrobenzofuran and l-hexanol totalling 7–9%, with 20 other constituents at less than 3% and the remainder below 1%. The furanone derivative, sotolon, is reportedly mainly responsible for the characteristic fenugreek odour. Arctander places it in the celery odour group. Its main use is as a liquid substitute for the powdered seed, especially in frozen dairy products, alcoholic and non-alcoholic beverages.

Oleoresin. The oleoresin is obtained by either hydro-alcoholic or petroleum ether extraction of seed with a yield of approximately 0.02%. The former produces a very dark, resinous extract, which is intensely sweet, the latter a lighter coloured and less sweet extract. The oleoresin is produced in small quantities almost wholly in user countries. It is an ingredient of imitation maple syrup, and substitutes for caramel, vanilla, butterscotch, rum and liquorice flavours. In the USA, the maximum permitted level of oleoresin in food products is generally 0.05%. The oleoresin is also used in soaps, perfumery, cosmetics and hair tonics with a maximum use level of about 0.2%. An absolute can be prepared by further extracting oleoresin, which is stated to be non-irritating, non-sensitizing and non-toxic to humans.

A monograph on the physiological properties of fenugreek oleoresin has been published by the RIFM. In the USA, the regulatory status generally recognized as safe has been accorded to fenugreek, GRAS 2484, and fenugreek oleoresin, GRAS 2486.

Medical uses

Fenugreek seed, oil and extracts have numerous uses in traditional medicine; seeds are emollient, laxative, a vermifuge, a paste is used to treat mouth ulcers and chapped lips, and poultices made from herbage are used to alleviate the pain of abscesses and wounds. Seeds are also considered to be a restorative, to ease menstruation, promote milk flow and have aphrodisiacal properties probably related to the trimethymine content. Fenugreek seeds and extracts are used to treat many afflictions in China, including abdominal pain, kidney ailments, hernia, arthritis and the debilitating disease beriberi. In India, it is a traditional anthelmintic, commonly used as a diuretic and used in the treatment of dropsy, heart disease, spleen and liver enlargement. Fenugreek has little place in modern Western medicine, although tablets are often available in health food shops, but is retained in veterinary formulae. In laboratory trials, leaves have been shown to lower blood glucose levels in diabetic laboratory animals and have analgesic properties. A water extract of seed reportedly had accelerating effects on the beat of isolated mammalian hearts.

Other Trigonella species

Trigonella caerulea (L.) Ser. (sweet trefoil) is cultivated on a minor scale in Switzerland and also in central Europe where leaves are used to flavour cheese, bread, soups and as a tea. *Trigonella corniculata* L., occasionally grown as a vegetable in Asia and elsewhere, also contains diosgenin. *Trigonella suavissima* Lindl. is an Australian species known as sweet fenugreek, and was used as a vegetable and flavouring by early settlers.

5

Myristicaceae

The order *Magnoliales* contains several major families including the *Myristicaceae* of 18 genera, with *Myristica* (L) Gronov. the largest with up to 120 species. The *Myristica* are mainly evergreen trees of tropical lowland rain forests from India and Sri Lanka eastwards through South-east Asia to Taiwan, the Pacific islands of Fiji, Tonga and Samoa to Australia, with other species occurring in the Americas. The centre of origin and distribution of the genus is considered to be Papua New Guinea with some 40 known species, 34 of which are endemic.

A number of *Myristica* species are cultivated on a small scale, but only *Myristica fragrans* Houtt. (nutmeg), *Myristica argentea* Warb. (the Papua nutmeg) and *Myristica succedanea* Reinw. ex Blume (the Halmahera nutmeg) are grown commercially and are later discussed, together with *Myristica* species that also produce a nutmeg-like seed, several of which could be marketed more widely to satisfy the considerable world consumer demand for a greater range of natural flavourings. The calabash or African nutmeg, *Monodora myristica* Dunal., of the *Annonaceae*, and native to West and East Africa, has seeds that resemble nutmeg in odour and flavour. Seed is collected for sale in local markets, especially Nigeria, and substantial amounts are involved nationally. It is also described later in this chapter for convenience. The term nutmeg without qualification in the text refers to *M. fragrans* and its products.

Nutmeg and Mace

Nutmeg as a food flavouring or medicine has long been used in its original home, and probably also in China, but few early references positively identify the spice. The Sanskrit *Susruta Samhita* about AD 600 named nutmeg *jai phal* and mace, which comes from the fleshy outer covering of the nutmeg, as *jai kosa,* and Hindu traders probably introduced it together with their religion to Java, where nutmeg is known as *jati-phala* and mace as *jati-kosa*, and thence to Malaysia as *buah pala.*

Authorities differ as to the extent nutmeg and mace were known and used by the Greeks or Romans, although Theophrastus (372–287 BC) in his *Enquiry into Plants* is credited with attributing the spice *comacum* to nutmeg because of its dual properties and differentiating it from cinnamon, which Pliny the Elder 350 years later supported in his *Historia Naturalis*, but more accurately called it a nut. Arab traders brought nutmeg to Europe from the Moluccas via India, and as with other spices concealed the true source to avoid competition; however, it was an Arab, Kazwini, who in AD 1300 was the first to publicly reveal the Moluccas as the source of nutmeg. The first authentic European record is from AD 540, and stated to be by Actius of Constantinople.

By the 12th century it was well known in Europe and when in 1191 Emperor Henry VI entered Rome for his coronation, the streets were fumigated with nutmegs and other strewing aromatics. In the 12th century, nutmeg was well known enough for the English chronicler Chaucer to write, 'And notemuge to put in ale, whether it be moist or stale', and in 13th-century England mace sold for 4s 7d per lb, equal to one sheep or half a cow. The search for a route to the Spice Islands resulted in Vasco da Gama's landfall on India's west coast in 1498. By 1512, the Portuguese had reached the Moluccas and they dominated the nutmeg trade for the next century. Garcia da Orta who visited the Moluccas in the early 16th century wrote that, 'The tree supplying nutmeg and mace is like a pear tree in both trunk and foliage. In these islands its fruit is sparse and wild, and they eat its fruit with betel leaf and make no other use of it' (da Orta, 1563).

The Portuguese were replaced in the early 17th century by the Dutch, who in turn monopolized the trade for another 200 years. The Dutch endeavoured to limit nutmeg production to Banda and Amboina by forcible destruction of all other trees and by 1650 had been generally successful. The programme was partially negated by fruit pigeons, which swallowed whole seeds and voided them on neighbouring islands and one nutmeg-producing island, Run, a British possession. To enforce their monopoly, the Dutch negotiated the exchange of Run for an island they owned off the North American coast, now known as Manhattan Island. In addition to limiting Moluccan production, the Dutch East India Company in 1735 burnt 570,000 kg of surplus nutmegs in Amsterdam to maintain a high price. Supplies of mace were also deliberately restricted, and the 1806 London price was 85–90s per lb plus import duty of 7s per lb.

The French were able to circumvent the Dutch monopoly when Pierre Poivre obtained 32 nutmeg plants and reached Mauritius, then Île de France, in December 1753 with five survivors. Later Captain Provost of the *L'Etoile du Matin* collected quantities of nutmeg seeds and nutmeg and clove seedlings on the island of Begy, and

returned with them to Île de France (Ly-Tio-Fane, 1958). Despite an export ban, plants were taken to the Seychelles and Réunion (Ile de Bourbon). A later governor of Mauritius, Mr Cere, made the important discovery that nutmeg was unisexual. According to a contemporary description, 'The first French nutmeg was picked in December 1778 in the presence of a distinguished assembly and Governor Cere afterwards gave a magnificent reception to celebrate the event.' Introduced to Zanzibar in 1818 from Mauritius or Réunion, descendants of these trees still grow on Zanzibar and Pemba. Trees thrived but nutmeg never became as important as cloves, although there was a small export trade by dhow to the Persian Gulf, which continued until the dhow trade ceased in the mid-1960s.

During the British occupation of the Moluccas (1796–1802), the Honourable East India Company sent their botanist, Mr Christopher Smith, to collect seedlings of nutmeg and clove to establish the spice in Penang and other countries under British control. By 1798, there were 600 nutmeg and some clove trees growing in Penang, augmented by Smith's subsequent shipments of 71,265 nutmeg and 55,265 clove plants. The company sold their spice gardens in 1805 including 51,000 nutmeg and 1600 clove trees, although Penang nutmeg and mace were considered superior to Moluccan products. Most trees in Penang and Singapore were destroyed by disease between 1859 and 1886, and commercial production virtually ceased early in the 20th century (Burkhill, 1966). Introduced into Sri Lanka in 1804, nutmeg flourished and continues to do so as a minor crop.

Nutmeg was taken to the Caribbean island of St Vincent in 1802, and then to Grenada in 1843 where cultivation expanded. The first crop large enough to influence the world market was exported in the mid-1860s, and Grenada continues to be a leading producer of nutmeg and mace. Mace is more expensive than nutmeg and there exist much-quoted anecdotes concerning British and Dutch colonial officials in Europe, who, unaware that both spices come from the same tree, requested growers in

Indonesia and Grenada to reduce nutmeg and increase mace production!

Nutmeg has long been used in herbal remedies and in medicine, but became unpopular in Europe when the toxic effects were widely published. An Ayurvedic name for nutmeg is *made shaunda*, or narcotic fruit, and a *Materia Medica* published in Bombay in 1883 stated that local Hindus ate *Myristica* as an intoxicant. Local women on Zanzibar and Pemba chewed nutmeg (*kungumanga*) as an alternative to smoking the local marijuana (*bhang*), with almost the same effects (E.A. Weiss, personal observation).

Botany

M. fragrans Houtt. (syn. *Myristica officinalis* L.f.; *Myristica moschata* Thunb.; *Myristica aromatica* Lamk.; *Myristica amboinensis* Gand.). The basic chromosome number of the genus is unclear but is probably $x = 11$. Nutmeg is diploid with $2n = 44$, although $2n = 32$ has been determined in India (Dhamayanthi and Krishnamoorthy, 1999). The genus name is said to be derived from the Greek for fragrant oil. The English nutmeg is from the Latin *muscus*, via French *mugue* and medieval English *notemuge*; mace is from *maccis*. In Portuguese, nutmeg is known as *noz moscado*, mace as *macca* or *clava*, in Dutch nutmeg as *notemuskaat*, mace as *foelie*, in Spanish nutmeg as *neuz moscada*; in German as *muskatnuss*; in Italian as *noce moscata*; in India as *jaiphal* generally, in Tamil as *sadhikai*, in Sri Lanka as *sadhika*; in Indonesian as *jati-pala* and in Malay as *buah pala*. The similarity of many South-east Asian vernacular names referring to blood is derived from the blood-red sap exuded from wounds or slashes to the trunk of nutmeg trees.

No varieties of *M. fragrans* are recognized, but many local cultivars exist. *Myristica* germplasm is maintained at the Research Institute for Spice and Medicinal Crops at Bogor, Indonesia, but is reportedly not used for breeding purposes. The germplasm collection at the Indian Institute of Spice Research, Calicut, Kerala, is being expanded (Krishnamoorthy *et al.*, 1997), while collections of superior trees as breeding stock exist in several countries. The variation between nutmegs from individual trees is frequently greater than the average variation between regions, indicating an almost unlimited reservoir of material for selection or breeding purposes, especially on Banda Island, Indonesia. Grafting *M. fragrans* scions to *Myristica dactyloides* Gaert. and *Myristica malabarica* Lamk. stock was successful in India.

Nutmeg is a spreading, normally dioecious, evergreen tree with dark green leaves, yellow flowers without petals and large yellowish fruit, whose dried seed is the nutmeg of the international spice trade and dried aril the spice mace. All parts of the tree are aromatic.

The roots are generally extensive, superficial, sometimes partially exposed, and on light soils trees are often unstable. In 1955, a hurricane destroyed about 90% of trees on Grenada, a mortality exacerbated by this shallow rooting. Trees are normally 10–15 m high, but can be up to 25 m, with one main trunk up to 30 cm DBH. The timber is red, soft and not durable, slash red and oozes a light red sticky sap, kino, which heals, although slowly, and must be treated on cultivated trees to prevent exhaustion and pathogenic infection. The kino produces a brown dye. The bark is greyish-black, becoming darker and fissured longitudinally with age. Steam-distilling bark yields an essential oil up to 0.15%, lacking any aldehydes and of no commercial importance. Branching is extensive beginning almost at the butt and wild trees are cone-shaped. The twigs are slender, glabrous, greyish-brown and carry a large number of leaves, which form a dense canopy.

The leaves are alternate, glabrous, exstipulate, with the petiole up to 1 cm long; the blade is $5–15 \times 2–7$ cm, elliptic or oblong-lanceolate, with the margin entire, the base acute and the tip acuminate. The leaves are coriaceous, shiny, medium to dark green above, light green or subglaucous below. The blade has 8–11 pairs of slender nerves curving out to the edge with indistinct anastomosis; reticulations are usually invisible above and distinct beneath, forming a lax network. The leaves are aromatic and steam distillation produces an oil similar to weak nutmeg oil, which is of no commercial importance. Dry leaves contain about 10% myristicin and 80% α-pinene.

The male and female inflorescences are similar being glabrous, axillary, carrying one to ten male or one to three female flowers in umbellate cymes; the main axis is 1.0–1.5 cm, and generally unbranched. The pedicels are pale green, 1.0–1.5 cm, with a minute caducous bracteole at the flower base. The flowers are fragrant, creamy-yellow, waxy, fleshy and glabrescent with sparse minute tomemtum; the perianth three-lobed. The calyx is bell-shaped, nectiferous at the base, with reflexed triangular lobes and petals absent. Female flowers are up to 1 cm, with the ovary superior, sessile, slightly hairy, one-celled, up to 7 mm, and surmounted by a very short, white, two-lipped stigma. Male flowers are smaller than female, with the androecium up to 7 mm, glabrous, stalk 2 mm, apex acute, and eight to ten stamens with anthers adnate to a central column and attached to each other by their sides. Anthesis occurs from 3–5 a.m. and insects, often moths, are primary pollinators; in some areas a specific insect may be pre-eminent, for example *Formicomus braminus* in Kerala, India. The floral biology and histology of the flowers has been studied in detail (Armstrong and Drummond, 1986). An essential oil of no commercial importance can be extracted from the flowers.

The tree is typically dioecious, but some bear male and female, and more rarely hermaphrodite, flowers; mature male trees occasionally produce female flowers and may become female. Determining the sex of seedlings produced from seed is difficult before flowering, but cloning and similar techniques have eliminated this problem. Trees normally begin flowering at 6 years but if well fertilized and managed, at 4 years. Trees may flower throughout the year, but normally in two peaks, for example, July and October in Kerala.

The fruit is a yellow, succulent, aromatic, fleshy drupe, usually pendulous, broadly pyriform, smooth, 6–9 cm long, with a circumferential longitudinal ridge and persistent remains of the stigma, containing one seed. The pericarp, about 1.3 cm thick, splits when ripe along the suture to expose the purplish-brown lustrous seed enveloped in a crimson, fleshy, lacinated membrane, the aril

(mace) (Fig. 5.1). In Indonesia, Malaysia, Singapore and India (Joshi *et al.*, 1996), the pulpy pericarp which accounts for 75–85% of total fruit weight, is used for jellies and preserves, and in Malaysia, piles of rotting fruit are used to grow the very popular *kulat pala* mushroom, *Volvariella volvacea* Bull. An essential oil can be distilled from the pericarp, which consists of 16 monoterpenes (60%), nine monoterpene alcohols (29%), eight aromatic ethers (7%), three sesquiterpenes (1%), and eight other aromatic components. Compared to nutmeg and mace oils, pericarp oil has a higher concentration of terpinen-4-ol and α-terpinol, and lower concentrations of sabinene, myristicin and safrole (Choo *et al.*, 1999).

The seed is ovoid or ellipsoidal, 2.0–4.0 × 1.5–2.5 cm, surrounded by a red aril attached to its base, and considerable variation exists in seed shape and size, aril thickness and colour (Fig. 5.2). In the Moluccas, trees bearing different types of fruit are identified by a suffix; *pala tidore* has pear-shaped fruits, *pala bali* large globoid fruit (Nitta, 1993). The seed surface is furrowed and longitudinally wrinkled, with the raphe grove extending from basal scar to apical depression; it is soft when fresh, becoming hard with age or drying, and is easily cut or scraped. A cut surface shows the pale brown endosperm marked by many veins containing the essential oil. The remains of the small embryo are visible when the seed is halved longitudinally. A detailed descrip-

Fig. 5.1. Ripe nutmeg fruit.

Fig. 5.2. Nutmegs. l to r: dried and wet nutmegs; aril; nutmeg enclosed in aril.

tion of the morphology and histology of the seed and aril has been published (Parry, 1962). Cut seed has a characteristic aromatic scent and a warm, slightly bitter taste.

The shelled seed is the spice nutmeg, and contains approximately: water 9%, carbohydrate 30%, fibre 11%, protein 7%, fixed oil (fat) 33%, essential oil 4.5%, and ash 2% containing Ca 0.1%, P 0.2% and Fe 4.5%, but as noted great variation exists. Analyses of Indian nutmegs gave average values of moisture 14.3%, protein 7.5%, ether extract 36.4%, carbohydrate 28.5%, fibre 11.6%, and mineral matter 1.7% containing Ca 0.12 mg, P 0.24 mg and Fe 4.6 mg per 100g seed (Gopalan and Zacchovia, 1971). A semi-solid yellowish-red butter can be obtained by hot-pressing nutmegs, with a yield of 24–30% containing trimyristin up to 75%, and 13% essential oil.

The essential oil varies in composition, and extracts from six Indonesian nutmegs had methyl eugenol at 0.3–18.0% of volatiles, myristicin at 0.2–15.0% and myristic acid at 0–11%, with no correlation between external appearance and volatile content (Sandford and Heinz, 1971). Indonesian and West Indian essential oils differ; the former are higher in myristicin, up to 14%, compared with less than 1% in the latter, while the latter are low in α-pinene and safrol. West Indian oils are considered of finer quality than Indonesian (East Indian). Nutmeg can also be hallucinogenic, and aromatic ethers are the most likely cause. This is discussed in the section Products and specifications.

The red pigment in mace is lycopene, identical to the red colourant in tomato. Mace contains about 22% fixed and 10% essential oil, although again very variable, plus approximately moisture 16%, carbohydrates 48%, phosphorus 0.1%, plus iron at 12.6 mg per 100 g. Indian mace yielded 21.6% ether extract with chloroform-soluble lipids accounting for 88% by weight, consisting of neutral lipids at 60%, mainly glycerides, free fatty esters and hydrocarbons, with free sterols at 1.7% and sterol esters at 0.9%; neutral lipids as glycolipids were 27% and phospholipids were 13%; the main fatty acids were palmitic and oleic (Prakashchandra and Chandrasekharappa, 1984).

A number of commercial grades of nutmeg are recognized in Indonesia and Grenada, with sound nutmegs selected by size, while mixtures of sizes are exported as sound unsorted. Sound nutmegs are used mainly for grinding into the powdered spice and to a lesser extent for oleoresin extraction in importing countries. Indonesian dried mace is exported either as whole or broken blades, and Grenadian cured mace in a number of grades (see also Products and specifications section).

Ecology

M. fragrans is considered native to Banda and Amboina islands in the Moluccas, Indonesia, although it is seldom found as a truly wild plant, and because of its popularity as a spice, nutmeg trees have been planted wherever conditions are suitable. Commercial production, however, is limited to the countries noted. Birds, especially fruit pigeons and hornbills, seek out fruiting nutmeg trees and are the main seed-dispersal agents. Thus, within a specific area, nutmeg trees grow in any suitable environment.

Nutmeg's natural habitat is the wet tropics and trees thrive with a high well-distributed annual rainfall, moderately high temperature and high humidity, with little seasonal variation. An annual rainfall of 2000–3500 mm is necessary for high fruit yield, but trees grow and fruit well on 1500–2500 mm with good management; below 1500 mm supplementary watering is usually necessary. At the higher rainfall levels, soils must be free-draining as nutmeg is very susceptible to waterlogging, thus trees are often planted on hillsides. In regions with a pronounced dry season, as on Grenada, fruiting is more seasonal than in Indonesia.

Nutmeg is seldom grown on a commercial scale above 500 m, at a maximum of 800 m above sea level, but little information is available on the effect of altitude on yield or fruit composition. Individual trees and small plots are grown at much higher elevations mainly to supply local or domestic needs. Sunny sheltered hill valleys often provide ideal sites. An open situation is preferred for mature trees, but shading seedlings promotes rapid growth and prevents sun-scorch, and thus young trees are often interplanted with fast growing species such as bananas.

A temperature of 25–35°C is the optimum, and although mature trees are little affected by higher or lower temperatures for short periods, flowering can be adversely affected by temperatures above 35°C, or a hot dry wind. Frost will cause extensive damage at any stage, and in areas where regular frosts occur, commercial production is hazardous. In exposed situations, lines of trees such as *Albizia* spp. or *Erythrina* spp. can provide intermittent shade and wind protection, and this reduces the adverse effect of nutmeg's generally shallow root system. Extremely high winds or cyclones can be devastating, as noted.

Soils

The rich volcanic soils of Banda and Amboina produce the highest nutmeg yields, and a similar soil type could be the reason why Grenada has become a highly successful producer. Nutmeg also flourishes in Tamil Nadu, India, and the neighbouring areas of Sri Lanka, whose climates and soils are also similar. However, trees grow well on the very common tropical laterite soils with good management. Reasonably deep clay loams, such as the Belmont Clay Loam on Grenada, when drained naturally or artificially, are most suitable, since nutmeg cannot tolerate waterlogging even for short periods, especially when young. Trees thus grow well in hill valleys but not on river flats subject to flooding. A neutral to slightly acid soil, pH 6.5–7.5, is preferable, but nutmeg will tolerate more acid soils provided these are well managed; saline and alkaline soils are unsuitable.

Fertilizers

To produce high fruit yield, nutmeg trees require a high level of natural soil fertility as occurs in Indonesia and Grenada, and on the island of Banda plantations exist that have received no fertilizer since they were established over 100 years previously. When available, well-rotted plant residues, animal wastes, bone meal, rock phosphate, residue from fish and food processing, or merely surrounding topsoil is placed in the pits into which seedlings are transplanted, and mulching is common until seedlings are established.

Efficient producers also place these materials in a trench round larger seedlings and young trees during growth and before the root system grows too large. Subsequently, such material should be spread within the canopy diameter; in India an annual application of 2.5 kg cattle manure or 40–50 kg plant residue per tree is recommended. Saplings and young trees thus treated grow faster, are larger and tend to flower before unfertilized trees, thus allowing earlier removal of unwanted males.

Since most existing plantations have a relatively low level of management and trees are usually planted on fairly fertile soil, little chemical fertilizer is applied, but its use can be beneficial. It is essential that the soil nutrient status is accurately determined before fertilizers are applied. The relatively low price of nutmegs also restricts fertilizer use. Little information is available on the effect of individual plant nutrients on tree growth, fruit yield, seed size or composition.

Reports from experimental stations and government farms indicate a substantial response to manure and fertilizers on less fertile soils, but there are few data from formal trials. Nutrient removal by nutmeg with an annual yield of 750 kg ha^{-1} in India, is shown in Table 3.4. The official recommendation in Kerala State, India, is NPK at 20 g, 18 g and 50 g, respectively in the planting hole, increasing gradually each year until at 15 years the rate reaches 500 g, 200 g and 1000 g per tree. Since a similar high K ratio was recommended for local pepper vines, this may indicate a substantial local K deficiency. However, the Grenada Agricultural Department recommended a 13:8:24 NPK mixture, similarly high in K, at 25 g per seedling increasing to 2.5 kg per tree at 10 years. Trials to determine whether nutmeg has a high K requirement could be profitable, since many tropical crops take up luxury amounts of this nutrient, which can be reflected in greater vegetative growth but no increase in yield. Trees receiving fertilizer are usually obvious by their greater size.

Cultivation

Nutmeg trees were initially planted in cleared areas in forests but the rapid decline in area of tropical forests has virtually eliminated this type of agriculture, and most nutmeg trees are now planted to replace those that have died or to replant neglected groves.

Seed remains the most common method of producing seedlings, but should be replaced by techniques that allow quicker sex determination, or produce seedlings of known sex. These include vegetative production from female trees selected for high fruit yield, resistance to a local disease or similar desired characteristic. Approach-grafting and marcotting are well-known techniques

easily applied by growers, although some types of grafting are more complicated (Haldankar et al., 1999). Clonal propagation and tissue-culture methods enable large numbers of plantlets from selected trees to be produced, and although these techniques are relatively expensive, they would quickly repay a cooperative or growers organization from the higher yields obtained.

Many smallholders continue to raise seedlings from seed, and seed nutmegs should be sound, less than 5 days old, with the testa intact and the aril removed. Seeds kept in wet moss or polythene bags remained viable for 15 days in India, but storing seed for more than a day or two is not recommended. Seed is sown in specially prepared beds preferably composed of approximately 3:1 sifted soil and sand, in natural shade or shaded. Fertilizers are unnecessary. Alternatively, one or two nutmegs may be sown in pots containing the same mixture. Some growers germinate nutmegs in moss, sawdust or chopped coir prior to planting in pots, but this is not necessary and provides no advantage. Nutmegs should be sown about 2.5–5.0 cm deep in rows 30 cm apart, and seedbeds kept moist but not wet, as this encourages foot and stem rots. A fungicide and insecticide watered on seedlings will give added protection. Snails can occasionally be a problem and should be picked off by hand.

Germination is slow, between 45 and 80 days and, despite widespread belief to the contrary, there is no difference in speed of emergence between females and males. Alternatively, seedbeds can be planted with cloned seedlings, which will be of known sex, and treated similarly. Such seedlings are now often available at a subsidized price through the local Agricultural Department or from nurseries maintained by grower cooperatives, and are the quickest method of raising the average yield of nutmegs per tree and cutting establishment costs.

Following emergence, unthrifty and diseased seedlings should be removed, and the shade progressively reduced to harden off seedlings before transplanting. It is essential when removing seedlings from seedbeds to ensure that the long taproot is not damaged, and sowing seed in biodegradable pots, including banana leaves etc., which can be placed intact into planting holes without root disturbance, is recommended; if polythene tubes are used, these must be carefully removed before planting.

Seedlings are ready for planting out after 6 months when around 15 cm high, although seedlings aged 18–24 months and 30 cm high are preferred in southern India. A common spacing is 6 m × 6 m with alternate trees removed after about 10 years to allow the remainder to expand as they mature. Alternatively, a spacing of 5 m × 5 m can be used, then removing alternate trees to give a 10 m × 10 m triangular spacing over the same period. Seedlings planted on hillsides should follow the contour to reduce the danger of soil erosion.

Seedlings require temporary shade, and banana or palm leaves are suitable. When planning a new grove in an exposed situation, shade and/or a wind break is desirable, and this should preferably be established the previous season; bananas are ideal as they give shade, mulch and an income to offset establishment cost. In Indonesia, *Canarium commune* L., the *kanari* or Java almond, is often chosen as it bears an edible fruit, the *pili* nut, which also produces a cooking oil.

Seedlings grown from seed are approximately half male, and as 10% are sufficient to ensure pollination, it was previously necessary to wait for flowering at 5–7 years to remove them, or to change their sex by grafting with scions from female trees. With the introduction of clonal material, this is no longer necessary and male and female seedlings can be planted out at the required spacing. Young trees that die should immediately be replaced with one of the same sex.

Trees raised from seed begin bearing at 5–9 years but those vegetatively propagated begin at 4–7 years. Fruit ripens 6–9 months after flowering, which normally has two peaks but may continue year-round. Trees reach maximum production in 15–20 years and continue bearing for decades. A study on the economics of nutmeg cultivation in the main producing state of Kerala, India, indicated that trees remained profitable for some 60 years (Ipe and Varghese, 1990). In Indonesia, fruiting trees around 80 years old are common, but the life of most cultivated trees is usually determined by factors other than genetic such as disease, mismanagement and natural disasters.

Trees are well branched, and to facilitate harvesting the lower branches should be carefully removed during early growth to allow easy access to fallen fruit and for weed control; very long branches that touch adjoining trees should also trimmed, since flowering takes place on the ends of branches. Water shoots, dead or diseased twigs and branches should be removed and burnt. Wounds on nutmeg are generally slow to heal and may allow the entry of pathogenic fungi, so should be treated with a fungicide or one of the many proprietary wound products.

Weed control

Weeds and grasses must be controlled in newly established plantations and a cleared area maintained around young trees to reduce insect damage. The dense canopy of mature trees normally suppresses most weed growth, but vigorous and persistent grasses such as *Imperata cylindrica* or the various tall *Panicum* spp. should be manually removed. Under-canopy vegetation should be cut or slashed to enable fallen fruit to be easily seen. Small livestock may be allowed to graze under mature trees, but goats and cattle are excluded.

Herbicides are used where profitable, and directed sprays of glyphosate and simazine are recommended in young plantations. Applying many commonly used orchard herbicides under mature nutmeg trees is hazardous, since the chemicals can be absorbed by, or damage, the superficial roots. Although not strictly weed control, parasitic plants including various *Loranthus*, *Ficus* and *Cassytha* species should be carefully removed (see remarks on wounds in previous section).

Irrigation

Plantations established from selected or clonal trees will require watering in dry seasons to maintain growth and yield, but in general irrigation is unprofitable in most existing plantations. Where water is available to supplement rainfall, it should be applied when the first flowers appear and as necessary while fruit is maturing. Watering should cease before fruit ripens to ensure the ground is dry under the canopy so that fallen fruit will not rot. Intercropping commercial plantations of young trees is possible with any local low-growing crop such as legumes or capsicum peppers, but in mature trees is not recommended as the largely superficial roots can be damaged.

Pests

The principal economic pests of nutmeg attack the fruit and seed, although a particular insect may cause major tree damage in a specific region or season. Chemical control is infrequent for two main reasons: first, the cost of the chemical and its application, as most nutmegs are produced by small growers who do not have the cash resources to carry out a regular spraying programme, or whose level of tree management is so low that pest control is irrelevant; and secondly, the degree of economic damage, with one major exception as noted below, is usually too low to warrant control. Pesticide residues have also been detected in both whole and ground nutmegs (Manirakazia *et al.*, 2000).

A most serious pest is *Phloeosinus ribatus*, a small, dark-brown weevil, 3 mm long, which bores through bark and cambium, and attacks both above- and below-ground plant parts. Numbers breed up very rapidly and can quickly cause dieback and death. This insect is blamed for the virtual collapse of nutmeg production in Singapore and Penang in the 1860s. Other damaging scolytid borers (ambrosia beetles) are *Xyleborus fornicatus* and *Xyleborus myristicae*, the former a well-known pest of many plantation crops including tea. *Stephanoderes moschatae* and *Dacryphalus sumatranus* can be very damaging in specific localities, and it is the larvae of these borers that are the 'worms' that attack the nutmeg itself, resulting in the Indonesian wormy nutmegs used to produce nutmeg butter and oil.

The larvae of a number of polyphagous tropical moths cause defoliation but frequently become heavily parasitized once numbers increase, and damage is often restricted to a few adjacent trees (Fig. 5.3). A scale, *Coccus expansium*, attacks leaves and young shoots in India and South-east Asia, but with no details of the degree of damage. *Coccus* species are widely distributed in the tropics, are pests of many tree crops and may occur more widely on nutmeg than has been reported. Superficial tree and seedlings roots

Fig. 5.3. Typical noctuid moth.

can be attacked by the larvae of several moths including *Agrotis* species, termites and nematodes, including the polyphagous *Meloidogyne javanica*. Snails, as noted, may cause damage in seedbeds or to very young seedlings, and are easily controlled by baiting or hand picking.

Diseases

The total damage caused by disease is much greater than that caused by insects, but the same remarks regarding chemical control apply. Probably the most obvious diseases are fruit rots, which either cause shrivelling, rotting or premature fruit fall. In India, *Diplodia natalensis* attacks and destroys half-ripe fruits, as does a related *Diplodia* spp. in Malaysia and Indonesia. In South-east Asia, premature fruit fall was associated with *Coryneum myristicae*, but the causal organism was not determined. In India, *Gloeosporium* spp. and *Fusarium* spp. were isolated from rotten fruits, but again it was not determined if they were the cause. In Kerala State, rots are a major source of fruit loss. A thread-blight due to *Corticium stevensii* occurs in the Caribbean, which can be very damaging in a particular season.

A serious disease where it occurs is wilt caused by *Rosellinia* spp., which also attacks the roots of many tropical trees; *Rosellinia pepo* is common in the Caribbean, especially on Grenada and Trinidad. Leaves of badly infected trees wilt, turn brown, fall, and the tree eventually dies; fruits appear wrinkled and desiccated. Root rots due to *Fomes* spp. commonly occur from Asia eastwards; *Fomes noxius* and *Fomes lamaensis* are most often recorded as causing severe damage and sometimes death of trees. Wilts caused by *Fusarium* spp. are frequently recorded, and are especially damaging on seedlings.

A leaf spot due to *Colletotrichum gloeosporioides* is common in India but the degree of damage has not been accurately assessed. A leaf spot, recorded mainly in the Caribbean, raises dark brown greasy lesions on leaves, and several organisms may be involved. Leaves with similar but less greasy lesions have been described from Asia and Indonesia, but the cause apparently remains unidentified. Mace scab reported from Grenada is not considered to be due to a pathogen but is physiological and characterized by the accumulation of calcium oxalate.

Harvesting

In non-seasonal climates, trees fruit year-round, but fruit is harvested only when it has split and the seed enclosed in its aril is clearly visible. Fruit may be allowed to fall but must be collected daily to avoid soil contamination and insect and disease damage. Picking is becoming more general to avoid these problems, but pickers paid by weight tend to be unselective and can also damage trees by climbing among branches. In Malaysia and Indonesia, high fruit is picked using long poles fitted with cutters and a basket.

Fruit yield per tree is very variable and dependent mainly on the standard of management, especially adequate pest and disease control. A healthy mature tree in full bearing in India yields 750–2000 fruits annually, but up to 10,000 per year can be harvested from one 25 years old. The Sri Lanka annual average is higher at 2000–4000 fruits per tree, although an individual yield of up to 12,000 fruits per tree is possible. In Indonesia, the yield of dry nutmegs averages 1250 ha^{-1}. In Grenada the annual average is 1500 fruits per tree, but is almost double in groves established from vegetatively propagated seedlings. The yield of mace varies similarly, but from sound healthy nutmegs averages 15–20% by weight.

Processing

Following collection the seed (nut) with surrounding aril is separated from the fruit, and the aril (mace) is detached. After drying, nuts are shelled and become the spice nutmeg. Practically all of Grenada's nuts are now mechanically shelled, as are an increasing proportion of Indonesian.

Natural drying of nutmeg. After removal of the mace, nuts are dried on large wooden trays, and turned daily with wooden paddles to promote even drying and prevent fermentation. Sun-drying for about 1 week is usual in Indonesia, or in special buildings heated by slow fires to prevent shells cracking and melting the fat, which forms a large part of the kernel. Grenadian nuts are air-dried for about 8 weeks in special buildings, losing about 25% by weight.

Artificial drying of nutmeg. Natural drying is lengthy and tedious, and the introduction of forced-air flatbed dryers has been shown to reduce drying time to 23 days using cold air, and to 7 days using air heated to 37°C, without affecting quality. Properly dried nuts rattle in the shell. A solar-heated tunnel drier in India reduced microbial contamination and drying time, and produced higher quality nutmegs (Joy *et al.*, 2000).

Mace. The detached fresh mace is carefully flattened by hand or between boards to avoid the breakage that reduces its quality and value. Mace is usually sun-dried on large trays or mats, and 2–4 hours is normally sufficient in Indonesia and Grenada to ensure mace retains its scarlet colour. Grenadian mace is then stored in darkness for up to 4 months, gradually becoming brittle and pale orange-yellow; it sells at a premium over Indonesian mace, which is usually exported in the red form, gradually changing to the reddish-orange of the retail product.

Distillation

When sound nutmegs are distilled, the fixed oil tends to retain some volatile oil thus reducing the yield, but lower grades including shrivelled nutmegs or those infested with weevil larvae, which consume most of the fixed oil, have higher yields (see section on Pests). Nutmegs should be comminuted to a coarse powder prior to distilling, then transferred immediately to the still, since once ground there is a rapid and substantial loss of volatiles. Distilling should preferably be by low-pressure live steam, and cohobation may be necessary; high-pressure or superheated steam is not recommended as it carries over small quantities of myristic acid from the fixed oil. About 80% of oil distils within 2 h, the remainder within 10 h, and oil yield should be 6–12% by weight, and average at least 8%.

High-quality oils are produced from defective nutmegs on Grenada; a single preliminary crushing in a roller mill is followed by water or steam distillation preferably without cohobation. A 45 kg charge of crushed nutmegs distilled for 6–12 h giving a condensate rate of $9–10\,l\,h^{-1}$ is the optimum.

Nutmeg contains 25–40% fixed oil obtained by expressing crushed nuts between heated plates in a steamer, or by solvent extraction. The expressed oil, nutmeg butter, is a highly aromatic, orange-coloured fat, with the consistency of butter at ambient temperature. The main constituents are trimyristin and a high proportion of volatile oil difficult to separate by steam distilling.

Transport and storage

Nutmegs and mace are normally sold by smallholders to traders, who further clean and grade them before local marketing or selling to exporters. Cooperatives or grower associations may carry out these operations and sell to exporters, or export direct. Stored nutmegs and mace are readily attacked by the usual pests of edible products, including the nutmeg weevil *Araecerus fasciculatus* (Fig. 5.4), also known as the coffee bean weevil, considered to be by far the most damaging in nutmeg-growing regions of Asia and the Pacific region. Storage pests are controlled by adequate and supervised storage. Indonesian nutmegs are sometimes white following treatment with lime wash as an insect deterrent.

In general, nutmegs and mace should not be ground until required, since in uncontrolled storage their organoleptic properties deteriorate fairly rapidly, mainly through loss of the volatile constituents. Incorrectly stored nutmeg oil may also undergo significant composition changes if exposed to high ambient temperature. Unprotected powders and oil can also absorb unpleasant odours. Powders, oils and oleoresins should be stored in full, sealed and preferably opaque containers until required. Provided the correct packages or containers are used, transport of nutmegs, mace and their derivatives is uncomplicated.

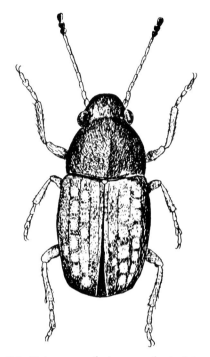

Fig. 5.4. Nutmeg weevil, *Araecerus fasciculatus*.

Products and specifications

Nutmeg trees produce three main products: nutmegs and mace used directly as spice; nutmeg and mace oils and oleoresins used as spice and flavourings; and leaf oil and other derivatives. While nutmeg and mace continue to be the most important domestic spice, oils and oleoresins have virtually replaced them for industrial uses. Leaf oil is normally produced only to order and is of no commercial importance (Weiss, 1997).

Nutmegs and mace traded on the international market must generally conform to recognized grades related to physical characteristics, i.e. size or colour. Individual countries also require both to conform to published national standards, which may include composition or end use, as noted in the Introduction. This also applies to their derivatives. The US regulatory status of generally regarded as safe is accorded to nutmeg, GRAS 2792, nutmeg oil, GRAS 2793, mace, GRAS 2652, mace oil, GRAS 2653, and mace oleoresin, GRAS 2654.

Nutmeg. Commercially accepted grades of export nutmegs are recognized, and those of the two major producers are summarized. Indonesian and Grenada nutmegs are internationally traded as East Indian and West Indian, respectively, although Sri Lankan nutmegs resemble West Indian in composition. Sound export nutmegs from Indonesia and Grenada are generally graded by size into (approximately) 170/180 or 240/250 nuts kg^{-1} (previously 80 or 110 nuts lb^{-1}), with large, uniformly coloured nutmegs the highest quality. Five other grades are based on the increasing numbers per kg, and mixtures of sizes as sound unsorted. In importing countries, nutmegs are sold as whole nutmegs or ground for retail sale as spice or in spice mixtures (Fig. 5.5).

Lower grade nutmegs from Grenada, known as defectives, are rejects from the grading process, plus badly bruised and broken pieces of sound nutmegs. The equivalent Indonesian is broken, wormy and punk (BWP) grade, and together with the sound shrivelled grade is used not only for grinding, but also for oil and oleoresin production in importing countries. These grades are also ground locally in India to provide nutmeg powder to be mixed with betel nut and other spices to make the chewing quid *pan*, and in Indonesia for snuff.

In addition to their well-known use as a spice, nutmegs also have a long history as a narcotic, but the first recorded hallucinogenic

Fig. 5.5. Dried spice nutmegs.

effect was by Lobelius in 1576, who in his *Plantarum seu Stiripium Historia* described a pregnant English lady who 'became deliriously inebriated after eating 10–12 nutmegs', apparently while trying to induce an abortion. A physiologist, Mr J.E. Purkinje, in 1829 ate three nutmegs, and described the effects as similar to *Cannabis* intoxication, including disorientation and hallucinations followed by a deep sleep. However, the response to nutmeg intoxication is extremely varied; some individuals experience a profound distortion of time and space, while others have visual hallucinations (Woolf, 1999). Whole or sliced nutmegs preserved in syrup were formerly popular in Europe, but fell into disfavour when their narcotic effects were publicized.

Mace. Grenada mace is exported as whole pale mace and No. 1 broken mace, both packed in cases for export, and No. 2 broken mace consisting of material that does not acquire the desired pale colour and is generally shipped in bags; unassorted and pickings are also shipped in bags. Indonesian mace is exported as whole or broken blades shipped in cases, double bags or corrugated cartons.

Nutmeg oil. Nutmeg oil is mainly obtained by low-pressure steam distilling second and lower grades of nutmeg in Grenada, India, Indonesia and Sri Lanka. A small quantity of oil is produced by companies in North America and the EU to satisfy local demand for a very high-quality product, and this oil is increasingly obtained by supercritical carbon dioxide extraction.

Distilled nutmeg oil is a pale yellow to almost water-white mobile liquid; its odour is fresh, warm-spicy and aromatic with a rich, sweet-spicy bodynote, and freshly distilled oil tends to have a rubbery topnote, which disappears with age. Nutmeg oil is placed in the fresh, spicy-woody group by Arctander (1960). Major components are monoterpene hydrocarbons at 61–88%, oxygenated monoterpenes at 5–15% and aromatic ethers at 2–18%. The main constituents of the monoterpene hydrocarbon fraction are α- and β-pinene and sabinene; myristicin is the main constituent of the aromatic ether fraction (Pino and Borges, 1999e; Lawrence, 2000c). Commercial steam-distilled oils also tend to differ from the natural oil present in spice or oleoresin extracts by containing a higher proportion of monoterpenes, particularly α- and β-pinene and sabinene, due to incomplete distillation of the oxygenated components. The main characteristics of nutmeg oils are shown in Table 5.1.

The organoleptic properties of nutmeg oil are directly related to the origin of the raw material; East Indian oils are higher in myristicin, up to 13.5%, compared with less than 1% in West Indian oils, which are low in α-pinene, safrole and myristicin, but higher in sabinene. The greater proportion of myristicin and safrole in East Indian oils and the different monoterpene component ratio probably gives the stronger nutmeg flavour. Oil from other countries generally resembles one or other of the two main types, due probably to the source of the original introduction. It is interesting, however, that Sri Lankan oil resembles West rather than East Indian oil, with a sabinene content of

Table 5.1. Physical characteristics of nutmeg oils.

Characteristic	West Indian[a]	Indonesian[a]	Sri Lankan[b]
Specific gravity	0.860–0.880	0.885–0.915	0.835–0.903
Refractive index	1.472–1.476	1.475–1.488	1.463–1.482
Optical rotation	+25° to +40°	+8° to +25°	+21° to +42°
Solubility (v/v 5% alcohol)	1:4 (90%)	1:3 (90%)	ng

[a] 20°C.
[b] 30°C.
ng, not given.

30–50% and mean myristicin and elemicin contents of 2.3 and 1.2%, respectively (Sarath-Kumara *et al.*, 1985). Indian oils, however, are intermediate between West and East Indian oils, with a sabinene content of 32%, α-pinene at 13.6% and β-pinene at 12.9% (Mallavarapu and Ramesh, 1998). Nutmeg oils can be adulterated with natural products high in monoterpenes, myristicin or synthetic materials, but there is little value in so doing.

Little information is available on the characteristics or composition of supercritical carbon dioxide-extracted nutmeg oils, and they are presumed to be similar to distilled oils, which they are rapidly replacing. Chromatographic analysis of such oils has confirmed that East Indian, West Indian and Papuan oils clearly differ in their composition. The major aromatic ether in East Indian oils is myristicin, in West Indian elemicin and in Papuan safrole (Ehlers *et al.*, 1998).

Nutmeg oil is mainly use as flavouring in a wide range of processed edible products, ketchups and soft drinks, but the monoterpenes have a pronounced tendency to polymerize and produce undesirable off-notes when heated, thus terpeneless oils are preferred for flavouring canned foods etc. Terpeneless oil is normally produced by counter-current solvent extraction to avoid heating. The oil is used in pharmaceutical products to alleviate bronchial troubles, in domestic aerosol sprays and increasingly in male toiletries, including aftershave lotions. The oil exhibits strong antimicrobial, antibacterial and antioxidant activity (Nolan *et al.*, 1996; Dorman and Deans, 2000; Dorman *et al.*, 2000a,b).

Oils from Indonesian or Asian nutmegs, including *M. fragrans* or *M. malabarica*, contain a high myristicin content of up to 14%, as noted, and 25% of this fraction is elemicin. On becoming ammoniated in the body, elemicin degrades to two potent hallucinogens, trimethoxy amphetamine (TMA) and 3-methoxy-4,5-methylenodioxy amphetamine (MMDA) (Emboden, 1979). However, other compounds may be involved, and toxicological data on the oil and its myristicin content have been evaluated (Hallstrom and Thuvander, 1997).

Mace oil. Distilled mace oil is a colourless to pale-yellow liquid, which partly resinifies and develops a turpentine-like odour upon exposure to air. The oil is placed in the fresh, spicy-woody group by Arctander (1960). The oil is very similar in its characteristics and organoleptic properties to nutmeg oil, but is produced in very small quantities (Lawrence, 2000b). Like nutmeg oil it is also now obtained by supercritical carbon dioxide extraction. Mace oil is mainly used as a flavouring in liquid products such as sauces, pickles, etc. and is generally a direct substitute for nutmeg oil. Cured Grenadian mace oil contains monoterpene hydrocarbons at 75–94%, oxygenated monoterpenes and sesquiterpenes at 4.7–17.6% and aromatic ethers at 0–5.9%. Nutmeg oil contains 85–93%, 6.6–12% and 0–3.5%, respectively. The composition of the monoterpene hydrocarbon and oxygenated monoterpene fractions is broadly comparable in the two oils from the same tree; the main difference is that aromatic ethers are usually slightly higher in the cured mace oil.

The commercially more important nutmeg oil has been more intensively studied than mace oil, but the two oils, when produced in the same geographical area, are usually very similar in quality and are generally interchangeable as a flavouring, although they differ in composition. Both oils have insecticidal, microbiological and antifungal activity, and this is increasingly discussed in the literature. Synthetic products purporting to be equal to true nutmeg oil in flavour and aroma have been marketed, but none has proved acceptable. Their use in food products is precluded in many countries by legislation defining nutmeg oil as obtained from the spice.

Oleoresins. Nutmeg and mace oleoresins are prepared by extracting comminuted spices with organic solvents, and contain steam-volatile oil, fixed oil and other extractives soluble in the chosen solvent. A high fat content is obtained with hydrocarbon solvents, a lower fat and resin content with polar solvents such as alcohol and acetone. The lower fixed-oil content of mace compared with nutmeg produces oleoresins containing less odourless material. Hexane extraction yields 31–37% oleoresin containing a substantial amount of odourless and flavourless material, mainly trimyristin, which is removed by washing with cold ethanol in which the fat is almost insoluble. Direct extraction of nutmegs with cold ethanol provides 18–26% crude oleoresin; on chilling, filtering and evaporating ethanol under slight vacuum, the yield is reduced to 10–12%. Mace extracted with petroleum ether yields 27–32% oleoresin containing 8–22% volatile oil. Hot-ethanol extracts yield 22–27% crude oleoresin, which after processing is reduced to 10–13%.

Nutmeg and mace oleoresins are considered to possess an odour and flavour nearer to that of the spice than the corresponding steam-distilled oils. Although the organoleptic properties of nutmeg and mace are similar, end-users consider that mace oleoresin has a finer, more rounded and fresh-fruity character. Oleoresins extracted by non-polar solvents with a relatively high fat content are preferred in flavouring foods, since they have greater tenacity and heat stability; perfumers prefer oleoresins extracted with solvents such as ethanol. The main characteristics of mace oils and mace oleoresin are shown in Table 5.2.

Nutmeg oleoresin is a pale to golden yellow viscous liquid, either clear and oily or opaque and waxy, becoming clear on warming to 50°C. Commercial nutmeg oleoresins are graded on volatile oil content (ml per 100 g) as 25–30, 55–60, 80 and 80–90. Mace contains little fatty oil or other odourless, flavourless substances soluble in hot ethanol, and ethanol-extracted mace oleoresin is one of the most concentrated forms of the nutmeg–mace flavour. Mace oleoresin is an amber to reddish-amber clear liquid, and graded on volatile oil content (ml per 100 g) as 8–24, 40–45, 50 and 50–56.

Nutmeg butter. Nutmeg butter, obtained by expressing sound nutmegs, is an orange-red to reddish-brown soft solid, with a warm, spicy, strongly aromatic odour and a burning, spicy taste of nutmeg. It has a melting point of 45–51°C, is partly soluble in cold alcohol and almost completely soluble in hot alcohol, and is freely soluble in ether and chloroform. It has specific gravity 0.9950–0.9990, saponification number 172–179, iodine number 40–52 and acid number 17–23. A saponification value of 196 and acid value of 9.0 have been recorded (Gopalam and Zacharia, 1989). Butter can be used directly in special formulations or alcohol-washed to remove the suspended solids, mainly glycerol myristate. Mace butter is produced by end-users, but contains 40% saturated and 60% unsaturated fats compared to 10% and 90% in nutmeg butter. Butters can replace their respective oils in many applications but their use is restricted, as oils and oleoresins are favoured by manufacturers.

Other nutmeg producers

Myristica argentea. *M. argentea* Warb. (syn. *Myristica finschii* Warb.) is known in English as Papua nutmeg, or long nutmeg, in Papua as *henggi*, in Indonesian as *pala irian* and in Malaysian as *pala bukit*. It probably originated in Irian Jaya, Indonesia, but now occurs wild only on the Bomberi peninsula; it is culti-

Table 5.2. Physical characteristics of mace oils and oleoresin.

Characteristic	Caribbean oil	Indonesian oil	Oleoresin
Specific gravity (25°C)	0.854–0.880	0.880–0.930	0.955–1.005
Refractive index (20°C)	1.469–1.480	1.474–1.488	1.469–1.500 (of oil)
Optical rotation	+20° to +45°	+2° to +30°	−2° to +45° (of oil)
Solubility (v/v % alcohol)	1:4 (90%)	1:3 (90%)	ng
Volatile oil (ml per 100g)	ng	ng	20–50

ng: not given. Essential Oil Association, USA, specifications.

vated in Irian Jaya, Papua New Guinea and neighbouring Indonesian islands. It can be distinguished from nutmeg by the rough twigs with numerous lenticels, larger leaves, which are silvery beneath, much larger fruit and a thinner, less divided aril.

Initially, in Papua New Guinea, fruit was gathered from wild trees but at the end of the 18th century high prices for the seeds encouraged cultivation, and trees were planted in abandoned cultivation and forest clearings. The area sown covered about 1500 ha, and although trees virtually grew untended except for clearing under-storey weeds prior to harvest or intercropping when young, private ownership was acknowledged. Production from Irian Jaya and Papua New Guinea is stated as some 500 t nuts and 60 t mace in the late 1950s, but storm damage in 1960 to plantations in both regions severely reduced production and many trees were not replanted. No recent production figures are available. The generally low price for nutmeg has reduced the international market for an inferior alternative, although it remains locally popular.

Papua nutmeg is a medium-sized evergreen tree, 15–25 m high, with a canopy width of 10–12 m, slash red. The bark is dark or blackish-grey with very small scales; older twigs are covered by conspicuous wart-like lenticels. Trees usually occur naturally below 700 m. The roots are superficial, as for nutmeg, and susceptible to similar storm damage. The leaves are chartaceous, the blade mostly elliptical-lanceolate, 10–20 × 4–6 cm, the base cuneate, the tip usually apiculate, and the lower surface appears silvery due to short, cobwebby hairs (indumentum); the midrib and 9–13 pairs of veins are sunken above, prominent below, and the petiole is 1.5–2 cm long.

The inflorescence is an axillary panicle with flowers in cymes. The male inflorescence is 2–5 cm in length, slender, simple or singly forked, bearing 3–5 flowers; the pedicel is slender, 1–13 cm long, and the perianth ellipsoidal, 7–11 mm × 5 mm, medium brown and subglabrous. The female inflorescence is 1–15 cm in length, and the main axis is usually simple, but sometimes forked. The pedicel is 8 mm long, the perianth ovoid-ellipsoidal, 8–10 × 5–6 mm, and the apex

beak-like; the ovary is superior, flask-shaped, 3–4 × 7 mm, light brown and tomentose. Pollination is by insects, mainly bees.

The fruit is ellipsoidal, 4.5–8.5 × 4.5–5.5 cm, slightly narrowed at both ends, glabrescent, yellow with some brown pustules, splitting when ripe into two, which are single-seeded. The seed is oblong-cylindrical up to 4 cm, broadening at base, shiny black-brown and harder than nutmeg, with an unpleasant rank flavour when fresh; the aril is thin, and red to orange. Mature cultivated trees yield between 2000 and 4000 fruits per year, and plantations with about 100 female trees ha^{-1} yield annually 1200 kg seeds and 335 kg mace. Mace is peeled from seeds, flattened and dried, and traded commercially as Macassar mace. Seeds are usually dried in a special smokehouse, and exported unshelled and ungraded. They are a widely used substitute and adulterant of nutmeg.

Seed contains about 30–40% fat lacking in aroma, and fat contains only 0.13–0.15% myristicin compared with at least 2.0% in nutmeg, but the safrol content is higher at around 0.5% compared with 0.3% in nutmeg; thus Papua nutmeg powder has a high safrole content. Seed yields 1.2–2.5% oil on distillation, stated to smell of sassafras, *Sassafras albidum* (Nutt.) Nees.; mace yields about 5.5–7.0% oil, said to smell of turpentine. Neither are produced commercially. Mace also contains a lignan, meso-dihydroguaiaretic acid, which is stated to be effective in the treatment of dental caries.

Myristica dactyloides. *M. dactyloides* Gaert. (syn. *Myristica laurifolia* Hook.f & Thoms.; *Myristica diospyrifolia* A. DC.) is native to Sri Lanka and probably also to neighbouring areas of India. It is a large tree with spreading branches and long, lanceolate, oval leaves. The fruit is an oblong-ovoid up to 4 cm in length. The seed is a poor nutmeg substitute and the orange-yellow aril is used as a food colouring. Neither are of commercial value, but may be used to adulterate nutmeg and mace.

Myristica malabarica. *M. malabarica* Lamk. (Bombay or Indian wild nutmeg) is native to the Malabar coast region of India (Ved *et al.*,

1998). The seed reportedly contains 40% fat, mainly myristicin, olein and free myristic acid. The mace contains 60% fat, similar in composition, and yields up to 3% essential oil. The seed and mace, Bombay mace, are of poor quality, have little aroma and are mainly used to adulterate true nutmeg and mace. Both are used in local herbal medicines. Bombay nutmeg is also used as a shade tree in cardamom plantations.

Myristica muelleri. *M. muelleri* Warb. (Queensland or native nutmeg – both names are also used for *Myristica insipida*) is a dioecious, evergreen rainforest tree up to 25 m tall, with an erect stem, and well branched, with a spread up to 5 m. The leaves are mid-green with prominent brown veins, narrowly elliptical, and up to 20 cm in length. The flowers are insignificant, and the fruit is an ovoid berry, used as a nutmeg substitute by early settlers to Australia.

Myristica succedanea. *M. succedanea* Reinw. ex Blume (syn. *Myristica radja* Miq.; *Myristica schefferi* Warb.; *Myristica speciosa* Warb.) is known in English as Halmahera nutmeg and in Indonesian as *pala patani*. It probably originated in the Moluccan Islands, where it is cultivated on a small scale and on Halmahere island. It can be distinguished from nutmeg by its stouter twigs, larger leaves and flowers, and smaller seed, and from *M. argentea* by the more coriaceous leaves.

It is a medium-sized tree of mountain forests, up to 8–15 m in height, with stilt roots and a pyramidal crown; the twigs are stoutish and slender towards the top. The leaves are subcoriaceous, with the blade broadly elliptical, elliptical-lanceolate, oblanceolate, 9–22 cm × 4–11 cm, shining dark green above, silvery and minutely hairy beneath. There are 18 pairs of veins, which are brownish or reddish-brown, and faint above, prominent beneath. The male inflorescence is brownish, tomentulose, usually bifurate, with a main axis 1–2 cm long; the pedicel is 0.7–1 cm long, and the flower fragrant, oblong, 7–10 × 4 mm, and medium brown. The female inflorescence is shorter, 0.5–1 cm, and simple or occasionally bifurcate; the pedicel is 1 cm long, the flower fragrant, ovoid and 7–10 × 5 mm, and the ovary is brownish and tomentulose.

The fruit is ovoid-ellipsoidal, 7 × 4 cm, and tomentulose becoming glabrous, with the hull 1 cm thick. The seed is broadly oblongoid, 3 × 2.5 cm, the aril red, and the endosperm aromatic. The seed reportedly contains about 40% fat and yields 5% oil on distillation, which is not produced commercially. The small fruit is a substitute and adulterant of nutmeg.

Other Myristica *species*. Myristica cinnamomea King is native to the Malaysian peninsula, Borneo and the Philippines; *Myristica elliptica* Wall. is known as the swamp nutmeg of Malaysia, and its seeds are locally collected and sold as nutmeg substitutes; *Myristica fatua* Houtt., native to the Moluccas and has long been cultivated in western Malaysia; *M. laurifolia* Warb. grows wild in India and Sri Lanka, and its seeds are collected and used as a nutmeg substitute; *Myristica womersleyi* Sincl. of Papua New Guinea has a very aromatic seed.

Unrelated species

The *Monodora* include some 20 African and Madagascan species, the most important being *M. myristica* (Gaertn.) Dunal. (syn. *Annona myristica* Geart.; *Monodora grandiflora* Benth.; *Xylopia undulata* P. de Beauv.), known in English as calabash or African nutmeg and in French as *muscadier de calabash*.

M. myristica is native to evergreen and deciduous forests of western Africa from Liberia to Cameroon and south to Angola, and is occasionally cultivated (Onyenekwe and Ogbadu, 1993). It occurs naturally in Ugandan forests, and west Kenya (Kakamega) where it is known as *lubushi*, but in neither are the seeds of any importance. It was transferred to the Caribbean in the 18th century by slave traders, where it became established on several islands especially Jamaica, and became known as the Jamaica nutmeg. First introduced to the Bogor Botanical Garden, Indonesia, in 1897, several cloned descendants remain, but although they flower regularly they fail to set fruit.

The tree is often grown as an ornamental and in arboreta for its large, scented and attractive yellow flowers, which are spotted with red and green.

The tree is up to 35 m in height, and up to 2 m DBH, with the trunk clear and the branches horizontal. The bark is thin and smoothish in West African types, but varies regionally to corrugated with vertical ridges in East Africa, slash greyish-white to brownish-white and vaguely layered. The timber is hard, easy to work, and used for house fittings, carpentry, joinery, stools and walking sticks. The leaves are drooping, alternate, thick, glaucous but paler below than above; the blade is obovate, oblong or elliptical, up to 45 × 20 cm, the apex obtusely acuminate, and the base rounded. There are up to 20 pairs of lateral veins, with secondary veins running parallel, and the midvein more prominent below; the petiole is purplish and thick, up to 1 cm. The canopy is thick and casts a heavy shade when in full leaf. An essential oil of no commercial importance is obtained from the leaves. About 20 compounds have been identified including β-caryophyllene, α-humulene and α-pinene (Fournier *et al.*, 1999).

The flowers are singular, appearing at the base of new shoots, with the pedicel to 20 cm long bearing a leaf-like bract one third of the length from the top; the flower is large, pendent and fragrant, and the calyx is over 2.5 cm long, with wavy edges, crisped and red-spotted. The six petals consist of an outer three up to 10 cm long with crisped margins, spotted red, yellow and green, and an inner three which are subtriangular, forming a white-yellow cone at the centre and spotted red outside and green inside. The carpels are whorled and uniting forming a one-celled ovary with parietal placentation. The flowers are protogynous and insect-pollinated.

The fruit is a smooth, green, spherical berry up to 20 cm diameter, becoming woody, and suspended on a long stalk, up to 60 cm, containing numerous seeds embedded in a whitish, fragrant pulp. The seed is pale brown, oblongoid up to 1.5 cm, containing 5–9% of a colourless essential oil not produced commercially. About 24 compounds have been identified in the seed oil; the major constituents are α-phellandrene, α-pinene, myrcene, limonene and pinene (Fournier *et al.*, 1999). The odour and taste are similar, but inferior to, nutmeg oil.

In western Africa, the seeds are collected in large numbers from wild trees and are a popular spice on general sale in most local markets, either whole, kibbled or ground, with a thriving inter-country trade. Powdered nuts are used as snuff, and seeds are made into scented necklaces. The bark is used to treat haemorrhoids, stomach-ache, febrile pains and eye diseases, while seeds are a stimulant, stomachic, used to treat headache and sores, and used as an insect repellent. Given sufficient incentive, this spice could become popular internationally as it has a distinctive and attractive aroma and flavour, and large quantities are available.

The Brazil nutmeg, *Cryptocarya moschata* Nees. & Mart. (*Lauraceae*) is used as a nutmeg substitute in South America; the Madagascan clove-nutmeg is obtained from *Ravensara aromatica* Gmel. (*Lauraceae*) (see Chapter 6); the Californian nutmeg is *Torreya californica* Torr. (*Taxaceae*).

6

Myrtaceae

The *Myrtaceae* consists of some 75 genera and nearly 3000 species of mainly tropical evergreen trees and shrubs. Chief centres of distribution are the American and Asian tropics, and Australia. A major revision transferred many members of the Asiatic *Eugenia* L. to the genus *Syzygium* Gaertn., now containing about 500 species mainly in the Asian tropics. The most important spice producer is clove, *Syzygium aromaticum* L. The genus *Pimenta* Lindl. contains 18 species native to tropical America and the Caribbean region; one produces the spice bay leaves, another an important essential oil. The genus *Myrtus* L. contains about 100 species mainly native to South America, but only *Myrtus communis* L. produces a commercial spice.

Clove

Clove and pepper probably played a greater role in world exploration and history than any other crops, and their exploitation changed forever the lands in which they grew. The clove is indigenous to the Moluccas, now part of Indonesia, but apparently the buds were little used locally as a spice or flavouring. The first recorded use was in the Chinese Han period 220–206 BC, when courtiers were required to sweeten their breath with clove buds in the Emperor's presence. This indicates that early

on China traded indirectly with the Moluccas, confirmed by later reports that in the first millennium BC, northern Chinese Yeah merchants obtained cloves via Mindanao, in the Philippines. The Chinese probably extended their knowledge of clove to regions with which they traded, as a later Chinese name *theng-hia* has been incorporated into many languages, for example, *makhak* in Persia. However, Burkill (1966) suggests that the habit of chewing cloves was general in the producing areas, and that the Chinese may have imported both.

When clove first reached India is uncertain, and its inclusion in very early Vedic and other writings may be hearsay, but it is more clearly identified in the *Ramayana* and *Susruta* of the 2nd century AD. A Sanskrit name, *kalikaphala*, is probably the origin of the Arabic *karanful*, itself the postulated origin of the Greek *caryophyllon*. The last is mentioned for the first time in writings by Pliny in the 6th century, implying a lost earlier Greek description, possibly by Serapion of Alexandria (200–150 BC). Another Sanskrit name for clove is *lavanga*, the origin of most modern Indian local names, but it is so similar to clove's contemporary name in countries of the ancient spice trade that it could be an import; however, in Tamil *lavanga* means cinnamon bark!

The various types, qualities and prices of cloves are discussed in the *Ain-i-Akbari* written in Agra at the end of the 16th century.

Cloves were then well known, although Rheede in his 17th-century *Hortus Malabaricus* makes no mention of clove cultivation. Clove trees were introduced to India toward the end of the 19th century, reportedly as seedlings from Mauritius. A major traditional Indian use of ground cloves is as an ingredient of the betel chewing quid, *pan pati.*

Cloves reached Alexandria, Egypt, in the 1st century AD, were regularly imported by the 2nd century, and in the 4th century were traded around the Mediterranean; the Emperor Constantine presented the equivalent of 70 kg of cloves to St Silvester, Bishop of Rome (AD 314–335). By the 8th and 9th centuries, cloves were known throughout Europe, probably through Jewish Radanite traders. The Alexandrian, Cosmus Indicopleustes, who visited India and Ceylon, described the clove trade in *Topographia Christiana* (about AD 548). Chilperic 2nd, King of the Franks, in AD 716 sent spices including 1 kg of cloves to the monastery of Corbie in Normandy. The Moorish physician and merchant, Ibrahim Ibn Yaacub, reported in AD 973 that cloves could be freely purchased in Mainz, and in AD 1265 the English Countess of Leicester's domestic records show payments of 10–12s lb^{-1} for cloves.

Venice was the leading European source of cloves and other spices in the 13th century and become tremendously rich. Trade was via Alexandria, itself supplied by Arab sailors who for centuries had a virtual monopoly of the seaborne spice trade until it was broken by the Portuguese in the 16th century. The origin of cloves became known in Europe following publication by Marco Polo in AD 1298 of his famous journeys, later resulting in Spanish and Portuguese searches for the Spice Islands, and brought Vasco de Gama to India in AD 1498. In less than 20 years, the Portuguese had also occupied the Moluccas, and trade in cloves and nutmegs became a royal monopoly for a century. A tree reportedly from the Portuguese period and said by the locals to be at least 350 years old survives on Ternate Island in a special enclosure; it is of enormous size for a clove and is still bearing.

The Dutch subsequently broke the Portuguese monopoly, instituting one of their own, and beginning in 1621 destroyed all clove trees except on Amboina and adjacent islands. This policy virtually ensured that clove trees were planted in other countries to circumvent the Dutch monopoly, finally giving rise to the pre-eminence of Zanzibar and Madagascar in the clove trade. Based on Batavia, now Jakarta, the Dutch clove trade is described in detail by Rumphius (*Herbarium Amboinense*, 1750). So effective was the Dutch policy of containing trees to certain islands where they could control their management that in 1710 directors of the Dutch East India Company (VOC) in their annual report 'noted with grief that production that season was likely to be 1.8 million pounds, to the detriment of the price! Future production should be limited to 500,000 pounds'.

The wholesale destruction of wild and cultivated trees by the Dutch resulted in a great loss of genetic diversity, a lack noted by Rumphius in 1741, when he described only three types of trees differing mainly in size and colour of ripe cloves. On Amboina island, also in a protected enclosure, is a clove tree believed to be at least 300 years old, a survivor of those planted by the Dutch.

Clove and nutmeg were introduced into Mauritius in 1770 from seeds smuggled out of the Moluccas for Governor Pierre Poivre, and the first crop of cloves was picked in 1776. Seedlings were later distributed to other Indian Ocean islands under French administration, including the Seychelles and Réunion (Sheffield, 1950). Few original seedlings survived and it is locally recorded that one tree on Réunion supplied seed from which are descended practically all trees on Réunion and Madagascar. Cloves were found growing on Papua New Guinea in the 1770s and were probably also introduced to neighbouring islands (Sonnerat, *Voyage en Nouvelle Guinee* 1776).

Clove was first planted in Malaysia in 1786 but died out; a second introduction in 1880 flourished and a small but thriving population remains in the Penang region. Seeds from Réunion were planted on Sainte Marie Island off the Madagascar coast in 1827, thrived and produced about 15 t of cloves annually by 1880. Plantings expanded after 1885, cultivation was extended to mainland

Madagascar in 1890, and that country became a major producer of cloves and exporter of clove leaf oil.

Possibly the most important introduction of cloves to any country was in the late 18th century to Zanzibar, an island off the East African coast opposite the present Tanzanian capital, Dar es Salaam. The most reliable account preserved in the Sultans' archives states that a local Arab, Harmali bin Saleh, was banished by the Sultan, took service with a French officer and obtained clove seeds in Réunion. These he presented to the Sultan, Sayyid Said Bin Sultan, and obtained his pardon, and the first seedlings were grown near the royal palace of Mtoni, 6 km north of Zanzibar town. The Sultan realized the economic potential of the crop and substantial areas were planted on Zanzibar and Pemba, helped by the large number of slaves, as described by Sir Richard Burton on his return from the historic journey to Lake Tanganyika (Burton, 1860). He wrote that cloves were first established on Zanzibar in 1818, and the American W.S. Rushenberger stated in 1835 that 'Clove trees had reached a good height and were bearing well' (Rushenberger, 1838). A review of contemporary Zanzibari seaborne trade with the Persian Gulf, the USA and Europe is contained in Nicholls (1971).

The 1872 cyclone caused widespread destruction to trees on Zanzibar, but the more important adjoining island of Pemba was not as seriously affected. The then Sultan, Sayyid Barghash bin Said, enforced replanting on Zanzibar while high prices encouraged additional planting on Pemba (Tidbury, 1949). In 1950, the Agricultural Department stated that there were approximately 4 million trees on Zanzibar and Pemba covering 20,000 ha, with Pemba supplying the majority of cloves exported (E.A. Weiss, personal communication). An interesting précis of the Zanzibar clove industry between 1920 and 1940 and the introduction of oil distillation is contained in Matheson and Bovill (1950).

Botany

The genus *Syzygium* Gaertn. includes some 500 species native to the Asiatic and African tropics and Australia. The most important

spice producer is the clove, long known as *Eugenia aromatica*, but following revision of the large *Eugenia* genus as noted, clove was reclassified and allocated to the genus *Syzygium* (Schmid, 1972). In addition to spices and essential oils (Weiss, 1997), members of the *Myrtaceae* are cultivated for their fruit, including *Eugenia michelii* Lam., the Surinam cherry, and *Psidium guajava* L., the guava.

S. aromaticum (L.) Merr. & Perry (syn. *E. aromatica* (L.) Baill.; *Eugenia caryophyllata* Thunb.; *Eugenia caryophyllus* (Sprengel) Bull. & Harr.; *Caryophyllus aromaticus* L.) has somatic number $2n = 22$. In English it is known as clove, in Dutch as *kruidnagel*, in Spanish and Portuguese as *cravo*, in Italian as *garofano*, in German as *gewurznelken*, in French as *Clou de girofle*, in Japanese as *choji*, in Indonesian as *cengkeh*, in Malaysian as *chengah* and in Zanzibar (KiSwahili), the tree is known as *mkarafuu*, the clove as *karafuu* and the clove stem as *kikonyo*. The modern English name of clove derives from the French *clou* meaning nail, probably from the Latin *clavus*, while an early Chinese name *ting hsiang* translates as sweet-smelling nail incense. In this text, clove will be used to designate clove tree, or harvested buds and the spice, the context ensuring no confusion.

The wild clove common on the forested lower slopes of the Moluccas and Papua New Guinea is considered to be cultivated clove's nearest relative. Wild clove has larger, less aromatic leaves and flower buds, and the essential oil content is lower with different characteristics, but a potentially valuable feature of these wild species is resistance to Matibudjang disease (see Diseases).

Soft-wood grafting on a number of *Syzygium* species was attempted in India but results were poor, with the exception of *Syzygium heyneanum* (Mathew *et al.*, 2000). Elsewhere, interspecific grafting on rootstocks of *Eugenia brasiliensis* Lam., *Eugenia uniflora* L., *Eugenia jambolana* Lamk. and *Syzygium cumini* L. were successful, as were approach grafts on rootstocks of *Syzygium pycnanthum* Merr. & Perry. and *P. guajava*. Hybrids of Indonesian and Zanzibar trees were superior to their parents, and hybrids between wild and cultivated cloves are fertile but have not been studied cytologically. It is unclear whether the wild clove is conspecific or a separate species, *Syzygium obtusifolia* Roxb.; chemotaxonomy will probably solve this question, and also the relationship between diverse and relatively isolated clove populations. A germplasm collection reportedly exists in Indonesia, but a collection of superior trees established in the mid-1950s on Zanzibar has apparently been destroyed.

Dr T.M. Meijer and his colleagues in Indonesia collected buds from wild trees and analysed their oils, which were found to be lacking eugenol but contained eugenone, eugenin, isoeugenitin and isoeugenital, with an odour quite different to clove oil (Meijer and Schmid, 1948). Meijer also wondered whether cultivated clove had evolved by selection to the present type, or if there were initially two quite different wild types. No true cultivars of clove are recognized in Zanzibar and Madagascar since virtually all existing trees are descendants of the original introductions, with little genetic variability. Greater variability exists in wild cloves and between trees growing in the Moluccas, as three main types are recognized in Indonesia: *Siputih*, *Sikotok* and *Bunga Lawang Kiri* considered to be a type reintroduced from Zanzibar with which it is identical. The three types differ in growth, leaf and clove size, but are heterogeneous; *Siputih* produces the largest cloves, especially prized as spice.

Cultivated clove is an evergreen tree up to 15 m in height, generally conical when young becoming more cylindrical with age, bearing glossy green leaves, fragrant red flowers and purple fruits. It is long-lived and can remain productive for 150 years.

The seedling has a pronounced but short taproot quickly replaced by two or three primary sinkers. In the first year, a mass of fibrous roots spreads out from the taproot to a radius of 50 cm and a depth of 25 cm; during the second year these fibrous roots thicken to become main horizontal laterals. Roots finally extend to a radius approximately equal to tree height, and when roots of neighbouring trees overlap, natural grafting can occur.

The trunk, up to 30 cm DBH, often forks near the base into two or three erect branches, sometimes simulated by planting two or three seedlings close together. The degree of branching varies between trees, being less in regions where trees are descended from a few imports. Since the inflorescences are terminal, the extent of branching directly affects flower numbers and thus cloves produced. The main branches are nearly upright with few side branches, giving a cylindrical appearance to mature trees. Smaller branches and twigs are also ascending, terete, numerous and forming a dense canopy, very brittle and greyish-white. During harvesting on Zanzibar, pickers frequently climb trees, damaging these brittle branches and leaving open wounds, allowing entry to the pathogen causing 'die-back' (see section on Diseases).

The bark is grey, smooth to rough depending on age, and slash on a healthy tree is white to rose-pink. The wood is very hard, and from a tree that has died from natural causes is ash-grey; a tree that has died from sudden death is yellow, while infection from die-back produces a reddish-brown discoloration.

The leaves are opposite, simple, coriaceous, exstipulate, glabrous and aromatic. The lamina is lanceolate or narrowly elliptic, sometimes narrowly obovate, 7–13 × 3–6 cm, gland-dotted, and densely and obscurely pinnatinerved. The apex is shortly or broadly bluntly acuminate, the base cuneate, and the margin wavy and recurved. The petiole is slender, 2–3 cm long, the base somewhat swollen and pinkish, and the leaf blade partly decurrent in the upper portion (Fig. 6.1). New bright pink leaves appear in flushes; the upper surface later becomes glossy dark green, the lower dull and paler. Crushed leaves emit the characteristic clove scent, and as the volatile oil also evaporates, clove plantations after rain are permeated with the rather cloying odour. Fresh leaves contain 5–7% essential oil whose main constituent is eugenol at 75–95%. The main characteristics of leaf oil from several origins are shown in Table 6.3.

The inflorescence is terminal, paniculate up to 5 cm, shortly pedunculate, shorter than the leaves and very variable in the number of flowers – from three on a simple three-forked peduncle to 50 or more on multiple

Fig. 6.1. Clove twig showing leaves and flower buds.

peduncles (Fig. 6.2). The angled peduncles and shorter 5 mm pedicels constitute the clove stems of commerce. Bracts and bracteoles are narrow, acute, up to 2–3 mm, and quickly shed. The flower is hermaphroditic with a fleshy hypanthium surmounted by sepals; the hypanthium is 1.0–1.5 cm, cylindrical, angled, green in the young bud but flushed pink at anthesis, becoming deep reddish after the stamens fall. The calyx is tubular, the four lobes fleshy, triangular, slightly incurved, 1–3 mm long, and easily observed in the spice. The four petals are imbricate, red tinged, rounded, 6 mm diameter, and shed as a hemispherical calyptra as the flowers open. The stamens are numerous, up to 7 mm long, the anthers pale yellow and ovate, and the style very stout, up to 3–4 mm, with the base swollen and pale green. The two-celled, multi-ovulate, inferior ovary is embedded in the top of the hypanthium (Fig. 6.3a). The morphology and histology of the clove flower and bud have been described (Parry, 1962).

The inflorescences are harvested when buds have reached full size but before they open; thus the petals together with enclosed stamens form the head of the dried clove. Clove oil is normally obtained from dried buds and published analyses refer to this oil type, but oil from fresh buds at different stages of maturity (i.e. 2, 3 and 4 months changes considerably in physical characteristics: specific gravity (27°C) from 1.0244 to 1.0395, refractive index (26°C) from 1.5215 to 1.5264, and optical rotation (23°C) from $-6.36°$ to $-9.04°$ (Gopalakrishnan *et al.*, 1982). Oil from fresh buds has a pleasant fruity odour, while oil from young buds is more mellow than that from mature buds.

Clove buds vary regionally in their composition, and contain cellulose, pentosans, proteins, tannins and minerals, plus 5–10% of a fixed fatty oil composed of resins and 94% saturated fatty acids, mainly stearic. Carbohydrates account for about 66% by weight. Dried buds also contain the following trace elements in p.p.m. (average of all samples): K

Fig. 6.2. Clove buds and flowers.

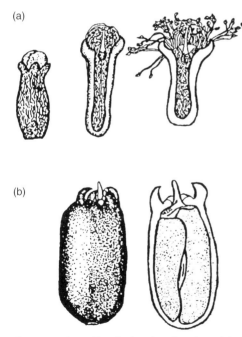

Fig. 6.3. (a) (l to r) Clove bud, and sections through bud and flower. (b) Clove fruit and section through fruit.

971, Ca 457, Ti < 40, Cr < 35, Mn 220, Fe 134, Zn < 25, Br < 15, Rb 58 and Sr 33 (Joseph *et al.*, 1999). In addition, buds contain an essential oil up to 22% but usually around 17% in good-

quality commercial spice samples, whose main constituents are eugenol at 70–90%, eugenyl acetate up to 17%, and caryophyllene sesquiterpenes at 5–12%. Physical characteristics of selected bud oils are shown in Table 6.1.

The fragrant flowers are attractive to and regularly visited by bees and other insects, which are probably the main pollinating agents. Maximum receptivity is reportedly on the 5th day following flower opening. In Indonesia, clove is considered to be self-pollinated but bagged inflorescences on Zanzibar did not produce viable seed, although flowers covered with polythene bags in Sri Lanka set 30% seed. Fallen flowers are used as a flavouring. The effect of growth regulators on clove has been little studied; a Zanzibar unreplicated trial using NAA applied in December caused a yield increase the following December.

Zanzibar has two flowering seasons, July–September and November–January; in Madagascar, the main flowering period is July–August, in India, September–October in the plains, December–January in the hills, and in Indonesia, October–February depending on the area. Bud initials appear about 6 months before they are ready to be harvested, and six stages with local names are recognized in Zanzibar. Thus, it is possible to make a reasonable crop estimate sometime

Table 6.1. Physical characteristics of clove bud and stem oil.

Characteristic	Bud oil		Stem oil	
	Zanzibar	Madagascar	Zanzibar	Madagascar
Specific gravity	1.051–1.056[a]	1.050–1.055[b]	1.040–1.065[a]	1.050–1.055[b]
Refractive index	1.530–1.539	1.5325–1.5357	1.530–1.538	1.5352–1.5357
Optical rotation	−0°32′ to −0°38′	−0°25′ to −3°36′	−0°30′ to −1°0′	−0°32′ to −0°36′
Total phenols (%)	91–95	91–95	ng	91–94
Eugenol (%) solubility (v/v 70% alcohol)	80–95	80–90	85–95	ng

[a]15°C.
[b]25°C.
ng, not given.
Figures in columns show the range and are the average of published figures. Specific samples may differ considerably.

before harvest. A profuse flowering leading to a bumper crop occurs every 4 years in Zanzibar and Indonesia.

The fruits, mother-of-cloves, are oblong fleshy drupes, shortly tapering at each end, reddish-purple, 2.5–3.5 × 1.2–1.5 cm in diameter, surmounted by four enlarged fleshy calyx lobes, and there are normally 350–375 ripe fruit kg^{-1} (Fig. 6.3). Distilling mother-of-cloves yields about 2% essential oil, whose main constituent is eugenol at 50–55%, with no commercial value except to adulterate other clove oils. Mother-of-cloves are also included in herbal medicines in South-east Asia, to treat a variety of ailments.

The fruit contains a single 2 mm oblong purplish seed, rarely two, rounded at both ends, with a 2–3 mm thick fleshy pericarp. Seed contains no endosperm, and the two large cotyledons lie parallel, inner faces folded and interlocking, attached centrally to the large hypocotyl. Seed from freshly harvested fruit is normally viable, should be planted as soon as possible, and germination is usually above 90%. Viability of stored seed is erratic and affected by air temperature and humidity, thus seed for sowing should only be stored for short periods (Hassounah *et al.*, 1984).

Ecology

The true clove is native to five small islands of the Moluccas, Bachain, Machian, Mutir, Ternate and Tidore now part of Indonesia,

with a tropical maritime climate. Although similar conditions exist in many countries to which clove has been introduced, few commercially viable plantations exist. Thus, the environmental conditions necessary for intensive clove production have yet to be fully established. For example, trees flourish on the inland Nilgiri Hills of Tamil Nadu State, India.

Clove occurs naturally as a second-storey component of lower montane forests, and in a plantation environment flourishes with partial shade from neighbouring trees. Isolated trees exposed to strong direct sunlight, as in some botanic garden specimens, frequently show foliar scorch and reduced growth and flower number. Young trees in plantations should be shaded initially with suitable local species; on Zanzibar, *Gliricidia* spp. and *Peltophorum* spp. were used but the most effective was bananas; in Indonesia, *Erythrina* spp. and *Leucaena* spp. are popular. These trees should be selectively pruned to provide the level of shade required by growing clove trees.

Clove is usually grown commercially below 300 m above sea level, and up to 600 m in the Seychelles and Madagascar, but some of the highest-yielding plantations in India are in the Western Ghats around 1000 m. However, trees will make healthy growth at higher elevations, as in the botanic garden at Entebbe, Uganda, at an altitude of 1200 m. Such trees normally make slow initial growth and the time to first harvest is considerably lengthened, but trees remain healthy and produce cloves for many decades. A temperature of 24–33°C is recorded in main clove-growing regions of Indonesia and Zanzibar, similar to that of the Moluccas. Lower mean daily temperatures occur in India and Madagascar, but there is no information on the effect of varying temperature on tree growth, flowering or fruit development.

The productive clove areas of Zanzibar and Madagascar have an annual rainfall of 2000–3000 mm, similar to the Moluccas with 2200–3600 mm, with a dry period coinciding with flowering, which also assists picking and drying. It would appear that a dry period is necessary with its related slowdown in tree development, to initiate the floral differentiation of buds. If this is so, then variation in the dry period, its length or indeed absence may be a factor influencing the highly variable annual yield. Further work on many aspects of clove's development is required. High humidity at flowering reduced yields in Singapore, Penang and certain areas of Indonesia. For leaf oil production a well-distributed rainfall of at least 3000 mm is desirable to ensure continuous regrowth.

Rainfall is considered to affect the type of clove produced, since in Indonesia cloves from wetter areas are less suitable for use in kretek cigarettes, as the smoke is less pungent with no crackling during smoking. Trees that produce the desired type of clove are said to require 3 months with a rainfall below 60 mm compared with trees producing spice, which require a minimum monthly rainfall of 80 mm. A very dry year can kill many young and also more mature trees, especially where soils are shallow or very free-draining, and trees often exhibit similar symptoms to dieback. Trees recover very slowly and those that are highly stressed frequently die; a similar phenomenon also occurred in Madagascar.

Young growth and branches are quite brittle, and high winds cause considerable damage. Wounded trees also appear to be more susceptible to infection by pathogenic fungi. On Zanzibar, this was considered mainly responsible for the spread of sudden-death disease. Although extensive structural damage can be caused by high winds, trees are seldom blown over or uprooted, cyclones excepted, as the extremely tough trunks resist breakage. Damaged trees generally recover, although regrowth is slow, and if trees are extensively damaged replanting is probably the quickest method of re-establishing clove production. The well-documented Zanzibar 1872 cyclone caused havoc and made extensive replanting essential to save the industry. When establishing new plantations in treeless or exposed situations, windbreaks are essential initially and permanent windbreaks may be necessary.

Soils

Deep, well-drained fertile loams, sandy loams or sandy clays are the most suitable, but soils with impeded drainage, or that are calcareous or markedly saline, should be avoided. Clove prefers a soil with a pH 5.5–6.5 but will tolerate acid soils to pH 4.5. Limits have not been determined and most records are taken from stands of established trees, which were originally planted with no regard to soil status. Trees flourished on well-drained soils derived from volcanic rock in Malaysia, but made poor growth on young volcanic soils in Madagascar. Heavy clay soils are generally unsuitable, but in Penang, clove thrives on such soils since the hillsides on which trees are planted ensure adequate drainage. Soils that become waterlogged for even relatively short periods are also unsuitable, and seedlings in Indonesia subjected to flooding resulted in the death of 60% after 5 weeks; death is hastened when *Phytophthora* spp. are also present. The most serious diseases of clove are the various forms of dieback, but similar symptoms can occur in uninfected trees and, as noted, is often due to unsuitable soil.

Fertilizers

The effect of fertilizers on clove has been little studied, and the small amount of information is conflicting, probably due more to varying conditions of climate and soils than to tree reaction. In general, fertilizer application, except in planting holes and to young saplings, is unprofitable. Spreading plant and animal residues under trees or mulching with grass can improve tree health and thus affect subsequent yield.

Placing compost or well-rotted cattle manure or castor pomace plus topsoil in planting holes is recommended in India, while 10–50 kg coconut meal per pit in Zanzibar resulted in substantial yield increase, but the response was not consistent, neither was superphosphate or an NPK mixture (Zanzibar vars). Fish and animal refuse is spread around trees and hoed in in Malaysia and Indonesia, and elsewhere rock phosphate, bone meal and similar material where available and cheap. Young trees in Indonesia receiving N plus K, but not singly, made improved growth and were larger with more leaves. There was little response to P (Wahid and Usman, 1984).

The nutrients removed by clove in India to produce 90 kg ha^{-1} annually are given in Table 3.4. Annual application per tree of 2–5 kg bone or fish meal is common, but in the first years after planting in Kerala State, the Department of Agriculture recommends (per tree) 20 g N, 8 g P and 42 g K, and to trees in bearing, 300 g N, 110 g P and 62 g K annually during the monsoon in May–June. Urea at 0.5 kg per tree, triple-superphosphate at 1.9 kg and potassium chloride at 0.6 kg applied to 14-year-old trees in Madagascar increased yield from 1036 to 1512 kg ha^{-1} but it was considered that potassium may not have been necessary although the total application was profitable (Dufournet and Rodriguez, 1972).

Yield increase was significant on mature trees in Indonesia only after 4 years' annual application of 900 g of a 12:12:17 NPK mixture plus 225 g urea per tree; applying 100 g of a 15:15:15 mixture improved growth of 3-year-old trees (Daswir, 1986). Little is known of the effects of minor elements on clove; data available are from laboratory or pot trials and may not be applicable in the field (Nazeem *et al.*, 1993).

Cultivation

Clove trees are grown primarily either to produce cloves used as spice or to produce leaves for leaf oil, since one generally precludes the other. Oils can be obtained from buds and/or stems and are basically by-products of the spice trade; leaf oil requires different cultural methods. Irrespective of the product required, new lands must be deep-ploughed or ripped to break up any hard layers, and all debris collected and burnt. The period prior to planting out seedlings is a good opportunity to destroy pernicious weeds, and obtain a quick-growing crop to offset the expense of land preparation.

Cloves may be propagated from seeds, layered, grafted, or more recently by clonal propagation, and large numbers of plantlets can be produced from selected parents using *in vitro* methods (Babu *et al.*, 1993). The usual method is to sow seeds, and special trees must be left in plantations to provide seed. High-yielding, regular-bearing trees should be selected and a single tree will yield about 2000 sound fruit in an average season. Self-sown seedlings are also transplanted from existing plantations. However, clove is normally cross-pollinated, thus seedlings from a single tree can vary widely in their characteristics, but variation is less where most trees are descendants of very few imports, as in Zanzibar and Madagascar.

Fruits for seed are allowed to ripen and fall on to a cleared area around the tree, then collected daily and kept moist for 2–3 days to soften and facilitate hulling, which should be carried out with care to prevent seed damage. Sound, large, olive-green seed from single-seeded fruit should be selected, with the radicle not blackened, and on Zanzibar 1000–1100 such seed weighed about 1 kg.

Immediately after hulling seed is normally sown either into nursery beds similar to those described for nutmeg (Chapter 5), commonly at a spacing of 20 × 20 cm, into biodegradable pots, which can be made from banana leaves, woven from palm fronds, or be proprietary products, or two to three seeds sown directly into prepared pits in the field. It is normally recommended that seed should be placed vertically in the soil, radicle down, and approximately one third exposed. However, trials in India have shown that seed sown

horizontally with the microphylar end facing sideways gave a significantly higher
germination percentage, and large seed was
more successful than small (Gunasekaran
and Krishnaswamy, 1999).

Sown seeds should be treated similarly to
nutmegs. Seeds normally germinate in
10–15 days, and emergence is usually above
90% and is epigeal. Seedlings are usually
ready for transplanting about 12 months
after sowing, when 25–50 cm in height,
sturdy, well branched, with well developed
roots. Seedlings should be lifted with sufficient soil to protect the roots fully, and the
taproot pruned. One or two seedlings are
planted in pits, shaded until established,
and seedlings that die should be quickly
replaced. A field spacing of about 8–9 m on
the triangle is recommended when grown
for clove production as this will allow full
development of trees. Pruning to reduce
height or spread of branches is not recommended as it invariably reduces yield, and if
severe can kill trees. For leaf production,
seedlings should be planted approximately
75–100 cm apart to produce a hedge, with
the row width up to 5 m to suit harvesting
machinery. For new plantations, shade and
windbreaks may be required (see Ecology
section), and should be established well in
advance of planting out seedlings. Where
possible, shade and clove tree rows should
also be orientated to reduce wind damage,
i.e. parallel to the prevailing wind direction.

Weed control

Weeding is essential for the first 2–3 years
after establishment: subsequently, ring-
weeding individual trees to 1–2 m diameter
and mowing or slashing is usually sufficient. Herbicides used in local plantations
are generally suitable for clove, for example, glyphosate, but vigorous grasses
including *I. cylindrica* should be removed
manually or spot sprayed, and several
applications may be necessary. Local trials
are essential to determine the effect of particular chemicals on clove, since in pot trials
2,4-D, dalapon and napropamide caused
abnormal development in seedlings (Friere
et al., 1985). Intercropping has been suggested as a method of controlling weeds,
and can be successful in newly established
plantations, but once trees have reached 3 m
their roots can be severely damaged by the
necessary cultivation.

Cover crops normally suppress only less
aggressive local weeds and are of little value
against persistent stoloniferous species.
Cover crops may also deplete soil moisture;
on Zanzibar, for example, tropical kudzu
(*Pueraria phaseoloides* (Roxb). Benth.) used
more soil moisture to 1 m depth than local
weeds. In Indonesia, a cover crop slashed
three times annually plus clean weeding
around the base of 3-year-old trees increased
the rate of tree growth and retained soil
moisture. However, unless it is important to
plant fodder or food crops due to a land
shortage, cover crops *per se* are generally
undesirable.

Although not strictly weed control, parasitic plants including *Ficus* spp., *Loranthus*
spp. and *Cassytha filiformis* L. should be routinely removed.

Irrigation

Where water is easily and cheaply available,
clove benefits from applications during prolonged dry periods and in years of erratic
rainfall just as trees begin to flower.
Plantations should not be established in
regions where irrigation is essential for tree
growth, since the environmental conditions
are generally unsuitable for commercial
clove production.

Intercropping and rotations

Newly established clove plantations can be
intercropped with maize or similar cereals in
the first and sometimes the second year, and,
depending on tree growth, for the next 2
years with low-growing annual food or fodder crops. After this, clove's superficial root
plate will be damaged by the necessary cultivations, and they should cease. However,
cloves may be gallery planted with two or
three rows of trees separated by intercrops.
This has occurred in Indonesia, not by intention, but to remove surplus clove trees!

Pests

Insects are seldom a serious problem in the main clove-growing areas although many species are reported to cause some degree of damage, but not sufficient to require control. Scarab larvae (cutworms) attack young plants in seedbeds (Fig. 6.4), while the larvae of various beetles bore shoots and stems causing varying degrees of damage. Although a particular species may be the main local insect pest, the extent of tree damage is usually minor. Borers causing damage in Indonesia include *Hexamitodera semivelutina*, *Nothopeus fasciatipennis* and *Nothopeus hemipterus*, but the most economically important is *Hindola striata*. Adults and nymphs cause serious damage to shoots and young stems in a particular season, but more importantly, the insect is a major vector of Sumatra disease (see Disease section). The most damaging borers in Malaysia are *Paralecta antistola* and *Chelidonium brevicorne*, and in Madagascar is *Chrysotypus mabillianum.*

Termites can kill newly planted seedlings by ring-barking the stem at ground level; *Bellicositermes bellicosus* is most often reported on Zanzibar. Roots may also be damaged by other insects but this is seldom serious enough to require control; the chafer *Idaecamenta eugeniae* has been recorded in East Africa. Scales are frequently reported but with little comment on the degree of damage, including *Chrysomphalus* spp., *Saissetia* spp., especially *Saissetia zanzibarensis*, and *Aspidiotus* spp., particularly *Aspidiotus destructor*, while unidentified scales have been reported as the most serious pest of clove nurseries in India. The red tree-ant, *Oecophylla longinoda*, known in Zanzibar as *maji-ya-moto* (firewater), is a serious nuisance rather than a pest. It makes large nests on branches and the very active, pugnacious workers swarm on pickers during harvesting, inflicting extremely painful bites. The ant is closely associated with the scale *S. zanzibarensis*, which it transfers from tree to tree.

Nematodes have been reported on clove roots in several countries but with no assessment of damage, including *Dolichodorus* spp., *Helicotylenchus dihystera*, *Rotylenchulus reniformis* and *Trichodorus* spp., with *Tylenchorhynchus* spp. the most frequently reported in India. Some termites are serious pests of other tropical crops, and when clove is planted in their proximity the potential damage is much greater. In Brazil, for example, *Meloidogyne incognita* is a serious pest of pepper and could require control in adjoining clove plantations.

Fig. 6.4. Scarab beetle larvae.

Diseases

A range of diseases has been recorded from clove, but most cause minor damage. The most economically important in terms of its effect on world production is sudden-death disease, which may have different causes but basically has the same symptoms. A major cause of tree death first recognized in western Sumatra in 1961 and named Sumatra disease has become widespread and is due to infection by *Pseudomonas syzygii*, whose major vector is the boring beetle *H. striata*. An attempt to introduce a degree of immunity into clove by grafting clove buds on to 27 immune or resistant species of *Myrtaceae* was unsuccessful (Jarvie *et al.*, 1986). So many trees died of the disease on Zanzibar and Pemba that clove timber became a major local source of fuel, and the disease continues to be the single most important factor affecting local clove production (Dabek and Martin, 1987). There are several causal organisms, probably the most serious being *Cryptosporella eugeniae*, which is always associated with injury particularly when branches are broken during harvesting. The branch slowly dies back, leaves turn brown and the fungus proceeds downwards. When a fork is reached, the unaffected branch above it dies suddenly.

A dieback in Indonesia whose primary cause is considered to be unfavourable soil conditions rather than the associated fungal infection is *Matibudjang*. The initial symptom is a general decay of the fine feeder roots, which quickly produces secondary leaf shedding and dieback, initially at the top of the crown and subsequently extending throughout the canopy. Trees slowly decline and usually die 2–3 years from onset of the symptoms. A similar disease in Madagascar, apoplexia, is also restricted to areas of poor soil.

Wilts may cause death of seedlings in the nursery and after transplanting; causal organisms vary regionally, and include *Cylindrocladium* spp., *Fusarium* spp., *Rhizoctonia bataticola, Phytophthora* spp. and *Colletotrichum* spp.: *C. gloeosporioides* in India also causes gleosporium leaf spot, twig blight and flower shedding; it has a range of alternative local hosts.

Leaf spots recorded from several countries include *Cylindrocladium quinqueseptatum, Gloeosporium piperatum*, which also attacks seedlings, *Mycosphaerella caryophyllata* associated with *Alternaria* spp., *Phyllosticta* spp., *Phomopsis* spp. and various sooty moulds such as *Capnodium brasiliense* and *Aschersonia* spp. A thread blight caused by *Corticium* spp. is also of minor importance.

The parasitic alga, *Cephaleuros mycoidea*, is most damaging to leaves in Malaysia but is seldom serious on healthy trees; it has also been reported from India. Phyllody is reported from Belarus in the former USSR, and is apparently widespread locally. Symptoms, which may be due to a virus, have been described from India and Indonesia, but not definitely confirmed.

Harvesting

Clove begins to bear after 4–5 years, attains full bearing at 20 years and can remain productive for a further 100 years. Although trees may begin to bear after 4 years, heavy picking is not recommended, since it not only physically damages young trees but adversely affects future growth. On Zanzibar and Indonesia, a 4-year cycle usually produces one poor, two medium and one excellent clove crop. The differences can be extreme; the Zanzibar/Pemba crop in the 1952/53 season was 2715 t and for the 1953/54 season was 20,018 t. A number of reasons have been advanced for this phenomenon, some published, some based on personal observations (Martin *et al.*, 1988). Since high or low yields occur generally over a region, climate must be a major factor, combined with its effects on clove development, particularly shoot and flower initiation. These aspects of clove's phenology are worthy of detailed study, to determine whether they can be modified to ensure a more regular bearing pattern, which would be of benefit to both trees and growers.

Clove clusters are picked when buds are full-sized, the calyx base has developed the characteristic pink flush, but no buds have opened or petals fallen to expose the stamens. Since not all clusters on a tree ripen simultaneously, it is usually necessary to

harvest each tree five to eight times in a season. Additionally, it is also necessary to
ensure that pickers paid by weight do not
include immature or opened buds, as their
removal is expensive and leaving them
downgrades the spice.

Two annual harvests are normal on
Zanzibar, the *mwaka* in July–October and
the *vuli* in December–January, although it is
uncommon for a tree to be harvested in
both, and about 1 month later on the adjoining island of Pemba; in India, it is January
in the plains and March–April in the hills.
Since clove is now grown in many regions
of Indonesia the harvest occurs at different
times: November–December on Ambon and
Pinang, and May–June on Java. In
Madagascar, harvesting is usually October–
November, with the main harvest in
November. In the 1950s and 1960s it was
estimated that some 10,000 persons trav-
elled annually to Pemba from Zanzibar to
work as pickers. The main harvest in
Indonesia is usually very short, 7–10 days,
and large numbers of pickers move from
island to island. In Indonesia and
Madagascar, an experienced picker paid by
the day harvests about 25 kg wet cloves, but
Zanzibar pickers paid by weight regularly
reap 55 kg.

Harvesting cloves has traditionally been
manual and will probably remain so, since
fragmentation of holdings, the need to provide work in rural areas, plus the generally
low cost of such labour, ensures mechanization is unlikely quickly to be introduced.
Tractor-mounted platform pickers as used
in modern fruit orchards would reduce the
heavy damage manual harvesting often
causes to trees, believed to be a significant
factor influencing subsequent yield.

To assist manual harvesting and to
reduce damage caused by pickers climbing
trees, topping at 3–4 m is practised in the
Moluccas, but in general pruning, as noted,
is not recommended. A less harmful
method is to use portable tripods, cutters on
long poles, or even to leave unpicked clusters on less accessible branches. The use of
chemicals to cause cloves to fall has been
tested in Brazil, and although successful
was more expensive than hand harvesting,

although the method became more finan-
cially attractive as plantation size and yield
of cloves rose (Araujo *et al.*, 1989).

Quoted regional average yields are basically an indication of the tremendous seasonal variation that exists for reasons still to
be accurately determined, as noted in the
Cultivation section. Dividing official production of cloves by the planted area under
trees produces ludicrous results; on this
basis the 1996 season in Indonesia gave an
average production of 1 kg cloves per tree!
(de Guzman and Siemonsma, 1999).
Additionally other factors may be important; on Madagascar a shortage of pickers in
specific districts affects annual plantation
yields. Thus, to compare one region with
another is of little value. Assessing the data
available, it appears that in virtually all
clove-growing areas, a long-term average of
4–5 kg cloves per tree was reasonable.
Improved agricultural and management
techniques becoming available have the
potential to increase this many times;
whether it will be profitable to do so is
doubtful given the present overproduction.

Similarly, the yield of individual trees
can range between 5 and 50 kg cloves per
season, but in a specific area the variability
may be less due to the lack of genetic diversity, as on Zanzibar. Since clove is open-pollinated, inter-tree yield comparisons are no
basis for selection. Clonal propagation is
probably the best current method of
increasing plantation yield, coupled with
more efficient management.

Madagascar is currently the only country growing clove for commercial leaf oil
production, and figures indicate that
2000–3000 kg green leaves ha^{-1} containing
about 1.5–2.0% oil, depending on the season, was obtained from hedge-planted
trees. When leaves from existing trees are
to be distilled, they are picked every 2–3
weeks, contain 5–7% oil and about 15%
moisture, and when dry equal 1–2 kg per
tree. Alternatively, fallen leaves can be collected, and in Indonesia about 1–2 kg per
tree per month is common. Oil yield from
distilling fresh, naturally dried fallen
leaves, or dried leaves is similar, but oil
characteristics differ.

Processing

Harvested clove clusters are taken to a central store or the homestead, where buds are carefully separated from peduncles and pedicels and dried. The peduncles and pedicels, usually called stems and will be so designated, account for approximately 25% of total dry weight of sound cloves. Buds are spread to dry in the sun on matting to prevent bruising and damage, which reduces oil content or quality; they are gently raked to ensure complete drying and prevent heating or mould formation. Concrete drying floors and even roadsides may be used, but mats under the buds are essential to produce a high-quality product. In showery weather, it is also easier and quicker to roll up mats to protect the buds from rain damage. Damaged and spoiled buds, pieces of twig, insects and other material are removed during drying.

With clear sunny conditions, drying is usually completed in 4–6 days. Buds lose about two-thirds of their green weight after drying and contain about 6% oil. Artificial dryers reduce the danger of spoilage due to inclement weather, and several solar-powered models suitable for cloves are available. Properly dried buds, now the spice clove, are bright brown, brittle, snap easily and are prickly when grasped in the hand. Smallholders then bag the cloves and deliver them to traders, while plantations that have their own equipment further clean, dry if necessary, grade and bag the cloves for export.

Stems are dried using the same technique, and lose about two-thirds of their weight. Stems may be sold as an inferior spice or distilled. Large quantities of stems were previously exported from Zanzibar, but as a significant proportion was distilled to be used to adulterate the more expensive clove bud oil, the then British administration prohibited stem exports for some 50 years.

Distillation

Clove leaf oil is usually obtained from foliage cut from trees originally planted to produce the spice, and currently only on Madagascar have cloves been planted specifically to provide leaves on a regular commercial scale, although the system is being tested elsewhere. The normal method in the Seychelles and Mauritius is to cut branches from existing trees, strip leaves and distil the oil. Since removing branches reduces the subsequent clove crop, choice is thus between a higher yield of cloves and no leaf oil, or a moderate yield of both. Trees regularly harvested for leaves are usually topped to produce a profusion of side branches for ease of cutting. Hedge planting of trees on Madagascar allows partial or full mechanization of all operations, thus reducing the labour required, as the frequent labour shortages are a major reason for low plantation yields of cloves. Fresh leaves contain about 5–7% oil and 15% moisture. In Indonesia, the abundant fallen leaves are collected by family labour, bundled and sold to leaf-buyers. Leaf-fall occurs throughout the year and 1–1.5 kg of naturally dry leaves per tree can be collected every 2–3 weeks.

Leaf oil is often obtained by hydro- or steam distillation in locally constructed stills. On Madagascar, many are locally made of sheet iron or iron drums, embedded in stone or mud hearths and heated by a wood fire, but are being replaced by the Deroy type made of copper. An average still charge is about 300–400 kg branchlets with a capacity of 1000–1500 l. Yield after 24 h continuous operation, including 5–7 h to boil the water, is about 1.5–1.8%. The oil obtained is dark due to the presence of iron eugenates, but plantations using modern stills imported from France produce a high-quality light-coloured oil.

Distilling in Indonesia is also generally a small-scale operation, although several large modern distilleries have been built. Smaller stills are usually vertical cylinders, 3 × 1.6 m diameter, charged and discharged through the lid. A long pipe connected to the lid is submerged in a ditch or concrete channel, where running water serves as condenser and cooler. The oil produced is sold to dealers for further processing. The usual charge is about 1000 kg dry leaves yielding 20–25 kg crude oil; from wet leaves the yield is 15–20 kg. Steam distilling takes 4–6 h, direct firing

6–10 h. The distilling method affects both oil yield and characteristics; combined water and steam distilling produces more oil than steam alone, as does an increase in distilling time (Nyrdjanah *et al.*, 1991), while storing leaves before distilling alters oil characteristics (Muchlis and Rusli, 1978; Rusli *et al.*, 1979; Nyndjanah *et al.*, 1991).

Clove bud oil is obtained by hydro- or steam distilling whole or comminuted buds, but the oils differ. Comminuted buds should be distilled immediately to reduce oil loss by evaporation; distilling whole cloves extends distillation time up to 24 h. A proportion of the eugenol acetate also hydrolyses, increasing the eugenol content of the oil. Steam distillation has the same effect and oils obtained by this method have the highest proportion of free eugenol. Clove stem oil was produced in commercial quantities at the government distillery on Zanzibar until the 1960s, with an oil yield of 15–17% from selected high-quality dried stems.

While steam or water distillation remain the traditional methods of oil production, supercritical fluid extraction, often with liquid carbon dioxide, is becoming more common in developed countries. Since a wide range of extraction plants with varying capacities are available, they can be installed wherever required. The product will be up to international standards.

Storage and transport

Bagged cloves can remain without deteriorating in clean, dry buildings for some time. They should not be in contact with a concrete floor, and stacks should not be too high or lower levels will be crushed. Sound, properly dried and stored cloves are relatively free from storage pests, but paths between stacks allow access for control or fumigation if necessary. Prolonged storage reduces oil content and may also dull the colour, both detrimental for sale as spice.

All clove oils should be stored in full, airtight, opaque containers, and remain sealed until required. Bulk oil is usually transported in 200 l metal drums. Changes in oil composition and the ratio of constituents can change in storage, and the rate and extent depends initially on the production method. Data from India indicated that in carbon dioxide-extracted oils, the terpene hydrocarbons were most affected (Gopalakrishnan, 1994).

Products and specifications

Clove trees produce buds, which are used whole or ground as spice, bud and stem oils and oleoresins, and leaf oil, used principally as a source of eugenol. The characteristic odour and flavour of cloves, clove stems and leaves is determined by the steam-volatile components of their essential oils. The major component of all is eugenol, but odour and flavour differ significantly due to the varying proportions of minor and trace components (Weiss, 1997) (Table 6.2).

Cloves. Good quality cloves should be of similar size, plump, of a uniform bright reddish brown, slightly oily, with no musty or mouldy flavour, and free of foreign matter, insect or other contamination (Fig. 6.5). Cloves can be used whole or ground, with the ground spice generally prepared where consumed. The whole spice is preferred in

Table 6.2. Major components of clove oils.

Component	Madagascan clove bud oil	Zanzibar clove stem oil	Madagascan clove leaf oil	Indonesian clove leaf oil
β-Caryophyllene	4.34	5.13	9.86	12.50
α-Humulene	0.54	0.60	1.10	1.36
Eugenol	88.95	92.35	86.89	84.77
Eugenyl acetate	5.54	1.25	1.59	ng

ng, not given.

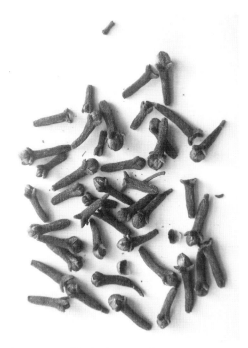

Fig. 6.5. Dried buds; the spice clove.

Clove oils. When clove oil was first produced is uncertain, and initial mention of an elixir of cloves was for the treatment of toothache, probably a direct result of sucking dried cloves, which not only sweetened the breath but acted as a palliative. The major constituent of clove oil was first described by Liebig in 1836, as *nelkensaure* or acid of cloves, subsequently eugenic acid, and in 1875 as eugenol by Johann Karl Tiemann of Berlin, who used it to produce synthetic vanillin (J. Chem. Soc. 1900:600). Clove bud oil was formerly produced in Europe or North America from imported cloves, but is now regularly produced in Madagascar, Indonesia and to a minor extent in India. Clove leaf and stem oils are placed in the dry woody, pronounced spicy group of oils, and bud oil in the warm, sweet spicy group by Arctander (1960).

Bud oil. Distilled bud oil is a clear, colourless to yellow mobile liquid, becoming more brown with age or contamination with iron or copper, with a strong characteristic sweet and spicy clove odour. The flavour is warm, almost burning and spicy. The composition and organoleptic properties of distilled bud oil differ somewhat from the volatile oil present in spice. Hydro-distillation reportedly provides the finest oils for perfumery and flavour use, with a eugenol content of 85–89%. The more common steam distillation produces oils with a eugenol content of 91–95% resulting from hydrolysis of eugenyl acetate. Distilling whole spice provides oil with a high eugenol content and a specific gravity above 1.06; comminuted spice has a lower eugenol content and a specific gravity below 1.06. Bud oil stores well in sealed light-proof containers, and although colour may darken, characteristics and quality are little affected. The finest UK oil containing 85–89% eugenol is from hydro-distillation, but in the USA, much clove bud oil is directly steam distilled. A recent extraction method using pressurized hot water is claimed to be very fast, with no loss of eugenol and eugenyl acetate at the high temperature of 300°C (Rovio *et al.*, 1999).

Bud oil contains three main components: eugenol at 70–90%, eugenyl acetate up to 17%, and caryophyllene sesquiterpenes (mainly β-caryophyllene) at 5–12%. Although

domestic cooking, and often removed before serving the meal. However, clove is a spice whose use must be controlled as it has a strong flavour, which can easily become overpowering. In England, cloves are a common ingredient of apple pies and tarts, in continental Europe of stews and soups, and widely studded into cooked hams and pork. Clove powder has less flavour and odour unless freshly prepared, and deteriorates further in storage. Buying the smallest amount available in a retail pack so that it is quickly used will ensure the powder is as fresh as possible; storage in an airtight, opaque container will retain odour and flavour. Powder is often added to baked products such as cakes, biscuits and puddings, and occasionally to mulled beer or wine. The main use of ground cloves in industrial food preparation includes processed meats, sauces, pickles, in some curry powders and bakery products. A major use is in Indonesian kretek cigarettes which contain up to 8% clove powder, and kretek manufacturers are virtually the only domestic users of Indonesian cloves.

bud oil composition (and stem and leaf oils) is well documented (Gopalakrishnan *et al.*, 1982, 1993; Gaydou and Randriamiharisoa, 1987; Lawrence, 1994b; Pino and Borges, 1999c), the relative importance of various trace components in determining odour is not clearly established; methyl-n-amyl ketone is said to be responsible for the fruity top-note. Adulterants of bud oil are usually clove stem or leaf oil, or clove terpenes remaining after eugenol extraction. Such adulteration may be difficult to detect analytically, thus organoleptic evaluation is necessary for use in flavourings or perfumery.

The major use of bud oil is in seasonings and processed food, perfumery, and to a lesser extent in pharmaceutical and dental products. It has minor properties as a stimulant, carminative and to relieve flatulence, and possesses antimicrobial, antioxidant and insecticidal activity (Barnard, 1999; Dorman and Deans, 2000; Dorman *et al.*, 2000).

Stem oil. Stem oil of good quality is a pale to light yellow liquid, but freshly distilled oil is almost colourless; its odour is strong, spicy and somewhat woody, similar to eugenol but coarser and more woody than bud oil. Redistilled stem oil contains 90–97% eugenol, but its odour is less sweet and floral than bud oil. Eugenol content of stem oil at 90–95% is higher than bud oil, but its eugenyl acetate content is low. Madagascan stem oil is now legally required to have a minimum eugenol content of 82%, Indonesian of 80–82%. Stem oil is used mainly in food flavouring and perfumery, a small amount to produce eugenol and its derivatives, and is a frequent adulterant of bud oil.

Production of clove stem oil was previously a monopoly of the Zanzibar government as noted, and the oil was one of the most uniform of available commercial essential oils. The plant consisted of 12 stainless steel stills each holding about 680 kg stems; steam distilling for 16 h yielded 5–7% of an almost water-white oil, which darkens with age to yellow, and is sometimes violet-tinted. Mother-of-cloves oil is produced by steam distilling ripe fruit only when there is spare capacity or a shortage of other material. Its main use is to adulterate other clove oils.

Leaf oil. Crude leaf oil is dark brown, often with a purple or violet tint, cloudy to some extent, and sometimes showing precipitation; its odour is harsh, woody, phenolic, slightly sweet and quite different to bud oil. Rectified oil is a clear pale yellow, with a sweeter, less harsh, dry woody odour closer to that of eugenol. The oil is obtained by steam or water distillation. In areas with a substantial production of leaf oil and little stem oil, or vice versa, the two are often bulked and sold on eugenol content. Eugenol content of leaf oil is 80–88%, lower than bud oil, with only a small content of eugenyl acetate, but a high caryophyllene content. Indonesian leaf oil may contain β-caryophyllene up to 18%, humulene up to 2% and eugenol up to 80% (Vernin *et al.*, 1994). Standard specifications for leaf oil are shown in Table 6.3.

Leaf oil was mainly used for eugenol production and its by-product caryophyllene. Rectified oil is used in less expensive perfumes, soaps and similar products. Soaps containing leaf oil are usually dark brown to mask the discoloration when combined with an alkali. Leaf oil is generally considered unacceptable in food flavourings since its harsher note does not reproduce the genuine clove flavour. The major competitor to leaf oil as a eugenol source is cinnamon leaf oil and should the price of clove leaf oil become too high for end-users, expansion of cinnamon leaf oil production becomes attractive, especially in countries with few other export crops.

Oleoresins. Bud oleoresin is a viscous brown liquid, which may deposit waxy particles on standing. The odour and flavour are superior to distilled oil and much closer to the natural spice, but can be tainted by the solvent employed. Extraction using organic solvents produces an oleoresin containing a volatile oil, a fatty oil and other constituents soluble in the particular solvent. In general, a yield of 18–22% is obtained with benzene (90–92% volatile oil) and 22–32% with alcohol, but the latter has a high resin content and is unsuitable for perfumery. Commercial oleoresins extracted with a hydrophilic solvent such as acetone normally contain 70–80% volatile oil.

Table 6.3. Standard specifications for clove bud, stem and leaf oils.

Specification	Bud BSS	Stem EOA 178	Leaf EOA 55	Oleoresin EOA 238[c]
Specific gravity	1.041–1.054[a]	1.048–1.056[b]	1.036–1.046[b]	1.04–1.06[a]
Refractive index (20°C)	1.528–1.538	1.534–1.580	1.531–1.535	1.527–1.538
Optical rotation (20°C)	0° to −1.5°	0° to −1°30′	0 to −2°	< −1°30′
Total phenols (%)	85–93	89–95	84–88	ng
Volatile oil (ml per 100 g)	ng	ng	ng	66–88

BSS, British Standard Specification; EOA, Essential Oil Association, USA.
[a]20°C.
[b]25°C.
[c]BSS is virtually identical.
ng, not given.
Figures in columns show the range.

Oleoresins extracted with alcohol are much darker than those from hydrocarbon solvents, and have a more powerful perfume and flavour, but are considered less refined or delicate in odour. Extraction using supercritical carbon dioxide produces an oleoresin considered closest in flavour and odour to the spice, and is becoming the oleoresin of choice for many food processors.

Extraction on an industrial scale is by specialist firms in major consuming countries, who are increasingly using supercritical fluid techniques. Oleoresin is normally dispersed on salt, dextrose, flour or rusk, and sold weight for weight equivalent in strength to the dry spice. The main advantages over dry spice are little risk of bacterial contamination, standard strength and quality. The oleoresin has gained a market share from both dry spice and bud oil especially in the food industry, mainly in meat and soup products, and is also used directly in perfumery.

Concrete and absolute. Concrete is obtained by solvent-extracting dried and comminuted buds with benzene or petroleum ether. Commercial concrete is a very viscous liquid or semi-solid mass, varying in colour from pale olive-green through yellow-brown to dark brown. The highest quality is usually a clear viscous liquid, olive to pale brown; its odour is refreshing, oily-sweet, intensively rich and spicy, closely resembling dry buds.

Bud absolute is prepared by extracting concrete using similar solvents; petroleum ether produces the finest, benzene a richer odour. Absolute is an olive-green, greenish brown to orange-brown viscous or oily liquid, semi-solid at low temperatures; it is soluble in alcohol in all proportions. Absolute differs from bud oil in containing only those constituents present in the unprocessed bud. The flavour strength of absolute is about twice that of average clove bud oil. Its odour is of clove flowers in full bloom: floral and refined, balsamic, sour-sweet and immensely rich with none of the unwanted off-notes and flavour found in bud oil (Arctander, 1960). Its main use is in perfumery and to a lesser extent in flavourings.

Eugenol. Eugenol is the principal constituent of clove oils, particularly leaf oil, and clove oils were initially the sole source of eugenol used in the production of vanillin. This use was greatly reduced with its replacement by artificial vanillin and synthetics (see Chapter 7). Eugenol is a colourless to light yellowish oily liquid, becoming browner on exposure to air, its odour and taste spicy, but less strong than bud and stem oils, although smoother than clove leaf oil. It is used mainly in inexpensive perfumes, soaps and similar toiletries, or where too definite a clove odour is undesirable.

Other Syzygium species

Syzygium polyanthum (Wight) Walpers. (syn. *Eugenia polyantha* Wight; *Eugenia nitida* Duthie; *Eugenia balsamea* Rid.) is known in English as Indonesian bay leaf, in Indonesia commonly as *salam*, in Malaysia as *samak* or *serah* and in Thai as *daengkluai*. It is native to Myanmar, Indo-China, Malaysia and Thailand to Indonesia.

It is a tall tree up to 30 m in height, and up to 60 cm DBH, with grey bark. The leaves are opposite, simple, glabrous and dotted with minute oil glands. The inflorescence is a 2–8 cm panicle, and the flowers white, small, fragrant and bisexual. The fruit is globose, up to 12 mm long, purplish when ripe, and single-seeded.

The aromatic leaves, either fresh or dried, are a common spice, freely traded and available at markets throughout the region, and equivalent in use to bay leaves from *Laurus* spp. in Western cuisine (Chapter 3). The fruit is edible but astringent. A considerable volume of leaves is collected from wild trees, but also from timber plantations. Dry leaves are distilled locally with an oil yield of 0.15–0.175%; the oil is dextro-rotatory (laurel is laevo-rotatory) and the main components are eugenol and methyl chavicol. The bark pounded in water yields a brownish-red dye used to tan fish nets and bamboo matting, and a bark extract is used to treat diarrhoea. The wood is pale to pinkish-brown, 540/790 kg m^2 at 15% moisture, and used for housing and furniture.

Unrelated species

A number of plants produce either a spice or essential oil locally called clove or clove oil, and two are well-known members of the *Lauraceae*.

Ravensara aromatica Gmelin., the Madagascar clove-nutmeg, is a medium-sized tree up to 12 m tall, with small leathery leaves and small round aromatic fruit, whose taste resembles a mixture of clove and nutmeg; oil distilled from its leaves is commonly an adulterant of true clove leaf oil.

Dicypellium caryophyllatum (Mart.) Nees, the Brazil clove, bears highly aromatic flower buds very popular locally as a flavouring, which are harvested and dried in a similar manner to cloves. Oil is distilled from wood and bark, known as clove-cassia or clove-bark oil, and exported to the EU, especially France. *Cinnamomum cultilawan* (L.) Kost.,

native to Indonesia, produces a clove-scented bark and oil (see Chapter 3).

Pimenta

The genus *Pimenta* Lindl. contains 18 species of aromatic shrubs and trees native to tropical America including the Caribbean. The genus name is derived from the Spanish *pimienta* (peppercorns), via the Latin *pigmentum*, and was originally applied to *Pimenta dioica* because its berries resembled the then more familiar spice. The genus *Pimenta* is closely related to *Myrtus* L. and *Eugenia* L., differing mainly in the flowers and fruits, hence the confusion in early identification. The most commercially important *Pimenta* spp. are *P. dioica* (L.) Merr. providing the spice pimento (allspice), and *P. racemosa* (Mill.) Moore., source of oil of bay, which should not be confused with bay leaf, a spice made from leaves of *L. nobilis* L. (Chapter 3), or with oil obtained from the Californian bay, *Umbellularia californica* (Hook & Arn.) Nutt.

Pimenta dioica

P. dioica is now indigenous to Central America and the neighbouring Caribbean islands although its original home is in dispute. It is widely cultivated throughout the region, primarily to produce dried berries known as pimento or allspice, with Jamaica a leading supplier. Use of pimento predates the arrival of the Europeans as the Maya, Aztecs and other peoples had an extensive knowledge of local herbs and spices both as food flavourings and in medicine, and the Maya also used the berries in religious ceremonies.

The first European mention of pimento is probably that of Columbus, who on 4th November, 1492, recorded in his journal that on being shown true pepper berries he had brought from Portugal, the local Cuban people said it grew there wild and in profusion. The berries were later found to differ from true pepper, and the tree was named *Piper tabasci* by the Spanish explorer Hernandez. However, the berries were widely known as *pimienta*, later anglicized as pimento.

Spanish explorers and later settlers on Jamica harvested and used the leaves and berries, and the island remained under Spanish control from 1494 until 1655, sending dried berries to the Royal Household. Thus, there has been continuous production of berries on Jamaica from about 1509 to the present day. The berries first reached London in 1601 as described by Clusius in his *Liber Exoticorum*, and plants were first cultivated in England in a hothouse in either 1732 or 1739. In his *Natural History of Jamaica* (1725), Sir Hans Sloane accurately described the tree and how the spice was prepared. By this time pimento was a cultivated crop, and many plantations (walks) has been established.

Botany

P. dioica (L.) Merrill. (syn. *Myrtus pimenta* L.; *Myrtus dioica* L.; *Eugenia pimenta* DC.; *Pimenta officinalis* Lindley) is known in English as allspice or pimento, in French as *piment jamaique* or *toute-espice*, in Portuguese as *pimienta da Jamaica*, in Spanish as *pimienta gorda* and in Dutch and German as *piment*.

The basic chromosome number of the genus is x = 11, and pimento is a diploid with $2n$ = 22. *P. dioica* has been successfully budded on to *P. racemosa* and vice versa. A germplasm collection is maintained by the Jamaican Department of Agriculture, in addition to breeding and selection programmes, which have produced named cultivars, and at the Indian Institute of Spices Research, Calicut, Kerala, India. *Pimenta* species from the Caribbean region have been described in a series of papers by Tucker *et al.* (1991). A number of unrelated species with aromatic fruit are incorrectly referred to as allspice, including Carolina allspice (*Calycanthus floridus* L.) and Japanese allspice (*Chimonanthus praecox* (L.) Link). Use of the words allspice and pimento in the text refer to the products of *P. dioica*.

Pimento is an evergreen tree up to 7–10 m tall, sometimes 15 m, with a slender trunk, branching from 1 to 3 m, with secondary branches profusely branched at the extremities. The bark is smooth, shiny, silvery pale brown and often shed in long strips. The wood is pinkish, close-grained, heavy, strong and durable. Woody shoots are frequently used as handles for brooms and other tools, and as walking sticks. The root system is extensive and generally more penetrating than nutmeg, and trees thus suffer less damage from high winds.

The leaves are simple, opposite, entire, thinly coriaceous, punctate, with pellucid glands, borne in clusters at end of branches and highly aromatic when crushed; the petiole is 1–1.5 cm long (Fig. 6.6). The blade is elliptical/oblong, 6–15 × 3–6 cm, the margin entire, the base tapering, the apex rounded and the venation pinnate; it is dark green above, paler below. Trees may bear a majority of large or small leaves. Although trees are basically evergreen, leaves are continually shed as they reach 2–2.5 years of age. Leaves contain an essential oil, and on distilling, fresh leaves yield 0.3–0.5%, and dried

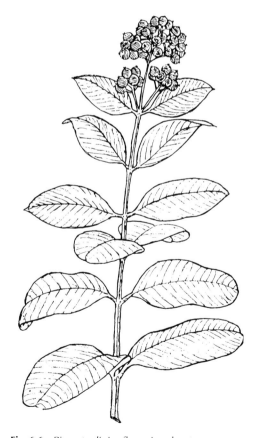

Fig. 6.6. *Pimenta dioica* flowering shoot.

leaves 0.5–1.25%; the oil is discussed at the end of this section.

The flowers are fragrant, white, small and on many-stalked cymes in the axils of upper leaves (Fig. 6.6). The flowers are structurally bisexual but functionally dioecious; thus, female and male flowers are borne on separate trees. The flowers are 8–10 mm in diameter, the pedicel about 1 cm long, pale green and pubescent. The calyx is tubular and projects above the ovary; the four lobes are rounded, 1.5–2 mm, creamy-white, expanded after anthesis, and persistent in fruit. The four petals are reflexed, rounded and 3–4 mm, falling early; the stamens are 5 mm, free, and there are 40–50 in functionally female flowers, 80–100 in male flowers. The filaments are white and slender, and the anthers are creamy, dehiscing by longitudinal slits. The ovary is inferior, two-celled, usually with one ovule in each; the style is 5 mm, white and pubescent, and the stigma is yellow.

Similar to nutmeg, no accurate method of visually sexing trees until they flower is known. Male flowers produce abundant pollen, and wind and insects are the main pollinating agents; pollen from female flowers is usually sterile. Male trees generally begin flowering before, and continue after, female flowering. The main flowering period in Jamaica is March–June according to district. In wild populations, the proportion of fruiting and barren trees is about 50:50.

The fruit is a sub-globose berry, 4–6 mm in diameter, containing two seeds with a spirally coiled embryo. Ripe fruit is a deep purple to glossy black, the mesocarp mucilaginous, sweet and spicy. On drying, the fruit is dark brown, rough and more aromatic (Fig. 6.7). Birds are the main seed-dispersal agents. The dried berry contains an aromatic essential oil, fixed oil, resin, protein, pentosans, starch, traces of alkaloids, pigments and minerals; the proportions vary considerably. Mature but green berries contain more essential oil than ripe berries. The aroma and flavour of the spice (and the oil) are due to steam-volatile compounds, the pungency to non-steam-volatile constituents.

Essential oils. Information on the chemistry of pimento is almost wholly from the Jamaican spice, berry and leaf oils, but considerable variation exists in chemical composition from different areas of the Caribbean and Central America (Green and Espinosa, 1989). The major constituent of pimento spice and leaf oil is eugenol (up to 90%), but the two oils are organoleptically quite distinct. The main characteristics of the oils are shown in Table 6.4. The oils are similar in general character to clove bud, clove leaf, tropical bay leaf and cinnamon leaf oils.

Substantial seasonal and regional variation in leaf-oil content and oil composition occurs; phenol content, for instance, can range from 50 to 90% in different districts of Jamaica. Leaf oil composition in a specific region tends to be similar, but the variation noted could indicate the existence of chemovars, although in general distilling techniques have a greater effect. The major constituents of berry oil are eugenol at 65–85%, eugenyl methyl ether up to 9%, plus cineole, α-phellandrene and caryophyllene.

Ecology

Pimento is native to the Central American/Caribbean region as noted and trees grow naturally in semi-tropical lowland forests with a mean average temperature of 18–24°C, a low of 15°C and maximum of 32°C. Pimento flourishes below 300 m where the incidence of rust is less severe, although trees occur up to 1000 m above sea level. An annual rainfall of 1500–1750 mm evenly spread through the year is desirable,

Fig. 6.7. Dried pimento berries; the spice.

Table 6.4. Physical characteristics of *Pimenta dioica* oils.

Characteristic	Berry		Leaf	
	BSS	EOA	BSS	EOA
Specific gravity	1.025–1.045[a]	1.018–1.048[b]	1.037–1.050[a]	1.018–1.048[b]
Refractive index (20°C)	1.526–1.536	1.527–1.540	1.531–1.536	1.532–1.536
Optical rotation	0° to −5°	0° to −4°	ng	+0°30′ to −2°
Solubility (v/v 70% ethanol)	1:2	1:2	1:2	1:2
Total phenols (% vol)	65	ng	80	80–90

BSS, British Standard Specification; EOA, Essential Oil Association, USA.
[a]20°C.
[b]25°C.
ng, not given.
Figures in columns show the range.

but trees grow well with a rainfall of 1200–2500 mm and a dry season of varying length. Pimento is a forest tree and thus seedlings and young trees benefit from shading until established. Although permanent shade trees are not considered necessary on Jamaica and neighbouring islands, they may be required in more exposed situations, in Brazil, for example.

Although pimento grows abundantly where it occurs naturally, attempts to introduce it as a commercial crop into many countries outside this region have failed for unspecified reasons. Thus, like clove, pimento may require quite specific environmental conditions to flourish. In the Peradeniya Botanic Gardens, Sri Lanka, for example, there is a well-grown healthy tree that produced only a small crop of fruit annually in July–August, and in East Africa there is a healthy, surviving tree from an introduction to Zanzibar (E.A. Weiss, personal observations). However, in some cases it may be that the original seed/seedlings were male and thus would not produce fruit.

Soils and fertilizers

Although pimento is planted on a wide range of soils, a well-drained fertile loam with a pH of 6.0–8.0 produces highest yields, and wild trees favour limestone soils. Soils with impeded drainage or with a compacted subsoil are unsuitable, as are acid and saline soils. However, soil type within fairly wide

limits is less important than management; thus, good berry yields are obtained on a very wide range of soils in central American countries and Brazil.

Little is recorded on the nutritional requirements of pimento as most trees are left to grow virtually untended once established. Well-rotted manure plus surrounding topsoil, or a small amount of phosphate is placed in planting holes, and young trees may be mulched. In established walks, grass and weeds are slashed and left as an annual mulch, plus foliage and twigs remaining after harvesting. This is probably the most profitable approach for most growers, as trees yield well and the price of berries is usually low. The Department of Agriculture, Jamaica, recommends 1 kg of a 10:10:10, or 15:15:15 NPK mixture spread around trees in two annual applications in February and September.

Cultivation

The cultivations required for nutmeg are suitable for the establishment of berry or leaf plantations. Many of the original pimento walks for berry production were established with self-sown seedlings transplanted from wherever they were found growing, and the excess males replaced until only the required number were left. Seedlings could also be produced from seed and the same system used once trees had flowered. Seed should be collected and treated as for nutmeg, and the seedbeds similarly. An improvement on the original system was the establishment of government nurseries, which supplied healthy seedlings at a subsidized price. However, the problem of sexing remained until replaced by vegetatively propagated seedlings from known parents.

It is doubtful if any of these methods can now be recommended although they will continue to be used, since various techniques including cloning, *in vitro* production of plantlets, etc. can produce seedlings of known sex from selected high-yielding parents. Whether it is profitable to do so depends on the level of plantation management and berry prices. Production of seedlings using these techniques should be undertaken by local Agricultural Departments or grower cooperatives, as all require specialized but relatively inexpensive equipment and trained personnel; the high-yielding, high-quality seedlings will, however, rapidly repay the cost.

Seedlings should be planted singly into pots or baskets and allowed to grow until about 30 cm, when they are ready for transplanting into prepared pits in the plantation. Pits about 60 cm deep and 30 cm in diameter may contain fertilizers as noted in the Soils section. Since seedlings will be of known sex, multiple planting is unnecessary, and one plant per pit sufficient. Seedlings which die should be quickly replaced. A common spacing for berry production is 6 × 6 m and the ratio between male and female 1:10. Planting out should correspond with the beginning of the rainy season, but additional watering may be required in a dry year. Seedlings require shade until established and any convenient material such as banana leaves is suitable, plus ring-weeding as necessary. Young plantations should be fenced against livestock. Trees begin to bear after 4–6 years depending on the standard of management and the origin of seedlings. Clonal material is usually quicker to flower and also comes into full bearing faster: 12–15 years instead of 15–25 years from unselected stock. Well-managed healthy trees can continue bearing for decades.

When pimento is grown for leaves to produce oil, sex is unimportant and trees are planted in hedges to allow mechanical harvesting as for laurel (Chapter 3). In-row spacing should be about 2 m, and row width to suit the machinery to be used. However, much leaf oil is still produced from leaves remaining after berry harvesting, or taken from trees after berries have been harvested, which usually reduces the subsequent berry yield.

Weed control

Weeds and grasses must be controlled around young seedlings, but after 2–3 years slashing once or twice annually is usually sufficient, and cut plants can be left as mulch, which will also assist in reducing the danger of soil erosion on sloping lands. Family labour is usually sufficient to maintain grass and weed control, assisted in mature plantations by small livestock but preferably not goats or cattle as both damage trees. Herbicides are seldom applied, but should their usage become profitable, those compounds, including glyphosate, used in other plantation crops will probably be suitable and should be tested. During weeding, parasitic plants should be removed from trees.

Irrigation

It is unlikely that pimento will be profitable if it must be grown as an irrigated crop, and thus water applications are generally a supplement to rainfall in seasons of low precipitation. Additionally, few plantations are established on lands level enough to allow controlled application rates per tree. Basin irrigation is practised where possible, but the amount of labour involved is seldom repaid by the extra berry yields obtained.

Intercropping

Young pimento plantations can be intercropped for 1 or 2 years with crops such as bananas and for a similar period with low-growing food crops such as pulses. Cattle can be herded to graze on the vegetation between trees, but cannot be left unattended.

Pests

The most important pests attacking *P. dioica* and *P. racemosa* are generally the same, varying only in regional importance. Leaves and young shoots are eaten by leaf-eating caterpillars, which occasionally occur in large numbers in a restricted area when the damage can be extensive. Caterpillars of the bagworm *Oeceticus abboti* and related species are most often recorded. Young leaves may also be badly damaged by whiteflies, *Aleyrodidae*, and the red-banded thrips, *Selenothrips rubrocinctus*. Adults of the weevils *Prepodes* spp. and *Pachnaeus* spp. also feed on leaves and their larvae damage roots. Also causing occasional damage to roots are larvae of unidentified scarab beetles, while other species have been recorded as feeding on flowers.

Borers are widespread but the damage is usually minor as severe infestations are rare; the greatest damage is due to secondary infection by pathogenic fungi. Most often noted are *Cyrtomenus*, *Cylindera* and *Neoclytus* spp. Scale insects, soft and hard, are frequently present on trees but normally do little damage. Also causing little physical damage but making harvesting unpleasant are black ants, *Crematogaster brevispinosa*, also responsible for transferring scale insects between trees.

Diseases

Similar to insect pests, the most serious diseases attack both *Pimenta* species, and also vary regionally in the degree of damage caused. Many diseases are recorded from *Pimenta* species in addition to *P. racemosa* and *P. dioica*, but in general the information is sparse, and thus the relative importance of the various diseases and control methods have been little investigated.

A damaging leaf rust caused by *Puccinia psidii* is most prevalent in areas where there are frequent fogs or heavy dews. The disease attacks young leaves, and shoots, and later inflorescences become covered by a bright yellow powdery mass of urediospores. Severe infection results in defoliation, and successive attacks may kill young trees. A wound parasite, which seldom affects young trees, is a dieback or canker locally known as fireblight, caused by *Ceratocystis fimbriata*. The disease is widespread, but can be very local; pruning and burning diseased branches is the most efficient control method. Microbiological contamination is not strictly disease control, but some of the pathogens concerned are initially contracted in the field, and thus good crop hygiene can be of major assistance in reducing the degree of infection.

Harvesting

Flowering usually coincides with the onset of the rains and fruit (berries) are picked about 3 months later, varying between seasons and regions. Harvesting is July–September in Jamaica, June–August in Central America and Mexico. Berries are ready for harvest when fully mature but green, and the state of maturity is important when intended for spice. Small or overripe berries should be removed as they detract from the appearance of a finished sample. When intended for distilling, appearance is unimportant, but too many small or overripe berries reduces the overall oil yield.

Harvesting remains basically manual with twigs bearing berry bunches broken off and collected in any convenient container; for piecework the ubiquitous 4 gallon kerosene tin is often used to assess payment. Berry clusters should not be placed in heaps on the ground awaiting collection, as soil contamination aids subsequent growth of microorganisms. Manual harvesting causes damage to trees and reduces subsequent berry yields, but no satisfactory alternative is available. Cutting bunches with secateurs or small hand-held mechanical pruners was rejected by pickers as it both prolonged the time required to fill a tin and reduced their earnings. Fruit harvesting platforms would obviate the damage caused by pickers climbing trees, but as for cloves (Chapter 6) are probably too expensive as more berries are produced in a normal year than are picked.

Berry yield not only varies widely between trees, but also between seasons and districts, and thus comparisons are of little value. In general, a healthy well-managed tree in a commercial walk will produce an average green yield of 10 kg annually over 10 years, varying between a few kg and 25 kg, as trees have a cyclical bearing pattern with about one good year in three. Trees from clonal or similar material reportedly yield more regularly and have a higher average yield, up to 25 kg, but little data is available from commercial plantations.

A considerable amount of leaf oil is produced from leaves remaining from berry harvesting, which are collected and delivered to independent distillers or cooperative stills. On specialized plantations, the leaves are mechanically harvested and loaded into trailers for delivery to the distillery. Leaf yield varies between 10 and 30 t ha^{-1}, and depends primarily on the standard of management, i.e. how many cuts can be taken per year. Oil yield from green leaves is about 1.0–1.25%.

Processing

Berries destined to be used as spice must be treated with care, as quality is assessed mainly on appearance, colour, flavour and essential oil content. Thus the emphasis in the Harvesting section on selecting mature, green berries is of great importance, as it is virtually impossible to turn poor berries into good spice. Berries destined for distilling require less care; retaining the highest essential oil content is most important.

On Jamaica harvested berry clusters are taken to a central store or drying shed and left in heaps or sacks for up to 5 days to ferment, according to local preference. Berries are then spread on outdoor drying floors, turned daily to ensure uniform drying, covered at night and if it rains, for about 5 days in sunny weather, up to 10 days if cloudy but dry. At the end of this period, moisture content is 12–14% and the yield from 100 kg green berries is 55–65 kg. The dry berries are then cleaned and bagged, and kept in a clean dry store.

In Central America, a similar system is used, but the periods of fermentation and drying are very variable since mainly wild trees are harvested and the spice is of lower quality. In Guatemala, berries may be blanched in boiling water for 10 min as this is considered to reduce the risk of contamination, but produces a paler, less attractive spice with a lower essential oil content.

It is during fermentation and sun drying when berries are exposed to the elements that microbiological contamination of the spice most often occurs, especially if the weather at these periods is wet or humid. This contamination is almost impossible to counteract, and a high microbial content results in berries being rejected for human use. They are quite acceptable for distilling.

Correctly dried spice should have a pleasant characteristic odour and as little microbiological contamination as possible. These objectives are most easily met by artificially drying the green berries under controlled conditions, but a large proportion of berries are still sun-dried and manually prepared.

Artificial drying is more common in areas where berries ripen in the wet season, and a number of simple wood-fired models are available. Solar energy dryers are more efficient, do not require a supply of solid fuel and are easy to dismantle and transport to convenient sites. However, such drying does not eliminate the possibility of contamination, and also requires a higher standard of post-harvest berry management to be effective.

Larger forced-air dryers are more suitable for permanent erection in the centres of production, and can be owned and operated by grower cooperatives or local government. Berries received can be handled under more hygienic conditions prior to drying, and data indicate that a temperature of 75°C is the maximum that can be used without loss of essential oil. When properly operated and supervised and the dried spice correctly stored, the danger of microbiological contamination is virtual eliminated. Since the international market for the spice is relatively static, artificial drying confers a major advantage as it allows the marketing of a guaranteed product with the minimum contamination.

Distillation

Leaves may be fresh, withered or dried and stored for 2–3 months prior to distilling with little effect on oil yield or phenol content. Yield from dried and fresh leaves is 0.5–3.0% and 0.3–1.25%, respectively. Modern stills in Jamaica take a charge of 1000–2200 kg, have perforated false bottoms upon which leaves are packed to prevent channelling, and distilling is by low-pressure steam for 4–5 h. Some distilleries run the distillation waters into large tanks where the oil settling out is dark but normal leaf oil, which is added to the other fractions. The combined oil is filtered and stored, preferably in stainless steel drums.

Berry oil is obtained by crushing berries immediately prior to loading, and distilling by direct steam for about 10 h usually with cohobation; total oil yield is 3.3–4.5%. Central American berries have a volatile-oil content of 1.5–3.0%, but commercial samples of Belize berries examined in London yielded 3.4–4.4% (average 3.5%). Oil from Central American berries varies considerably in organoleptic properties due mainly to origin, but there is little reliable information on differences in composition (Green and Espinosa, 1989). Local variation in Jamaica is high, with methyl eugenol content ranging from zero to greater than eugenol, and a number of chemovars apparently exist. This is supported by oil distilled from pimento collected in Belize, whose methyl eugenol content ranged between 0 and 10%. Information on oil composition of wild pimento from other sources in Central America is limited as this material is seldom distilled. Oil from Jamaican undried green berries is similar in composition to that from dried berries, but has a higher monoterpene content (Ashurst *et al.*, 1972).

Storage and transport

Whole correctly dried fruit can be stored in sacks or in small-volume bulk bins before recleaning if required, then graded and packed for export. Essential oils should be stored in full, sealed, opaque containers until required for use or shipment.

Products and specifications

The three main pimento products are the berry, whole or ground, berry oil and leaf oil. An estimate of world end-use of berry products gave domestic spice at 5–10%, the food industry at 65–70%, berry oil at 20–25% and oleoresin at 1–2%. Leaf oil is used almost wholly in the food industry.

Berry. The dried fruits, berries, become the pimento spice of commerce, and the best-quality Jamaican, which set the international standard, should be between 6.5 and 9.5 mm in diameter, with an uneven surface, approximately 13 fruits g^{-1}. They should be a uniform, unblemished, medium to dark brown colour, with a pleasant odour, characteristic of the spice, with no mustiness; the taste is spicy, never bitter, and described as a combination of cinnamon, cloves and nutmeg, hence the name allspice. Jamaican spice is considered superior, with spice from other islands, Central and South America less favoured, and this is reflected in the wholesale price. Little loss of essential oil occurs when spice is correctly stored, but prolonged storage is detrimental to both oil content and flavour, and can also allow mould growth.

An analysis of the ground spice gave (per 100 g) water 8.5 g, protein 6.1 g, fat 8.7 g, carbohydrates 50.5 g, fibre 21.6 g, ash 4.6 g containing Ca 661 mg, Fe 7 mg, Mg 135 mg, P 113 mg, K 1.0 mg, Na 77 mg and Zn 1.0 mg. In addition, the ground spice contained (per 100 g) 40 mg ascorbic acid, 0.1 mg thiamine, 0.06 mg riboflavin, 3 mg niacin, plus vitamin A 540 IU, 61 mg phytosterols, and had an energy value of approximately 1100 kJ. An oil expressed from the spice reportedly contained 54% unsaturated fatty acids; the principal acid components were linoleic 41%, stearic 36%, oleic 12% and palmitic 9%.

The major uses of ground pimento are as a flavouring in processed meats and bakery products, in pickles and in some fish products particularly raw marinated herring, but whole berries are preferred in prepared soups, gravies and sauces. It is a common ingredient of many eastern Mediterranean dishes, but seldom used in curries. For domestic culinary use, pimento is often supplied mixed with other ground spices, or labelled allspice, but to obtain the best flavour, berries should be stored whole in small sealed containers and ground when required. A major adulterant of ground spice is ground clove stems.

Whole ripe berries are an essential component of the local Jamaican drink Pimento Dram and an ingredient of the classic liqueurs Chartreuse and Benedictine. Berries are also used in alternative medicine as a carminative and stimulant, or as a thick paste to provide a plaster to alleviate rheumatism and neuralgia. A water extract of berries is used to treat flatulence.

Berry oil. The oil was formerly distilled only in the main consuming countries, but is now produced in Jamaica and other producing countries on a minor scale. Berry oil obtained by distilling dried Jamaican berries in copper or mild steel stills is yellow to brownish-yellow, but oil from stainless steel stills is much lighter. The odour is warm-spicy sweet, with a peculiar fresh and sweet top-note. The oil is placed in the warm, sweet spicy group by Arctander (1960). The main characteristics of some berry oils and specifications are shown in Table 6.4. About 60 constituents have been detected including three phenols, 11 monoterpene hydrocarbons, five oxygenated hydrocarbons, 12 sesquiterpene hydrocarbons and four oxygenated sesquiterpenes. The principal components are usually eugenol 65–90%, methyl eugenol up to 10%, β-caryophyllene 4%, humulene 2–3% and terpinen-4-ol 4,5-cineole 2–2.5% (Lawrence, 1999a; Pino, 1999c). The main constituents affecting taste and flavour are the abundance and ratio of 1,8-cineole and α-phellandrene. Berry oil is placed in the warm, spicy sweet group of oils by Arctander (1960).

Berry oil is similar in composition and odour to the ground spice and oleoresin, but lacks their flavour, and is normally blended with ground spice in processed meat products of all kinds. It is used as a flavour enhancer in jams and other fruit-based products, and the maximum permitted level of berry oil in food products is about 0.025%. The oil is also used in soaps and perfumery, especially men's aftershave lotions and colognes. Very small quantities are used in medicines to relieve indigestion and related ailments. The oil also has bactericidal, fungicidal and antioxidant activity (Kikuzaki *et al.*, 1999).

Berry oleoresin. Oleoresin is produced in major importing countries by specialist firms, and is prepared by solvent-extracting crushed spice then evaporating the solvent. The oleoresin is a brownish to dark-green oily liquid, and two grades are normally available based on volatile oil content: 40–50 and 60–66 ml per 100 g. A US specification requires a minimum of 60 ml per 100 g. Fat content depends on the solvent used and is highest with hydrophobic solvents such as petroleum ether. For flavouring purposes, oleoresin is dispersed on a neutral base, salt, dextrose, flour or rusk, and sold by weight equivalent in strength to dry spice, and the small quantities prepared are used in meat processing and canning. The oleoresin has the considerable advantage of avoiding risk of bacterial contamination and its strength and quality are more consistent. Some perfumers consider odour quality of solvent-extracted oleoresin superior to distilled berry oil and also regard water-distilled as finer than steam-distilled oils.

Leaf oil. Pimento leaf oil is produced almost wholly in Jamaica by distilling fresh or dry leaves. The oil is brownish-yellow liquid, the odour dry-woody, warm-spicy aromatic, although odour and flavour are coarser than berry oil with which it is frequently blended. It can also be mixed with clove stem and leaf oils to produce a relatively inexpensive berry oil substitute. Eugenol content of leaf oil is somewhat higher than berry oil at 65–96% (Pino and Rosado, 1996); prolonged distillation increases the eugenol content. The main characteristics of the oil are shown in Table 6.4. The oil is mainly used in meat products and confectionery as a partial replacement for berry oil, but is much more expensive. A less expensive substitute is a mixture of leaf oil with clove stem and leaf oils. The oil is placed in the dry woody, pronounced spicy group by Arctander (1960).

Monographs on the physiological properties of berry and leaf oils have been published by the Research Institute for Fragrance Materials. In the USA all pimento products have been given the regulatory status generally recognized as safe: pimento, GRAS 2017, berry oil, GRAS 2018, oleoresin, GRAS 2019, and leaf oil, GRAS 2901.

Pimenta racemosa

This is commonly called the bay or bay rum tree and the latter is to be preferred, since the former may be confused with the spice from *L. nobilis* (Chapter 3). In the Caribbean it is known as *ausu, malagueta,* wild cinnamon and bay-berry, in Cuba as *pimenta de Tabasco,* in Haiti and other French-speaking islands as *bois d'Indie,* in the Dominican Republic as *ozua* or *canelillo,* and in Surinam as bay boom. *Malagueta* is incorrect and more accurately applied to the unrelated *Aframomum melegueta* (Rose.) Schum. (*Zingiberaceae*), source of melegueta pepper, native to West Africa and introduced by the Portuguese to their South American colonies (Chapter 8). *P. racemosa* is cultivated to provide leaves that are distilled to produce an essential oil known as bay leaf oil or bay rum (Weiss, 1997). Since only a minor amount of the oil is used as a flavouring, *P. racemosa* will not be discussed in detail.

Botany

P. racemosa (Miller) J.W. Moore (syn. *P. acris* (Swartz) Kostel; *Amomis caryophyllata* (Jacq.) Krug & Urban; *Myrcia acris* Swartz.) has a number of local varieties or races that have been described, although most are of doubtful authenticity. One locally known as ausu and named *P. racemosa* var. *grisea* (Kiaersk) Fosberg has dense, finely grey or white hairs on the undersides of leaves, young twigs, flower stems and fruits. The leaves yield oil high in geraniol, methyl eugenol and isoeugenol. Another designated *P. racemosa* var. *citriodora* (syn. *P. acris* var. *citriodora*) coexists in certain areas; the leaves yield a strongly lemon-scented oil containing 33% neral and 54% geranial. A believed chemovar

of *P. racemosa* produces an anise-scented oil containing 32% methyl chavicol and 43% methyl eugenol but only small amounts of eugenol and chavicol. Indiscriminate harvesting and distillation of either with true bay rum produces inferior-quality oil.

The bay rum tree is native to the Caribbean and north-eastern South America, and has been successfully introduced into south-eastern USA, Sri Lanka, West and East Africa and Indonesia. It is frequently planted in botanic gardens and arboreta, indicating a much wider potential range, but is grown commercially for oil principally in Dominica and Puerto Rico.

Bay rum is a small to medium-sized forest tree up to 15 m, occasionally 20–25 m, with a single trunk up to 25 cm DBH, often slightly angled or grooved; the bark is greyish, sometimes scaly, and often shed giving the trunk a mottled appearance. Timber from mature trees is used for general carpentry, posts and is an excellent fuel. The leaves are large, 5–13 × 3–5 cm, dark shiny green, simple, opposite, entire, elliptic to obovate, based tapering, with the petiole short and reddish (Fig. 6.8). The leaves are leathery, glandular on the lower surface, and highly aromatic when crushed. The inflorescences are terminal on branchlets, the small fragrant white to greenish-white flowers on much branched cymes or corymbs (Fig. 6.9).

The oil content of fresh mature leaves is 1–3%, but can be 5% from specific trees. Harvested leaves can be distilled immediately, or stored for up to a week in suitable conditions with no oil loss. Leaves are normally finely chopped or comminuted, and distillation may be by water, steam or a combination of both plus cohobation. Two types of distillate are produced, one lighter and one heavier than water, later combined to produce crude oil of acceptable phenol content.

Products and specifications

The only important product obtained from *P. racemosa* is the leaf essential oil, bay leaf or myrcia oil. The oil is yellow to dark brown depending mainly on the still used; stainless steel produces the lightest coloured oils. Crude oil has a sweet rather penetrating

Fig. 6.8. *Pimenta racemosa* flowering shoot.

odour, strongly phenolic with a fresh, spicy, sweet-balsamic undernote, and a pungent, warm-spicy, somewhat bitter taste. The oil is placed in the dry-woody, pronounced spicy group by Arctander (1960). Approximately 80 compounds have been detected, including 20 monoterpenes, 19 sesquiterpenes, six aromatics (predominantly eugenol and chavicol) and 23 aliphatic compounds. The main constituents are eugenol up to 40%, myrcene up to 32% and chavicol up to 11%. The main characteristics are shown in Table 6.5. A major use for leaf oil is in hair lotions, especially Bay Rum hair dressing, with minor amounts in perfumery, and as a food and confectionery flavouring. Bay leaf oil is frequently adulterated with clove leaf, bois de rose, lime or synthetic terpenes, and oils with a phenol content above 65% are usually adulterated.

Terpeneless bay leaf oil is crude oil with low-boiling-point monoterpenes removed leaving the sesquiterpenes. It is pale straw to brownish-orange colour, with an intensely sweet, mellow, spicy-balsamic scent, and read-ily soluble in alcohol. Bay leaf oil absolutes are very pale oils produced from steam-distilled de-terpened oils, or by alcohol-washing crude oil, and lack terpenes and sesquiterpenes. Oleoresin has been produced in small quantities for use in flavourings and fragrances.

Myrtle

The genus *Myrtus* L. contains about 100 species, but only *M. communis* produces a commercial spice and essential oil. The ancient Persians regarded myrtle as a holy plant, and in Pharaonic Egypt women wore its blossoms together with pomegranate and lotus on festive occasions. To the Greeks it was known as *myrtos* and sacred to Aphrodite, by the Romans to Venus, and to the Jews as a symbol of love and peace. The Greek physician Pedanius Dioscorides of the 1st century AD prescribed extracts of the leaves or berries macerated in wine to treat lung and bladder infection, and it was also considered, incorrectly, as an aphrodisiac. Dried whole or ground berries were also considered a pepper substitute.

An extract of the leaves is used locally in north Africa to alleviate coughs and chest complaints, especially in children. In 16th-century Europe the leaves and flowers were the major ingredients of a skin lotion known as *Eau d'ange* (angels' water). Myrtle was also a component of the famous Hungary Water made for the Queen of Hungary in 1235. Around the Mediterranean, lamb and goat meat was, and still is, cooked over dried myrtle branches with their leaves to impart the flavour. Wrapping or stuffing hot meats with myrtle leaves after cooking has the same effect. In this section, myrtle refers to *M. communis* and its derivatives.

Botany

M. communis L. is known in English as myrtle, in French and German as *myrte*, in Italian as *mirto* or *mortella* and in Spanish as *arrayan* or *mirto*. Myrtle is an evergreen shrub or small tree, generally 3–7 m in height, and well branched; the branches are stiff and young branches and twigs reddish. The leaves are opposite, entire, ovate to lanceolate, 3–5 cm long, and a dark glossy green dotted with transparent oil glands. Fresh leaves are a common flavouring for Mediterranean meat dishes. The flowers are fragrant and white or pinkish. The fruit is a round, reddish-blue to violet, aromatic berry. Its major constituents are α-pinene up to 80%, 1,8-cineole up to 70% and limonene up to 20%, the ratios depending on berry origin. Berries are used to flavour bitters and certain liqueurs. Dried berries may also be used to flavour food but have little distinctive taste compared to leaves. Wood from larger trees is hard and tough, and used for tool handles, small domestic items and walking sticks.

Ecology and soils

Myrtle is native to the Mediterranean region and western Asia, frequently on hillsides, widely cultivated elsewhere as a garden plant, and is especially common in Algeria, Italy, Morocco and Tunisia (Anon., 1984), while on Sicilian hillsides it can account for over half of the vegetation. Experiments in Portugal showed that leaves maturing in full sunlight were larger, thicker and different in chemical composition to those partially shaded (Mendes *et al.*, 2001). A basic requirement is that soil must be free-draining, and if this is fulfilled myrtle will grow on a wide range of soil types. No information is available on the nutrient requirement of plants nor the effect of fertilizers on plant growth.

Cultivation

Most spice is obtained from the large areas of wild shrubs that are the basis of local leaf and oil production in the Mediterranean region. Trials to establish it elsewhere are generally on experimental stations, since cultivated myrtle is unlikely to compete with production from wild plants. In Israel the season when cuttings were taken significantly influenced the time and extent of rooting (Klein *et al.*, 2000). *In vitro* propagation on a laboratory scale, by various methods,

Table 6.5. Physical characteristics of *Pimenta racemosa* oil.

Characteristic	EOA specification	Commercial samples
Specific gravity (25°C)	0.95–0.99	0.957–0.988
Refractive index (20°C)	1.51–1.52	1.5075–1.5170
Optical rotation (20°C)	0° to −3°	−4°4′ to −2°56′
Solubility in 70% alcohol[a]	1:1–2	1:1
Phenol content (%) (mainly eugenol and chavicol)	55–65 (max)	55–65

EOA, Essential Oil Association, USA.
[a]Fresh oil only (solubility declines rapidly with age).

has been successful, as in Italy (Lucchesini and Mensuali-Sodi, 2000).

Pests and diseases

Published data on pests and diseases, their relative importance and degree of damage is lacking, mainly because the extensive areas of wild plants ensures an abundant supply of leaves, and any local damage due to either pests or disease is unimportant.

Harvesting and processing

Leafy branches are harvested from late European spring to early autumn, and the effect of various harvesting techniques on wild Sicilian bushes, including cutting at various heights, concluded that an 80 cm cut was most productive in the long term (Mulas et al., 2000). The spice is prepared by drying leaves in the shade. Highest quality leaves are pressed between boards to keep their shape.

Oil yield from local stills using chopped branches with leaves is between 0.25 and 0.55%, but twice this when only fresh young leaves, flowering shoots and twiglets are harvested and distilled; this material also yields the highest quality oil. The main constituents are normally α-pinene, 1,8-cineole and myrtenyl acetate (Lawrence, 1996c). Analysis of Russian material gave limonene, linalool, α-pinene and 1,8-cineole, in that order, as major components (Shikiev et al., 1978), but regional samples from Turkey differed from both (Akgul and Bayrak, 1989). Albanian oil had α-pinene and 1.8-cineole up to 22%, linalool up to 17% and myrtenel acetate up to 18% (Asllani, 2000).

Products and specifications

All plant parts are aromatic, but fresh leaves are the main product and dried for use as a culinary spice, or distilled to produce an essential oil used as a flavouring.

Leaves. The spice is made from dried leaves, although they lack the characteristic scent of fresh leaves. Good-quality leaves are whole and flat, a uniform greenish with a brownish tinge but not brown; the odour is aromatic and free of any mouldy notes and the taste spicy, pleasant and astringent. Arctander (1960) places myrtle in the fresh, spicy-woody group of spices. Myrtle is not a commonly used domestic spice and is mainly added to meat dishes including pork, but especially lamb when a Mediterranean flavour is required, or to soups and stews. Dried powdered berries with a slight juniper taste may also be used as spice and to flavour beverages.

Leaf oil. Oil distilled from fresh leaves is known as oil of myrtle or the preferred myrtle oil. The oil is a yellow to greenish-yellow liquid, with a fresh camphoraceous, floral-herbaceous odour, later slightly resinous, and a warm spicy taste; it is used mainly in the food industry as a flavouring for processed meats and prepared soups, and in alcoholic drinks. The main characteristics of myrtle oil from selected countries are shown in Table 6.6 (Mazza, 1983). Myrtle oil also provides the unsaturated primary alcohol myrtenol, occurring as the acetate. Myrtenol has a warm woody odour, and is used in perfumery and toilet preparations.

Table 6.6. Physical characteristics of myrtle leaf oil.

Characteristic	Spanish	Moroccan
Specific gravity (15°C)	0.911–0.930	0.900–0.912
Refractive index (20°C)	1.466–1.470	1.466–1.467
Optical rotation	+21° to +26°	+23° to +24°40′
Solubility (v/v 80% alcohol)	1:2	1:7
Acid number	2.7 (max)	ng
Ester number	63–92	ng
Ester number (after acetylation)	80–117	ng

ng, not given.

Unrelated species

Myrica pensylvanica Mirb., the wax myrtle or bayberry, is the source of bayberry tallow in the eastern USA, and *Myrica gale* L., sweet gale or bog myrtle, previously an important herb in northern Europe, is used to flavour beer; the berries of both can be toxic. Other plants with the prefix myrtle include Tasmanian myrtle or beech (*Nothofagus cunninghami* (Hook) Oerst.), red myrtle (*Eugenia myrtifolia* Sims.), grey myrtle (*Backhousia myrtifolia* Hook & Harv.) and Australian myrtle (*Backhousia citriodora* F. Meull), which produces an essential oil.

7

Orchidaceae

The *Orchidaceae* with about 700 genera and some 20,000 species is probably the largest family of flowering plants, widely distributed throughout the world with the greatest number in the tropics. Members may be terrestrial, climbing, epiphytic or saprophytic, with a great variety of floral forms due mainly to restricting pollination to particular insects. The *Orchidaceae* contains two subfamilies, the *Cypripedioideae* and the *Orchidiodeae*. The latter contains 110 species, and although a large number are grown commercially for their exotic flowers, *Vanilla* Mill. is the only genus whose members produce a spice, primarily *Vanilla planifolia*, vanilla.

Vanilla

Vanilla is indigenous to Central America from south-eastern Mexico through Guatemala to Panama. Bernal Diaz, a Spanish officer with Hernando Cortes, first recorded its use by the Aztecs when he saw the Aztec leader Montezuma drinking *chocolatl* (hence chocolate) prepared from pulverized cocoa seeds (*Theobroma cacao* L.), which are also native to the region, and ground vanilla bean. In Nahuatl, the Aztec language, vanilla is called *tlilxochitl*, and its seeds collected from wild plants were considered among the most important tributes paid to the Aztec leader. A Franciscan friar, Bernhardino de Saghun,

wrote about Mexican vanilla in the early 16th century but his work was not published until 1829–1830. He is quoted as saying that the Aztecs used vanilla seeds (pods) in cocoa drinks and as a medicine, and sold them in their markets. It is unlikely the Aztecs knew how to cure vanilla pods, and most contemporary illustrations examined appear to be of dried or partially dried pods, although some may have been fresh (E.A. Weiss, personal observation).

The Spaniards soon began shipping vanilla beans to Spain, and in the second half of the 16th century chocolate flavoured with vanilla was produced commercially. In his *Rerum Medicarum Novae Hispaniae Thesaurus*, first published in Rome in 1651, Dr Francisco Hernandez, the celebrated physician to the Emperor Philip II, described and illustrated many orchids under the collective name *Tzautlis*, including vanilla. He translated the Aztec name for vanilla, *tlilxochitl*, as black flower, but noted that it referred to the pod not the flower. Later writers were not so accurate! He added that the Aztecs used the 'seeds' as a tonic, for headaches, and as an antidote to bites of poisonous insects. It is unclear whether he was referring to fresh or dried pods, as both were available and had different uses.

Cultivation and processing of vanilla in Mexico is currently confined to higher ground inland from the Gulf of Mexico including part of central Veracruz and northern Puebla, with the main growing districts and principal trading/curing centres within a radius of about 65 km around San Jose Acateno. Vanilla cultivation was traditionally by Tototnec Indians on small plots, but few cured their own vanilla and the majority sold their crop to traders who specialized in this operation. Later production was on larger plantations, but vanilla production steadily declined from 1990, and fell to an all time low of 10 t in 1999.

Europeans also used vanilla in medicine as a nerve stimulant, as an aphrodisiac (Madame du Barry's favourite aromatic in her relationship with King Louis XV of France) and as a flavouring ingredient in tobacco and snuff. An apothecary to the English Queen Elizabeth I, Hugh Morgan, suggested vanilla as a flavouring in its own right, and in 1602 gave some cured beans to Carolas Clusius, the Flemish botanist, who described them in his *Exoticorum Libri Decem* of 1605. The English navigator William Dampier noted vanilla growing in 1676 in the Bay of Campeche, southern Mexico, and in 1681 at Bocotoro in Costa Rica. Vanilla plants were apparently taken to England prior to 1733 but subsequently lost. Reintroduced by the Marquis of Blandford at the beginning of the 19th century, plants flowered in Charles Greville's collection at Paddington in 1807, and he reportedly supplied cuttings to the botanic gardens in Paris and Antwerp.

The Indonesian vanilla plantings originated from two plants sent in 1819 from Antwerp to Buitenzorg, Java, but only one survived; it flowered in 1825 but did not fruit. However, subsequent introductions flourished, and in 1846 Mr Teysmann, Director of the Buitenzorg (now Bogor) Botanic Gardens, began systematic cultivation. West and central Java became the main producing areas and remain the most important, later joined by South Sumatra and Sulawesi, with cultivation almost exclusively by smallholders.

Rooted vanilla cuttings were also taken to Réunion, and production was by smallholders on plots up to 1 ha centred on the north-east coast. Cuttings were subsequently sent to Mauritius in 1827 and about 1840 to the Malagasy Republic. Failure of the Réunion sugarcane crop in 1849–1856, and discovery of a satisfactory method of hand pollination, increased vanilla cultivation there and on Madagascar, where cultivation is concentrated along the north-east coast around the trading centres of Antalaha, Andapa, Sambava and Vohemar. Most vanilla is grown by smallholders, few of whom cure their own beans, instead selling to specialist firms in local towns. Final sorting and packaging of the cured beans is carried out by exporting firms in Antalaha, one of the major vanilla shipping ports. The 1960s saw the introduction of comprehensive government legislation, mainly to guarantee producers a minimum price and protect them against monopoly buying, speculation and other malpractices.

Cuttings were introduced to the Seychelles in 1866 and became one of the island's earliest agricultural industries (Lionnet, 1958a,b). Exports began in 1877, expanded rapidly to reach 70 t in 1901 worth £70,000, then declined to 20 t in 1929, and is now an important cash crop only on the small islands of Praslin and La Digue. Collapse of the vanilla industry concentrated local farmer's attention on cinnamon, but tourism, as on Réunion, is now the major local industry. Cultivation began in the Comoro Islands in 1893 and it is now a leading world producer. Production, originally undertaken by Europeans, is now carried out by numerous smallholders whose individual annual output is generally below 200 kg of green beans, with curing carried out by traders. Three-quarters of production is on Grande Comoro with minor contributions from Anjouan, Moheli and Mayotte. Vanilla was cultivated as early as 1839 on the West Indian island of Martinique, and probably about the same time on Guadeloupe. Introduced to Tahiti from Manila in 1848, its cultivation developed into an important local industry, but later declined, again in favour of tourism.

Sri Lanka supplied plants to Uganda in the early 1920s but these apparently remained in the botanic garden at Entebbe. Additional cuttings were obtained in 1930, and a small plantation established.

Following the Second World War, a commercial plantation was planted near the capital, Kampala. High-quality pods were produced, but production languished during the Idi Amin era. In the 1980s the crop was re-established with official support and from 1992 other plantations added with aid from the USA. In 1999 it was officially stated that some 2000 smallholders produced about 300 t green beans, cured by six local processors. Exports could be rapidly expanded as production cost is low by world standards (see Harvesting section).

Botany

V. planifolia H.C. Andrews (syn. *Vanilla fragrans* (Salisb.) Ames.; *Vanilla viridiflora* Blume; *Vanilla mexicana* (P.Miller); *Epidendrum vanilla* L.) is frequently referred to in the literature as *V. fragrans*. In English it is known as vanilla, in French and German as *vanille*, in Italian as *vaniglia*, in Spanish as *vainilla*, in Portuguese as *baunilha* and in Indonesian as *panili*. All are basically derived from *tlilixochitl* and *tlilli* of the Aztec language Nahuatl. The genus name is from the Spanish *vaina*, little pod. The use of vanilla in the text without qualification refers to *V. planifolia*, its products and derivatives.

The basic chromosome number of the genus is x = 16, with *V. planifolia* a diploid with 2n = 32, together with *Vanilla pompona* Schiede. and *Vanilla tahitensis* J.W. Moore. Some 110 *Vanilla* species distributed in the Old and the New World tropics have been described. No cultivars of *V. fragrans* are known, and the vanilla plantations of Réunion, Mauritius, the Seychelles and the Malagasy Republic basically derive from a single cutting introduced into Réunion from the *Jardin des Plantes* in Paris (Lionnet, 1958a, b). In Indonesia, however, a number of local types are distinguished and named, with cultivar Gisting giving high pod yields and resistant to *Fusarium* spp. infection.

Crosses of *V. fragrans* × *V. pompona* were successfully grown from seed in Puerto Rico and the Malagasy Republic, and *V. fragrans* crossed with *Vanilla phaeantha* H.G. Reich. in an attempt to increase resistance to root rot caused by *Fusarium* spp. The major germplasm collection is held at the Centro Agronomico Tropical de Investigacion y Ensenanza (CATIE), Costa Rica, and includes accessions from Central America, and India is expanding its germplasm collection at the National Research Centre for Spices, Calicut, Kerala.

Vanilla is a fleshy, herbaceous perennial vine, climbing trees or other supports to 10–15 m by adventitious roots; in cultivation it is trained to a height that allows easy hand-pollination and harvesting. The long, whitish, aerial, adventitious roots, about 2 mm in diameter, are produced singly opposite leaves and adhere firmly to the support; the basal roots proliferate in the mulch, which contains an endotrophic mycorrhiza. The valeman, the outer parchment-like sheath found on tropical epiphytes, is poorly developed.

The long, succulent, cylindrical, monopodial stems, up to 2 cm in diameter, may be simple or branched, flexuose and brittle; the internodes are 5–15 cm long. Stems are generally dark green, and are photosynthetic, with stomata. An anatomical study of the stem has been published (Zhao and Wei, 1999). In the wild, stems grow to a considerable height up supporting trees, but when cultivated are pruned to a convenient height for picking pods. The leaves are large, fleshy, flat, subsessile, carried on short, thick petioles, and canalized above (Fig. 7.1). The leaves are alternate, oblong-elliptic to lanceolate, 8–25 × 2–8 cm, with the tip acute to acuminate and the base rounded. The veins are numerous, parallel, and indistinct on fresh but more obvious on dry leaves. A method of determining leaf area has been developed (Krishnakumar *et al.*, 1997), and leaf permeability and transpiration rate (Schreiber and Riederer, 1996a,b).

The short, 5–10 cm long, axillary inflorescences borne towards the end of the branches, are usually simple, rarely branched, stout, bearing 6–15 but up to 30 flowers, opening from the base upwards, and generally with only one to three flowers open together. The rachis is stout, often curved, 4–10 mm diameter, and the bracts are rigid, concave, persistent, and 5–15 × 5–7 mm at the base.

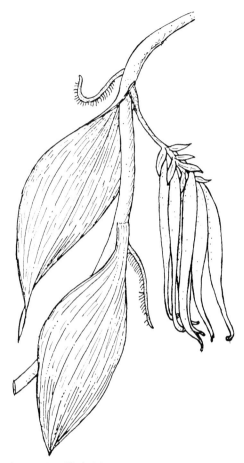

on the inner surface, bearing at its tip the single stamen containing the two pollen masses covered by a cap. Below is the concave sticky stigma separated from the stamen by the thin flap-like rostellum. Thus, self-pollination is made impossible by the rostellum, which separates the stamen from the stigma; flowers, however, can be hand-pollinated.

Vanilla normally flowers only once a year over a period of about 2 months; in Mexico this is usually April–May, in the Malagasy Republic October–December, in Réunion December–February, in the Comoro Islands November–January, and in India March–April. The flowers open early in the morning, are receptive for about 6–8 h and wither the following day. Fruit-set is highest when flowers are hand-pollinated as early as possible on a bright morning

Fig. 7.1. Vanilla fruiting stem.

The large, fragrant, pale greenish-yellow waxy flowers are about 10 cm in diameter and last 1 day; the pedicel is very short (Fig. 7.2). The ovary is inferior, cylindrical, tricarpillary, often curved, and 4–7 cm × 3–5 mm in diameter. The three sepals are oblong-lanceolate, obtuse to subacute, slightly reflexed at the apex, and 4–7 × 1–1.5 cm. The two upper petals resemble the sepals but are slightly smaller; as in most orchids the (third) lower petal is modified as a trumpet-shaped labellum, 4–5 × 1.5–3 cm at its widest, and attached to the column it envelops. Longitudinal verrucose darker-coloured papillae form a crest in the median line and a tuft of hairs in the centre of the disc. The column is 3–5 cm and attached to the labellum for most of its length; it is hairy

Fig. 7.2. Vanilla flower.

following rain. Flowers remain on the rachis if pollinated, or fall in 2 or 3 days. Nectar is secreted at the lip base, and in the wild, flowers are pollinated by bees, mainly *Melapona* spp., while humming birds may be secondary agents. Natural pollination is reportedly 1–3% in Central America, and thus hand-pollinating is essential for cultivated vanilla.

The Belgian botanist Mr Charles Morren of Liège, Belgium, first hand-pollinated vanilla in 1836, but it was Mr Edmond Albius of Réunion who in 1841 developed the technique still in use. The Albius technique uses a sliver of bamboo or similar material about the size of a toothpick. The flower is held in one hand and the labellum pushed down with the thumb releasing the column. The stamen cap is removed by the bamboo held in the other hand to expose the pollinia. The flap-like rostellum is then pushed up under the stamens with the bamboo and, by pressing with the thumb and finger, the pollinia are brought into contact with the sticky stigma to which the pollen mass adheres. Alternative methods of fertilizing flowers have been only partial successful, including the application of growth regulators, such as IAA, IBA, 2,4-D and dicamba.

The fruit or pod, incorrectly but commonly known as the vanilla bean, is aromatic on drying. The pod is pendulous, narrowly cylindrical, faintly three-angled, very variable, and $10–25 \times 0.5–1.5$ cm in diameter (Fig. 7.1). When ripe it contains a mass of minute, round, black seeds, shed by the pod splitting longitudinally. Seeds are difficult to germinate, require special conditions, and were originally considered sterile (see Cultivation). Mature, fresh, green pods contain about 20% water, and per 100 g of dried pod contain, on average, protein 3–5 g, fat 11 g, sugar 7–9 g, fibre 15–20 g, ash 5–10 g, vanillin 1.5–3.0 g, a soft resin 2 g, and an odourless vanillic acid.

The period to pod maturity is variable, depending mainly on region, from 4–9 months, and maturing pods change colour from green to yellow, or yellowish-brown, depending on regional strain. In commercial production, the pods are harvested before they are quite ripe, when they have none of the familiar vanilla aroma or flavour, which develops on curing (see Harvesting). However, vanillin content is often highest when pods are fully ripe. Pods contain vanillin plus other aromatic compounds derived from the reaction between an enzyme and a glucoside during curing, with vanillin the most abundant. Research on suspension cultures of vanilla indicated that cinnamic, but not ferulic acid is a precursor of vanillic acid.

The flavour of pods from different origins varies due to climate, soil, degree of ripeness at harvesting and method of curing. Indonesian cured pods contain approximately 2.75% vanillin, Mexican 1.75%, Sri Lanka 1.5%, Tahiti 1.7%, and the last also contain heliotropin, which gives the distinctive flavour. Production of secondary vanilla metabolites, particularly vanillin, from *V. planifolia* cell suspension cultures developed from various plant parts using biological technology remains experimental (Havkin-Frenkel *et al.*, 1997). The epicuticular waxes contain β-dicarbonyl compounds, not previously identified (Ramaroson-Raonizafinimanana *et al.*, 2000).

Ecology

Vanilla has been widely introduced into other tropical countries where climatic conditions are suitable, but has seldom become commercially important. Although plants generally grew well in the Old World tropics, fruits were often not produced because the special natural pollinators were absent (see Botany). Attempts to develop vanilla plants adapted to cooler climates by using cell cultures have not been productive, although genetic manipulation could be more successful.

Wild vanilla is usually found in tropical lowland forests between 20°N and 20°S of the equator, generally below 600 m but up to 700 m in Indonesia. Partial shade is necessary, since dense shade and prolonged direct sunlight adversely affect growth and suppress flowering. Shade is usually provided by trees supporting the vines, or trellis rows can be interplanted with bananas or maize, which also act as windbreaks to shelter the brittle vines.

Vanilla thrives in hot, moist, insular climates, with a temperature range of 21–32°C, averaging around 25–27°C. A rainfall of 2000–2500 mm annually is usual, evenly distributed except for two drier months to check vegetative growth and induce flowering, and a dry period at harvest is an advantage. Regions with prolonged heavy rains or a dry season should be avoided. The monthly rainfall and maximum and minimum temperatures in the Malagasy Republic are given in Table 7.1 (Bouriquet, 1954).

Soils

The most suitable soil for vanilla is a friable clayey loam covered with a thick layer of humus or mulch in which roots can proliferate. Adequate drainage is essential as plants are very susceptible to waterlogging for even short periods at all growth stages. Thus, gently sloping lands are an advantage for commercial cultivation. A neutral to slightly acid soil reaction is generally preferred, but vanilla will grow well in soils with a range of pH 6.0 to pH 7.5; lower than pH 6.0 may require liming to counteract the acidity and this is frequently not possible. The effect of soil amendments on open-grown and protected vanilla has been little studied; on Réunion, those tested had no significant effect on pod aromatic compounds, but highly significant differences in plant growth and flowering (Odoux *et al.*, 2000).

Fertilizers

A major requirement is to maintain a thick mulch around plants, and its decomposition is the most important source of nutrients. Incorporating locally available material in the mulch, including rock phosphate, guano, bonemeal, fishmeal, etc. is considered beneficial. It is also believed that animal residues or farmyard manure (FYM), especially chicken manure, should be stored for a period and not applied 'raw' as this adversely affects plant growth and pod aroma.

Little data is available on the effects of chemical fertilizers on vanilla. These should only be applied if determined by formal field trials, since Mexican growers state that there was little or no increase in pod yield from their use, and where trees were the supports, encouraged a spreading and too dense canopy (E.A. Weiss, personal observation). However, an NPK mixture or superphosphate applied to holes prior to planting cuttings is becoming common in all vanilla-growing regions and also a top dressing of ammonium sulphate after cuttings begin to grow. However, acidic fertilizers can decrease the soil pH and adversely affect plant growth. Application of Ca and K reportedly improved yield and pod quality, but no supporting data are available.

Table 7.1. Rainfall and temperature in the Malagasy Republic (from Bouriquet, 1954).

	Rainfall (mm)	Maximum temperature (°C)	Minimum temperature (°C)
January	215	32.2	22.5
February	204	31.8	23.1
March	249	31.4	23.0
April	235	30.2	22.3
May	110	29.0	21.1
June	140	27.6	19.5
July	136	26.4	18.8
August	113	26.7	18.6
September	85	27.5	19.1
October	50	29.0	20.0
November	79	30.4	21.3
December	181	31.3	22.4
Total	1797		

Cultivation

Smallholder vanilla was usually grown in semi-cleared forest with vines trained up individual trees, simulating the plants natural habitat, but demand expanded cultivation and most vanilla is now produced on special plots (Fig. 7.3). Although some vanilla was produced on relatively large plantations these are now uncommon, and direct overseas or local investment in vanilla production is basically confined to the various types of artificial processing. Smallholders also generally have a second occupation, as there is a time-lag of several years between initial planting and first harvest, and thus growing vanilla is not a full-time occupation.

Cultivated vanilla requires support, either living or dead, and wooden posts or trellises must be chemically treated as protection against termite and fungal attack. Living supports should be planted well in advance of vanilla cuttings. The ideal tree has been described as quick-growing from large cuttings, providing light, chequered shade, having a rough bark and many low branches so vines can be trained to give easy access to pods, and being well rooted to support vines in strong winds, as well as being easily pruned. It is an advantage if the tree chosen can also provide a saleable product, and the following have been either utilized or suggested: Liberian coffee (*Coffea liberica* Hiern.), avocado (*Persea americana* Mill.), annatto (*Bixa orellana* L.); hog plum (*Spondias mombin* L.) and white mulberry (*Morus alba* L.). The two trees most commonly used in the Malagasy Republic are the physic nut (*Jatropha curcas* L.), which grows rapidly from seed, and *Casuarina equisetifolia* L. In Uganda (and elsewhere) *Gliricida sepium* (Jacq.) and *Bauhinia* spp. are favoured. Other trees include calabash *Crescentia cujete* L., *Croton tiglium* L., *Albizia lebbeck* L., *Dracaena* spp., *Erythrina* spp. and *Pandanus* spp.

Commercial vanilla is propagated by stem cuttings although seeds are used for special purposes or breeding crops, as noted later. Cuttings should be from healthy vigorous plants, may be from any part of the vine, and should preferably be taken during dry periods. For production of cuttings, a small plot of mother vines is recommended, which should be protected against pests and diseases and not allowed to flower.

Length of cuttings is usually determined by the amount of planting material available. Cuttings 90–100 cm long are preferable, as cuttings of 30 cm take 3–4 years to flower and fruit. Some growers use cuttings up to 3.5 m long with their tops tied to supports,

Fig. 7.3. Smallholder vanilla plantation – Uganda.

since such cuttings will fruit in 1 or 2 years. Cuttings are succulent and can be stored for up to 2 weeks in cool shade. Basal leaves should be removed and the cutting inserted into the existing layer of soil and humus, or into a prepared pit covered with a thick mulch. When using short cuttings, at least two nodes should be above ground. Cuttings should be tied to supports until aerial roots have obtained a firm hold. Cuttings can also be started in nursery beds or pots and planted out when rooted. In India, a mixture of vermicompost or well-rotted coir was the best potting medium (Siddagangaiah et al., 1996). With healthy material and care, the strike rate is usually high. When transplanting, care is necessary to avoid damage to roots or the lower stem, since such wounds provide entry to pathogenic fungi. Shoots appear on cuttings 4–6 weeks after planting.

Vanilla can also be grown from seeds, but like all orchids, this is a long, complicated process and not recommended unless specialized facilities are available. Full details are available in the literature. Briefly, seeds are germinated, which usually occurs after 4–8 weeks, in flasks containing a sterilized mixture, and the minute plantlets initially transplanted into larger flasks and again every 2 months as they develop. Seedlings can be planted out after hardening off in about 6–8 months. Exposing seeds to greenhouse conditions before incubating in the dark at 32°C results in a higher percentage germination, and treatment with various growth regulators in India accelerated germination and seedling growth (Mary et al., 1999). Laboratory data indicates that dormancy in vanilla seeds may be several years depending on origin and treatment of seeds. Since most Old World vanilla plantings originated from a single clonal source there is little variation, and this discouraged conventional breeding methods. However, the techniques currently available for growing orchid seedlings allow the interchange of regional vanilla strains, and it is thus relatively easy to introduce new genetic material. Many are already in use (Ganesh et al., 1996; Mathew et al., 1999; Pett and Kembu 1999). Plantlets have also been produced by several in vitro techniques (George and Ravishankar, 1997), and a simple method developed in India (Geetha and Shetty, 2000).

Cuttings are planted about 3 m apart at the foot of the supporting trees or poles, although a closer in-row spacing of 1.5 m in 3.0 m rows may be chosen. For some commercial crops, row width is regulated by the cultivating machinery. Under good conditions plants can grow from 50 to 100 cm per month, and vines must be trained to grow horizontally at a convenient height for pollinating and harvesting. Vines are usually carefully twisted round the longer branches of supporting trees or top of the trellis so they hang down. The top 10 cm is usually pinched out 6–8 months before the flowering season to encourage inflorescences on hanging branches. Vanilla usually begins flowering in its third year after planting, earlier if large cuttings are used, and a small crop can be harvested. Maximum flowering occurs in 7–8 years and may continue at this level for 4–5 years before gradually declining, and after a further decade vines are usually unprofitable and replaced.

Flowers are hand-pollinated as described, but only those flowers on the lower side of the raceme are pollinated to ensure that fruits hang vertically to produce straight pods. Experienced operators, mainly women, can pollinate between 1500 and 2000 flowers daily. Usually only one flower per day opens in each inflorescence and is receptive for about 8 h; thus, most pollination is in the morning. If pollination is successful, the perianth withers, but remains attached to the ovary, which then increases in size. Unpollinated flowers fall next day. Pollination continues until the required number of fruits has set, up to 3 months, and the remaining buds are then clipped off.

The number of inflorescences and flowers, and the number that are pollinated and allowed to produce mature beans, depends upon vine growth. On vigorous plants, eight to ten flowers on 10–20 inflorescences are pollinated, but only four to eight pods on each raceme allowed to mature. Damaged or malformed pods are removed. Pods reach maximum length about 6 weeks after pollination, and take another 7–9 months to mature. Well-grown second year but usually third-year vines are allowed to bear 50–100 pods, and around 150 for the fourth and subsequent years. More than this usually reduces pod size, final weight and value of cured pods.

Weed control

The superficial roots of vanilla plants preclude normal manual or mechanical cultivation, and in most plantations slashing weed growth as necessary is considered sufficient, with residues added to the mulch. Removing climbing weeds is also recommended, since their often fast and rampant growth can choke young vanilla plants. Herbicides are normally used only on larger commercial plantations or where vanilla is ridge-planted. A major problem is the very superficial root system, which can quickly absorb the compounds applied.

Irrigation

Water use by vanilla has been little investigated, and those growers who do irrigate their crops mainly rely on their own experience. A thick mulch together with adequate shade is usually sufficient, and to grow vanilla wholly under irrigation is unlikely to be profitable at present prices and commercial yield of around 800 kg ha^{-1}. Neither are likely greatly to increase in the short-term. If vanilla cannot be rain-grown, then in general the environment is unsuitable.

Where water is available, vanilla benefits from applications when rainfall is insufficient, especially during the reproductive and fruit maturation phases, or to maintain humidity at flowering. Water is generally supplied by surface irrigation via furrows between rows, or by a trickle system, which conserves water and prevents the overwatering so detrimental to vine growth. Overhead irrigation is feasible if the system can be used on other crops, thus spreading the cost. An advantage is that water applications can be carefully adjusted for amount and duration or the system can be used to distribute pesticides and fungicides, particularly if computer controlled.

Intercropping and rotations

Intercropping vanilla other than with the crops earlier noted as windbreaks is usually impractical, and the total return from intercropping vanilla is frequently less than from vanilla alone. However, smallholders may sow catch crops of low-growing beans or vegetables for domestic use, which can mature while vanilla stems are elongating. Planting vanilla to grow up coffee, mango, avocado or similar fruit trees may be successful, but the agricultural operations necessary to maintain the tree plantations can cause damage to the vines.

Since the life of a vanilla plantation can be 20–25 years, lands are often left fallow between replanting, as an intervening crop must be suitable for planting between the rows of supports. The fallow period also allows the use of herbicides, or deep interrow cultivation to bury persistent weeds, and reduces the incidence of root diseases in the subsequent vanilla crop (see Diseases).

Pests

The most important pests damage flowers and fruit, with those attacking roots and seedlings more easily controlled. The emerald bug, *Nezara smaragdula*, cosmopolitan throughout the tropics, lays its eggs on the leaves and stalks of vanilla and the larvae suck the sap of buds and stalks. Another bug, *Memmia vicina,* has been reported as damaging vanilla in the Malagasy Republic, while *Trioza litseae* is a major pest in Réunion, where it punctures buds and flowers, producing spots of decay.

Two weevils attack flowers, buds and other plant parts in the Malagasy Republic, *Perissoderes oblongus*, and *Perissoderes ruficollis*, which also occurs in the Comoro Islands. An ashy-grey weevil, *Cratopus punctum*, which bites holes in the flowers and can destroy the column, also occurs in the Malagasy Republic, Réunion and Mauritius; in Puerto Rico, a black weevil, *Diorymerellus* spp., causes injury to vanilla shoot tips. The small lamellicorn beetles, *Hoplia retusa* and *Enaria melanictera*, also damage flowers in the Malagasy Republic. Another beetle, *Saula fer-*

ruginea, can cause considerable damage to vanilla leaves in India.

The caterpillars of two moths cause damage in Réunion, *Conchylis vanillana* and *Phytometra aurifera*, which has a wide tropical distribution and host range, and a woolly bear caterpillar, *Ecpantheria icasia*. Insects causing minor injuries are a small leaf-tyer, *Platynota rostrana*, in Puerto Rico, an aphis, *Cerataphis lataniae*, which also occurs in the Seychelles and Réunion, an earwig, *Doru* spp., and unidentified mites reported from vanilla roots, which in Uganda were considered a possible vector of *Fusarium* root rot.

Snails and slugs are frequently serious pests of vines, feeding on roots, leaves and apical shoots and cause most damage during warm damp weather; for example, *Thelidomus lima* occurs in Central America generally but especially in Puerto Rico, and the widespread *Achatina fulica* with a wide host range can become a problem in the Malagasy Republic and Uganda. Unidentified snails, which appear annually in large numbers at the beginning of the rains in Indo-China, cause little damage, but are a major factor in the spread of the fungus *Nectria* spp., as they transfer and deposit its spores in wounds caused when feeding on foliage. A slug, *Veronicalla kraussii*, is damaging in Central America generally; it proliferates in the mulch and damages plants by eating buds, leaves, shoots and immature fruits.

Chickens are a pest to smallholder crops near villages by scratching among the mulch and exposing and breaking roots, thus allowing entry of pathogenic fungi, while the ubiquitous goats eat the lower foliage and cause more damage by pulling down vines.

Diseases

Vanilla is infected by a range of diseases with many causing minor damage. Only the most important are described, although a specific pathogen may be damaging in a specific crop or year. A most serious disease is anthracnose caused by *Glomerella vanillae*, which occurs in all vanilla-growing areas. Initial symptoms are small lesions on the stem apex and leaves; subsequently plants wilt, fruits turn black at the tips and mid-section and quickly fall. In some instances, roots are attacked initially causing plants to wilt and die. Prolonged humidity or rain, poor drainage, excessive shade and close planting favour the disease. Sunburn can occur on poorly shaded vines and is often mistaken for damage due to fungal infection. Tip blackening and dieback, without the other symptoms, are often due to a soil moisture shortage.

A root or stem rot due to infection by *Fusarium oxysporum* f. sp. *vanillae* occurs in many vanilla-growing regions and can be so damaging that it limits vanilla production, for example, in Puerto Rico and parts of Indonesia. The most obvious symptoms are a cessation of shoot growth and the increased production of many aerial roots, which frequently die on reaching the soil surface. Stems and leaves become flaccid, begin to shrivel and eventually droop and desiccate; on examination, roots are brown, dying or dead. Water stress, lack of major nutrients, too little shade and too many fruits induce the disease and exacerbate the damage. Heavy mulching and adequate shade during the dry season materially reduce the incidence of the disease. *V. pompona* and *V. phaeantha* are reportedly resistant and hybridize with vanilla. Another stem and shoot rot is due to *G. vanillae* and is reportedly troublesome in Central America, Mauritius and Sri Lanka. The initial symptom is dieback of young shoots.

Probably the most damaging leaf disease is brown spot caused by *Nectria vanillae*. Leaves become covered in brownish-black spots, and finally the whole plant can be infected and die. Overmature and unhealthy plants are most susceptible. Another damaging leaf disease is due to *Phytophthora* spp.; *Phytophthora jatrophae* is found in the Malagasy Republic where the physic nut *J. curcas* L., a support tree for vanilla, is a major alternative host, while *Phytophthora meadii* has recently been identified in India (Bhai and Thomas, 2000). A less-damaging regional leaf rust is *Uredo scabies* in Central America, especially Colombia. Mildew due to *Peronospora* spp. commonly occurs, particularly downy mildew caused by *Peronospora parasitica*, and is most damaging following periods of high humidity.

Diseases due to viral infection have become more important, probably due to more efficient detection methods rather than a general increase in their incidence. The most often reported in Réunion is the cymbidium mosaic virus and potyvirus, which had infected all shade houses, but the level of damage varied between areas (Benezet *et al.*, 2000). Two potyviruses isolated from vanilla on Tonga demonstrated an ability to improve resistance to vanilla necrosis potyvirus (Pearson *et al.*, 1999).

The following diseases have been reported as damaging vanilla but some may have been incorrectly identified: *Atichia vanillae* in Tahiti, *Macrophoma vanillae* and *Pestalotia vanillae* in Brazil, *Physalospora vanillae* in Java and *Vermicularia vanillae* in Mauritius.

Harvested beans are attacked by moulds due to infection by *Penicillium* and *Aspergillus* spp., and although this may occur in the field on beans harvested before they are mature or awaiting transport, infection generally occurs during conditioning and storage. The two types are distinctive; one is initially white later becoming green, the other is black. Infection begins at the base and spreads to the tip, when the whole bean becomes wrinkled and dry, with an undesirable smell. This odour is almost impossible to eliminate and downgrades beans. To reduce damage, beans must be kept dry following killing, and rigorously inspected during curing and storage to remove those infected. Sweating and sun-drying increases the incidence of moulds, as does a general lack of cleanliness and ventilation in curing rooms. Mould-damaged beans can be salvaged by cutting out the infected portions and the remainder sold as cuts.

Harvesting

Pods mature in 9–12 months depending mainly on the degree of shade, seasonal conditions, etc., but the length of time is reasonably constant in a specific area. In Uganda pods matured in 9 months, which coincided with the beginning of the dry season, reduced the danger of pod rots and facilitated picking. Harvesting, depending on the season, is usually November–January in Mexico, in Madagascar June–October, in Comoro April–June, in Réunion June–August and in Tahiti April–July.

Pods are ready for picking when they are fully grown and tips turn yellow, and as it is essential to harvest at this stage to obtain the highest quality pods, daily picking during the season is normally necessary. Pods are carefully cut or clipped from the vine without damaging the skin as this causes problems during curing. Theft of mature pods is common and can be a serious source of loss to smallholders (see also pepper, Chapter 8); some growers scratch beans with their own marks to establish ownership. Vanilla plants contain calcium oxalate crystals, which cause dermatitis in the susceptible.

Pod yield is very variable mainly depending on the skill of the grower, and one report into the industry stated, 'The chief cause of pod loss by smallholders and thus a low average yield, is incompetence and mismanagement, producing 200 to 300 kg ha^{-1} of cured pods when the potential average was at least twice this.' Commercial plantations have achieved an average of 1000 kg ha^{-1}, with highest yield from the fourth to the eleventh year after establishment. However, a relationship exists between high initial yields and longevity, with high yield correlated with a shorter plantation life, probably due to overcropping vines.

Following harvest, old and diseased stems are removed, vines pruned of unwanted growth, and supports given the necessary maintenance. Shade trees should also be pruned to induce branching at the required height for vine support, and the canopy thinned to allow about 50% shade. Where a cooperative curing facility exists or a nucleus plantation system operates, smallholders are normally required to deliver pods to the plant within 4–6 days of picking. Picked pods should be cured within a week.

The cost of establishing 1 ha of vanilla containing 1500 plants in Uganda in 1998–1999 was between 50,000 and 60,000 Uganda Shillings (USh) per plant (roughly $30–35), including planting material, supports and initial heavy mulch. Excluding labour (whose opportunity cost was virtually nil), total production cost averaged USh50 kg^{-1}, total processing and packaging cost USh25,000 kg^{-1}, depreciation and other fixed costs about USh 2000 kg^{-1} and the export FOB price was USh 30,000 kg^{-1} ($16–18 kg^{-1}).

Processing

Methods used to cure vanilla pods, hereafter called beans, are basically similar and use the following steps:

1. Killing or wilting. This stops vegetative development in the fresh bean and initiates enzymatic reactions responsible for aroma and flavour; beans become brown.
2. Sweating. The temperature of killed beans is raised to promote enzymatic reactions and initial drying to prevent harmful fermentation; beans become a deeper brown, supple and the aroma is detectable.
3. Drying. This involves slow drying at ambient temperature, usually in the shade, until the beans are reduced to about one-third of their original weight.
4. Conditioning. Beans are stored in closed boxes for at least 3 months to permit full development of the aroma and flavour.

The various traditional procedures for curing vanilla beans continue in use among smallholders, but large producers use more modern methods. The most important traditional methods developed in Mexico and used on the Indian Ocean islands producing Bourbon vanilla are described below; others were basically local variants.

Mexican processing. When the present system of processing was first practised is unclear, but the traditional method of sundrying was partially replaced by oven-wilting in the mid-19th century. Both methods are used by specialist curing firms, and processing by either takes 5–6 months before the product is ready for market. Since Tahiti vanilla pods are indehiscent, the beans are not killed artificially but are harvested when the tip of the bean turns brown. Curing is carried out by specialist firms using a very similar method to the Mexican.

SUN WILTING. Following arrival at the curing house, fresh beans have the peduncle removed, and are sorted by maturity, size, whole or splits into grades, which cure at different rates. Beans that are already beginning to darken are removed, wiped with castor oil and cured separately. Beans are killed the day following sorting by spreading on dark blankets resting on a cement floor or wooden racks. In the afternoon when beans have become too hot to handle, they are covered by folding the blanket. Later in the afternoon, while still hot, the thick ends of the beans are laid towards the centre of the blanket, rolled up, and taken into a shed and placed in blanket-lined, air-tight mahogany boxes to undergo their first sweating. Blankets and matting are placed over the sweat boxes to prevent heat loss. After 12–24 h the beans are removed and inspected. Most will be a dark-brown colour indicating a good killing, while those that remain green or are unevenly coloured are separated and oven-wilted.

Killed beans are then alternately sunned and sweated. Beans are sunned on blankets for 2–3 h during the hottest part of the day, then stored on wooden racks in a well-ventilated room. To ensure fairly rapid initial drying, beans are sunned virtually every day with overnight sweatings until they become supple, in 5–6 days. Subsequently beans are sorted approximately into the various grades and unsuitable beans rejected, followed by further but not daily sunnings, as too rapid drying is considered to reduce their quality, with 4–8 sweatings. These processes usually take 20–30 days after killing, when the majority of beans become very supple, resemble the final product, and are ready to be dried very slowly under cover. This takes approximately 1 month, during which beans are regularly inspected and those considered dry enough are removed for conditioning. Depending on the weather, sweating and drying operations may take up to 8 weeks from when beans are killed.

Beans ready for conditioning are straightened by drawing them through the fingers, which also spreads the oil exuded during curing and imparts the characteristic lustre. Beans are then tied with black string into bundles of about 50, and the bundles wrapped in waxed paper and placed in metal conditioning boxes also lined with waxed paper. Conditioning lasts for at least 3 months, during which beans are regularly inspected and those not developing the required aroma returned for additional sunnings and sweatings.

The regular inspections during conditioning also allow treatment of any beans developing mould (see previous Disease section). A common treatment is to wipe affected beans with 95% alcohol; treatment with dilute formaldehyde solutions is banned by the US Food and Drug Administration and prevents entry to that market. Alternatively, slightly mouldy beans are treated and then wiped with the oily secretion of mature beans; very mouldy beans are often immersed in hot water for an hour.

OVEN WILTING. This takes place in a specially constructed brick or cement room, the *calorifico*, measuring about 4 × 4 × 4 m with a small access door, solid floor, wooden racks along the walls and a furnace stoked from outside. The beans to be killed are made into bundles of about 1000, rolled in a blanket, then covered with matting to form a *malleta*. These are moistened, placed on shelves in the *calorifico*, and more water is poured on to the floor to maintain a high humidity, the door closed and the furnace lit.

The temperature in the *calorifico* steadily rises for about 24–28 h when it reaches 70°C, and is maintained at this level for a further 8 h when the *malletas* are removed. If the temperature cannot be raised above 65°C then the period is extended to 48 h. After removal, the matting is stripped from the *malletas*, the blanket-wrapped beans placed in sweating boxes, and 24 h later the beans are removed and inspected. Beans are then treated as described under sun wilting. If the weather does not permit outside treatment for 3 days, beans are returned to the *calorifico* for a further processing.

During sweating by whatever method, some beans develop a creosote-like aroma, which is impossible to eliminate. This aroma generally becomes apparent early in the sweating process and results from an abnormal fermentation due mainly to improper storage of fresh beans prior to killing. The aroma can be avoided by keeping fresh beans in small heaps in well-ventilated storage, and curing as soon as possible.

Bourbon processing. Bourbon vanilla is the trade name of beans from the former French Indian Ocean colonies, processed by a technique first developed on the island of Réunion (Isle de Bourbon). Vanilla production of this type is now carried out by specialist firms, primarily on Madagascar and Comoro, with a smaller production on Réunion, Vanuatu and Tonga. Bourbon vanilla usually has a higher moisture content than other origins, frequently with a white crystalline coating. A study of the traditional drying system on Réunion found that vanillin content rose during the curing process until drying began, remained steady, then decreased during conditioning (Odoux, 2000). Strict control of these processes is necessary to retain a high vanillin content and thus high quality.

Since most Bourbon type is now produced on Madagascar, the local method is described. Following arrival at the curing factory, beans are sorted according to maturity, size, whole and split beans. Perforated cylindrical baskets are loaded with 25–30 kg beans, and plunged into water heated to 63–65°C. Those beans that will become the top three grades are immersed for 2–3 min, smaller and split beans for less than 2 min. Beans are then drained, wrapped in dark cotton cloth and placed in cloth-lined sweating boxes for 24 h, then removed and those not killed are rejected.

Killed beans are spread on dark cloth resting on raised, slatted platforms, and after 1 hour of direct exposure to the sun, the edges of the cloth are folded over the beans to retain the heat. The covered beans are then left for a further 2 hours and then rolled up and returned to the store. This procedure is repeated for 6–8 days until the beans become quite supple. Beans are then dried slowly on racks in a well-ventilated room for 2–3 months, inspected regularly, and sufficiently dried beans are removed for conditioning. Where, or when, sun-drying is not possible, curing ovens fired to 45–50°C are used. Bean conditioning requires about 3 months, and the Bourbon process 5–8 months. On average, 3.5–4.0 kg green pods give 1 kg beans.

These processes are labour- and time-consuming, and subsequently more efficient and rapid methods were investigated. The initial trials were carried out at the US Department of Agriculture Federal Experiment Station, Mayaguez, Puerto Rico. The Station de la Vanille of the Institut de Recherches Agronomiques, Antalaha, Malagasy Republic, also investigated artificial drying methods and later introduced a hot-air drying system capable of treating 3 t fresh vanilla per day, and producing 40 t dry vanilla in one season.

A completely different system to produce artificially dried cut vanilla has been developed and patented by the American spice firm, McCormicks and Co. of Baltimore. This involves cutting fresh beans into 1 cm lengths (cuts), killing and sweating on trays in an oven at 60°C for 70–78 h; cuts are then put through a rotary hot-air drier at 60°C to reduce moisture content to 35–40%. Cuts are then forced air-dried in layers about 10 cm deep on perforated trays at ambient temperature down to 20–25% moisture content. The equipment in contact with cuts is stainless steel and the oven heated by a hot-water jacket. In 1965, McCormicks set up a plant in Uganda in cooperation with the Uganda Company. The curing operation takes about 4 days, and five men handle 1 t green beans per week. Treated beans are shipped in sealed drums to the USA for extraction. McCormicks later established a similar cooperative venture in Java, whose beans have a high vanillin content and are frosted.

An increasing amount of beans entering consuming countries is used to produce vanilla extract, and the appearance of the beans is thus of no great importance. Larger extracting firms are encouraging production of artificially dried cut vanilla possessing good aroma and flavour plus a high vanillin content. Some producers now prepare their entire output to suit these requirements, rather than continuing the traditional methods. As extract is apparently gaining market share, this trend could accelerate.

An allergy previously noted, vanillism, sometimes occurs in persons working with vanilla especially during curing and conditioning. The allergy is caused by contact with the latex exuded from cuttings or stems, which causes skin inflammation and more severe injury to eyes. During curing, symptoms include skin eruptions, headaches, fever and stomach disorders. Similar symptoms have been recorded in persons handling synthetic vanillin, but it has been determined that vanillin is not the principal active agent. The effects usually disappear within 2–3 days and resistance appears to build up as a person becomes used to handling vanilla. Very susceptible persons should not handle the pods or be involved in processing.

Storage and transport

Following arrival at a store, beans are inspected, as they often become infested by mites during conditioning or transport, which produce small holes in beans. Minor attacks can be cured by sunning, or more serious infestations by fumigation. Clean beans are then sorted and graded to local export requirements, and the following is a general description of the procedure as applied to Bourbon beans; details of all regional procedures is considered unnecessary, but are fully described in the literature.

Madagascar, Comoro and Réunion vanilla is classified into five main grades of whole and split beans, plus a bulk category, cuts. Grades of whole and split beans are also sub-divided according to size, and average length for the top five grades in Réunion is 18–20 cm with a maximum of 26 cm, for mainland Madagascar is 16–18 cm, maximum 23 cm, while for Nossi-Be and the Comoro Islands it is 14–16 cm, maximum 22 cm.

Beans are first sorted to separate those below 12 cm, then whole and split beans are classified into grades according to aroma, moisture content and appearance. The five main categories for whole beans are:

- Extra. Whole, supple, unsplit beans, free of blemishes, possessing a uniform chocolate-brown colour and an oily lustre; aroma clean and delicate.

- First. Similar to Extra grade but less thick and uniform appearance.
- Second. Somewhat thinner beans with a chocolate-brown colour but with a few skin blemishes and good aroma.
- Third. Thinner, more rigid, beans with a slightly reddish chocolate-brown colour; aroma fair.
- Fourth. Rather dry beans with a reddish colour and numerous skin blemishes; aroma ordinary.

Splits are sorted into categories corresponding to those for whole beans. Foxy splits are thin, hard and dry short types with a reddish-brown colour. Bulk refers to the beans remaining after sorting and grading and consists of cuts, broken and very short beans, and those discoloured and lacking aroma.

Bourbon vanilla may be also be sold in bulk as a head-to-tail lot, and consists of 50% of top qualities. Bourbon vanilla tends to have a somewhat higher moisture content than the corresponding Mexican grades and it is often frosted with vanillin crystals.

After sorting, beans used to be tied in bundles containing 70–100 beans weighing 150–500 g; 20–40 bundles (about 8–10 kg) were packed in boxes lined with waxed paper and six boxes were fitted into a wooden case for shipment. While this system of bundling has been retained, wooden boxes have generally been replaced by double, corrugated cardboard cartons. Bulk beans are not bundled before packaging.

Beans for export may also be sorted into internationally recognized grades, distinguished by size, colour, appearance and odour. Individual countries may have their own grades, but in general the highest quality are similar, although lower and mixed grades may differ. Excepting very specialized end use or customer preference, the better grades are generally interchangeable and most end-products can utilize beans from any of several origins with little obvious difference. To discuss these grades in detail is unnecessary, since they do not directly affect producers, and traders are fully aware of the requirements.

Products and specifications

The vanilla vine is source of a number of products, but the most important is the cured pod, the vanilla bean of commerce; it also produces an extract, an oleoresin and various derivatives. Vanilla is notable among spices for its rapid acceptance and popularity following its first introduction to Europe, and it remains one of the most widely used flavouring materials. While most world usage of vanilla flavour is satisfied by the relatively inexpensive synthetic vanillin, natural vanilla remains in demand. In the UK and USA, extracts are preferred by industrial and domestic users, but other European domestic consumers prefer beans or powdered vanilla.

Beans. High-quality beans are long, fleshy, supple, very dark brown to black, somewhat oily in appearance, strongly aromatic and free from scars and blemishes (Fig. 7.4). Low-quality beans are usually hard, dry, thin, brown or reddish-brown, with poor aroma. Moisture content of top-grade beans is between 30 and 40%, but around 10% in the lower grades. Frosting, a surface coating of naturally exuded vanillin crystals, is no indication of quality, as high-quality Mexican beans are not frosted. While a high vanillin content is desirable, it is not directly related to the overall aroma/flavour quality. The most important characteristics of cured beans, which basically determine their value, are aroma and flavour, then general appearance, flexibility and length. Beans are normally tied in bundles and packed in waxed paper-lined tin boxes, 40 bundles in rows of ten; boxes are marked with the grade and length of beans, and four to six tin boxes are packed in a cardboard container for shipment. Cuts are shipped in bulk according to buyer requirements. Beans are seldom used as such, but are raw material for a number of derivatives.

Vanillin is the major flavour component and is usually between 1.5 and 3.5%, but about 150 other aroma constituents have been identified, and the relative importance of a single attribute depends on the end use. Appearance, flexibility and size are important as a fairly close relationship exists between them and aroma and flavour. The aroma and flavour of beans and thus their organoleptic properties

Fig. 7.4. Dried vanilla pods.

are basically determined by the volatile constituents, primarily vanillin, but the influence of minor and non-volatile constituents is still unclear (Werkhoff and Guntert, 1997). Resins possess no aroma but a pleasing taste and are apparently important in fixing the volatile aromatic constituents in solvent extracts. Aroma and flavour are very subjective assessments, but an accepted classification is that of Arctander (1960) as follows:

Bourbon vanilla (from Madagascar, Reunion and Comoro) has an extremely rich, sweet, tobacco-like aroma, somewhat woody and animal, possesses a very deep balsamic, sweet-spicy bodynote, but vanillin scent is lacking. The odour varies considerably according to moisture content; very moist beans have a stronger vanillin character while the higher-boiling aromatic constituents are more perceptible in drier beans.

Mexican vanilla has a somewhat sharper and more pungent aroma than the Bourbon type. Java vanilla beans frequently have a very heavy/woody flavour. Tahiti vanilla has an aroma almost perfumery sweet and not tobacco-like, not very deep woody, nor distinctly animal. Guadeloupe vanilla has a peculiar anisic-like floral sweet, liotropin/iosafrol type, and it is more perfumy than all other vanillas. It is considered a poor flavouring.

The occurrence of numerous non-vanillin aroma and flavour components present in minor or trace quantities in beans, as noted,

is the reason for their organoleptic superiority over synthetic vanillin and blends. Natural vanilla has a delicate, rich and mellow aroma, while that of synthetic material tends to be heavy and grassy. The natural extract has a delicate and mellow aftertaste, while that of synthetics is less pleasant.

Importing countries are gradually introducing legislation to ensure a standard analytical method is used to determine vanillin content for pod imports, to be quoted on retail vanilla products, including the EU (Ehlers *et al.*, 1999). Additionally the identification of genuine vanilla can become important for the same reasons, since bulk shipments may be adulterated; identity profiles are being developed (Kaunzinger *et al.*, 1997). Minor constituents can also be used to determine the origin of vanilla pods, and this may also become a statutory requirement (Overdieck, 1998; Ramaroson-Raonizafinimanana *et al.*, 1998).

Vanilla powder. This may be prepared from whole ground beans, but this operation is difficult owing to the presence of moisture and fats. Beans are usually ground together with sugar, to which is later added food starch and a natural gum, such as gum acacia. Thus, vanilla powder is a blend of the ingredients used. It should not be confused with vanillin powder, which is vanilla powder fortified with synthetic vanillin.

Vanilla extract. Vanilla extract was initially prepared by soaking chopped beans for up to a year in a solvent bath, the menstruum, which included glycerine, sugar and other additives considered necessary by the operator. This produced a dark brown extract whose colour depended on the ethanol content of the menstruum; about 60% was considered the optimum. The extract varied in its constituents, flavour and odour according to the origin of the beans, and beans of various origins were blended to produce an extract to suit particular end-users. This time-consuming method is retained to provide a small amount of extract for users who consider it superior to modern extraction methods.

Extract is currently produced mainly by ethanol percolation in a vertical stainless steel container fitted with trays or baskets of chopped beans, and the solvent, menstruum, is automatically returned to the top of the vessel to continue the cycle for up to 5 days. Three cycles are necessary, the first using 60%, the second 30–35% and the third 15% ethanol, and finally beans are washed in cold water. The four extracts are then combined and standardized to provide an ethanol content of about 35%. Extracts are then aged for a minimum of 3 months, but for 6 months if an enhanced flavour and aroma is required. Storage should be in stainless steel or glass containers to prevent the development or absorption of unwanted odours or flavours. The production of extracts from cell cultures is possible, and several patents covering the techniques have been filed, but no commercial plants are operating.

The average composition of extract (per 100 ml) is vanillin 0.19 g, ash 0.32 g, soluble ash 0.27 g, lead number (Winton) 0.54. the alkalinity of total ash is minimum 30, maximum 54, and of soluble ash 30, both at N/10 acid per 100 ml extract; total acidity is 42, acidity other than vanillin 30, both at N/10 alkali per 100 ml extract. The major constituents of commercial extract are vanillin up to 85%, 4-hydroxybenzaldehyde up to 8.5%, all others less than 1% (Klimes and Lamparsky, 1976). Extracts are usually commercially available as singlefold or twofold strengths, with fourfold being the maximum. The extract is the main source of natural vanilla flavour, and used in ice creams, yoghurt, chocolates, confectionery, custards, biscuits and other baked products, beverages, the liqueur Vanille, and added to less-expensive alcoholic drinks including brandies and whiskeys to give a simulated smoothness and bouquet. The highest average maximum use level in baked goods is reportedly 9.642 p.p.m. Vanilla tincture for pharmaceutical use is an extract containing a minimum of 38% ethyl alcohol. A very recent use is as an appetite suppressant. Patches impregnated with vanilla extract attached to the person were claimed by a London company to assist in weight loss, and they supplied data in support. Vanilla extracts are currently prescribed in pharmacopoeias to flavour unpleasant tasting medicines, and may be included in pharmaceutical products for the same reason.

Adulteration of vanilla flavours and extracts is common, but methods of detection are becoming more sophisticated, more accurate and more easy to operate (Remaud *et al.*, 1997). In mid-2001, the French company Alpha Mos of Toulouse developed an electronic analyser that can fingerprint the compounds dissolved in a liquid sample. This 'electronic tongue' can then accurately determine whether a sample of vanilla extract has been adulterated, or determine the origin of an unadulterated extract.

Vanilla oleoresin. This is misnomer as it is not an oleoresin but a resinoid. Solvent-extracting chopped beans (as for extract), then removing the solvent, ethanol or isopropanol, under vacuum produces a dark brown, viscous extract. Yield of extract varies according to bean origin, between 30 and 50% on a wet basis. Flavour and odour of oleoresins are inferior to extracts due to loss of some aromatic constituents during solvent stripping. The major use of oleoresins is as a flavour base for synthetic vanillin, and is commonly available as one-, two- or tenfold strengths. Several proprietary oleoresin products are available containing more or less natural oleoresin, mainly for specific end-users.

A monograph on the physiological properties of vanilla extract has been published by the Research Institute for Fragrance Materials. The regulatory status for beans, extracts and oleoresins in the USA is generally regarded as safe, beans GRAS 182.10, extracts GRAS 182.20, and oleoresins GRAS 169.3.

Vanilla absolute. This again is a misnomer as it is not a true absolute but a concentrate, usually solvent-extracted from the oleoresin, which is soluble in ethanol and essential oils used in the fragrance industry, its main user. It is the most concentrated form of the vanilla aroma, up to 12 times that of the bean, but a little harsher. It is available in a number of formulations and aroma levels.

Essential oil. Dry cured vanilla pods can be steam distilled to yield a small amount of essential oil of no commercial importance. Tahiti pods generally have the highest essential oil content.

Vanillin. Vanillin was first isolated from vanilla in 1858, and was first produced artificially in 1874 from the glucoside coniferin, occurring in the sapwood of certain pine species. Vanillin is now produced synthetically, and is much cheaper than natural vanilla. Raw materials are waste sulphite liquor from paper mills, coaltar extracts and eugenol from clove oil, and while they are virtually identical chemically, they differ in flavour and odour due to minor constituents in each. Java vanilla with its full-bodied flavour is frequently blended with synthetic vanillin as an enhancer. However, the flavour of natural beans is far superior to synthetic vanillin, due to the presence of other flavour compounds. Synthetic vanillin normally accounts for about 90% of the US vanilla flavouring sector, at about 1% of its cost.

Other *Vanilla species*

Vanilla abundiflora. *V. abundiflora* J.J. Smith is known as the Borneo vanilla. It is indigenous to Borneo, and found in fairly dense, swamp forest. It is a succulent climber up to 20 m; the leaves are elliptical, 25 × 9 cm and the inflorescence is a dense raceme of large white flowers. The fruit is a yellow-brown compressed cylinder, 25 × 2 cm. It is an inferior substitute for true vanilla.

Vanilla gardneri. *V. gardneri* Rolfe is the Brazilian or Bahia vanilla. It is similar to *V. pompona* but with smaller leaves and shorter fruit. It is another poor substitute for true vanilla, and also an occasional adulterant.

Vanilla phaeantha. *V. phaeantha* H.G. Reichenb is known as the Florida vanilla. It is indigenous to south-eastern USA and the Antilles islands. It generally resembles *V. planifolia* but has much larger flowers and shorter fruits. It is harvested and occasionally cultivated for its aromatic fruit. It is important for its resistance to *Fusarium* spp. root rot of vanilla.

Vanilla pompona. *V. pompona* Schiede is the West Indian vanilla, which occurs wild over a larger area than *V. planifolia* from south-eastern Mexico, Central America, northern South America to Trinidad. It resembles *V. planifolia* but the leaves are larger, 15–30 × 5–12 cm. The greenish-yellow flowers are also larger and more fleshy, with perianth lobes up to 8.5 cm. The lip has a tuft of imbricate scales instead of hairs in the centre of the disc. Wild and cultivated vines may flower and fruit over a long period. The cylindrical pods are shorter and thicker, 10–17.5 × 2.5–3.5 cm in diameter.

It has been introduced as an alternative into countries where vanilla will not flourish, Singapore and Malaysia, for example. *V. pompona* is commercially cultivated on a small scale only in Guadeloupe, Martinique and Dominica. A major constraint to increased production is that the fruits are more difficult to dry naturally than vanilla. Vanillons is the trade name for *V. pompona* pods, which have a low vanillin content and possess a characteristic aroma. These pods are the most widely used substitute and adulterant of true vanilla, used in the cosmetics and tobacco industries, and are raw material for the extraction of heliotropin. Vanillons tend to provide gummy aqueous alcohol extracts. The typical heliotropin aroma was considered to be due to the piperonal content but a German analysis did not detect this compound among those identified, and noted the chemical composition more closely resembled that of Tahitian vanilla rather than vanilla pods (Ehlers and Pfister, 1997).

Vanilla tahitensis. *V. tahitensis* J.W. Moore is the Tahitian vanilla. It is indigenous to Tahiti, and is cultivated there and on Hawaii. This species is less robust than *V. planifolia*; the stems are more slender and the leaves narrower, 12–14 × 2.5–3.0 cm. The lip is shorter than the sepals. Pods are 12–14 cm × 9–10 mm wide, tapering towards each end, and indehiscent. Pods have a lower vanillin content than *V. planifolia* but with a finer scent, and are often preferred in cosmetic products. Tahiti vanilla beans are generally less popular for flavouring due to a somewhat rank flavour and relatively high volatile-oil content, which can result in cloudy extracts. Tahiti vanilla rarely frosts.

Unrelated species

C. culitlawan (L.) Kosterm is native only to Indonesia, specifically the Moluccas, Ambon and adjacent islands (see Chapter 3). The bark oil (lawang oil) is rich in eugenol and a raw material for synthesis of vanillin.

Liatris odoratissima Willd. (*Compositae*), known as Deer's tongue, is a North American species whose leaves are considered to resemble a deer's tongue, hence its local name; it is also known somewhat fancifully as wild vanilla or vanilla leaf. The fleshy leaf bases contain coumarin, and its main use is in traditional medicine. Other *Liatris* spp. are used as herbs or are ornamentals.

8

Piperaceae

The *Piperales* comprises three families, the *Chloranthaceae*, the *Saururaceae* and the *Piperaceae*, which includes 10–12 genera with *Peperomia* and *Piper* accounting for most species. The genus *Piper* L. contains about 1000 species of tropical shrubs or woody climbers, including the commercially important pepper vine, *Piper nigrum* L. Less significant are cubeb pepper, *Piper cubeba* L. and the long peppers, *Piper longum* L. and *Piper retrofractum* Vahl., while *Piper betle* L. is cultivated for its leaves. Fruits of other *Piper* species used as a substitute or alternative to pepper are mentioned later. Fruit of *Aframomum melegueta* (Rosc.) K.Schum. (*Zingiberaceae*), the melegueta or alligator pepper, is used as a spice, and the species is included in this chapter for convenience. All references to pepper in this text refer to *P. nigrum* and its derivatives unless otherwise stated.

Black Pepper

The pepper vine is native to the hills of western India but dispersed by trade throughout the world. So important was pepper that the search for its source and thus control of the trade was a significant factor influencing world exploration and history. The long pepper, *P. longum*, is the *pippali* of Sanskrit from which the genus, Greek, Latin and most European names are derived. It was probably the first pepper to reach the Mediterranean and was originally more highly regarded in the Classical World than *P. nigrum*. Two peppers were recognized by the Greek Theophrastus (372–287 BC), black pepper and long pepper. The Greek physician Pedanius Dioscorides of the 1st century AD mentions black and white pepper, which he believed were produced on different plants. In the Roman Empire, pepper was well established as an article of commerce, shipped to Red Sea ports by Arabs using the Arabian Sea monsoons, then on to Alexandria. The Emperor Marcus Aurelius imposed a customs tax on long pepper and white pepper arriving at Alexandria, but exempted black pepper.

Cosmas Indicopleustes visited India and Ceylon and accurately described the harvesting and preparation of pepper on the Malabar coast in *Topographia Christiana* (*c.* AD 545). By the Middle Ages, pepper was of great importance in Europe to season or preserve meats, and to overcome the odours of rancid food. Peppercorns were very expensive and accepted in lieu of money in dowries, taxes and rents, and this survives in the term peppercorn rent, which now means virtually free – exactly the opposite! The statutes of King Ethelred in 10th-century England required German spice traders to pay tribute, including 10 lb pepper, to trade with London merchants. A pepperers' guild of wholesale merchants founded in London in 1180, subsequently incorporated into a spicers' guild, was succeeded in 1429 by the present Grocers' Company. The original pepperers and spicers were the predecessors of apothecaries, emphasizing the role spices formerly occupied in medicine. Pepper was no exception, and concurrent with its use as a spice, pepper and later pepper oil were included in most early European herbals and medical treatises.

Pepper was probably taken to Java by early Hindu traders, and its widespread cultivation dates from at least AD 500. Marco Polo visited Java in 1280, and described pepper cultivation and the Chinese junks loading up to 6000 baskets, and later the abundance of pepper on the Indian Malabar coast in his *Description of the World*, 1298. In the 13th century, the great wealth of Venice and Genoa was derived mainly from the spice trade, and records from the end of the 15th century show that Venice annually imported the equivalent of 500,000 kg pepper from Alexandria to partially satisfy a European demand estimated at 0.75–1.0 million kg (Wake, 1979).

Portugal's desire to obtain a share in this lucrative trade was mainly responsible for Vasco da Gama circumnavigating Africa and arriving at Calicut on the Malabar coast of India on 20 May, 1498. At that time Malabar produced about two-thirds of the pepper available to world trade. Da Gama subsequently established in 1503 an export price for pepper at Cochin of 2.5 cruzados (11.5 g gold) a quintal FOB, a price the Portuguese maintained for decades. Portugal subsequently gained control of the region and a virtual monopoly of its spice trade, which lasted for a century, and almost ruined Venice, Genoa and Alexandria. Lisbon became the most important European trading centre for oriental spices, with entrepôts in Goa and Malacca, which also exported pepper to China. In the century 1500–1600, Portugal imported from Malabar the equivalent of about 2 million kg annually (Kieniewicz, 1986), while European usage increased to around 3 million kg.

The value of pepper varied substantially according to the season, wars, etc. and in the early 16th century the original and final selling prices for the equivalent of 1 kg was 1–2 g silver at Malabar, in Alexandria 10–14 g, in Venice 14–18 g, in Lisbon 16–20 g and to the final consumer 20–30 g. Trading via Cairo in the 15th century was hardly less profitable; a load of pepper bought in Malabar for 2 dinars, was worth 10 in Mecca, and sold in Cairo for 20 dinars. However, in 1430, an order from the then Mamluk Sultan raised the price for sale to European traders to 80 dinars, causing consternation to Venetian and other merchants.

Garcia da Orta gives an excellent and accurate description of pepper in the 46th of his colloquies, *Colloquies on the Simples and Drugs of India*, published in Goa in 1563, although he believed white pepper was a different species (Fig. 8.1). The Italian traveller, Varthema (1465–1519), gave in his *Itinerario de Ludovico de Varthema Bolognese* of 1510 a vivid picture of the Calicut plantations in the 16th century, while Barbosa in his *Coasts of East Africa and Malabar* gives a detailed account of the plant and the Calicut pepper trade.

Early in the 17th century following the successful voyages to Indonesia by Captains de Houtman and van Neck, the Dutch acquired pepper-producing areas near Bantam, Java, and Lampong, Sumatra (more correctly Lampung but Lampong is too well established in the trade to change). Although the Dutch subsequently controlled much of the pepper trade, they never succeeded in maintaining a virtual monopoly as they did with clove and nutmeg. The end of Dutch dominance coincided with entry of the USA into the Far Eastern spice trade, and in 1797 Captain J. Carnes of Salem, Massachusetts, sailed the schooner *Rajah* into New York harbour carrying pepper from Sumatra worth $100,000. During the next 50 years, ships from Massachusetts played an important role in world pepper trade, which reputedly produced in Salem some of America's first millionaires (Putnam, 1924). Early in the 19th century, the British organized pepper planting in Malaysia and later Sarawak, which subsequently became major producers (Burkill, 1966).

Fig. 8.1. Sketch of pepper vine by Garcia da Orta.

Botany

P. nigrum L. (syn. *Piper aromaticum* Lamk.) is known in English as pepper, in Dutch as *peper*, in Italian as *pepe*, in Spanish as *pimienta negra*, in Portuguese as *pimenta*, in Arabic as *filfil*, in India, Hindi and Urdu as *kali mirch*, in Tamil as *milagu*, in the Philippines as *paminta*, in Malaysia as *lada*, in Indonesia generally as *lada* and black pepper as *lada hitam*, in Japanese as *kosho*, and in Chinese as *hu-jiao*.

Cytology of the genus *Piper* is confused; Darlington and Wylie (1955) give the basic chromosome number of the genus *Piper* as x = 12, 14, 16, and somatic number of *P. nigrum* as 2n = 52 or 128; Mathew (1974) gives 2n = 52 for six cultivars and four wild types, and 2n = 104 for two other wild types; Samuel and Bavappa (1981) reported chromosome numbers from nine *Piper* species including 2n = 26 for *P. longum* and 2n = 52 for *P. betle*. Jose and Sharma (1984) state three *P. nigrum* cultivars, *Piper attenuatum*, *Piper boehmeriae-folium* and *P. longum* had 2n = 52, one *P. nigrum* cultivar and *Piper chaba* had 2n = 104, and *Piper peepuloides* had 2n = 156; *P. cubeba* (2n = 24) with large chromosomes and high total chromatin length was regarded as primitive. Most Old World species are thus polyploids and New World species diploids. It has been suggested that species with fused or partly fused ovaries and fruits should be moved from the genus *Piper* into a separate (unnamed) genus.

The main germplasm centre for *P. nigrum* is India, where extensive collections are maintained at a number of Institute of

Spices Research sites. Chemotaxonomy has established the relationships between south Indian peppers (Ravindran and Babu, 1994), and with populations elsewhere. Smaller but important germplasm collections exist in Indonesia and Sarawak, while those in Brazil include many *Piperaceae* species. Breeding programmes, including hybridization, based on these collections have mainly concentrated on introducing disease resistance. The very large numbers of local cultivars also provide plant breeders with a choice of material when selecting a specific plant characteristic. A very important cultivar is the hybrid Panniyar-1, developed at the Taliparamba Research Station, Cannanore, southern India, which has four times the yield of the best local variety and is early bearing.

P. nigrum is a perennial, glabrous woody climber up to 10 m, sometimes taller, very variable in physical characteristics and virtually dioecious. Cultivars are generally hermaphrodite. The root system is extensive with large roots up to 4 m, but the major portion remain within 1 m of the surface, often the first 40 cm, in a circle round the stem. Root anatomy and characteristics have been described (Hu *et al.*, 1996), as has the root essential oil composition (Hu *et al.*, 1995). Cultivated pepper grown from cuttings produces initially one main climbing stem, but others rapidly appear and branch profusely, and mature vines have a bushy, columnar shape, generally 3–5 m × 1.0–1.5 m (see Fig. 8.2).

Young climbing stems are green tinged with purple, becoming dark green with age; mature stems are 4–6 cm in diameter, woody, with thick, flake-like bark, the internodes 5–12 cm with short adventitious roots at the nodes, which support the vine. Horizontal fruiting branches are produced from these stems, which may again branch. The pungent principle of pepper, piperine, is present at 0.2% in mature fully differentiated shoots, but only as a trace in young leaves and shoots.

The leaves are alternate, simple, 8–20 × 4–12 cm depending mainly on cultivar, and the petiole is 2–5 cm (Fig. 8.3). The lamina is ovate, entire and coriaceous, with five to seven main veins; the base is oblique, obtuse or rounded, and the tip acuminate. The leaves are usually dark shiny green above, paler and densely glandular below. A direct relationship exists between leaves produced and fruiting spikes, thus cultural techniques in commercial plantations aim for healthy, well-foliaged vines. The leaves contain an essential oil differing from berry oil, with a scent resembling lime. Piperine is absent from mature leaves and stem.

Fig. 8.2. Pepper plantation – Sarawak.

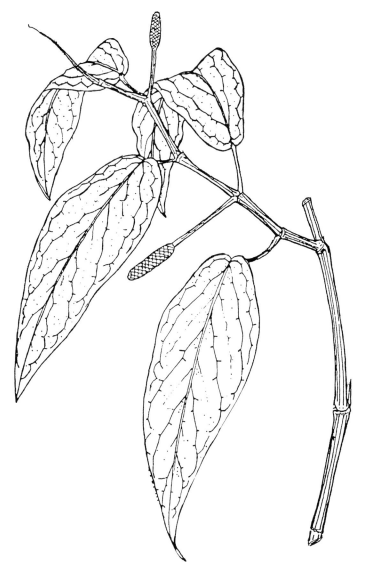

Fig. 8.3. Fruiting stem.

The inflorescence is a pendant spike on lateral branches, 3–15 cm long, bearing 50–150 very small yellowish-green to whitish-yellow flowers. Flowers may be unisexual, but frequently bisexual in cultivars, and in general the higher the percentage of bisexual flowers in a particular cultivar, the higher the berry yield. Flowers lack a perianth, and the two to four minute stamens borne laterally on the ovate sessile ovary are surmounted by three to five somewhat fleshy stigmas. Cross-fertilization is usual in the wild and wind is the main agent, but self-pollination appears to be general in hermaphrodite flowers. The main flowering period varies seasonally, mainly depending on the onset of the rainy season; in Kerala, India, it is May–August with a July peak, in the Philippines July–September, in Sarawak September–November, in Indonesia October–December and in Brazil November–December.

The fruit is a single-seeded, sessile, inde-hiscent, globose drupe (the berry), 3–6 mm in diameter, and the pericarp pulpy, with 50–60 fruits per spike (Fig. 8.4). Unripe fruit is green, becoming red when ripe and black after drying. When dry, the black wrinkled skin encloses a single whitish seed, known as a peppercorn. The morphology and histol-ogy of pepper seeds has been described (Parry, 1962).

An analysis of dried berries gave an aver-age content (per 100 g) of: moisture 9–12 g, protein 11–13 g, starch 25–45 g, fibre 9–17 g and ash 3–6 g (see also Table 8.1). Berries nor-mally average 4.5–5.5 g per 100 depending on cultivar, and in India where litre cans are often used as a measure in markets, weight varied from 426 to 850 g per can, with Tellicherry and Baliapatnam the highest. The seed is globose, 3–6 mm in diameter contain-

ing mainly starchy perisperm, a small endosperm and minute embryo. A detailed description of the physical structure of berries determined by scanning electron microscopy has been published (Arnott, 1999).

Berry composition. The berry contains an essential oil (steam-volatile), a fixed (fatty) oil, alkaloids, resins, proteins, cellulose, pen-tosans, starch and minerals. Indian berries also contain the following trace elements (average content in p.p.m.): K < 50, Ca < 45, Ti 70, Cr 36, Mn < 30, Fe 96, Zn < 25, Br 19, Rb 19; Sr 33 (Joseph *et al.*, 1999). The fatty oil, which constitutes 2–9% of the berry, contains 19–38% saturated and 51–83% unsaturated acids; the major components are palmitic 16–23%, oleic 18–31%, linoleic 25–35% and linolenic 5–19%. Together with other data quoted this indicates the very wide differ-ence in berry composition, and thus tables and analyses quoted herein are only gener-ally applicable.

Numerous cultivars of *P. nigrum* are known, especially in India where nearly 100 have been described, and within a producing country each region has its favourite. These cultivars produce a wide range of berry types, varying in physical characteristics and constituents (Kanakamany *et al.*, 1985; Sumathikutty *et al.*, 1989; Gopalkrishnan *et al.*, 1993; Pruthy, 1993: Menon *et al.*, 2000). A method of differentiating groups of cultivars, graph clustering, developed in Kerala, India, is expected to provide a more accurate sys-tem for selection and hybridization pro-grammes (Mathew *et al.*, 2001). Considerable data on berry composition from specific cul-tivars has been published, but composition can be significantly affected by seasonal influences and growers' methods of harvest-ing and storage, as noted later. New geno-types also produce oil with differing or unusual component ratios, and thus vary in their taste and odour (Gopalkrishnan *et al.*, 1993).

An analysis of 23 berry types from Kerala, Coorg and Assam gave the following values (range): moisture 8.7–14.1%, total N 1.55–2.60%, N in non-volatile ether extract 2.70–4.22%, volatile ether extract 0.3–4.2%,

Fig. 8.4. Ripening berries.

Table 8.1. Berry composition during maturation in Sri Lanka.

Age (months)	Mean dry weight (mg)	Starch (g per 100 g)	Volatile oil (ml per 100 g)	Total extractables	Piperine[a]	g per 100 g[b]
3	9.0	1.0	3.5	16.0	12.5	12.5
4	12.8	13.0	5.0	17.7	14.1	13.9
4.5	13.2	18.0	5.5	17.9	14.2	14.1
5.25	16.2	29.3	8.6	15.8	11.5	11.0
6	18.5	30.6	8.0	16.7	9.7	9.4
6.75	25.3	42.1	4.9	14.3	8.7	9.0
7.5	28.5	53.8	6.3	14.0	7.5	7.1

[a] tlc-uv.
[b] tlb-densitometry.

non-volatile ether extract 3.9–11.5%, alcohol extract 4.4–12.0%, starch (by acid hydrolysis) 28.0–49.0%, crude fibre 8.7–18.0%, crude piperine 2.8–9.0%, piperine (spectrophotometrically) 1.7–7.4%, total ash 3.6–5.7% and acid insoluble ash (sand) 0.03–0.55%.

The effect of time of picking (maturity) on physical characteristics and constituents of Sri Lankan berries is shown in Table 8.2, while Table 8.1 indicates the difference in piperine levels recorded by various analytical methods; Indian data are similar. Pungency level, aroma and flavour of the spice and its volatile oil are cultivar characteristics, and can be improved by selection. The most important non-genetic factor is maturity at harvest, as the oil content of immature berries reaches a maximum relatively early and then diminishes while piperine content continues to increase. Thus, suitability of the final product for a particular end use is largely governed by berry maturity at harvest.

Sri Lankan berries generally have maximum oil content at 5.0–5.5 months after flowering, with highest piperine content at 4.5 months; Indian cultivars often continue to increase their piperine content to maturity, but frequently have a final piperine content less than half that of Sri Lankan berries. Commercial consignments of black pepper usually contain 0.6–2.6% essential oil, white pepper 0.5–1.0%. The piperine content of both is usually 4–7%, and a quoted higher level is suspect. Commercial samples of pepper oil in the USA are required to meet the following characteristics: specific gravity (15°C) 0.890–0.900,

Table 8.2. Effect of berry maturity on some selected oil constituents in Sri Lanka (as percentage of total volatile oil).

Constituent	Time of harvest (weeks)						
	13	17	18	22.5	26	29.5	33
α-Pinene	9.3	9.8	7.5	11.1	10.1	10.1	12.5
β-Pinene	11.1	12.2	10.1	12.1	10.3	10.7	11.1
Sabinene	17.1	15.1	14.0	19.2	18.4	18.5	23.6
α-Phellandrene + carene	6.4	5.3	5.0	4.8	5.1	5.4	3.9
Limonene	19.0	20.0	18.0	19.5	14.5	17.1	17.0
Terpinen-4-ol	2.1	2.1	3.4	1.5	1.2	1.7	0.8
β-Caryophyllene	11.2	10.6	10.4	12.5	18.5	14.3	15.3
β-Bisabolene	3.2	1.8	2.6	2.2	3.6	3.2	2.0
Unidentified	4.2	5.9	8.3	3.7	1.9	2.3	0.5

refractive index (20°C) 1.4935–1.4977, optical rotation −3° to −5°, insoluble in water, soluble in about 15 vols 90% alcohol.

Ecology

Pepper is native to wet tropical forests with its centre of origin the south-western Ghats of India, where it can be found wild in an area from North Kanara to Kanyakumari, and also in Assam and Borneo, most probably as an escape from early cultivation. Kerala State is India's main pepper-growing region accounting for about 96% of production, Tamil Nadu, Pondicherry and Karnataka the remainder. In Indonesia, main producing regions are Lampung, East and West Kalimantan, and Bangka island. The vine has been introduced wherever there was the slightest possibility it would grow, so popular is the spice! A pepper vine is easily grown in a large pot, and if trained as a bush or up a suitable support will provide a plentiful supply of fresh berries for many years.

The adaptability of *P. nigrum* is demonstrated by the large number of local cultivars; there are nearly 100 in India alone, as noted in the Botany section. Popular cultivars are Balancotta with large, light green leaves and high, regular yields, Kalluvally with narrower, dark green leaves, high yield, and drought and wilt resistance, Cheriy Kaniyakadan with small, elliptical leaves, regular and heavy yields, and wilt-resistance (unsuccessful when planted in Sarawak) and Uthirankotta, with tough, leathery, light green leaves and very hardy.

The five major Indonesia cultivars differ mainly in leaf shape and size, internode length, branching, flowering and fruit yield: Lampung cultivars are Kerenci, Belantung and Jambi; and on Bangka island, Lampung and Bangka. In Sarawak, the large-leaved, dense-branching high-yielding Kunching is most popular, although very susceptible to foot rot; there is also the smaller-leaved Sarikei. In a 7-year variety trial in Sarawak including the highest yielding local, Indian and Indonesian cultivars Kuching and Jambi gave the highest yield at 20,260 and 19,960 kg ha^{-1}, respectively.

Commercial pepper production is basically limited to a band from 20°N to 20°S of the Equator; production in Indonesia is within a few degrees of the Equator, in Sri Lanka and Malaysia about 10°N, in southern India 12–15°N, Madagascar 15°S and Brazil to 15°S. Pepper flourishes and produces excellent berry yields from sea level to 1500 m; in the major Indian producing region of Kerala most commercial plantations are between 1000 and 2000 m. Berry yield or quality is believed to deteriorate above 2500 m but no comparative data for a specific cultivar are available to demonstrate the effect of altitude.

A rainfall of 2000–3000 mm annually uniformly distributed over the year is the optimum, although up to 4500 mm is tolerated on hillside plantings, while good berry yields can be obtained at 1500 mm provided this falls mainly in the growing and flowering season, and soil moisture is conserved. Under seasonal rainfall conditions a dry period of 2–3 months occurs while berries are maturing.

Warm, humid and cloudy conditions promote growth, with an average day temperature of 25–35°C, and a relative humidity of 65–95%. Long periods of strong sunlight will scorch unshaded plants, and the north slopes of hills are preferred for pepper plantations in India and Sri Lanka. Light intensity is important in determining growth and thus berry yield, and in addition to a general reaction to this factor, cultivars also differ. Vines, especially seedlings, benefit from intermittent or partial shade but heavy shade adversely affects flowering and fruit set. Prolonged very cloudy or misty periods at this time also have a similar but less damaging effect.

The minimum temperature in the main pepper-growing regions seldom falls below 15–18°C, and low temperatures depress reproductive more than vegetative growth. Very hot dry wind inhibits growth, may kill young plants and adversely affects pollination. In hilly regions of India, easterly facing slopes are preferred. Windbreaks are recommended in exposed situations, particularly when vines are not grown on living supports, which provide partial protection plus shade.

Soils

Pepper grows naturally on the humic loams of its forest habitat, where the upper 50 cm contains a wealth of plant nutrients. This soil type is becoming increasingly scarce following destruction of tropical forests, but other well-managed soils are equally suitable. In India, loams, tropical red earths and laterite are used; in Indonesia, Sri Lanka, the Philippines and Sarawak, loams are the first choice, and in Brazil, loams and the red *terra rosa* soils are utilized. Commercial plantations are established on preferred soil types, but smallholders grow vines where they can, from heavy clay in Sarawak to the sandy soils of Zanzibar. The intensive treatment given to smallholder vines frequently compensates for the disadvantages of an unsuitable soil, but does not indicate larger plantings will be successful.

A neutral to slightly acid soil is preferable, but forest soils of pH 5.5 in the Philippines grow excellent crops. A pH of around 6.5 is favoured in India and Sri Lanka, but there are few data on the effect of soil reaction on plant growth or berry yield. Peat soils, saline, alkaline or very acid soils are unsuitable. Although pepper roots should be kept moist, poorly drained soils are unsuitable as pepper is very susceptible to waterlogging even for relatively short periods, and 7–10 days can cause death of young plants. Wet soils also increase the damage caused by root rots. Sloping valley alluvia and river banks are quite suitable provided they do not become flooded. A plantation must be kept weed-free, thus adequate soil conservation measures should be routine on soils subject to erosion.

Fertilizers

It may be obvious to state that to obtain high annual berry yield it is essential vines are well nourished, but anyone who has visited smallholder plots cannot fail to notice the often poor growth and generally unthrifty appearance of vines. The earliest plantations were fertilized mainly by mulching with available plant and animal wastes, or burnt earth and broken-up termite mounds. While all these materials are still used, their availability has decreased with the diminished area of forest available. It has been calculated that approximately 2 ha of jungle was necessary to provide fuel and topsoil to support 1 ha pepper. The effectiveness of mulching to increase yield has frequently been demonstrated; green berry yield rose from 4.5 to 7.3 kg per vine over 3 years in Indonesia when mulch was applied directly after planting. In Sarawak, mounding soil around plants and mulching with *I. cylindrica* increased average green berry yield from 15,500 to 18,250 kg ha^{-1}, providing a substantially increased return to growers from additional labour only. Similar responses have been reported elsewhere. The incidence of disease is also often lower on mulched plots.

Organic manures were formerly extensively used, but such material must now be easily and cheaply available and application cost low to be profitable; in Kampuchea, berry yield tripled following application of shrimp waste at 10–12 kg per vine. A general recommendation of the Indian Department of Agriculture is for an annual application of 10 kg cattle manure or compost, 500 g ammonium sulphate, 1 kg superphosphate and 100 g potassium chloride applied in a trench round each vine; in Kerala State, 5 kg FYM, 0.5 kg each of neem cake and bone meal, plus 140 g N, 18 g P and 116 g K per vine per year increased yield 2.5 times (Sandanandan *et al.*, 1993). Fertilizer is normally applied at the end of the monsoon, although organic manure may be applied separately when the rains begin. Pepper is mainly a smallholder crop in Sri Lanka, usually an intercrop, and seldom manured separately to the main crop.

Published data show that mature vines can remove annually (ha^{-1}) 233 kg N, 17 kg P, 172 kg K, 18 kg Mg and 75 kg Ca, contained in 11,500 kg ha^{-1} dry material, while a healthy stand of 700 vines removed the following annually: 102 kg N, 13 kg P, 91 kg K, 27 kg Ca and 9 kg Mg (de Waard, 1964; de Geus, 1973). Nutrient removal by pepper vines to produce 200 kg ha^{-1} berries is given in Table 3.4. In Brazil, nutrient extraction over time was, in decreasing order, N > Ca > K > Mg > P, while fruits extracted 11.22 kg N,

6.15 kg K, 3.84 kg Ca, 1.18 kg Mg and 1.07 kg P (Veloso and Carvalho, 1999). Leaf sampling indicated the following percentage of dry matter below which deficiency symptoms will occur in Indonesia: N 2.7%, P 0.1%, K 2.0%, Ca 1.0% and Mg 0.2%; and in India: Ca 1.51%, Mg 0.913% and S 0.121% (Nybe and Nair, 1987). A significant correlation can exist between nutrient status of leaves and potential yield; in India this was the leaf P index as calculated by the formulae P/(N+P+K+Ca).

In the major pepper-growing region of Kerala, an annual application per vine of 50 g N, 80 g P and 110 g K raised average berry yield from 1340 to 2000 kg ha^{-1}, with over half the increase due to potassium. Indian fertilizer companies supply specially formulated pepper mixtures containing the correct local balance of NPK and microelements. In Sarawak, a special pepper mixture contains 12:3:4:1.5 NPK and Mg, plus a compound containing (approximately) Fe 20%, Cu 3%, Zn 5.5%, Mn 9%, B 7% and Mo 0.22%. Local NPK mixtures are used in Brazil, and are frequently high in phosphate, moderate in potash and moderate to low in N with a recommended rate of 2.5–3.0 kg per mature vine annually, but micronutrients are essential on many regional soils. Castor or coffee processing residue is also used as a mulch around individual vines, especially in northeastern regions.

Inorganic fertilizers generally increase berry yield, but other factors probably have greater influence on berry quality, i.e. time of picking and drying method. An interaction between uptake of fertilizers and the type of support is noted later. Thus, a relatively small improvement in plantation management will probably result in greater return than purchasing expensive fertilizers. The effect of fertilizers on berry composition and oil content has been little investigated, and where a particular type of berry is required this is often obtained by introducing a suitable cultivar. Data indicating the response to minor nutrients are irregular and often conflicting, and in general they should not be applied until shown to be necessary by foliar or soil analysis, as noted previously.

Deficiency symptoms induced in greenhouse-grown vines in India were described as: calcium, small brown necrotic spots or chlorotic areas, near leaf margins, which later enlarge to form necrotic areas; magnesium, oval interveinal chlorotic areas which become black necrotic areas; sulphur, uniform yellowing with brown necrotic spots (Nybe and Nair, 1987). Major nutrient deficiency symptoms noted in Indonesia are: nitrogen, uniform yellowing of leaves, becoming orange-yellow; potassium, tips of mature leaves become necrotic, brittle and light grey. The yellow disease in Indonesia is mainly a nutrient deficiency, and effectively controlled by application of a balanced fertilizer, and in Malaysia, leaf spotting was due to a manganese deficiency.

Cultivation

To establish a plantation the land must be cleared and all debris burnt and the ash spread. Topsoil disturbance should be kept to the minimum to retain soil nutrients, as the bulk of pepper roots remain near the surface. A major decision is whether vines are to be grown up living or dead supports and factors to be considered are: (i) is the plantation on an exposed or protected site; (ii) the extent shading will be required; and (iii) availability of quick-growing suitable trees or timber for supports. In Bangladesh, Indonesia and the Philippines, smallholder vines may be grown up existing forest trees and the surrounding area kept clear of undergrowth.

In Sri Lanka, supports made from eucalyptus timber gave higher berry yield than vines trained up living trees (Bavappa *et al.*, 1981), but living trees as supports are becoming more common, as in Indonesia. Concrete posts or circular frames made from stone blocks are used in addition to timber supports in Vietnam (Pohlan and Heynoldt, 1986), and also in India (Reddy *et al.*, 1991). Suitable trees for live supports, which in general must be quick-growing and withstand heavy pruning, are *Erythrina variegata* L. and *Erythrina indica* Lam., *Garuga pinnata* Roxb., *Grevillea robusta* Cunn., *Gliricidia* spp. and *Leucaena* spp., in addition to existing coconut, areca nut, kapok or rubber plantations, while in

Madagascar vines are often trained up shade trees in coffee plantations.

The type of support also influences fertilizer uptake, as vines on living or dead supports differed in their uptake of phosphate at the same rate and method of application (Geetha *et al.*, 1993). Height of stakes is also important; in Brazil, 2.5 m gave a higher average berry yield than 1.5 m at 3760 and 2890 kg ha^{-1}, respectively (Kato and Albuquerque, 1980), but in India and Sarawak, stakes are usually much higher at 4–6 m. Spacing between supports depends on the method of weed control, and if this is mechanical then 3–5 m must be left between rows to allow passage of a small tractor with implements. Spacing on non-mechanized estates is usually 2–3 m on the square, with timber supports up to 4 m in Sarawak, up to 9 m in India, while living supports are pruned of branches to this height. The cost of providing supports has steadily increased, and by 1995 the Malaysian Agricultural Research and Development Institute (MARDI) reported the expense was unprofitable for growers with less than 2 ha of vines.

On sloping land, rows should be aligned on the contour to reduce danger of soil erosion; on steep slopes, contour bunding is essential, since most plantations are clean-weeded and thus very susceptible to erosion. Alternatively, low-growing plants such as *Calopogonium mucunoides* Desv. or *Mimosa invisa* Mart. will provide an effective soil cover and assist in preventing soil erosion, and their foliage can be used as a mulch.

On loam soils with a high plant nutrient status, the preparation of planting pits is usually unnecessary, and surrounding topsoil mixed with compost or manure is drawn up into a mound around the support base. On heavier clays or lateritic soils, two or three pits filled with a similar mixture are prepared around the support. No pit is prepared on the south side north of the Equator and on the north side south of the Equator, to reduce seedling loss from sun-scorch. Six or more unrooted or two well-rooted cuttings are planted per support, but when using large existing trees the number should be doubled.

Cuttings are the simplest and cheapest method of propagation, but pepper can be grown from seed, grafted, marcotted, or plantlets grown from single leaf cuttings. Plants raised by micropropagation require skilled and specialized treatment, but cloning allows large numbers to be raised from a selected parent. Shade is very important for plantlets, and in Sri Lanka 50–75% shade for 12 weeks produced larger and healthier plants than under no or complete shade (Seneviratne *et al.*, 1985). Should it be necessary to raise plants from seed, mature, ripe berries should be collected, dried and the mesocarp carefully removed. Seed can be sown in pots or seedbeds 1.5–2.0 cm deep, kept moist but not wet, and seed should germinate in 2–3 weeks.

If cuttings are the method selected then the type is important; in India stoloniferous shoots or cuttings are popular, in Indonesia and Sarawak cuttings from fruiting shoots, and in Brazil both methods are used but aerial-shoot cuttings preferred. A system of rapidly producing a large number of cuttings has been developed in Sri Lanka. Rooted cuttings are tied to bamboo supports, and when vines reach the top the growing point is excised. Two weeks later the vine is severed near the base and divided into single node cuttings. One vine can supply 50 cuttings per year. A similar number of cuttings was obtained in Brazil.

In India, runner shoots at the base of selected high-yielding vines are coiled round a forked stick to prevent premature rooting. Runners, or stolons, are separated from the parent vine in February–March, cut into lengths of three to five nodes and planted in nurseries, or directly into the field. A simpler and cheaper modification of the Sri Lankan method has been officially introduced, but is not proving popular, as although it produces more cuttings per vine, additional labour is required plus the cost of plastic tubes. Government farms in the main pepper-growing areas of Panniyur and Kerala maintain nurseries and supply rooted pepper cuttings to growers at a subsidized price. If planted at two cuttings per pit at a spacing of 3 × 3 m, about 2000 rooted cuttings per ha are required. In Sarawak and Indonesia,

60 cm cuttings with five to seven nodes are normally taken from terminal shoots of vines less than 2 years old. The terminal bud of the selected shoot, leaves and small branches between the third and seventh node from the apex are removed, and when the terminal bud has regenerated, the shoot is carefully cut below the seventh node.

Cuttings should preferably be rooted in containers or in a nursery, and treatment with a rooting compound usually results in more rapid root growth; pre-rooting also ensures quick growth in the field and reduces expensive infilling. Cuttings should be planted in the nursery at least 4–5 months before planting out. Alternatively, cuttings can be planted directly into mounds, but large numbers are required, which may not be available, and losses are usually high. Cuttings are usually planted at 45° with three or four nodes below the surface and not deeper than 10–15 cm. If planted direct into mounds, leaves and branches of the top three nodes are left and trained up a temporary stake; however, in Indonesia, cuttings treated with a rooting compound and placed horizontally in the soil produced more roots more quickly than obliquely planted cuttings.

Cuttings should be planted out at the beginning of the rainy season; in India this is May–June or July–September, in Sarawak and Indonesia September–October, in Sri Lanka October–November and in north-east Brazil January–February. When plants have made sufficient growth they are tied to supports, three climbing stems retained, and regularly pruned to encourage development of lateral fruiting branches.

Stems are pruned to 15–25 cm when eight to ten nodes have developed and leaves removed; when a further nine to ten nodes have developed, stems are cut back to within three to four nodes of the previous cut. Vines are thus pruned seven to eight times until they reach the top of support posts. The terminal growth is pinched out periodically to prevent further growth, and low side branches and stolons removed. The main object of pruning is to remove unproductive nodes without lateral branches, leaving only stems with nodes each bearing a branch. Thus, a mature vine will have the maximum number of fruiting branches with profuse leaf development, as a flower spike can only develop in a leaf axil. During the first 2 years, flower spikes are removed, and this may be combined with selective leaf removal to encourage branching.

Main flowering occurs in July in the Philippines, in September in Sarawak and in September–October in Indonesia. The period from flowering to fruit maturity is a cultivar characteristic; in Sri Lanka it is 7–10 months, in India 4–5 months, in Sarawak 5–6 months and in Indonesia 6–8 months. The natural variation in time to maturity is also reflected in berry composition and volatile oil constituents. Fruit development can also be affected by growth stimulants, but whether this is profitable or indeed desirable is debatable, and their use on pepper has been reviewed (Kandiannan *et al.*, 1994). To meet the growing worldwide demand for organically grown products by consumers, trials with organically grown pepper have been shown to be successful in India (Krishnakumar *et al.*, 1999).

Weed control

Pepper vines grown by smallholders are normally interplanted or trained up existing trees, and little cultivation is necessary other than maintenance of a weed-free area round the base of the vines. Weed control in commercial plantations is essential as the shallow roots will be in direct competition with weeds for plant nutrients. Although mechanical weed control is possible, vine roots are easily damaged by too-deep cultivation; where a ground cover is desirable, mowing or slashing is effective. In most pepper plantations hand-weeding is normal and this too must be shallow; deep-rooted grasses such as *I. cylindrica* are easily and cheaply eliminated with a herbicide-wand or spot spraying.

Mulching can also greatly reduce weed growth in addition to conserving moisture, and significantly increases berry yield. Mulching with rice husks, hay or sawdust at 3 t ha^{-1} in Brazil proved that sawdust was the most effective, with a seed yield of 3600–7800 kg ha^{-1} compared to rice, second at 3500–7300 kg ha^{-1}. Herbicides are little used, and as pepper is very shallow rooted, soil inactivated compounds are to be preferred, glyphosate, for example. All herbicides should be tested on a small plot before general use, as pepper cultivars reportedly differ in their reactions to a particular compound, especially the newer high-yielding cultivars.

Irrigation

Little information is available on the water requirements of pepper other than the optimum rainfall and its distribution. In India and Indonesia, any prolonged dry period in the annual monsoon results in lower yields. The effects of moisture stress on pot-grown pepper plants indicated a direct slowing in vegetative or reproductive growth depending on when the stress occurred (Krishnamurthy et al., 1998). Stressed plants seldom regain the loss in growth or yield, but substantial differences in cultivar resistance to moisture stress exist. The possibility of using certain plant reactions to water stress as a method of determining drought resistance has been investigated in India (Krishnamurthy et al., 2000).

Compounding the problem of water application is the frequently hilly lands on which the crop is grown. If additional water is necessary to ensure vine growth and fruiting, then the environment is generally unsuitable for commercial pepper production, and it is doubtful if pepper could be profitable as an irrigated crop at current or foreseeable prices.

Intercropping and rotations

Intercropping pepper vines in commercial plantations of other crops is generally not recommended as there is frequently a conflict between management techniques, except in specific circumstances such as coffee in India (Korikanthimath et al., 1997; Damodaran, 1998). Cocoa, for example, requires a fairly heavy shade, which is inimical to pepper, while mango is host to insects that can devastate pepper vines (Korikanthimath et al., 1994). Interplanting in plantations of jackfruit (Artocarpus heterophyllus Lam.), areca palm (Areca catechu L.) and kapok (Ceiba pentandra Gaertn.) is more successful.

Pepper can be interplanted with suitable local crops in the year of establishment, including capsicum, groundnuts, soya and other beans, low-growing fodder crops for livestock, and less often tobacco and ginger. It is recommended that 1, preferably 2, years of alternative crops should be grown before replanting vines, which will greatly reduce the incidence of soil-borne diseases and pests.

Pests

Insects attacking flowers, flowering shoots and berries are the most damaging; foliage eaters and insects damaging roots are of lesser importance, except in specific locations in a particular season. An additional advantage of pruning vines is that the flowering and fruiting period is condensed. Thus, the time spikes are exposed to insect attack is also reduced, and the opportunity for effective control enhanced. The number of insect and related pests reported from pepper in a specific country is large, but those causing economic damage or requiring control is much less – in India about 40, in Indonesia and Malaysia 35, with about a dozen accounting for the majority of damage.

In Asia generally the pepper flea beetle Longitarsus nigripennis is a major pest, especially in India where it is known as the pollu beetle and associated with hollow berry disease (see Disease section). The small, shiny blue and yellow adults bore fruits and deposit eggs, which hatch in 5–8 days. The pale yellow larvae eat their way out and then bore into and eat out other fruits leaving a hollow shell. A single larva can damage a number of fruits before dropping to the ground to pupate below the surface. Damage can reach 30–40% of a specific crop. Fruits of high-yielding cultivars are reportedly more susceptible. Altitude affects the degree of infestation in India, where plantations at or near sea level were severely attacked, damage becoming less with altitude until at 900 m it was virtually nil, but this factor has not been mentioned elsewhere. The beetle is most damaging during periods of heavy rain, which hampers control measures.

Heavy shade also increases the rate of infestation and total berry damage.

Pepper weevils, *Lophobaris* spp., are serious pests in the Indonesian-Sarawak region, but less so in other countries; most damaging are *Lophobaris serratipes* and *Lophobaris piperis*, eating holes in developing fruits and causing premature shedding. A severe, uncontrolled infestation can result in massive berry loss. In India, *Eugnathus curvus* is most damaging.

A top-shoot borer very damaging in India is *Cydia* (*Laspeyresia*) *hemidoxa*, a tiny crimson and yellow moth whose larvae bore terminal and other shoots, causing dieback. The pest is most common in August–October when new shoots appear, and control measures are then essential as damage to 50% of shoots has occurred. Mirids also cause malformation and death of shoots, young stems and leaves, for example, in India and Sri Lanka, the slender dark brown *Disphinctus maesarum* and *Helopeltis antonii*, which also attacks tea, and numbers can increase very rapidly since the development period is very short. The most important pest in Cambodia is a small black chafer, *Elasmognathus nepalensis*, which feeds by sucking flower stalks, which then wither and fall. Damage varies seasonally and in a year of severe infestation, vines are virtually devoid of flowers.

The bugs *Dasynus piperis* and *Diplogomphus hewitti* commonly attack fruits in Indonesia, Sri Lanka and India. The first-named is also considered a most damaging pest of pepper in the Indo-China region, especially Kampuchea. Of minor importance in the region is the beetle *Pelargoderus bipunctatus* but *H. antonii* is also a problem in Sarawak. Thrips may be a pest, including the gall-producing thrips, such as *Gynaikothrips chavicae* generally, but the most important in India is *Liothrips karnyi*. The pepper gall midge, *Cecidomyia malabarensis*, is also a minor pest from West Asia to India.

The greyish-green larva of the small brown moth, *Thosea sinensis*, can cause serious damage to foliage, and has been reported on pepper from India through Indonesia, South-east Asia to China. Larvae of another moth *Cricula trifenestrata*, the mango hairy caterpillar, can become a major

pest, especially where mango trees are used to support or shade vines, which can be completely defoliated. Caterpillars of both these moths bear urticating hairs, and when present in numbers make berry picking almost impossible. Stem borers may cause damage but are seldom of major importance; most often reported in India are *Pterolophia annulata* and *Diboma procera* whose larvae bore the main stem of mature vines.

Scale insects and mealy bugs debilitate vines and can cause severe damage if not controlled. Most often mentioned are the cosmopolitan *Ferrisia virgata* and *Saissetia coffeae*, the pepperscale *Lepidosaphes piperis* in India and Sri Lanka, and *Planococcus citri* in Indo-China, especially Kampuchea.

The roots and stems of young plants can be damaged by the larvae of scarab beetles (Figs 8.5 and 6.4). Roots are also frequently host to various eelworm species including the root-knot eelworms, *M. javanica* and *M. incognita*, and the burrowing nematode *Radopholus similis*. The extent of damage has not always been accurately determined, nor the exact relationship between root disease and local eelworm species. An association between *R. similis* and incidence of vine wilt was established in India, and control of the

Fig. 8.5. Scarab beetle.

nematode increased vine vigour. A relationship between nematodes and incidence of pepper yellows caused by *F. oxysporum* has also been observed in Indonesia (Sitepu, 1993). In Brazil, treatment of cuttings with a nematicide significantly increased growth, and different mulches had varying effects on the growth of vines and soil nematode populations; a combination of both is recommended. Nematicides also controlled soil nematodes in India and in Indonesia, but some were toxic to vines.

Diseases

A number of pathogenic fungi have been reported from *P. nigrum*, but most are of minor importance, and the potential damage from fungal diseases can be reduced by good plantation management, and many are readily controlled by fungicides.

Probably the most economically important are *Phytophthora* species, which occur in pepper-producing areas worldwide. A disease caused by *Phytophthora capsici* (*Phytopthora palmivora*) is known as quick wilt in India and Sarawak and sudden death disease in Indonesia, because of its effect on vines. Control is difficult as the pathogen has a wide host range. Overall loss from this disease in Kerala State, India, was estimated at 9% of total yield (Prabhakaran, 1997), and responsible for the loss of 120 t pepper in one district in the same state (Nair and Sasikumaran, 1991).

Initial infection in the field begins in minor roots a few centimetres below soil level, then spreads to major roots and stems. When an appreciable portion of the underground stem has been attacked wilt symptoms appear, usually 50–60 days after infection. The first above-ground symptom is a slight droop over the whole vine, followed by yellowing of leaves, which usually fall within 7–14 days leaving branches bare, and the vine dies. In wet weather, leaves may be directly attacked causing brown circular lesions with a fimbriate edge; abscission occurs before infection can spread to the stem. Artificial inoculation of root cuttings produced typical symptoms after 6–9 days, indicating the ease and rapidity with which

the pathogen can spread. Snails and nematodes are vectors.

The damage caused by quick wilt in India is substantial as noted, and the disease is now the major constraint on expansion of pepper production in Indonesia. Control of the disease or limiting its spread is difficult, because most pepper is grown by small farmers, few of whom apply either pesticides or fungicides. In Brazil, the disease can be devastating on susceptible varieties; plantings in Bahia Province were reduced by half prior to the routine application of fungicides (Tsuda, 1989). In Brazil and Puerto Rico, outbreaks of root rot caused by *P. capsici* usually occurred when vines began fruiting, but the incidence was low in areas where pepper had not previously been planted. Resistance at various levels may exist in some cultivars (Mozzetti *et al.*, 1995), and a biological control programme has proved of value in India (Anandaraj and Sarma, 1994; Jubina and Girija, 1998).

Resistance to *P. capsici* varies substantially, with *Piper* spp. from South-east Asia generally highly susceptible, but American species resistant or slightly to moderately susceptible; *Peperomia* spp. are considered resistant. *Piper colubrinum* was resistant to *P. palmivora* and *Fusarium solani* ssp. *piperis*, and used as rootstock for susceptible *P. nigrum*, but grafts deteriorated after the fourth year. Research into control of *P. capsici* is a major project at the MARDI Biotechnology Centre, which is also producing pathogen-free plants by *in vitro* culture.

Root rots or wilts caused by the following are locally damaging: *Fusarium* spp., *Diplodia* spp., *Rosellinia bunodes* and *Sclerotium rolfsii* (basal wilt, Fig. 8.6.) in various regions of India; *F. lignosus*, *Ganoderma lucidum* (red root rot) and *R. solani* in Sarawak; and various *Pythium* spp. in Indo-China. Root rot due to *F. solani* f.sp. *piperis* is the most damaging disease in Brazil, and control with arbuscular mycorrhizal fungi is being investigated (Chu *et al.*, 1997). The importance of root nematodes in spreading root and other diseases has still to be determined.

Diseases that can originate in the canopy are often due to *Corticium salmonicolor* (pink disease) or *Marasmius scandens* (thread

Fig. 8.6. Sclerotina base rot.

blight), and both may cause almost complete necrosis of leaves and branches. They can be distinguished from root rot as symptoms differ. In pink disease, white pustules occur on bark or lateral branches, followed by the pink encrusting mass of the fungus. In thread blight, cobweb-like strands grow over leaves, petioles and stems. A parasitic alga, *Cephaleuros parasiticus* (black berry disease), may cause serious loss in Sarawak and to a lesser extent in western Malaysia. The algae attacks young fruits, whose skin turns black and the fruit falls prematurely. Leaf anthracnose caused by *Colletotrichum* spp. is damaging to some extent in Malaysia and China, but apparently of little importance elsewhere. In India, *C. capsici* can be very damaging; a severe infection causes

defoliation, a lesser results in some leaf drop hollow or shrivelled berries. Varietal resistance to the disease may exist. *Xanthomonas beticola* causes water-soaked lesions on leaves in Sri Lanka. *Cochliobolus geniculatus* and *Nectria haematococca* commonly infect berries in Malaysia.

An unnamed virus attacks vines in Indonesia and Sarawak, and a mycroplasma-like organism or virus has been noted from Brazil and India. The piper yellow mottle virus (PYMV) found on *P. nigrum* and *P. betle* in South-east Asia, Indonesia and the Philippines is apparently mechanically transmitted by the citrus mealy bug, *P. citri*. The pepper yellow disease in Indonesia is, however, due not to disease but to a nutrient deficiency, and in Malaysia a manganese deficiency produces dark brown to black spots on leaves. The condition known as *pollu*, meaning hollow in India, may be due to several causes: physiological, insect attack mainly by the pollu beetle *L. nigripennis*, or infection by *Colletotrichum* spp. It is thus not strictly a disease but a combination of factors, which results in hollow berries and spike abcission. Loss from *pollu* can reach 20% of total yield in a particular year.

Fungal contamination commonly due to *Aspergillus* spp. and *Penicillium* spp. can become a problem during berry drying and initial storage. To reduce the incidence of infection it is imperative to dry berries as quickly as possible to less than 11% moisture, and maintain this level as long as required.

Harvesting

Pepper vines are normally harvested 18–20 months after planting out, but other factors may locally affect time of harvest; in Sri Lanka, for instance, berries are often picked earlier than the optimum time to reduce loss through theft! Picking in India is January–March in the hills, November–January on the plains, in Sarawak it is May–August with the heaviest yield in July–August, in Indonesia August–November. The main harvest in Sri Lanka is October–January with a minor crop in June–July, and in peninsular Malaysia in August–September and March–April.

Smallholder vines are usually harvested year-round. Time of picking (i.e. maturity of berry) affects berry composition (Table 8.1) and berry oil (Table 8.2).

Picking for white pepper begins when one to three berries on a fruiting spike turn red, for black pepper when berries are still green, and picking continues every 2–3 weeks until the end of harvest. The application of growth hormones to spikes to accelerate maturity had little effect, although berries tended to ripen more evenly and be heavier.

Tripod ladders are used to reach higher branches and avoid damage to vines. When all fruiting branches have been removed on intensively managed plantations, branches are stripped of leaves except for one or two terminal leaves. Damaged and diseased branches are also removed. The yield of green berries per vine varies widely, from a few grams to a very high 10 kg from specially selected and treated plants and indicates the potential yield possible. In general, any improvement in cultural techniques by smallholders quickly increases yield and may also extend vine life, which can be less than 10 years of a potential twice this.

Green berry yield from well-managed plantations in Sarawak averages 8000–9000 kg ha^{-1} from the first picking rising to 18,000 kg ha^{-1} at the sixth to seventh harvest, which is maintained to the tenth year. In India, yield is generally lower at 100–400 kg ha^{-1} since pepper is a smallholder crop, with a vine life of 25 years; efficiently managed plantations, however, produce 900–1000 kg dried berries per 1000 mature vines (approximately 1 ha). In India, high-yielding selections under trial showed a yield variation in alternate years greater than lower yielding types (Kumar *et al.*, 1999).

The average yield from smallholders is (ha^{-1}) Kampuchea 1450 kg, Sumatra 1350 kg, Sri Lanka 1350–2500 kg, Malaysia 1200–1500 kg from mature vines in full bearing, in Brazil 600–4000 kg depending on variety, location and standard of management (Milanez *et al.*, 1987), and in the Philippines an average 3000 kg from well-managed mature vines (Malijan, 1982).

Little published information is available on the establishment cost of a commercial pepper plantation. In Brazil, 1 ha at a spacing of 2 × 2 m cost $12,700 and returned $15,867 over a 16-year period. Costs were recovered in the tenth year at 2 × 2 m spacing, in the eleventh at 2 × 2.5 m, in the ninth at 2.5 × 2.5 m, and in the eighth at 3 × 3 m (de Brandao *et al.*, 1978). In Indonesia, a dwarf vine planted at a close spacing and producing berries in its second year is considered to be potentially more profitable than normal vine plantations (Wahid and Zaubin, 1993). The economics of pepper production in India was calculated in detail by the Central Plantation Crops Research Institute (CPCRI). In 1983 and based on 10 years of average crops, the full cost of establishment will be recouped in the seventh year. Cost of production was calculated at Rs 9.25 kg^{-1}. The life of a commercial plantation in India depends directly on the standard of management and at least 12 years of full bearing from planting out should be the objective with a further 3 years of declining yields; currently yield usually declines from the ninth to the twelfth year.

Processing

Fruits on a spike do not mature together, and when a number are ripe a spike is harvested; unripe, green berries are processed to become black pepper, ripe berries white pepper. The two operations are dissimilar. Unripe fruits are separated, sun-dried usually after soaking in hot water, although they may also be smoke-dried, and produce black pepper. Ripe fruits are soaked in running water, the pericarp removed, and the dried berries ground to form the spice, white pepper. Black pepper produced from unripe fruits is the more pungent, due to the presence of various resins and a yellow crystalline alkaloid, piperine (the *trans–trans* form of N-piperoylpiperidine). The proportion of the crop made into white or black pepper usually depends on the price differential at the time, except in India, which produces almost exclusively black pepper.

The following details of berry processing in India were kindly supplied by the Pepper Board of Kerala. Similar methods are used in the majority of producing countries where smallholders predominate. More efficient and modern methods, which greatly reduce the risk of microbial contamination, are described later. A locally made power-operated pepper thresher to be used by cooperatives or larger traders has been developed in India, which greatly increases the throughput per worker (Kumar *et al.*, 1998).

Black pepper. When the majority of the crop is required to produce black pepper, spikes are harvested when berries are mature but still green. Harvested spikes are left in heaps for several hours to initiate slight fermentation, which browns the berries, which are then stripped from spikes, spread on mats to dry, and turned regularly to ensure uniform colouring and prevent fungal infection. Bamboo mats are preferred but thick black plastic is becoming common, and is also used to protect berries from rain and dew. Sun-drying requires up to 14 days, depending on the season, to reduce moisture content to around 12%, and during drying, up to 20% volatile oil can be lost. The objective is to produce berries with a uniform dark-brown to black colour, free of mould, and with the characteristic pungent aroma. About 35 kg black pepper is produced from 100 kg newly picked green pepper. The majority of the Indian crop becomes black pepper, as does that of the Philippines and pepper grown in Lampung, Indonesia.

This system is also common in Sarawak, but in Indonesia, berries are usually left on spikes for the first few days of sun-drying, then detached. Immersing berry clusters in boiling water for about 10 min prior to sun-drying, which hastens browning and drying, is common in Indonesia and Sri Lanka, but prolonged immersion can deactivate the enzymes responsible for browning. This method allows a large volume of berries to be quickly processed.

Solar energy is a simple and cost-effective method whereby smallholders or their cooperatives can partially or wholly replace sun-drying, and eliminate many of its problems (Kamruddin and Wenur, 1994). A range of solar-powered dryers of varying capacity are now commercially available, which partially or completely dry fresh berries. Cheap locally produced solar dryers have been developed in India, Malaysia and Indonesia, similar to those in Fig. I.2 (see Introduction). Continuous-flow hot-air dryers are now more common and suited for a range of spice crops, including pepper. A machine imported from The Netherlands by the Spice Board of Kerela dried and winnowed up to 40 t in 8 h, but caused social problems as it replaced the large number of workers formerly employed. In addition to the major advantage of virtually eliminating microbial contamination, as for vanilla, dryers are also an advantage where harvesting may overlap with a rainy season. Dryers are equally effective for black or white pepper, provided the latter is fully decorticated.

The original Brazilian hot-air drum dryers had a capacity of 9 t fresh green berries, heated to about 80°C for two periods of 4.5 h with a 6 h interval, temperatures very high in comparison with those recommended for most other spices. These temperatures have since been reduced following complaints from importers of loss of pungency; in addition, the depth of berries in horizontal dryers has been reduced and the drying periods varied to take into account the moisture content of berries received. The drying operation must also be followed by efficient storage and handling procedures, as berries rapidly reabsorb moisture when ambient humidity is high.

White pepper. When most of the crop is required to produce white pepper, spikes are not harvested until a majority of berries have turned red. In India, berries are stripped from spikes, packed in sacks, and placed in clean, running water (increasingly difficult to find naturally and thus a potential source of contamination) for up to 14 days, which softens and loosens the pericarp. The period for which sacks are left in the water depends on the maturity of the berries and experience of the grower. Bags are then placed in a tank of water and trampled slowly but firmly to remove any adhering pericarp and pulp; finally bags are replaced in running water until all discoloration disappears. The buff-coloured berries are then sun-dried as for black, except that if the weather becomes cloudy berries are submerged in water until the sun reappears. Thus, the drying process can be prolonged and the risk of contamination increased. The objective is to produce berries with a uniform buff colour, free of mould, and with a mild but characteristic aroma. About 25 kg white pepper is produced from 100 kg fresh green berries. Discarded hulls, approximately 25% of dry berry weight, can be distilled to produce pepper oil.

While this method is followed by many smallholders, improved processes, which are wholly or partially mechanized, have been introduced by the National Research and Development Corporation. Their 'Improved Process' involves passing dried black berries through a special mill to remove the pericarp, and automatically grading the white pepper produced. This is then packed in airtight containers for export. The CFTRI (Central Food Technological Research Institute) Process involves steaming black berries for 10–15 min, then passing through a pulper, which removes the softened pericarp, followed by drying. Yield is around 20%. This method is fast and greatly reduces bacterial contamination. A fully automatic continuous flow pulper/dryer is under test, to be used in conjunction with a still to produce oil from separated skins and pulp.

Sarawak berries are processed into white or black pepper, generally depending on the ruling price differential, but pepper from Bangka Island, off Sumatra, Indonesia, known in the trade as Muntok, is predominantly white. Indonesian white pepper is produced in a similar method to Indian; fully ripened black berries are retted, abrasion-peeled and sun-dried. The operation is performed by large traders to whom growers deliver their berries, and has recently been partially mechanized.

Distillation

Oil is mainly distilled from black berries and although white berries can be used, they produce a very similar oil at a lower yield, with no commercial advantage. Distillation of pepper berries is uncomplicated, most modern stills are suitable, with no special equipment or technique necessary. Whole berries can be distilled but the process is time-consuming, although it has an advantage to the unscrupulous in producing countries; spent berries can be redried and sold to the unsuspecting as spice! Berries are normally reduced to a coarse powder and on steam distilling yield 1.0–4.0% oil, but up to 9% has been obtained under controlled conditions from specially selected cultivars. The majority of pepper oil distilled in consuming countries was from Indonesian (Lampong) or Indian (Malabar) black berries; the latter usually has a slightly higher oil content. In Comoro, Madagascar and Indonesia, good-quality oil is distilled from siftings, low-grade berries, and husks from white pepper production.

Pepper oil can also be obtained by vacuum distilling oleoresin, which is closer in composition, aroma and flavour to the natural oil present in the spice, and thus superior to steam-distilled berry oils. However, oil and oleoresin are increasingly obtained by supercritical fluid extraction using carbon dioxide as the solvent, which produces an oil with very similar characteristics to the natural berry oil. The process is adaptable to both small and large units, and is simple and relatively inexpensive (Prulhy, 1993). Complete plants with varying throughputs and ready to be installed are available from specialist manufacturers. Oil and oleoresin producers in consuming countries prefer berries or grades from a known origin, as noted, sometimes from a specific cultivar or district, since cultivar differences exist in oil content and characteristics.

Storage and transport

Pepper is mainly produced by smallholders who leave the final cleaning, grading and bagging of the dried spice to local merchants or exporters. However, if the berries delivered to traders are insufficiently or carelessly dried, as frequently occurs due to laziness, inclement weather or a pressing need for money, no amount of recleaning and redrying will improve the quality. Stored berries are often attacked, most commonly by *S. panicea*, *P. farinalis* and *L. serricorne*. Application of berry oil to pepper and other whole spices slows the rate of population growth and inhibits larval development. Washing and redrying to reduce microbiological contamination is also effective in removing most surface insect contamination, but fumigation may be necessary in serious infestations.

Mould contamination is essentially restricted to the berry surface, and washing berries in water containing an approved sterilant and redrying is normally effective, inexpensive and does not affect berry quality. Chemical treatment may cause residue problems and ionizing radiation affect spice quality by reducing the pungent principle piperine; however, ozone treatment oxidized some oil constituents but did not affect oil in whole berries (Zhao and Cranston, 1995).

Following cleaning and treatment, the berries are mechanically sorted by size into internationally recognized grades. These differ substantially by origin, but remain constant from a given origin, and may be required to meet national export specifications. For example, the composition of Indian pepper grades are shown in Table 8.3; grades are also discussed in the next section.

Exporters generally store bagged berries in godowns prior to shipment, and prolonged storage under these conditions can cause significant deterioration; up to 20% of volatile oil can be lost after 6 months' storage at ambient temperature and humidity, thus reducing pungency of the spice and its value. Stacking must ensure ventilation without the weight of filled sacks crushing berries, and stacking to about 3 m high with a space of 75 cm between stacks is recommended in government warehouses in India. Ground berries must be stored in sealed containers or specially lined sacks and shipped with the minimum delay. Changes in berry composition, black and white, while in storage can produce off-flavours in both berries and extracts. Compounds affecting these changes have been determined (Jagella and Grosch, 1999). Methods of detecting irradiated berries have been developed (Satyendra *et al.*, 1998).

Storage also includes preserving seed for later planting, and investigations in India showed that berries harvested with a high moisture content then dried to 12% retained

Table 8.3. Composition of pepper grades in India.

Grades	Average berry wt (g)	Bulk density (g l⁻¹)	Density (g ml⁻¹)	Moisture (%)	Oil (%)	Piperine (%)	NVEE (%)	Starch (%)	Crude fibre (%)
Pin heads	0.004	280	0.571	13.0	0.6	0.8	7.1	11.5	27.4
Light pepper	0.010	240	0.476	13.0	2.9	4.1	13.5	14.6	27.8
Malabar garbled 1	0.042	540	0.952	13.0	3.7	5.0	12.3	39.7	11.8
Tellicherry garbled extra bold	0.057	544	1.000	13.0	2.2	4.4	9.1	39.7	10.8
Tellicherry garbled special extra bold	0.061	592	1.053	10.0	3.2	4.9	10.3	40.9	9.2
Malabar ungarbled	0.039	540	0.952	12.0	2.8	5.0	11.4	41.8	12.5
Tellicherry ungarbled	0.034	528	0.952	12.0	4.0	6.3	13.5	39.3	11.0
High range ungarbled	0.040	600	1.053	12.0	2.6	4.0	11.1	41.8	10.5

All figures are on a dry weight basis.
NVEE, non-volatile ether extract.

their viability, but below 12% steadily lost viability in normal storage; if preserved in liquid nitrogen at −196°C seeds retained up to 50% viability depending on the period of storage (Chaudhury and Chandel, 1994).

All traditionally harvested and dried pepper berries are liable to a high degree of microbial contamination, as noted, and many importing countries have introduced legal standards governing the degree and type of such contamination or imposed statutory treatments. Reducing fungal infestation at producer level with the help of specially trained extension workers and introduction of appropriate technology is thus important (Ahlert *et al.*, 1998). Methods of reducing dangerous contaminants, including aflatoxin, on berries and in derivatives are under continual review, including steam decontamination of berries before sale or processing, but these may also reduce the quality of the end product. Irradiation is also effective but requires special equipment and skilled operators; cobalt-60 at 10 kGy protected berries for 3 months in India at an ambient temperature of 28–30°C. Techniques for determining whether berries have been irradiated have also been developed.

Marketing

Pepper is a major crop in the main producer countries and export marketing is usually under some kind of official supervision, if only to ensure grading standards are enforced. However, primary marketing by growers is generally unregulated, but usually follows a similar pattern from grower to exporter. The traditional system that has evolved in India is typical. Smallholders normally sell their produce to village merchants, who then sell on at local markets to larger merchants, who bulk their purchases for sale to wholesale traders, themselves usually exporters' agents. Cooperative societies operating in pepper-producing areas also serve as collecting agencies for their members, and can obtain higher prices by selling directly to wholesalers or exporters. Large-scale pepper growers or commercial planters sell directly to traders or wholesalers, but if large enough may sell directly to exporters or overseas customers. Most pepper finally arrives at export markets in Cochin, Calicut and Alleppey, where commission agents sell the produce to wholesale traders and exporters. Exporters also trade in Pepper Futures on the various local exchanges, or the recently established Dollar (US) Trading Section, which quotes FOB. In mid-2000 an official on-line market was established, although individual exporters were already using this medium.

Products and specifications

Pepper vines produce a berry, berry oil and oleoresin. The most important product is the berry, which provides black pepper from the whole berry and white pepper from black berries after removal of the mesocarp. All products are derived from berries, thus their quality basically determines that of any derivative. The major constituent governing berry pungency and taste is the alkaloid piperine, described at the end of this section, together with other alkaloids concerned.

The major producing countries have introduced national standards governing the quality of berries exported, and to these may be added grades from pepper-producing regions. It is not necessary to detail the many different grades of berries from exporting countries, as current specifications are easily available on the websites of national spice organizations; the Indian Agmark grades are given as an example:

> Whole pepper;
> Malabar Garbled (MG Grades 1 and 2) Black
>   Pepper
> Malabar Ungarbled (MUG Grades 1 and 2)
>   Black Pepper
> Tellicherry Garbled Black Pepper Special Extra
>   Bold (TGSEB)
> Tellicherry Garbled Extra Bold (TGEB)
> Tellicherry Garbled (TG)
> Garbled Light Black Pepper (GL Special, GL
>   Grades 1 and 2)
> Ungarbled Light Black Pepper (UGL Special,
>   UGL Grades 1 and 2)
> Pinheads (PH Grade Special and PH Grade 1)

Black Pepper (Non-specified (NS) Grade X)
Ground black pepper;
Standards and General Grades (2 grades only)

Indian berries from Malabar and Tellicherry, Kerala, are highly regarded in the trade as they have large, evenly coloured black berries with excellent flavour and pungency. Lampong, the highest quality Indonesian, has smaller more pungent berries than Indian, which are favoured by distillers and extractors. Sarawak (Malaysia) berries have a less pungent, milder flavour and are mainly used as a flavouring and domestic spice. Sri Lankan berries are large, bold, with high oil and extractive content, favoured by oleoresin manufacturers. The grades and origins of white pepper are less important since most is ground for spice, and white berries are less pungent and flavoursome than black from any origin.

Black berries (Fig. 8.7) are used mainly as a spice and as raw material for the production of pepper oil and oleoresin. Berries for domestic spice may be whole, cracked, coarsely or finely ground, and virtually all dishes benefit from adding pepper, other than sweet foods. In the food-processing industry, black pepper is usually ground and added to an infinite variety of meat and similar foods.

White pepper (Fig. 8.8) is less pungent, has a more mellow flavour and is low in fibre but high in starch. This pepper may be used in foods and food processing similarly to black, but is especially useful in sauces, mayonnaise and similar light-coloured products or dishes where the dark specks of black pepper are undesirable. The majority of white pepper is sold as powder in importing countries. The best grades of Indian and Indonesian Muntok are preferred in the trade for retail sale; Brazilian is lighter, very uniform in colour and less pungent; Sarawak is often included in blended powders.

Whilst whole dried black and white berries account for most of the pepper traded, a range of berry products are now available to satisfy consumer demand; for example, decorticated black pepper is an alternative or substitute for white pepper. This is prepared from dry black berries milled to remove the pericarp (outer skin) and tissue, which is a skilled operation and machinery must be accurately adjusted for each batch of berries. As one operator commented 'It looks like white but tastes and smells like black!' It has strictly limited uses.

Green pepper (Fig. 8.9) is made from mature but green berries, and either dehydrated or preserved in vinegar, brine or other liquids, canned or bottled, and is mainly used as a domestic and table spice (Mathew, 1993). The canned spice owes its popularity to its attractive green colour, freshness, aroma and mild pungency. For use by food processors it is exported in drums containing up to 35 l. The NRDC (National Research and Development Corporation) are currently investigating more efficient and faster methods of processing green pepper as

Fig. 8.7. Dried black pepper.

Fig. 8.8. Dried white pepper.

Fig. 8.9. Dried green pepper.

it is a very profitable value-added pepper product. This research may also require development of specific cultivars to extend the period green berries are available.

Dehydrated green pepper is prepared in modern continuous-flow dryers or dehydrators under controlled conditions. The product's main advantage is that it is light, with much lower transport costs, although having less aroma and pungency than preserved green pepper when reconstituted. Its main use is in food processing because of its lower delivered cost. Freeze-dried green pepper is superior to all other green peppers and is considerably more expensive, but has a growing market in the EU. Similarly, frozen green pepper has high-priced niche markets in some Western countries.

Pink pepper is prepared by harvesting ripe red berries which are cleaned, graded, pickled in brine, and marketed in bulk or in retail bottles. Pink pepper from *P. nigrum* should not be confused with pink pepper made from the artificially coloured berries of *Schinus molle*, the pepper tree, also marketed as pink pepper.

Pepper oil is produced almost wholly from black berries, and is a colourless to pale green free-flowing liquid becoming viscous with age. Oil odour is similar to freshly prepared spice, but the flavour is rather flat and lacking pungency since the alkaloids involved are not soluble. A number of pepper oils are available commercially, differentiated by the method of distilling or raw material. The main commercial oils conform to the specifications in Table 8.4. Two grades are produced on Madagascar by fractionation: light oil or forerun, and a heavy oil consisting of higher fractions. White pepper oil is occasionally produced using the same methods as for black, but is normally more expensive due to the higher cost of raw material and lower yield. Little difference can be detected between the two oils when produced from similar-quality material.

The aroma and flavour of steam-distilled oil is determined mainly by monoterpene hydrocarbons plus smaller amounts of sesquiterpene hydrocarbons; oxygenated compounds are relatively minor constituents but important in determining organoleptic properties; pungency is due to non-steam-volatile alkaloids, principally piperine, as in Cuban oils (Pino and Borges, 1999c). However, oils produced by supercritical fluid extraction differ from distilled oils as

Table 8.4. Published standards for black pepper oil.

	ISO 3061	BS 2999/12	EOA[a]	AMC-EO
Apparent density (20°C)	0.870–0.890	0.868–0.907	0.864–0.884[b]	0.872
Refractive index (20°C)	1.480–1.492	1.480–1.492	1.4795–1.4880	1.4835
Optical rotation (20°C)	−16° to +4°	−15° to +4°	−23° to −1°	−7.2°
Ester value (max)	11	11	ng	4
Solubility (v/v 95% ethanol (20°C))	1:3	1:3	1:3	1:2.8

ISO, International Standards Organization; BS, British Standard Specification; EOA, Essential Oil Association, USA; AMC-EO, Analytical Methods Committee – Essential Oils.
[a] Mainly Indonesian berries.
[b] 25°C.
ng: not given.
Figures in columns shown the range.

the method preserves some important heat-sensitive components, and is considered most accurately to reflect the pungency and aroma of the berry.

The chemistry of *P. nigrum* berries, oil and other extracts and their non-food uses has been extensively investigated, and for comprehensive details and methods employed, the relevant literature should be consulted since it is outside the scope of this book; descriptions of the oil and oleoresins have been published (Purseglove, 1981a; Weiss, 1997). Over 100 constituents have been identified in pepper oil and oleoresin, not all found in a particular sample. Constituents include 15 monoterpene hydrocarbons (70–80%) predominantly pinene, β-pinene and limonene, 21 sesquiterpene hydrocarbons the major being β-caryophyllene, 32 oxygenated monoterpenes, four phenyl esters, 12 oxygenated sesquiterpenes and 20 miscellaneous compounds (Pino, 1990). The composition of black pepper oil can vary considerably according to origin and method of preparation (i.e. harvesting and drying). The greatest variation is within the monoterpene hydrocarbon group as follows: limonene (0–40%), β-pinene (5–35%), α-pinene (1–19%), α-phellandrene (1–27%), β-phellandrene (0–19%), sabinene (0–20%), δ-3-carene (trace to 15%), myrcene (trace to 10%). Sesquiterpene hydrocarbons and oxygenated compounds also vary: β-caryophyllene (9–33%), α-humulene (2–6%), α-bergamotene (1–4%). A monograph on the physiological properties of black pepper oil has been published by the Research Institute for Fragrance Materials.

The main use for pepper oil is in flavourings where pungency is unnecessary, but is increasingly being replaced by oleoresin. Pepper oil is often adulterated with other oils or synthetics, including phellandrene, limonene, *S. molle* oil, or sesquiterpenes from clove or *Eucalyptus dives* oils. Poor quality is frequently confused with adulteration, due to distilling low-grade material (Rusli and Nurdjannah, 1993).

Berries and oil are considered to have very similar properties in traditional medicine and herbal remedies. Both are described as being, to varying degrees, stomachic,

carminative, an aromatic stimulant, antibacterial and insecticidal. Diluted oil or seed extract may be used as a gargle and externally as a rubifacient.

Oleoresins are normally prepared by solvent-extracting black pepper berries, but can also be obtained by supercritical extraction using carbon dioxide. Berries are normally comminuted to flakes 0.05 mm thick or coarse powder 0.2–0.3 mm in diameter, and solvent-extracted with acetone, ethanol or dichloroethane. Oleoresins can also be obtained by solvent-extracting pepper oil or distilling residues. Fresh oleoresin is a dark green or dark brownish-green viscous liquid, containing piperine, several other minor alkaloids and may contain caryophyllene. Crystals of piperine appear on standing and the oleoresin requires mixing before use to ensure uniformity.

Oleoresin prepared by extracting with organic solvents is very similar to the odour, flavour and pungent principle of the original material, although supercritical-extracted oleoresins are considered closest to the pepper berry. A comparison of solvent-extracted oleoresins is shown in Table 8.5. Oleoresin yields of 10–13% are obtained from Indian Malabar pepper, the highest using Malabar Light (Mathew, 1990). Commercial products contain 15–20% volatile oil and 35–55% piperine (see below). One kilogram of oleoresin obtained from 8 kg of black pepper and dispersed on an inert base can replace up to 25 kg of the spice for flavouring purposes. The oleoresin is available as a liquid, dissolved in an edible solvent (e.g. polyoxyethylene sorbitan fatty acid esters), emulsified with gum arabic and dispersed on an edible dry powder carrier (salt or sugar), encapsulated in gum arabic or gelatine. The principal use for oleoresin is in processed and canned meats, pickles, dressings of all kinds and a variety of other food products where the spice enhances their flavour. A decolorized oleoresin is available for special use. An extract of black pepper, ultrasonic, is normally free of harmful solvent residues and commercially available for use in processed foods. The maximum permitted level of the oil or oleoresin in food products is about 4%. Extracts of ground fruits have considerable

Table 8.5. A comparison of solvent-extracted oleoresins.

Origin	Volatile oil (% w/w)	Piperine (% w/w)
Brazilian	29–33	35–46
Lampung	29–34	43–48
Malabar	32–44	42–53
Sarawak	31–37	43–47

insecticidal activity and proved toxic to the common housefly and a number of crop pests in India, where research continues to use the oil for biological control of pests.

The regulatory status in the USA of generally regarded as safe has been accorded to black pepper, GRAS 2844, white pepper, GRAS 2850, black pepper oil, GRAS 2845, white pepper oil, GRAS 2851, black pepper oleoresin, GRAS 2846, and white pepper oleoresin, GRAS 2852.

The alkaloid piperine, present at between 4 and 10% in the crude form in berries, crystallizes in columns, has a melting point around 129°C, is insoluble in water, soluble in alcohol and intensely pungent, being at first tasteless but developing a burning aftertaste; it hydrolyses to piperic acid and piperidine. The amount of pure piperine is determined by analysis, mainly to detect adulteration. Chavicine, also present, is a resinous isomer of piperine and considered the most biting ingredient; it hydrolyses to isochavicinic acid. The isomers of piperine and relative pungency have been stated as: piperine (*trans-trans*) 5; isopiperine (*cis-trans*) 1; isochavicine (*trans-cis*) 1; chavicine (*cis-cis*) 2; (*trans-cis*) 1. Interestingly the pungency of piperine is only 200,000 Scoville units (SU), compared to 30 million SU for capsaicin from *Capsicum* species (Chapter 9).

Other Piper *species*

The species shown in Table 8.6 also have fruit used as a spice. Fruits of *Piper aduncum* L. are a pepper substitute. Leaves of *Piper pinnatum* Lour. are used to scent clothes in Indo-China, and *Piper bantamense* Blume. similarly in Malaysia. Leaves, roots and fruit of *Piper excelsum* Forst. have a variety of medicinal uses in the south Pacific region and Australasia. The pounded roots of *Piper methysticum* Forst. blended with coconut milk make the mildly intoxicating Polynesian drink kava-kava, and roots of *Piper medium* Jacq. are similarly used in central America. *Piper porphyrophyllum* N.E.Br. with its attractive purple and silver leaves is cultivated as an ornamental. *Piper hispidinervium* Jacq. and *Piper callosum* Mt., native to Brazil, are being investigated as future sources of natural safrole oil. Comparisons of berry composition, oils, etc. from some of these species have been published (Orjala *et al.*, 1993; Singh and Blumenthal, 1997; Amvam Zollo *et al.*, 1998; Martins *et al.*, 1998).

Unrelated species

S. molle L., the pepper tree, has pungent red berries, which are a poor substitute for true

Table 8.6. Other *Piper* species with fruit used as a spice.

Species	Fruit	Country
P. clusii DC.	Wild pepper	West Africa
P. guineense Schum.	Ashanti pepper	West Africa
P. longifolium R. & P	Wild pepper	South America
P. saigonense C.DC[a]	Vietnam pepper	Indo-China

[a]Probably a P. nigrum cultivar.

peppercorns; red or chilli pepper is the ground fruit of various *Capsicum* spp., discussed in Chapter 9; Japanese pepper is obtained from *Zanthoxylum piperitum* DC. and Jamaican pepper from *P. dioica* (L.) Merr. (Chapter 6). *Xylopia aethiopica* A. Rich., of West Africa and a pepper substitute, is considered to be the *Piper aethiopium* of 17th-century Europe. *A. melegueta*, grains of paradise, is described later.

Cubeb Pepper

P. cubeba L. (tailed pepper) is indigenous to Indonesia and Malaysia where its fruits have long been used by local peoples as a condiment and in herbal medicines. These uses were maintained by Arabs who traded with the region and named the fruit *kabab*. In the 13th century, it was being shipped from Java to China. It was one of the first peppers to reach Europe and was used mainly as a spice by the Greeks and Romans although it later became more widely used in medicine. By the end of the 17th century in Europe long pepper was both uncommon and expensive. *P. cubeba* enjoyed a brief revival in the 19th century as a medicine but has since almost disappeared. Ground berries steeped in wine or brandy produced a drink believed to be an aphrodisiac.

Cubeb pepper was regularly shipped from Indonesia to Europe and the USA in the early 20th century, but the trade gradually diminished to an average of 135 t annually, and virtually ceased after 1940. Trade is now mainly between neighbouring countries in South-east Asia, but most cubeb pepper produced is used in the country of origin. However, as with many other minor spices, demand could be stimulated in the developed countries by promoting it as an unusual flavouring and by emphasizing its organic nature (due mainly to neglect!).

A number of related species are locally substituted for *P. cubeba* or used to adulterate dried fruits; most common is *Piper crassipes* Korth. and *Piper mollissimum* Blume. In this text, cubeb refers to *P. cubeba* and its derivatives.

Botany

P. cubeba L.f. (syn. *Cubeba officinalis* Raf.) is known in English commonly as cubeb or tailed pepper, in French as *poivre cubeba*, in Arabic as *kabab* or *kubaba*, in Malaysian as *lada berekur* or *kemukus,* in Indonesian as usually *kemukus* and in Hindi as *kabab chini*. A number of somewhat ill-defined cultivars exist in Indonesia with local names; only two, *rinu katuncur* and *rinu cangke*, are considered to be genuine.

P. cubeba is a perennial, woody climber, the stem greyish, smooth, flexuous, and up to 15 m but generally less when cultivated, and the twigs glabrous. The leaves are coriaceous, sub-glabrous, entire, the base rounded, the apex acuminate, oblong to ovate-oblong and obliquely cordate, 8–15 × 2–9 cm; the lower surface has numerous small sunken glands and the petiole is 0.5–2.0 cm long. The inflorescence is a spike, 3–10 cm, bearing clusters of sometimes 50 or more dioecious flowers, with the female spike shorter than the male. The female flower has an oblong bract up to 5 × 8 mm, and three to five stigmas. The male flower has an oblong-ovate bract up to 2 × 1 mm, and three to five stamens. The fruit (berry), on a 3–15 mm stipe, is a small 6–8 mm diameter, reddish-yellow drupe drying to black, with a coarsely reticulated surface. The base of the pericarp extends into a pedicel-like formation and the globose seed is usually free from the pericarp. Berries separated from the spike retain the pedicel (stalk), hence the name tailed pepper, and this is a distinguishing feature of the fruits. The histology of the fruit and seed has been described (Parry, 1962). Dried fruit, or cubebs, contain about 1% cubebic acid, the colourless crystalline cubebin, and up to 20% essential oil.

Commercial samples of steam-distilled cubeb oil in the USA usually have the following characteristics: specific gravity (25°C) 0.905–0.925; refractive index (20°C) 1.4800–1.5020; optical rotation −20° to +15°; insoluble in water, soluble in 10 vols alcohol, very soluble in chloroform. The main constituents of cubeb oil are the sesquiterpenes cadinene and dipentene.

Ecology

Cubeb pepper is indigenous to tropical monsoon forests of Malaysia and Indonesia, where it is an uncommon minor crop, and also in India. Although introduced to other countries of the region, it is seldom cultivated, except on a very minor scale in India, although vines can often be found as an escape from cultivation. On Java, *P. cubeba* plants are abundant on the landward boundaries of mangrove forests, where it is cultivated from sea level to 700 m, and it prefers partial or intermittent to full shade.

Soils and fertilizers

Cubeb pepper grows naturally on the humic soils of tropical forests, and clay loams when cultivated. Plots are mulched to provide ground cover and nutrients. No data are available on the effects of any fertilizers.

Cultivation

Cubeb is wholly a smallholder crop using similar methods to black pepper, but vines are less robust then pepper and require more attention initially. Cubeb pepper can be grown from seed or more commonly from cuttings taken from basal shoots. These are grown in pots and when rooted are planted out at the base of a tree and regularly weeded. When the stem begins to elongate, basal shoots are removed and the stem trained up support trees. After about 12 months, stems are some 1.0–2.0 m, and subsequently receive little attention other than the occasional slashing of surrounding undergrowth. Once vines have flowered in about 12–14 months and their sex has been established, one male is left for every nine females.

Unlike black pepper there are no cubeb plantations, and vines are grown as a supplementary crop up shade trees including kapok, *C. pentandra* Gaert., and the silk tree, *Albizia chinensis* (Osb.) Merr., in established plantations of coffee or cocoa. Pests and diseases are generally as for pepper.

Harvesting

Fruiting spikes are harvested when berries are mature but unripe and turning yellow; too early picking of mature but green fruits results in shrivelled and off-coloured berries. Spikes must be picked and handled carefully as bruised berries do not become the desired dark colour after drying. Berries are sun-dried as for pepper, and lose about two-thirds of their weight; correctly dried fruit is very hard. The annual yield per vine of dried fruit is about 0.5 kg. Vines produce a small harvest after 1 year, come into full bearing after 3–4 years, and can remain productive for up to 60 years.

Distillation

An essential oil can be obtained by steam distilling dried, crushed fruits, and yields 4–30%, and this very wide range indicates not only differing distillation methods but considerable regional variation in fruit oil content. Indonesian fruit locally distilled seldom yields above 10% oil, but up to 20% can be obtained from Malaysian and Indian fruits distilled in Europe. Distilling time affects not only yield but specific gravity, optical rotation and acid value; prolonged distillation in Indonesia increased oil yield and raised values of the three characteristics mentioned, and weight of charge affected acid and saponification values.

Products and specifications

The major products of the cubeb vine are the dried fruit, tailed pepper and cubeb oil.

The dried fruits resemble peppercorns, and good-quality fruit has a pleasant aromatic odour and bitter, pungent taste resembling allspice. Whole or ground fruits are the domestic spice and food flavouring, but are seldom available outside South-east Asia. Despite their very hard exterior, the berries are hydroscopic, and must be stored in airtight drums to avoid deterioration. Genuine cubebs are distinguished from other peppers by the adhering stalks (pedicels), and the powder by sprinkling it into a container of sulphuric acid, whereupon a crimson colour appears due to the presence of cubecic acid in the powder. This test is not wholly accurate and if pure cubeb powder is required then analysis is necessary.

The dried fruits are still used in herbal medicines and are stated to be antiseptic, diuretic, a stimulant, an expectorant and a carminative; they were formerly used in the production of asthma-alleviating cigarettes in the USA. In Malaysia, powdered dry fruits are widely used to treat amoebic dysentery and on Java as a popular aphrodisiac often added by wives to their husband's meals!

The main use for cubeb fruits outside producing countries is as raw material for distilling cubeb oil, produced mainly on demand in Europe and North America from imported material. Small amounts are also distilled in Indonesia and India, and frequently adulterated with similar but cheaper oils, most commonly clove leaf oil (especially in Indonesia), *S. molle* seed oil and oil of the false cubeb, *P. crassipes* Korth.

Cubeb oil is a viscous, pale greenish to bluish-yellow, occasionally colourless liquid, depending to a considerable extent on distilling time and temperature; blue indicates a very high temperature. The six main constituents are (approximately) β-cubebene 11%, copaene 10%, cubebol 10%, δ-cadinene 9%, α-cubebene 7%, α-humulene 5%. The oil has a camphoraceous, dry-woody, spicy but not peppery odour, with a bitter but not pungent or peppery taste. The oil is placed in the woody, warm-peppery group by Arctander (1960). To maintain its quality and flavour, oil must be kept in sealed opaque containers in a cool store. The main use of cubeb oil is in perfumery and soaps, certain medicines, some beverages and bitters, but little as a food flavouring. The maximum level in the USA for oil use as a fragrance component in soaps, creams and alcoholic perfumery is 0.8% and in food is 0.004%. A monograph on the physiological properties of cubeb oil has been published by the Research Institute for Fragrance Materials.

Cubeb oleoresin is obtained by solvent-extracting crushed fruits with a hydrocarbon solvent or ethyl alcohol, and is produced in small quantities almost wholly in user countries, with a very minor amount in India. The dried fruits are coarsely ground and extracted with a hydrocarbon solvent or ethyl alcohol; technically the product is a resin absolute. Oleoresin is a clear viscous liquid, free of bitter cubebin, wax and other extracted material, with a warm-spicy, peppery somewhat camphoraceous odour and a spicy rather peppery but not pungent taste. Its main use is in flavouring, especially sauces and pickles, and in certain pipe tobaccos. In the USA, the status of generally regarded as safe has been accorded to cubeb pepper, GRAS 2338, and cubeb oil, GRAS 2339.

Indian Long Pepper

Indian long pepper, *P. longum* L., was probably the first known pepper in the Mediterranean region; in Pliny's Rome it was twice as expensive as black pepper and more highly esteemed. Since then its importance has declined, and today long pepper usage is confined basically to producing countries and their immediate neighbours. In India, the medicinal properties of fruits and sliced, dried roots and stem are well known, and are often locally more important than the spice. As an ingredient of curry and similar mixtures it is often referred to by the Indian name of *pipel*.

Indian long pepper does not enter international trade, but is confined to the Indian subcontinent and to South-east Asia, and the amount is unquantified. It is becoming available in retail packs in developed countries with an Asian population.

Botany

P. longum L. (syn. *Piper latifolium* Hunter; *Chavica roxburghii* Miq.) is known in English as long pepper, in French as *poivre long*, in Italian as *pepe lungo* and in Malaysian as *chabai*. Some authorities wrongly consider *Piper sarmentosum* Roxb. ex Hunter as identical to *P. longum*. The two are distinct species, do not occur naturally in the same regions, and *P. sarmentosum* is not considered a spice, nor a substitute for *P. longum*.

P. longum is a perennial, dioecious herb with a woody rootstock and slender prostrate or ascending shoots. The leaves are thinly membranous, dull and gland-dotted, the blade ovate, 4–7 × 2–3 cm, the base cordate, and the apex acute or acuminate. The petiole is up to 1.2 cm, longest on lower leaves, with the upper leaves almost sessile and amplexicaul. The fertile branches are erect, up to 30–60 cm, bearing erect inflorescences; the peduncle is 1–2 cm and the bracts peltate and orbicular. The female spike is 1.4–3 cm × 3–8 mm and the flowers close together, each with three to four stigmas; the male spike is 3–6 cm × 1.5–2 mm, and the flowers close together, each with two stamens. The berries form a thin, fleshy, slender, cylindrical spike-like cone, which contains about 4–6% piperine, 1.0–1.5% cadinene, and about 1% essential oil of no commercial importance. The whole spike consisting of the very small fruits embedded in a pulpy rachis becomes the spice when dried.

Ecology

P. longum is native to the rather dry forested foothills of the north-eastern Himalayas through Assam to northern Myanmar, and although introduced to other countries in the region, is a cultivated crop only in the drier regions of India and Sri Lanka. The vines flourish on soils similar to those of their natural habitat, but when cultivated are grown on a range of well-drained loams to clay loams. No information is available on the effect of fertilizers on vine growth or fruit composition.

Cultivation

Long pepper is usually grown from stem cuttings and treated as for cubeb pepper. Supports are normally trees growing naturally around the homestead, or shade trees in plantation crops. Vines receive little attention once established. Little information is available on the pests and diseases of long pepper, but are generally as for pepper. The main flowering period is towards the end of the rains and spikes are harvested when most fruits are ripe, usually December–January. The fruits are sun-dried as for cubeb. The yield from cultivated long pepper is about 250 kg ha^{-1} in the year of establishment, rising to 1250 kg ha^{-1} in the third year. After harvesting, vines usually wither or are cut to

ground level; some roots are also harvested and dried, and those remaining are protected by mulching during the hot season. The average vine life is 4–6 years.

Products and specifications

The fruits are not separated from the spikes, which are sold whole in markets as spice. The main use is as a food flavouring with a milder taste than pepper. The essential oil steam-distilled from the spikes is light green, somewhat viscous, aromatic and pungent, with a slight ginger odour. It is placed in the woody, warm-peppery group by Arctander (1960). The seven main constituents are (approximately) pentadene 18%, β-caryophyllene 17%, β-bisabolene 11%, tridecane 7%, heptadene 6%, α-zingiberene 5%, and germacrene D 5%. The oil is mainly used as a food flavouring, but is seldom produced or available outside India.

In Asia and South-east Asia generally, *P. longum* fruit is said to be effective against coughs, asthma, dyspepsia and paralysis, with laxative and carminative properties, and aged plant products are said to be more effective than fresh; in China the pulped fresh or powdered dry root is given to pregnant women to speed up delivery.

Java Long Pepper

Piper retrofractum Vahl. and Indian long pepper, *P. longum*, are often treated as identical products in Western countries, although in the early 20th century Java pepper was exported to both Europe and the USA. Trade is now confined almost entirely to South-east Asia and China, and unquantified.

Botany

P. retrofractum Vahl. (syn. *Piper chaba* Hunt.; *Chavica retrofracta* (Vahl) Miq.; *Piper officinarum* (Miq) C.DC.) is known in English as Javanese long pepper, in French as *poivre long de Java*, in Indonesian commonly as *cabe jawa*, in Malaysian as *chabai jawa*, in Thai generally as *dipli* and in Laotian as *sali*.

P. retrofractum is a perennial dioecious

herb, recumbent, or climbing with adhesive roots, the stem up to 10 m and soft woody. The leaves are glabrous, firmly coriaceous, with the underside with many sunken gland dots; the blade is ovate to oblong, 8–20 × 3–13 cm, the base cordate, obtuse or cuneate, and the apex tapering or acuminate. The petiole is 0.5–3 cm long. The flowering spikes are erect and spreading; the peduncle is 1–2 cm and the bracts broadly ovate and 1–2 mm. The female spike is 2–3 cm, with two to three stigmas, which are short and persistent; the male spike is 2.5–8.5 cm, with two to three very short stamens. The infructescence is cylindrical, 2–4 cm × 4–8 mm on a 1 cm stalk. The berries are connate and adnate to the stalk of the bract, and broadly rounded, hard and pungent when green; they are soft, sweet and bright red-brown when fully mature, often covered with a grey powder. The seed is globose, 2–2.5 mm in diameter, and white and mealy inside. The fruit contains piperine, resin, fibrous material at 10–15%, starch at 44–49%, ash at 8%, fixed oil and essential oil. The fruit resembles that of Indian long pepper, not true pepper.

Ecology

P. retrofractum occurs naturally in mainly deciduous forests from Myanmar through northern Malaysia to Thailand, Indo-China and on to the Moluccas. It is cultivated principally in Peninsular Malaysia and on Java, Indonesia, and introduced to the Ryukyu Islands of Japan where it is now a common escapee. Vines can be found growing naturally from sea level to 600 m, and cultivation is usually confined to these limits.

Wild vines frequently occur and are cultivated on less-fertile soils on Java, but this may be because the fertile soils are utilized for more important food crops, as occurs with similar minor crops throughout the region. In general, sandy clays and loams are often chosen, and vines flourish when underplanted in coconut plantations on coastal sands. Smallholder vines may be planted on whatever soils are available. Mulching is common, but no data are available on the effect of chemical fertilizers on vine growth or fruit composition.

Cultivation

A substantial amount of fruit is collected from wild vines, which are numerous in specific regions. When cultivated, Java pepper is established from cuttings as for Indian long pepper, and planted at the base of trees in established plantations. Special shade trees are planted on Java round the borders of cultivated plots in two 8–15 m rows at an in-row spacing of 1.5–2.5 m. Superphosphate fertilizer is spread around vines about a month after planting, weeded four to five times per year, and earthed up at least twice. Java pepper is a vigorous climber and must be cut back to around 5 m, when trees are also pruned to reduce shade density. A second thinning may be necessary. Vines must be pruned regularly to induce flowering, which occurs year-round. Fruit thus ripen continually and spikes are harvested when mature. Pests and diseases are generally as for pepper.

Harvesting

Spikes are picked when the majority of fruits are mature and becoming reddish at the apex, and must be quickly sun-dried as fruits easily decay. Drying normally takes at least 5 days, but in cloudy periods spikes can be boiled in water, excess water removed, then dusted with wood ash to hasten drying. Dry whole spikes are sold in markets as spice. Yields are slightly higher than for *P. longum*, and vines may continue in bearing for many years, depending on the care they receive.

Products and specifications

The main product of Java pepper is the dried spikes used as domestic spice. The taste is pleasant and aromatic, but more pungent than black pepper and Indian long pepper. The spice is used to flavour a range of local foods and is also an ingredient of pickles and curries. Steam-distilled dried fruit yields about 1% of a light green, viscous essential oil, with the slight odour of ginger. It is seldom produced and of no commercial importance.

In South-east Asia generally, the ground

fruit is used to treat a variety of skin diseases, digestive and intestinal disorders, and haemorrhoids. A tincture is used in childbirth to accelerate expulsion of the placenta. In Indonesia, a leaf extract provides a mouth wash and is used to alleviate toothache; in the Philippines the root is chewed to treat colic (de Guzman and Siemonsma, 1999).

Betel Pepper

P. betle L., betel pepper, is native to Malaysia, Sumatra and probably also Java, and does not produce a culinary but a masticatory spice, as the plant is grown almost exclusively to provide fresh leaves to flavour the quid known as 'pan' in India. The 'pan' quid is prepared by spreading on the upper surface of a fresh betel leaf a thin layer of slaked lime paste, a sliver of areca nut, *A. catechu* L., and a variety of other seasonings according to local taste. This quid is then chewed and the liquid discolours teeth and stains saliva, mouth and lips red. The addictive taste is a combination of the essential oil contained in the fresh betel leaf and the alkaloid, arecoline, from the nut. The modern geographical distribution of *P. betle* and *A. catechu,* the other main constituent, is identical since the presence of one led to the introduction of the other, or both where neither occurred naturally as in East Africa.

Chewing *pan* is ancient in Asia, especially India, and although the copious saliva generated results in continual spitting, which is a menace to health, the arecoline swallowed reduces the incidence of internal parasites. This slight benefit is far outweighed by the adverse effects of habitual use, including tooth loss, serious disorders of the digestive system and mouth cancer. Many countries have attempted to ban chewing *pan* in public.

India has the largest area of betel under cultivation, about 40,000 ha, mainly in Orissa and Tamil Nadu, and Bangladesh with about 13,000 ha, although plantations exist wherever a local demand exists and the environment is suitable, for example, on Yap, Federated States of Micronesia (Stone, 2000).

Botany

P. betle L. (syn. *Chavica betle* Miq.) is known in English as betel vine, in Malaysian and Indonesian commonly as *sireh*, in Indian (Hindi) *tambuli* and in Arabic as *tambula*. Betel is a perennial dioecious climbing vine, with angled stems bearing large cordate, alternate, shining green leaves; the flowers are borne on lax panicles.

Cultivation

Betel vines do not produce a spice *per se* and cultivation is not described in the text, but in detail elsewhere (Chattopadhyay and Maiti, 1990; Balasubrahmanyam *et al.*, 1994). Betel is grown for its leaves only by smallholders who may own from one to several hundred vines. Successful cultivation of vines requires close and constant attention and this is only possible because of the strong and sustained demand for fresh leaves. Vines are trained up shade trees, and leaves are picked year round as they become large enough; vines are replaced every 2 years. Cultivars from India have a higher leaf-oil content and more pungent taste, and for these reasons have been introduced to many countries in preference to Malaysian or Javan cultivars. A major constituent is generally eugenol imparting a clove-like taste, although in some cultivars it is *E*-anethole giving a liquorice flavour.

Products and specifications

The only other product of importance is the essential oil, betel oil, obtained by steam distilling fresh, sometimes partially dried leaves with a yield of 0.5–2.0%; young leaves give the highest yield. The oil is usually a light to medium brown, occasionally bright yellow, depending on the type of leaf used, i.e. young or old, fresh or dried. The odour is sharp and distinctly phenolic; the taste is acrid, bitter and burning. The oil is placed in the bitter-herbaceous, phenolic or medicinal group by Arctander (1960). Considerable variation in the abundance of the main constituents exists in regional cultivars, probably due to selection to satisfy a local preference, and this is reflected in oil composition. At least 22 compounds have been identified in the oil including chavicol up to 98% in some cultivars (Lawrence, 1993b; Garg and Rajshree, 1996). The main characteristics of Indian betel oils are: specific gravity 0.9408–1.0482; refractive index 1.5048–1.5088; optical rotation +0.9° to +9.3°; soluble in 80% alcohol. The range indicates the wide regional variation.

Betel oil is produced on a very small scale and seldom used by European or North American manufacturers. The main use is in Asia and supplies the betel taste and flavour where leaves are not easily obtained. Refrigerated fresh leaves and the oil are now available in Western countries with a large Asian element. The oil has significant fungicidal and nematicidal activity. The leaves and oil are traditionally used in herbal medicines in Asia and South-east Asia. To increase the range of value-added betel products and thus the total market for the 20,000 local growers, Sri Lanka's Industrial Technology Institute has developed an instant betel quid, betel pellets and a range of personal products including betel tooth paste, shampoo, face and hand creams.

Melegueta Pepper

The genus *Aframomum* K.Schum. (*Zingiberaceae*) contains between 40 and 50 species of perennial herbs, native to Africa. Several provide a local spice, but only *A. melegueta*, source of melegueta pepper, is of importance. In this text *A. melegueta* will be designated melegueta for brevity.

Melegueta is native to West Africa, and cultivated in Ghana, Guinea, the Ivory Coast, Nigeria and Sierra Leone. Today, Ghana and Nigeria are the major producers and consumers. Melegueta pepper was first taken to Europe via the great caravan routes across the Sahara, the Golden Trade of the Moors, to the port of Mundibarca on the coast of Tripoli. Following the Portuguese maritime expeditions to West Africa and their discovery of the origin of melegueta pepper, it was exported to Europe from that part of West Africa known to the Portuguese as the 'costa da Malagueta', especially from Senegal. The English knew

the region as the Windward or Grains Coast, roughly Liberia and the Ivory Coast. Records of the Portuguese trade indicate that in 1506 a quintal of melegueta pepper cost 8 cruzados in Lisbon compared with black pepper at 22 cruzados. The Portuguese later carried melegueta to their South American colonies, particularly Brazil, and a thriving export trade to Portugal developed. From Brazil, melegueta spread to many other countries of the region including Surinam and Guyana, and while it is nowhere important, it is often cultivated by smallholders for their own use. Dutch traders from Amsterdam followed the Portuguese, and the records of their voyages illustrate how the trade was effected. One shipmaster recorded he returned with 'eight lasts (*c.* 16,000 kg) of malaguetta at which the Herren Meesters took great satisfaction and were very pleased' (la Fleur, 2000).

An early record is from a festival held at Treviso, Italy, in 1214, and early European physicians included melegueta in their prescriptions. The spice was well known in Europe by the 14th and 15th centuries, and may also be the *P. aethiopium* of 17th-century Europe. Melegueta was in demand as a spice, a stimulant, a carminative, and popular with Queen Elizabeth I of England (1533–1603). It was used for spicing wines and for strengthening beer, a custom prohibited by law in the reign of King George III (1738–1820), and heavy fines were prescribed. Later an Act of Parliament was passed that no brewer or publican should have any 'grains of paradise' in their possession, or use it in making beer, under a penalty of £200, and any druggist selling it to a brewer was fined £500. Melegueta was carried on ships to treat dysentery and other stomach disorders. However, following the increased availability of black pepper, melegueta as a spice gradually declined in popularity. Today melegueta is virtually unknown in Europe, although available in specialized retail outlets.

During the 18th and 19th centuries, interest in the spice waned in Western countries. Exports from Ghana were: 86,719 kg in 1887, the UK taking 38,818 kg; 28,567 kg in 1872 valued at £10,303; and 68,909 kg in 1875 valued at only £912. Exports from West Africa virtually ceased after the First World War, and

melegueta is now of no international importance, little traded and will not be discussed
in detail. However, melegueta pepper is commonly available in West African markets, the
trade is locally important, and demand would
rapidly increase the amount available. The
characteristic spicy flavour and a marketing
drive could revive exports to Western countries, whose consumers are seeking new, naturally produced food flavourings.

Botany

A. melegueta (Roscoe) K.Schuman (syn. *X.
aethiopica* A. Rich.) is known in English as
melegueta pepper, or grains of paradise, in
French as *malaguette*, in German as *malaget-
tapfeffer*, in Spanish and Portuguese as
malagueta, in Ghana, Twi as *famu wisa* and in
Nigeria, Bini as *chie ado*. The taxonomy of the
genus *Aframomum* remains confused, and

many references to particular species in the
literature are probably unreliable.

A. melegueta is a perennial herb with
short rhizomes from which arise distinct
leafy shoots, 1.5–2.0 m, typical of the
Zingiberaceae. Descriptions of the plant differ in detail according to the authority and
origin of specimens.

The leaves are sessile or subsessile, mostly
glabrous, and glossy green; they are lanceolate with the base rounded, the apex acuminate, 18–22 × 1.8–2.5 cm, with numerous
parallel veins (Fig. 8.10). The inflorescences
are borne at the base of the leafy shoots on
very short peduncles with reddish bracts. The
delicate trumpet-shape flower has one erect
lanceolate lobe, two white and pale violet linear lobes, and a large, spreading fan-shaped
lip or labellum (a petaloid staminode), which
is white/pink with a red or yellow blotch at
the base; there is a single fertile stamen (Fig.

Fig. 8.10. Upper leaves – *Aframomum melegueta.*

8.11). The fruits are ovoid or pear-shaped berries, 5–10 cm, red or orange when ripe, with a persistent calyx, containing a sweet, white, edible pulp in which are embedded 60–100 brownish seeds, 3–4 mm, which are aromatic and pungent; when dried these are the grains of paradise (Fig. 8.12).

Cultivation

Melegueta is native to West Africa and occurs naturally and is cultivated in the wet tropics from Guinea, Sierra Leone to the Republic of Congo (Zaire) and Angola, although reportedly known only under cultivation in Ghana. It is grown by small-holders over most of this area, but mainly in Ghana and Nigeria (van Harten, 1970; Lock *et al.*, 1977).

Products and specifications

The major product is the spice, which is the dried fruit containing the seeds that constitute the pepper; seeds alone are the grains of paradise. In West Africa the fruit pulp of this and other species of *Aframomum* are chewed as a refreshing stimulant, and the seeds used for seasoning foods. A very small amount of dried fruit, estimated at about 10,000 kg, is imported by Western countries for use in cosmetics and perfumes.

An essential oil can be steam distilled from the fresh or dried fruits to yield

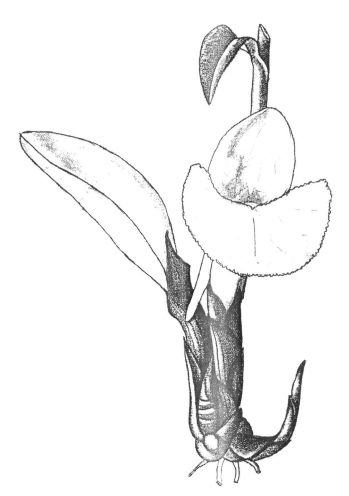

Fig. 8.11. Flower and rhizome – *A. melegueta.*

0.5–0.75% oil. The oil is yellowish to brownish-yellow, with a sharp, spicy taste and odour of the crushed pods. The oil is placed in the woody, warm-peppery group by Arctander (1960). The main constituents of the oil are β-caryophyllene up to 9%, α-humulene at 32%, together with their epoxides at 18% and 28%, respectively. The main characteristics of the oil are: specific gravity (15°C) 0.8970, optical rotation −3°10′, refractive index (20°C) 1.4916, soluble in 1 vol. 95% alcohol. The oil is produced on demand by specialist companies, but is not commercially available.

Its main use is in cosmetics, as noted, processed speciality foods, flavourings and vinegars as a replacement for the ground fruits. In West Africa and South America, to which it was introduced, various plant parts are used to treat a variety of diseases, internal parasites and as an emollient.

Other Aframomum species

Many *Aframomum* species were either confused with or used as substitutes for grains of paradise. The most important include: *Aframomum angustifolium* (Sonn.) K. Schum., which is native to swampy areas of East Africa, the Malagasy republic and other Indian Ocean Islands, is not cultivated and fruits are collected from wild plants; *Aframomum corrorima* (Braun) Jansen, which occurs wild and is cultivated in Ethiopia, and small amounts are exported to neighbouring countries; *Aframomum sceptrum* (Oliv. & Hanb.) K. Schum. (syn. *Aframomum granum-paradisi* (L.) K. Schum.), which is native to West Africa, and a distinct species and not the true grains of paradise; *A. daniellii* (Hook.f.) K.Schum. and *A. sulcatum* (Oliv. & Hanb) K.Schum., which are both native to the Cameroon and their seeds collected and used as spice. Various species of *Amomum* and *Alpinia* may be locally substituted for melegueta.

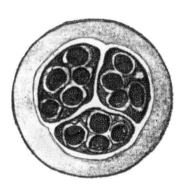

Fig. 8.12. Fruit and cross section showing seeds – *A. melegueta.*

9

Solanaceae

The *Solanaceae* contains about 90 genera and some 2000 species of mainly tropical herbs, shrubs and small trees. The genus *Capsicum* contains at least 20 species, most native to South America but two, *Capsicum annuum* var. *glabriusculum* and *Capsicum frutescens*, extend through Central America to the southern USA. The *Solanaceae* also includes a number of important food and drug plants; *Solanum tuberosum* L., the potato, *S. melongena* L., the aubergine, *Lycopersicon esculentum* Mill., the tomato, and *Nicotiana tabacum* L., tobacco.

Capsicum and Chilli

In pre-Colombian Mexico and South America, spices, particularly pepper, occupied an important place in Aztec and Inca cookery and medicine, but had been utilized by local peoples for many thousands of years. The oldest known evidence attributed to *C. annuum* is from the pre-agricultural Ocampo culture in caves near Tamaulipas, apparently dating from 7000 BC, and similarly dated material has also been found from human-occupied sites elsewhere in Mexico; but by 5200 BC and 3400 BC local people were cultivating peppers. The earliest South American records date from 6000 BC in Bolivia and are of *Capsicum pubescens*, while cultivated forms of *C. baccatum* var. *pendulum* were identified from Ancon and Huaca

Prieta, Peru, about 2500 BC, as the fruits are non-deciduous and include orange forms unknown in the wild. An interesting description of the ethnobotany of peppers is contained in Bosland and Votava (2000).

The Aztec ruler, the Great Speaker, maintained arboreta, pleasure and kitchen gardens to provide the palace with fresh vegetables, herbs and spices, and regularly added to the variety of plants by sending collecting teams throughout the empire (Pico and Neuz, 2000). Their subjects were no less knowledgeable about local herbs and spices, which were freely available in the markets. Thus, the first Europeans were confronted with an array of culinary spices and herbs, especially the many varieties of peppers commonly known in the Aztec language Nahuatl as *axl* or *axi*, although there were separate names for the various types, degree of hotness and culinary use. Many of these names remain in current use in the region. The Spaniards pronounced *axi* as chili, now the common pronunciation for hot peppers generally. It is frequently spelt chillie in English, but chilli or chili in North and South America.

When Columbus first landed on Hispaniola in 1492, capsicum was grown and used in virtually the whole of the Caribbean, Mexico, Central and northern South America, although the greatest number of types of what later became *C. annuum*

190

var. *annuum* was in Mexico. He recorded in his journal, 'Also there is much axi, which is their pepper, and it is stronger than pepper, and the people won't eat without it, for they find it very wholesome. One could load 50 caravels a year with it.' Eighty years later, Dom Nicholas Monardes, a Spanish physician, wrote of this pepper, 'A certain kind of long pepper, which has a sharper taste then the pepper of the Oriente and it does bite more, and it is of more sweet taste and better smell than that of Asia. I have caused it to be put in to dreste meats in place of the Oriental pepper, and it giveth a gentle taste'.

Capsicums were taken back to Europe by Columbus, and then shortly afterwards to India and South-east Asia on Portuguese trading voyages. The French botanist Charles de Lecluse (Carolus Clusius) in his *Rariorum plantarium historia* of 1601 mentions *Capsicum brazilianum* as being brought to India from the Spanish West Indies by the Portuguese. Many early writers wrongly identified the *siliquastrum* of Pliny as a *Capsicum* rather than *Cardamomum*, and incorrectly stated that chilli peppers had been grown in India from antiquity. Supporting the later introduction is the fact that the common Indian name for chilli pepper of *achar* is directly derived from the Portuguese *achi*, another pronunciation of the Aztec *axi*. By the end of the 17th century, *C. annuum* var. *annuum* and *C. frutescens* were being grown in most of the world's warmer regions. The fruits are attractive to birds who widely disperse the viable seeds, thus some *Capsicum* species became naturalized in many tropical countries, especially *C. frutescens*, as did another New World export, the Aztecs *tomatle* (tomato) for similar reasons.

It was Columbus's reputed finding of capsicum or chillis that became the New World's most important contribution to the family of spices, for capsicums subsequently spread throughout the tropics and warm temperate regions of the Old World. It is now almost impossible to imagine the dishes of Asia and the Pacific region without chilli peppers, while the traditional African sorghum or maize porridge (*ugali*) would be tasteless without them. It is the national spice of Ethiopia, an essential ingredient of the viscously hot *wat*, and as one historian

commented, 'Without capsicum pepper one cannot imagine a food, almost not even an Ethiopian!' Initially it was the flavour that ensured spice capsicums very wide dispersal, but more recently the huge range of fruit shapes of bell and spice capsicums and their attractive colours has produced what is virtually a pepper cult, whose devotees collect, photograph, discuss, write about, use and try to extend the culinary or medicinal uses for peppers! A similar cult exists for garlic.

The genus name *Capsicum* is believed to be derived from the Greek *capsicon* via the Latin *kaptein*, to bite, apparently in reference to the fruits' pungency, and the name bell peppers from the Latin *capsa,* boxlike. *C. annuum* var. *annuum* species have a variety of forms, mainly expressed in the huge number of cultivars, from the large bell or sweet peppers with the mildest flavour and little pungency and mainly consumed fresh, to the smaller chillis whose pungency ranges from mild to extreme. *C. annuum* var. *annuum* is further divided into chilli pepper (small, pungent) and paprika (larger, mild), while bell pepper does not qualify as a spice and will be mentioned only where necessary.

Cultivated peppers were, and remain, difficult to classify as species or by common names. Species are extremely variable, particularly in fruit characteristics, and many so-called species are locally adapted strains or cultivars. Five cultivated species are now recognized and those starred are the most commercially important: *C. annuum* var. *annuum**, *C. baccatum* var. *pendulum*, *C. chinense**, *C. pubescens* and *C. frutescens**. Classification was summarized by Purseglove (1981a), who quite correctly doubted whether some botanical separations were justified, as they intercross and intermediates occur. His key to cultivated *Capsicum* species remains valid as shown in Table 9.1.

Cultivated and wild species are diploid (x = 12, 2*n* = 24), and sterility barriers between species are generally well developed; *C. pubescens,* for example, will not cross with any other species. Hybrids were successful between *C. annuum* and *C. frutescens, C. baccatum* var. *pendulum* and *C. chinense*, but the progeny were usually highly sterile. Thus, a sound genetic basis exists for recognizing the

Table 9.1. Key to cultivated *Capsicum* species (Purseglove, 1981).

A	Corolla lobes purple; seeds black	*C. pubescens*
AA	Corolla lobes white or greenish-white, rarely purple; seeds light in colour	
B	Corolla white with yellow or tan markings on throat; anthers yellow	*C. baccatum*
BB	Corolla without yellow markings on throat; anthers light blue to purple	
C	Corolla usually clear or dingy white; pedicels usually solitary at a node	*C. annuum*
CC	Corolla usually greenish-white; pedicels usually more than one at a node	
D	Pedicels usually two per node, erect at anthesis, without distinct constriction with calyx	*C. frutescens*
DD	Pedicels usually 3–5 per node, usually curved, with distinct circular constriction with calyx	*C. chinense*

five distinct cultivated species. However, the discovery in 1958 of cytoplasmic male sterility and of genetic male sterility in 1969 allows the production of hybrid seed, and hybrid plants are highly uniform and usually higher yielding. More recently, biotechnological techniques are developing new cultivars and mapping genomes to assist breeders.

Self-incompatibility is rare in cultivated forms, which can thus be grown in small areas without loss of identity, but is reported in some wild species. However, both self- and cross-pollination occurs, the latter ranging from 16–20%, with Indian cultivars recording up to 70% natural cross-pollination, and is higher in flowers whose stigmas protrude beyond the stamens. Increased self-pollination is due to styles being similar in length to anthers. Heterosis in capsicums is expressed by earlier fruit maturity, more uniform and increased numbers of fruit, less flower abscission, and heavier and more viable seeds. Red skin colour and pendent mature fruit characters are apparently dominant over yellow and erect fruits.

Wild capsicums were initially known as *C. annuum* var. *minimum* but later became *C. annuum* var. *glabriusculum* (Dunal) H & P, which is not cultivated but collected fruits are locally sold in markets. Common names are chilitepin and bird pepper, although the latter is also used for *C. frutescens*. Variety *glabriusculum* is probably the wild progenitor of the cultivated var. *annuum*, with which it readily crosses. In South America, *C. annuum* may be replaced by *C. pubescens* at higher altitudes, *C. baccatum*, *C. pendulum* and *C. chinense* at lower altitudes. Authorities state that *C.*

annuum and *C. frutescens* no longer occur as truly wild plants, but as escapes or have become naturalized, especially the latter.

Pepper germplasm collections are located at the Southern Plant Introduction Station, Griffin, Georgia, USA; the Asian Vegetable Research and Development Centre (AVRDC), Tainan, Taiwan; Centro Agronomico Tropical de Investigations y Ensenanza (CATIE), Turrialba, Costa Rica; at the National Repository of Plant Genetic Resources, Idukki, and other centres in India; the Central Institute for Genetics and Germplasm, Gaterleben, Germany; and the Centre for Genetic Resources, Wageningen, The Netherlands. An international cooperation programme to promote paprika breeding was begun in Hungary in 1996, based on the Research Station at Szeged (Somogyi, 2000).

Capsicum var. *annuum* has been greatly modified to suit grower preferences, initially through selection and then deliberately under cultivation, especially in the fruit, which is the most important plant part. Thus emerged the great variety of fruit shape, size, colour and degree of pungency. In the last 20 years the number of cultivars has increased enormously, with new ones apparently released monthly! For example, the Japanese Sakata Seed company released in mid-2000 a baby red bell capsicum named Il Bello Rosso. It is only 3–4 cm deep compared with the usual 10–12 cm, the flesh is 6 mm thick, with a sweet yet mild flavour, and fresh fruits weigh 80–100 g. It is now extensively planted in Queensland, Australia. Thus, cultivar names will generally be ignored unless they have acquired the status of a type, for example, Tabasco pepper. Horticultural cultivars have also been developed and are very popular as pot plants. Peppers from *Capsicum* species should not be confused with black and white pepper from *P. nigrum* and long pepper from *P. longum* (Chapter 8), Jamaica pepper, pimento or allspice from *P. dioica* (Chapter 6), and Melegueta or Guinea pepper from *A. melegueta* (Chapter 8).

Botany

Capsicum annuum group. C. annuum var. *annuum* L. (main syn. *C. frutescens* L.; *C. abyssinicum* Rich. and many more) comprises various types, which each have their own names in most languages, and are too numerous to mention; for example, there are at least 200 in Mexico. The most likely ancestor of *C. annuum* is *C. annuum* var. *aviculare*, the wild chiltepin pepper of Central and South America. *C. annuum* var. *annuum*, hereafter var. *annuum* for brevity, may be classified by fruit characteristics, i.e. size, shape, colour, pungency, flavour and use, into many types, but in the text will be referred to as bell, chilli, paprika, etc. Figures in the following description indicate the range occurring over cultivars, not the range of a single cultivar.

Var. *annuum* is an erect perennial herb or subshrub usually grown as an annual (Fig. 9.1), while *C. frutescens* is a perennial. The strong brownish taproot is frequently damaged or loses vigour on transplanting; later numerous mostly superficial branched laterals develop, extending to 1 m. Data from the USA indicate that root weight in modern cultivars is approximately one tenth of total plant weight. An association with mycorrhizal fungi increases nutrient uptake, and where this does not occur naturally, young plants can be inoculated.

The stem is erect, 45–100 cm, often becoming woody at base, sparsely to densely tomentose, rarely glabrous, green to brown-green often with purplish spots near nodes, and irregularly angular to subterete. The stem is much branched, the main shoot radial, but later branches are cincinna; the subtending bract or bracts are adnate and carried up a lateral shoot to the node above. A variety of growth regulators, including giberellic acid, triazols and ethephon applied to young pepper adversely affected subsequent growth, flowering or fruit set. The leaves are alternate but upper leaves are almost opposite, and simple; the petiole is light green and angular, up to 10 cm. The lamina is thin, variable 10–16 × 5–8 cm, broadly lanceolate to ovate, the margin entire, and serrulate to papillate; the base is cuneate or acute, and the tip acuminate. The leaves are light to dark green, paler green beneath, subglabrous, and tomentose on the veins.

The inflorescence is terminal, but due to the form of branching may appear axillary; the flowers are usually single but there may be up to five; the pedicels are erect or pendent, slender or thick, up to 1.5 cm. The calyx is campanulate, shortly five-dentate and ten-ribbed, up to 2 mm, normally enlarging and enclosing the fruit base. The corolla is usually dull, light green, dirty white, or yellowish white; it is campanulate to rotate, 5–11 mm, usually with five lobes, sometimes more. The five stamens are inserted near the base of the corolla, the free filament is 2.5–6.0 mm, and white to purple; the anthers are 2.5–4.0 × 1.0–2.0 mm, dark green to blackish, dehiscing by two longitudinal slits. The ovary is conical, normally two-celled but multiple-celled in some cultivars; the style is simple, white or purple, and the stigma capitate, and light green to yellow. The number of flowers is a cultivar characteristic and can be 100 per plant, far more than fruit set for reasons still to be determined, since it is not due to obvious seasonal influences. In general, once fruit begins to set, flowering decreases. It appears that fruit that develop from early flowers are usually larger, brighter red, and from chilli types more pungent.

Pollination is basically by bees, although thrips, flies, pollen beetles and ants have been recorded as visiting flowers. In some African countries where the local bee population is low, placing hives in the vicinity of crops, including sunflower and chillis, gives excellent results (E.A. Weiss, personal observation). The majority of flowers open between 5 and 6 a.m., with pollen shedding between 9 and 11 a.m., with 10 a.m. on the

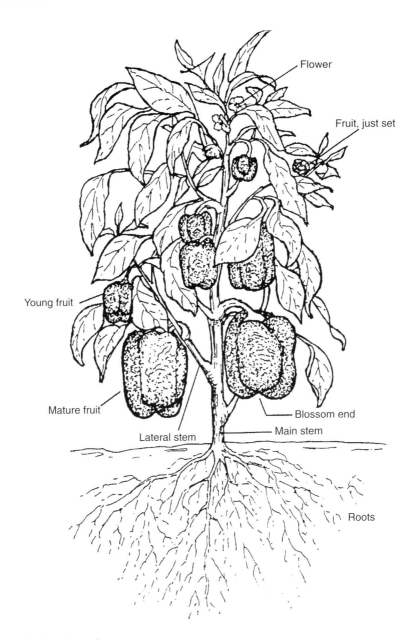

Fig. 9.1. Typical capsicum plant.

day a flower opens the optimum for hand-pollination. The pistil remains viable for 5–9 days after emasculation at the white bud stage. Optimum fruit-setting and number of seeds per fruit is obtained from pollen collected immediately after anther dehiscence, with pollination 3 days after emasculation.

Pollen retains viability for 8–10 days at 20–22°C and 50–55% humidity. Stigma receptivity can be preserved for 5–7 days.

The fruit is a pendulous or erect, indehiscent many-seeded berry, usually borne singly at nodes. The size is very variable, but generally in the range 1–19 × 0.5–4.5 cm diameter, although *The Guinness Book of Records* includes a monster 32.5 cm long! The fruit shape is linear, conical or globose, the apex acuminate to blunt, and the base rounded to obtuse, with the persistent calyx surrounding the base (Figs 9.2 and 9.3). Immature fruit is light to dark green becoming yellow-orange and finally red to bright red when ripe. Fruit varies greatly in the degree of pungency, and numerous 'hot' cultivars of var. *annuum* are recognized in Mexico, India, the USA and elsewhere. Bell peppers are generally lacking in pungency, except the Hungarian paprika type with its longer, pointed fruits. Fresh capsicum fruit is very rich in vitamin C (ascorbic acid), whose isolation won the Nobel Prize in 1937 for Dr Szent-Gyorgyi, the Hungarian scientist. His work showed that capsicum fruits are one of the richest available sources of vitamin C. Juice extracted from crushed fresh fruits has antibacterial properties. Further details of fruit characteristics and constituents are given at the end of this section. The chemical composition of fruits has been described in detail (Bosland and Votava, 2000).

The fruits are attractive food for birds who are the main seed distribution agents, as noted. So efficient are they that in the Azande region of the southern Sudan, plants from bird-distributed seed are so numerous that all the locals need to do is weed around the plants and collect the fruit, which is their second most important cash crop. Despite steady prices and demand, chilli peppers have never been a truly cultivated Azande crop, as is cotton. As one local stated to the author 'If cotton appeared in the same way we would not cultivate that either!'

While fruit is usually seeded, seedless forms exist. It appears that seed number affects fruit growth and development, with high seed number per fruit depressing both (Marcelis and Hoffman-Eijer, 1997). Sections through two types of fruit are shown in Fig. 9.4. Seed number is variable, usually numerous, and they are orbicular, flatted at the hilum, slightly rugose, 3.0–4.5 mm diameter and about 1 mm thick, and usually yellow; 1000-seed weight is 6–8 g. The embryo is white, strongly curved and embedded in copious grey endosperm. Data from India showed that about 60% of the dried weight of local chilli pepper fruits was seed.

Pepper seeds have been studied extensively, and a considerable literature on their composition and characteristics is available. The seed oil content of Indian chilli pepper

Fig. 9.2. Bell pepper fruit.

Fig. 9.3. Chilli pepper fruit.

cultivars is 12–26%, and the unsaturated fatty acids included linoleic 70.6%, palmitic 16.3%, oleic 10.9%, stearic 2.2% and myristic 0.2%. Analysis of tonnes of seed in the USA gave the following average chemical composition: oil 26%, moisture 6.25% and dried extracted meal 67.7%. Analysis of the meal gave N-free extract (carbohydrates) 36.4%, fibre 29.1%, protein 28.9% and ash 5.6%. Some characteristics of the seed oil were (at 24°C): specific gravity 0.918; refractive index 1.4738; acid number 2.18; Hanus iodine number 133.5; acetyl number 7.0; unsaponification number 1.7.

Seed generally remains viable for about 3 years, although no data are available on the longevity of species or types. Dormancy has been reported in some cultivars but in others fresh seed germinated well. Should it be necessary to overcome dormancy, treatment with potassium nitrate is successful. A reason that dormancy may not have been reported as affecting germination of commercial crops is that most seed sown is at least 6 months old. For example, sun-dried var. *annuum* seed retained in Kenya from the previous crop and stored in brown paper bags at ambient temperature and humidity for up to 7 months usually had a germination percentage of at least 90% (E.A. Weiss, personal observation).

Fruit composition varies considerably, not only between cultivars, but also due to seasonal conditions and maturity at harvest. Fresh capsicum fruits contain 0.1–2.6% of steam-volatile oil, fixed (fatty) oil at 9–17%, pigments, pungent principles, resin, protein at 12–15%, cellulose, pentosans and minerals. The fruits are one of the richest known plant sources of vitamin C, as noted, and an important source of vitamin A (Howard *et al.*, 2000).

The pungency is mainly due to capsaicin, and associated amides, present at 0.8–1.2%, but in most commercial samples at about 0.5%. The associated amides include dihydrocapsaicin and nordhidrocapsaicin, with small amounts of homocapsaicin and homodihydrocapsaicin. Inheritance of these constituents controls fruit pungency, and has been partially determined (Zewdie and Bosland, 2000). A rapid method of determining the level of capsaicin and its analogues has been developed (Manirakiza *et al.*, 1999; Sato *et al.*, 1999), but simple or standard methods are acceptable (Thomas *et al.*, 1998; Perucka and Oleszek, 2000). Carotenoids are responsible for fruit colour; red fruits contain capsanthin as the most important, although also capsorubin and carotene, but no lutein; yellow fruits contain lutein and violaxanthin. Since the red pigment is most important for paprika powder, time of picking was found to be the most influential factor controlling fruit pigment colour in Bulgaria (Todorova *et al.*, 1999) and in Hungary (Markus *et al.*, 1999; Deli *et al.*, 2001), while fruit colour is a most important selection factor in India (Verma and Joshi, 2000). Treatment of harvested fruit with ethylene had no effect on colour or pungency in Australia (Krajaklang *et al.*, 2000).

Fixed oil content varies between 9 and 25% according to species and cultivar, and gradually increases from the green to the ripe red stage. The oil includes about 60% tryglycerides, mainly linoleic and other

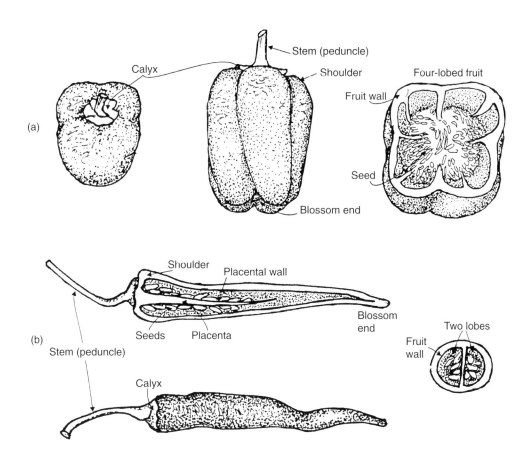

Fig. 9.4. Sections through fruits. (a) Bell peper; (b) chilli pepper.

unsaturated acids, and it may be possible to distinguish chilli from paprika oil by the amount of certain fatty acids. The volatile oil is generally regarded as being of lesser importance than colour and pungency of the dried products in determining quality and value (see also Products and specifications section).

Capsicum baccatum *group. C. baccatum* L. is native to South America and distinguished from the other *Capsicum* species by the yellow, brown or dark green markings in the corolla throat, and yellow anthers. It is difficult to distinguish from *C. annuum* when fruiting, as it has many similar fruit forms, although mainly ovate to almost round fruits predominate. The wild *C. baccatum* var. *baccatum*, which gives fertile hybrids when crossed with cultivated var. *pendulum*, is probably the wild progenitor and reported as occurring in Paraguay, northern Argentina, southern Brazil and south-eastern Bolivia. The greatest diversity is found in Peru, Ecuador and Chile. In South America generally it is less common than *C. chinense*.

The cultivated *C. baccatum* var. *pendulum* (Willd.) Eshbaugh (syn. *Capsicum pendulum* Willd.) occurs in Argentina, Bolivia, Brazil, Chile, Ecuador and Peru, and also Costa Rica and Hawaii. It grows naturally from sea level to 1500 m in the Andes, is very popular in Peru, and remains of *Capsicum* at very early archaeological sites in the region are probably this species. It has been introduced into the USA, and in trials matured early and produced heavy yields.

Capsicum cardenasii. *C. cardenasii* is native and was apparently restricted to the Bolivia antiplano, and was classified only in 1958. It is locally known as ulupica and sold regularly in the markets of local towns. The very small, aromatic and very pungent fruits are gathered mainly from wild plants, although it may also be planted near homesteads for domestic use.

Capsicum chinense. *C. chinense* Jacq. is probably the most commonly cultivated and widely distributed species in northern South America and the West Indies, but its wild progenitor is uncertain. Capsicums discovered in prehistoric Peruvian sites have been identified as *C. chinense*. Hybrids with *C. frutescens* lack viability. Plants are up to 45–75 cm; the stem and leaves are essentially glabrous, rarely densely short pubescent. The leaves are ovate to ovate-lanceolate, up to 10.5 mm broad, smooth or rugose, and light to dark green. There are generally three to five flowers per node; the pedicels are declinate, rarely erect at anthesis, and relatively short and thick. The calyx is without teeth and usually with a distinct constriction at the base; the corolla is usually greenish-yellow, 0.5–1.0 cm, the lobes not spreading; the anthers are blue to purple. The fruits are 1.0–12.0 cm varying from spherical to elongate, smooth or wrinkled; when mature they may be red, pink, orange, yellow or brown. The seed margins are usually wavy and rarely smooth.

C. chinense resembles *C. frutescens* to which it is closely related. The inflorescence can be distinguished by the usually shorter, thicker, curved pedicels, three to five per node, with a distinct circular constriction between the pedicel and calyx. In *C. frutescens*, the pedicels are long, slender, erect at anthesis and lack the constriction. The corolla lobes are not, or only slightly, spreading; those of *C. frutescens* are usually spreading and frequently recurved.

Capsicum frutescens *group. C. frutescens* L. (syn. *Capsicum minimum* Roxb) is commonly known as bird pepper. It is found as a weed or wild plant in tropical climates from Florida, Mexico and the West Indies to north-ern parts of South America, and as large-fruited cultivars from southern Mexico to Costa Rica. Its current wide distribution obscures its American centre of origin. It was taken in post-Columbian times to the tropical Pacific, Africa and South-east Asia, where it quickly became naturalized. As a cultivated plant, the species is much less variable and has a more restricted distribution than *C. annuum*, partly due to the long period required for its fruit to mature.

C. frutescens is naturalized in the Equatoria Province of the Sudan and its fruits collected and exported. It is also common in Uganda, Tanzania and Kenya, renowned for their very pungent chillis known most aptly in KiSwahili as *pilipili hoho*, and sometimes exported to the EU, the Middle East and Gulf States as Mombasa peppers. Forms of *C. frutescens* are grown and have become naturalized in many other parts of the tropics. Indian botanists discovered in Assam a local type, *naga jolokia*, whose very small red fruits were stated to be the most pungent (hottest) yet known at 855,000 Scoville units (Mathur *et al.*, 2000). Scientists at the Indian Defence Research Laboratory, Tezpur, are reportedly studying methods of using the very high capsaicin content in various sprays.

C. frutescens is a rather woody perennial subshrub up to 1.5 m, similar in structure to var. *annuum*. It has an inflorescence of several flowers. There are usually two fruits at fruiting nodes, but can be more; they are small, erect, conical 2–3 cm, usually red when ripe and very pungent. The only cultivar grown commercially in the USA is Tabasco, mainly in Louisiana, where it was introduced from Mexico. Tabasco is a cultivar with small, very thin-walled, extremely pungent fruits, about 3.7 × 1.2 cm in diameter, and used to make the famous Tabasco Sauce (see Products and specifications section).

Capsicum pubescens *group. C. pubescens* Ruiz & Pavon (syn. *Capsicum guatemalense* Bitter) is known as apple chilli. It is native to the Andean highlands from Colombia to Peru, limited to altitudes between 1500 and 3500 m, and known only as a cultivated plant. This species was apparently well known in the Inca empire, remains very common today, and is generally known as *rocoto*. The species was apparently taken to the highlands of Costa Rica, Guatemala and southern Mexico at an unknown date, and has also been introduced to Indonesia and cultivated above 1400 m on West and Central Java. However, outside its specialized environment it seldom flourishes.

C. pubescens is a perennial with a life span of 2–3 years; it is a herb or subshrub, up to 0.5 m, or climbing shrub to 3 m, differing from other cultivated peppers by its overall pubescence, but similar in structure to *C. annuum*. The stem is striate and branched and the nodes purple. The leaves are conspicuously pubescent, alternate, ovate, rugose, the margin entire, and occasionally ciliate. The flowers are fragrant and blue or purple rather than white or greenish; the pedicels are frequently in pairs, usually erect at anthesis. The calyx is truncate or slightly dentate, six-lobed up to 1 mm; the corolla is campanulate to rotate, greenish-white, waxy or shiny, and 6–10 mm; the stigma is green.

The fruits may be pendent to erect, very thick, fleshy and usually small, but larger fruits occur; they are 0.7–3.0 × 0.3–1.0 cm, green and yellow when immature, red-orange or brown when ripe, and late maturing. Fruits are very variable in shape, elongate to spherical and can have a pronounced neck; they are usually very pungent, but pungency is variable regionally. The seeds are 2.5–3.5 mm in diameter, wrinkled and black instead of yellow.

Since most *Capsicum* species have basically similar agronomic requirements and behave similarly when cultivated, only *C. annuum* var. *annuum* will be described in detail, with any important deviations mentioned.

Ecology

Many types of small-fruited peppers occur naturally from the southern USA, through Mexico and Central America to northern South America, although central Mexico has the greatest diversity of cultivated forms with a secondary centre in Guatemala. Capsicums are grown from sea level to 2000 m, but to 3000 m in South America and Ethiopia, although at these altitudes fruit may not fully ripen or be difficult to dry. In India, eight ecological/geographical zones are recognized from humid to semiarid with suitable cultivars recommended for each. It also appears that cuticular water permeability is an acquired characteristic evolved by plants to suit a specific environment (Schreiber and Reiderer, 1996).

A rainfall of 600–1250 mm is suitable; above 1500 mm soils must be free-draining as in general plants cannot tolerate waterlogging, even for relatively short periods, especially as seedlings. Rainfall should ideally fall over the main growing period and cease when fruits are maturing. Heavy rain at full bloom adversely affects pollination and reduces fruit yield, and on nearly ripe fruits exacerbates fungal damage.

Capsicums flourish in warm, sunny conditions, and require 3–5 months with a temperature range of 18–30°C; below 5°C growth is retarded, and frost kills plants at any growth stage. Proprietary chemical sprays to limit potential frost damage are available; if overhead irrigation is installed a thin film of water will also protect plants. A seedbed temperature of 20–28°C is the optimum for germination, which is slowed at 15°C and ceases at 35°C. Height of plants in Korea was affected more by day than night temperature, stem length by the difference between night and day temperatures, and a day temperature of about 24°C produced compact healthy plants (PakHeng Young *et al.*, 1996). Temperatures above 30°C without adequate soil moisture will adversely affect growth, and should this occur at flowering, may cause substantial flower drop. High temperature was shown to be the main factor inducing abscission of flowers and fruit in Israel (Huberman *et al.*, 1997). Fruit set is adversely affected when mean temperatures remain below 16°C or above 32°C, or by prolonged periods of cloudy weather. Fruit colour and the pungency of some cultivars is also affected by temperature. The optimum range for promoting a uniform red is 18–24°C; above this fruits acquire a yellow tinge; below 13°C red colour development is retarded and fruits remain green. Pungency is generally increased at high and reduced by low temperatures during maturation.

Maturing fruits turning from green to red and exposed to high radiation levels, i.e. direct strong sunlight, can develop sun-scald expressed as bleached or white spots of varying size, often with a yellow halo. Some scald can be reduced by selecting cultivars that are light rather than dark green, as the former absorb less heat.

A strong dry wind also adversely affects flowering, and high temperatures immediately prior to main blooming will suppress flowering, but if the temperature quickly drops then flowering usually resumes, often with a lower fruit yield. However, local cultivars are often highly resistant to local weather conditions, and cultivars of *C. frutescens* are more tolerant of hot weather. Outside the tropics, capsicums often grow well in regions where grape vines flourish, especially the bell peppers.

The effect of light intensity or shade on growth and yield of capsicums remains to be fully determined, while information on photoperiodic sensitivity of specific cultivars is also lacking. It appears that most cultivars produce the highest number of flowers with a minimum day length of 10 h and a maximum of 16 h. Fruit on some cultivars is susceptible to sunburn, especially paprika and bell peppers, and maintaining a full canopy to provide shade is important or fruit will be badly damaged. However, the presence of shade trees or interplanted tall crops generally reduces fruit yield.

Soils

The range of soils on which capsicums are grown reflects the popularity of the spice, and soil type *per se* is generally less important in many areas than the standard of management. Sandy and clay loams are generally preferred with a minimum depth of 30 cm, slightly acid with pH 6.0–7.0. Saline soils should be avoided, as pepper is classified as being moderately sensitive, and is also sensitive to saline irrigation water (see also Irrigation section). *C. chinense* is reportedly more salt-sensitive than *C. annuum* in the USA.

Smallholders grow the crop as and where they can, but commercial crops require a more selective approach. Highest chilli yields in

Kenya were achieved on a sandy clay derived from granite, but good crops were also produced on coastal sandy/coral rag soils and on Zanzibar (E.A. Weiss, personal observation). In Uganda, most chillis are grown on the loams around the eastern shores of Lake Victoria. Sandy clays suitable for irrigated crops are favoured in the Sudan and the rich riverside loams of the Nile are popular in Egypt. In India, red or black soils and clayey loams are utilized. In the USA, Mexico and Central America, chillis are grown commercially on the truck soils used for vegetable production and in southern Europe, soils favoured for vines are suitable for chilli. The adaptability of capsicums can thus be appreciated, and it is again emphasized that crop management is generally as important as soil type.

Fertilizers

Most smallholder crops receive no fertilizer except crop and animal residues when available, or night soil in China. Prior to construction of the Aswan High Dam in Egypt, excellent crops were grown on lands fertilized by the annual Nile floods. In trials with FYM, biogas slurry and vermicompost in India, the last gave the highest fruit yield (Shashidhara *et al.*, 1998). However, some organic manures can increase susceptibility to insect attack as in Nigeria, where plants receiving poultry manure suffered the greatest damage from the pepper fly, *Atherigona orientalis* (Ogbalu, 1999).

Data on the nutritional requirements of capsicum indicate the very significant variation in cultivar response to applied fertilizers, especially in respect of N as a top dressing. Where large commercial crops are involved, formal fertilizer trials to establish the correct nutrient level, ratios and type of fertilizer are essential for optimum profitability. In Nigeria, for example, ammonium nitrate produced higher fruit yield over ammonium sulphate (Owusu *et al.*, 2000); in Thailand, a mixture of NPK at 30:30:15 kg ha^{-1} gave the highest fruit capsiacin content, but 15:15:15 kg ha^{-1} produced fruit with the highest ASTA Colour Units (Yodpetch, 2000).

Sap testing is an efficient method of quickly establishing the nutrient status of plants, and a grower can easily carry out the necessary tests using commercially available kits. Nutrient status of leaves during growth can be a guide to the effectiveness of a fertilizer programme, and indicate future improvements. In Queensland, the optimum nutrient levels of the youngest fully mature leaf (dry weight) taken when first fruit matures is as shown in Table 9.2. The most important nutrient is usually nitrogen (N), with great variability in the ratios of phosphorus (P) and potassium (K). Although compound NPK mixtures are applied to most commercial crops of chillis and bell peppers for the fresh market, it is often the standard mixture used on all local vegetable crops. Where a local deficiency exists, these mixtures contain additional amounts of that element, plus any minor nutrients necessary, and are advertised as 'chilli mixtures'. Soils in capsicum-growing regions of Queensland, Australia, generally require boron (B) and zinc (Zn), and this may be sprayed on the seedbed prior to planting or applied through the irrigation system. Elsewhere calcium (Ca), magnesium (Mg) and sulphur (S) and may be required, and single superphosphate with its higher S content may be sufficient on many tropical soils. Copper ore tailings (COT) are a by-product of copper mining, and in India contained significant amounts of micronutrients considered a cheap replacement for pure compounds (Basavaraja *et al.*, 1998). In Guyana, ground sea shells were an excellent substitute for

lime, and increased pepper yields over the local phosphate lime (Brandjes and Lauckner, 1997).

A general recommendation in India is for 60–100 kg N ha^{-1}, 40–60 kg P ha^{-1}, and 20–40 kg K ha^{-1} varied according the zone, as noted in the Ecology section, with half the N in the seedbed. In the USA, a 5:8:8 NPK mixture at about 600 kg ha^{-1} is recommended for clay soils; on more sandy soils about 1000 kg ha^{-1} of a 5:10:10 mixture, plus on both a top dressing of 140 kg ha^{-1} ammonium sulphate or its urea equivalent when fruit begins to set. The amount of N applied may in fact depend on a grower's preference, since application rates vary from 50 kg N ha^{-1} to 700 kg N ha^{-1}, some applied in the seedbed and as split top dressings. However, N applied after fruit has set may affect pungency by reducing the capsaicin level. Seedbed N produces sturdy, quick-growing plants but not necessarily significantly higher yields, but may do so as di-ammonium phosphate (DAP). In Kenya on a rather sandy granitic soil, 100 kg DAP ha^{-1} in the seedbed and 100 kg ammonium sulphate top dressing ha^{-1} were applied when first flowers appeared, as recommended by the local Agricultural Department (E.A. Weiss, personal observation). Two common NPK mixtures applied in Queensland are 5:6:5 at 900–1000 kg ha^{-1} and 13:15:13 at 350–400 kg ha^{-1}. On soils with a high P content, either natural or residual, a 15:4:12 mixture is used.

Some deficiency symptoms in pepper plants have been described, but may not be generally applicable: P-deficient plants were stunted and the leaves narrow, becoming a glossy greyish-green; K-deficient plants developed bronze leaves, which were later shed; Mg-deficient plants were stunted, developed pale green leaves followed by interveinal yellowing, then leaves were shed; Ca deficiency is associated with blossom-end rot and spotting of fruit; B-deficient plants produced fruits with cracked skins; on molybdenum-deficient plants older leaves become mottled, may curl and die back from the tip; with a Zn deficiency leaves become yellow between the veins and may be smaller than normal.

Table 9.2. Optimum nutrient levels in young, fully mature leaves (dry weight) – Australia.

Nutrient	Normal level
Nitrogen	3.0–5.0%
Phosphorus	0.3–0.6%
Potassium	3.0–5.5%
Calcium	1.0–3.5%
Copper	10–200 p.p.m.
Zinc	20–100 p.p.m.
Manganese	26–300 p.p.m.
Iron	60–300 p.p.m.
Boron	30–100 p.p.m.
Molybdenum	0.5–2.0 p.p.m.

Cultivation

Lands for smallholder crops should receive a thorough cultivation producing as level a seedbed as possible. If organic manure or rock phosphate is applied, it should be spread and ploughed in as deeply as possible before planting. Seed may be sown in drills and harrowed in, but transplanting seedlings is most common. In India about 1 kg seed is broadcast on 200 m^2 of nursery beds, sufficient to plant 1 ha, but in East Africa in similar conditions using local seed, abut 1.5 kg was required and beds were fenced against small wild animals. Irrespective of whether seed is sown direct or in seedbeds, it should be protected against fungal infection, and sodium hypochlorite is cheap and efficient.

Germination is epigeal, and seedlings should be large enough, 2.5–5.0 cm, for transplanting at 6–8 weeks. Older seedlings are to be preferred since they generally have a greater survival rate, as in Taiwan (Shih *et al.*, 1999) and Hungary (Somogyi *et al.*, 2000). However, young plants are usually raised under cover in colder countries, and then transplanted into the field. On outdoor hotbeds in the USA about 60 seeds m^{-1} in rows 20–40 cm apart is common, and a temperature of 21–24°C is recommended pre-emergence, 18–21°C post-emergence, and seedlings must be hardened off before transplanting. A common in-row spacing of 45 cm in 100 cm rows requires about 30,000 plants ha^{-1}, with larger cultivars at an in-row spacing of 60 cm. In the Gangetic area of India, capsicum is a cold-weather crop, transplanted in September and harvested in January–February. In parts of the Punjab, it is sown towards the end of the cold season in March–April to avoid frost, and harvested in September–December. Around Mumbai, seeds are sown in nurseries in June and July, the seedlings transplanted in August–September and the crop harvested after 3–4 months.

In general, cultivars are spaced according to local experience and conditions; for example, in India this varies between 25 and 30 cm between plants in 25–60 cm rows. With any cultivar, wider spacing generally produces the highest number of fruits per plant, but close spacing produces the greatest number ha^{-1}. Most capsicum crops in East and Central Africa are manually grown by smallholders, frequently intercropped, and little attention paid to either plant populations or correct spacing. Commercial crops grown in western Kenya (E.A. Weiss, personal data) were mechanically transplanted into 90 cm rows with plants 25 cm apart – an arbitrary spacing to use available equipment. A row spacing of 45 cm would probably have given a higher yield. Recent plantings in Zimbabwe of paprika peppers were in 50 cm rows with 25 cm between plants. A system similar to that adopted by grape growers is used in Korea, parts of China and Japan, with raised beds or wide ridges planted with two rows of plants supported on twine or wire strung between wooden stakes (Fig. 9.5).

A most important factor to obtain uniform growth and fruit quality of commercial capsicum crops is to ensure that only pure, certified seed of a specified cultivar is obtained from a reputable producer. Some seedsmen may also state the guaranteed germination percentage. A variety of treatments can encourage or hasten germination, including soaking in chemical solutions, but results are conflicting and such treatments unnecessary in commercial crops.

Capsicums can be sown directly in the field using coated seed to facilitate precision planting, and although 1 kg of certified seed will normally suffice for 1 ha, most growers in the USA sow 2.5–3.5 kg seed ha^{-1} to achieve a final stand after thinning of around 100,000 plants ha^{-1}. However, the seed of some new cultivars is so expensive (A\$9000 kg^{-1} in 2000) that this may not be possible. Alternatively, seedlings can be raised in trays to ensure uniform growth for mechanical planting, or, as is common for large-scale commercial growers, be obtained from specialist seedling producers; between 30,000 and 40,000 seedlings ha^{-1} are required.

Cup transplanters require seedlings to be at least 12 cm high; waterwheel planters have no height requirement other than that seedlings must be tall enough not to be trapped under the plastic strip (Fig. 9.6). A major difference between direct seeding and

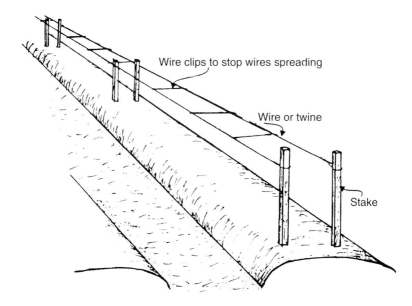

Fig. 9.5. Support trellis for plants on raised beds.

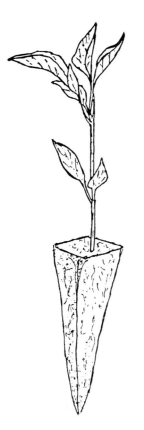

Fig. 9.6. Seedling ready for mechanical transplanting.

transplants is that the former have a strong penetrating taproot and a larger root system, while transplants give shorter more sturdy plants, often with a higher fruit yield. A common spacing is 1 m rows with 30–45 cm between plants depending on cultivar. Time to flowering and duration is a cultivar characteristic, and varies from about 10–12 weeks after transplanting for chillis to 12–15 weeks for larger varieties. Reducing the duration of flowering to produce fruit over a shorter period and thus more even ripening has allowed mechanical harvesting to become more efficient.

Capsicums are also grown in transparent plastic tunnels covering one or two rows, a system common in cooler regions such as China and Japan to obtain a second, protected crop following harvest of the main crop. Capsicums are also produced in greenhouses with plants growing in soil, trays, or in troughs filled with hydroponic solutions. Production in permanent greenhouses is expensive and in developed countries is normally located near towns to take advantage of a high-value market. Plastic greenhouses can be erected virtually anywhere; one producing bell peppers in northern Japan was heated by a thermal spring and surrounded by over 1 m of snow (E.A. Weiss, personal observation). The area under plastic greenhouses worldwide totals many thousands of hectares; for example, some 15,000 ha in Spain. A higher degree of management is required for both types, but very high-quality fruit is produced.

Weed control

Most crops are hand- or mechanically weeded, which generally causes some damage to plant roots. Herbicides do least damage at the recommended levels but are generally too expensive for most smallholders. The following chemicals have been recommended for use pre-sowing or pre-transplanting: chlorothal, diphenamid, glyphosate, trifluralin; and post-transplanting, directed sprays of chloramben, chlorothal, diphenamid and glyphosate. Commercial growers also use plastic strips through which seedlings are planted, and these strips can reduce the damage of herbicides to seedlings, as in Spain (Cavero *et al.*, 1996). Plastic strips also act as a mulch and retain soil moisture in drier regions such as Israel, Mexico and the eastern USA. Fumigation with methyl bromide before laying the plastic will prevent weeds piercing the sheeting and also growing through the planting holes.

Irrigation

Capsicums can be grown as a rain-fed or irrigated crop, and different soil types are generally chosen for each. In India, the former is grown on the fertile black cotton and clay soils, but under irrigation sandy clays or alluviums are preferred. In Egypt before the construction of the Aswan High Dam, plants were sown on river verges as floods retreated and produced excellent crops. Today the Nile no longer floods and chilli crops require artificial irrigation, which may not be available or affordable. It is believed in areas such as India and East Africa where rain-grown and irrigated crops are grown, that naturally withered fruit from the former tend to be slightly more pungent than from irrigated plants.

Unless water is easily and cheaply available, very high yields are obtainable, or a high value market can be supplied, growing capsicum crops wholly under irrigation is seldom profitable. An exception is when portable systems already exist for use on other crops, or trickle irrigation is possible. The increasing cost of water is now limiting irrigated paprika production in Hungary (Somogyi *et al.*, 2000). Brackish water from bores is frequently used for irrigation in arid areas, and although water with a salt content above 1200 microsiemens cm^{-1} (μS cm^{-1}) is unsuitable for young pepper, it is tolerated by mature plants with up to 50% yield loss, while 2500 μS cm^{-1} will generally be fatal.

Consumptive water use by capsicums in general remains to be fully determined, although some data are available from the USA, Israel and India. The total amount of water required to produce an irrigated pepper crop in the USA is stated as 60–75 cm ha^{-1}, with most water being removed from the top 30 cm of soil. Moisture stress during the period of rapid vegetative growth and at flowering reduces yield by up to 50%, depending on cultivar; watering during intervals of the monsoon in India increased fruit yield by between 50 and 90%, depending on the length of the dry period. In general, when plants wilt in the early afternoon a watering is necessary. The watering interval varies with the soil type and local climatic conditions from 7 to 10 days, seldom longer, until fruits are nearly mature; in Queensland 30–40 mm per week is recommended. On a sandy clay in a low rainfall region of Tanzania, alternate rows were watered to increase the planted area with no significant loss of yield (E.A. Weiss, personal data).

Pepper plants are very susceptible to a moisture stress at flowering and over watering at any stage, and thus accurate determination of root zone soil moisture level is essential, either manually based on experience or by sensors, known as tensiometers. Two tensiometers are necessary, one placed about 15–20 cm deep to indicate when to water and the other at about 45 cm to indicate the amount required to cover the root zone (Fig. 9.7). Where no accurate data are

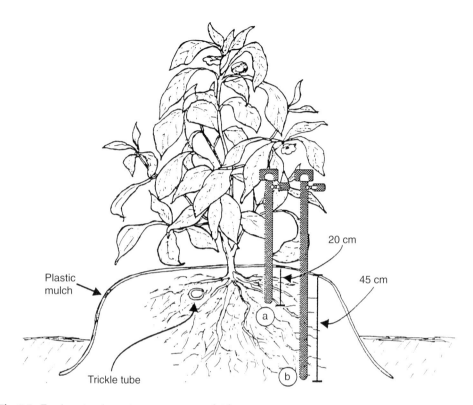

Fig. 9.7. Tensiometers in root zone on protected ridge.

available, tensiometer readings must be empirically related to local conditions for future use. In arid regions, drip or trickle irrigation optimizes water use, and most peppers in Israel and Australia are watered by these methods. Fertilizers can be applied via irrigation systems, known as fertigation, and highly soluble solid or liquid nutrient solutions are available.

Intercropping and rotations

Many species of *Capsicum* can be interplanted by smallholders with a variety of crops including beans, onion, okra, pearl millet, groundnut and chickpea, or stripped-cropped with cotton, maize, castor and occasionally tobacco. Strip-cropping rows of chilli and dwarf beans in Sri Lanka did not affect bean yields, but fruit yield of chillis was greater than chilli in pure stands (de Costa and Perera, 1998).

Chilli may also be underplanted in cocoa, coffee, areca nut and pepper planta-tions, but in plantations such as rubber, underplanting is possible only in the first year, sometimes the second, as subsequent weeding is either mechanized or by overall herbicide spraying, and both preclude further chilli cultivation.

A rotation should allow at least 3–4 years between pepper crops, and not include other solanaceous crops such as tomatoes, potatoes, aubergines or tobacco, as they are affected by many common diseases and pests. In India, capsicums are often grown in rotation with wheat, maize, vegetable crops, okra and curcubits; elsewhere in Asia also with groundnuts, beans, oilseed rape, rice, leguminous cover and fodder crops. On an irrigated tenant farming scheme under the author's supervision in Tanzania, one-cash crop rotation was sunflower, groundnuts, cotton and chilli. In Europe and North America, rotations include cereals, oilseed rape, safflower, root crops and non-solanaceous vegetables.

Pests

Capsicums and chillis (hereafter capsicums) are attacked by a range of insect and related pests; some are cosmopolitan, others are limited in their geographical range or to a particular country. Only the most important are noted, since many are of minor importance. Where a number of solanaceous crops are grown in a district, pests frequently move from one to another, thus increasing total damage. Residues of pesticides may also adversely affect plant growth, DDT in India, for example (Mitra and Raghu, 1998).

Commonly damaging seedlings are mole crickets, *Gryllotalpa* spp. generally, *Scapteriscus* spp. in the southern USA and Caribbean, *Brachytrupes* spp. in Africa, Myanmar and Indo-China, grasshoppers, *Zonocerus* spp. mainly in Africa, *Chrotogonus* spp. in Asia, and *Melanoplus* spp. in North America, while locusts can devastate crops in a specific season. A major cause of gappy stands are cutworms, *Agrotis* spp. generally, *Amsacta* spp. in Asia, *Feltia* spp. and *Peridroma* spp. in North America, and scarabs, especially *Schizonycha* spp. in Africa (see Fig. 6.4).

Aphids are a most common and widely distributed pest of capsicums, particularly the cosmopolitan green peach aphid, *M. persicae*, cotton aphis, *Aphis gossypii*, and potato aphid, *Macrosiphum euphorbiae*. Aphids are also vectors of viruses, including tobacco mosaic, leafcurl, and potato banding virus (see also Disease section). Whiteflies, *Aleyrodidae*, can occasionally become so numerous as to defoliate plants, and some species are vectors of mosaic virus. Flea beetles, *Phyllotreta* spp. are common polyphagous pests and can be very damaging to young plants by 'shotholing' foliage. Flower-eating beetles can occasionally cause damage in a particular crop, but are seldom a major pest (Fig. 9.8).

The ubiquitous corn borers, especially *Ostrinia* spp. and *Elasmopalpus* spp., and bollworms, particularly *Helicoverpa* spp. and *Spodoptera* spp., damage young pods and fruit. Larvae of hawk moths, *Sphingidae*, are damaging locally by eating foliage, especially the hornworm, *Manduca* spp. in the USA, and a pyralid moth, *Sceliodes cordalis*, whose larvae bore fruit in Australia. A major pest in the warmer regions of North America is the pepper weevil, *Anthonomus eugenii*, which feeds on buds and pods and can cause such extensive fruit damage that fields may be abandoned as not worth harvesting. Also damaging is the pepper maggot, *Zonosemata electa*, which feeds within the developing fruit causing it to decay or fall. Various bugs attack fruit, including *Lygus* spp., *Acrosternum* spp., *Calidea* spp. and *Nezara* spp., with the leaf-footed bug *Leptoglossus phyllopus* important in the USA and Caribbean.

Fig. 9.8. Typical flower-eating beetle – *Centoniid* spp.

Pests that may cause economic damage in a particular season are: thrips, especially *S. dorsalis* from West Asia to India, and *Frankliniella tritici* in the USA; stem borers; leaf hoppers, particularly *Circulifer tenellus* in the USA; scale insects; fruit flies, especially *Bactrocera tryoni* in Australia and *Atherigona orientalis* in West Africa, which may cause heavy loss of fruit; leaf miners and mites, especially *Tetranychus* spp. Nematodes are usually present in capsicum fields, but damage to plants is usually patchy. Use of resistant cultivars and rotations are common methods of reducing damage, while plant extracts from castor (*Ricinus communis*) and neem (*Azadirachta indica*) reduced nematode population on pepper in India (Akhtar and Mahmood, 1996).

Diseases

Insects are serious pests of capsicum and chillis but the greatest damage is caused by disease. However, as both types are grown worldwide, the importance of a particular pathogen differs in specific regions. Those diseases described are the most frequently recorded and most damaging. Many are soil-borne and remain infective in the soil for long periods, thus a rotation that includes non-susceptible crops is essential to minimize their incidence.

Damping-off, caused by *Rhizoctonia* spp., especially *R. solani* and *Pythium* spp., particularly *Pythium aphanidermatum*, are especially damaging in nurseries; seed may rot or seedlings be killed before emergence. After emergence, stems can also be attacked; symptoms are softening and shrivelling of stems, which collapse and die. The disease is most damaging on very moist soils. Root rot or phytophthora blight due to *Phytophthora* spp. occurs in most pepper-growing areas, and is seed- and soil-borne. Infected plants are usually girdled at ground level, followed by sudden wilting and death; alternatively leaves show dark green spots, which become dry, and dark water-soaked patches on fruits, which become coated with a white growth; fruits eventually wither and remain attached to plants.

Collar rot caused by *Sclerotium rolfsii* is widespread in the tropics and subtropics, and in the USA it is known as southern blight. The cultivar Tabasco has some resistance. The main symptom is girdling of the stem at ground level, which eventually kills plants. Uprooted plants show the white feathery mycelium with embedded pale brown sclerotia on roots.

Fusarium wilt due to *Fusarium* spp. is widely reported, especially *F. solani* in the capsicum areas of the southern USA. The initial symptom is drooping of the lower leaves and later all leaves wilt and sunken brown areas appear on the stem at ground level, which is finally girdled; roots are also attacked. The fungus is soil- but apparently not seed-borne, and is most damaging on heavy, poorly drained soil and at temperatures above 27°C. Verticillium wilt due to *Verticillium dahliae* is more damaging in temperate and warm-temperate regions; symptoms are very similar. No current cultivars are resistant to either of these wilts. Another locally damaging wilt in Peru is caused by *Phytophthora citrophthora*.

Bacterial spot, *Xanthomonas campestris* pv. *vesicatoria*, is probably the most serious bacterial disease of peppers worldwide, and several local races are known. No commercial cultivars are resistant. Initial symptoms are yellowish-green spots on young leaves, but on older leaves dark water-soaked lesions, which later develop straw-coloured centres with dark margins; severely affected leaves become yellow and fall. Small raised spots develop on fruit, which later become brown and warty in appearance. Severe defoliation occurs due to this bacterial disease in India and Australia. Another damaging disease, especially in the USA, is bacterial canker caused by *Corynebacterium michiganense*. Scabby spots, which later coalesce, form on pods, which become unsaleable. In Israel, the disease is reportedly most frequent in protected cropping and glasshouses. Bacterial soft rot, *Erwinia* spp., and the widespread bacterial wilt, *Ralstonia* (*Pseudomonas*) *solanacearum*, can be damaging in specific regions or seasons.

Frogeye leaf spot due to *Cercospora capsici* often causes serious defoliation. The large circular to oblong spots develop light grey centres with dark brown margins; heavily infected leaves desiccate and fall. Fruit may also become infected, especially when wet and humid conditions prevail. Another leaf spot is due to *Colletotrichum nigrum*, which gains access via injuries or following insect damage. A leaf spot also called frogeye in the USA, ripe rot elsewhere, is due to infection by *C. capsici* and mainly attacks fruit. Small yellowish spots appear on the ripe fruit, which increase in size during damp weather and become sunken and soft. The same fungus causes dieback of plants in India.

Fruit can be infected by the widespread and damaging *Glocosporium piperatum*. The fruits develop large, dark, circular sunken spots, which occur on immature and ripe fruits. The disease spreads rapidly during wet weather, when dark, raised bodies occur coated with masses of pink spores. Anthracnose caused by *Colletotrichum* spp. is widely reported as attacking ripening fruits (Fig. 9.9). Other fruit rots including *Alternaria* spp., *Glomorella cingulata* and *Phomopsis* spp.

Fig. 9.9. Anthracnose of chilli fruit.

have been reported as causing economic damage in specific regions. However, small bleached areas or white spots on developing fruit may be due to sun-scald, as noted in the Ecology section.

Powdery mildew caused by *Leveillula taurica* is most damaging in the dry tropics, and *Oidiopsis taurica* in Europe, Mauritius and North America, and a severe attack can result in almost complete defoliation. Downy mildew, *Peronospora tabacina*, which is a major disease of tobacco, also attacks capsicum, especially in seed and nursery beds in south-eastern USA. Pale spots occur on leaves, which later become covered with a light blue coating of spores. Grey mould due to *Botrytis cinerea* and white mould caused by *Sclerotinia sclerotiorum* can be damaging in very wet seasons.

Numerous viruses have been recorded from capsicums and several may infect a single plant, and they are most serious in tropical regions, Ethiopia, for example (Hiskias *et al.*, 1999). A general symptom is stunted plants, yellow mottling and curling of leaves, and some viruses may also cause yellow and dark green raised spots on fruit. Capsicums frequently have a higher degree of viral infection when planted in the vicinity of more favoured hosts, wild and cultivated. Capsicums are apparently mutagenic, and such aberrant plants can be mistaken for those affected with a virus. A method of identifying a new tospovirus has been developed by Australian CSIRO scientists using DNA fingerprinting techniques, which should allow much faster progress in the breeding of resistant cultivars.

The viruses most commonly recorded are potato virus Y (PVY), pepper mottle virus (PMV) and tomato spotted wilt virus (TSWV). Other viruses include alfalfa mosaic (AMV), beet curly top virus (BCTV), cucumber mosaic virus (CMV), tobacco etch virus (TEV) and tobacco mosaic virus (TMV). In addition, a number of what are designated geminiviruses have been described from Central America, Mexico in particular. Common symptoms are stunting of plants, leaves becoming bright yellow and curling, distorted fruit, and greatly reduced yield (Bosland and Votava, 2000). Viruses can be transmitted mechanically by infected equipment, TMV on hands of smokers, or by insects, including the ubiquitous green peach aphid *M. persicae*, the potato aphid *Macrosiphum euphorbiae*, a thrip *S. dorsalis*, the whitefly *Bemisia tabaci*, and a leafhopper *Hyalesthes obsoletus*.

Physiological disorders

Blossom-end rot produces water-soaked spots initially appearing on the blossom end of fruits, which later become light brown and papery. Damage is increased when soil dries out after a period of high soil moisture when plants are making rapid growth; the damage is increased by high N availability. Another watery type of spot is due to oedema, occurring as numerous small raised areas on the underside of leaves, and is often due to over-watering. Greenish or greyish-brown spots on fruit affects only certain cultivars and regions, and is apparently due to a calcium deficiency or unsuitable cultivar. Sun-scald occurs on unshaded or partially shaded fruit, and is most severe on plants defoliated by leaf diseases. Fruits show spots that are initially soft, light-coloured, sometimes slightly wrinkled, later becoming sunken, white and papery.

Herbicide damage distorts or turns leaves yellow, but later growth is unaffected. The ever-increasing level of air pollution can also cause foliar and fruit damage, especially high levels of sulphur and nitrates. Symptoms of pollution damage are necrotic areas on leaves or leaves with reddish or brownish discoloration on otherwise healthy plants. Severe air pollution can stunt growth of young plants. High salt content in soil results in seedlings dying back at the soil surface, and young plants may die when heavy rain lifts subsoil salt into the root zone.

Harvesting

The period from transplanting to first harvest varies with the type and cultivar; bell peppers are usually harvested about 60–75 days from transplanting, when fully grown fruit is green, yellow or red. Chilli peppers are ready for picking in 100–115 days but up to 150 days to the fully ripe red stage, with Tabasco being very late maturing. In New Mexico, USA, fruit for canning are picked 90–110 days after transplanting, when fully ripe but still green. The first fruit from initial sowings in India raised in nursery beds and transplanted are usually picked green and sold fresh, as this is considered to stimulate further flowering and fruiting. Ripe fruits are later harvested every 14 days over 3–5 months. In Hungary, diseased fruits are picked after the first harvest to reduce the spread of disease to ripening fruit; between 60 and 65% of the total is normally harvested at the first picking.

The most important factors affecting pungency are the species and cultivar grown, season and environment, but a major modifier is fruit maturity at harvest. Bell and chilli peppers are affected differently, but only the latter is described in detail, although paprika generally behave as chilli. Capsaicin levels gradually increase as fruit matures; they are lowest in green and highest in bright red fruit, while pungency and colour of fruit is closely related to maturity at harvest. However, fruit left on the plant to ripen and partially wither are superior to those picked when fully coloured but succulent, and withered fruit commands a premium when marketed on pungency content. Withering the fruit has a greater influence on the quality of the final product than any subsequent drying or storage method. Thus, chilli growers in the USA harvest only fully mature fruit including partially withered, while unripe fruits are left to mature and harvested in a second picking.

Pungency levels vary in different fruit components as fruit matures; the capsaicinoid content of the whole fruit and the hull gradually increases, that of the enclosed seeds and placental attachments decreases, although the rate and final content varies with the cultivar. These findings support empirical experience that fruit withered on-plant has greater pungency.

Most peppers are harvested manually in less developed regions because labour is less expensive than machines and in more developed countries because it produces a higher quality fruit, since pickers can select only whole unblemished fruits, and include fewer leaves and twigs (Fig. 9.10). Some chilli fruits

Fig. 9.10. Harvesting smallholder chilli – Zimbabwe.

can cause the hands to blister and gloves may be required. In Japan and Korea where fields tend to be small and top-of-the-plant bird peppers popular (Fig. 9.11), hand-held mechanical tea-cutters can be used. Mechanical harvesting is under investigation, but no currently available machines are wholly successful; the major defect is fruit damage (Palau and Torregrosa, 1997; Bosland and Votava, 2000). Harvesters under development will be either fully automatic or labour-assisted, similar to pineapple or lettuce harvesters. Another approach is to develop suitable chilli cultivars for mechanical harvesting with the majority of their fruits at or near the top of rather flat-topped

Fig. 9.11. Bird peppers – Japan.

plants (Fig. 9.11), and paprika cultivars that produce fruits over a short period to allow once-over harvesting. The reduction in yield per plant is compensated by a higher plant population. Defoliants or desiccants, including sodium chlorate and ethephon, can assist harvesting in wet seasons, and the latter can also enhance the red colour of the fruit.

The average yield of dried chillis from an Indian rain-fed crop varies from 500 to 1000 kg ha^{-1} depending mainly on how the crop was treated by the grower, but a commercial irrigated crop can yield 1500 to 3000 kg ha^{-1}. In East and Central Africa, smallholder crops yielded from 200 to 500 kg ha^{-1} depending on the skill of the grower, but most chilli crops are interplanted to some extent. In the USA pepper growing is highly developed and an irrigated crop can produce up to 15,000 kg chilli fruits ha^{-1} depending on cultivar; in Europe 10,000 kg ha^{-1} is considered good.

The capital cost of establishing a fully mechanized 20 ha capsicum or chilli plantation in Queensland, Australia, in 2000 was about A$250,000, which included all machinery, irrigation equipment, cool rooms and packing sheds. Annual average costs including wages of two full-time workers and production of 4000 × 8 kg cartons of fresh capsicum or 6000 × 3 kg cartons of chillis were estimated as shown in Table 9.3.

Processing

Bell pepper growers usually market their produce as soon after picking as possible,

before the fruits lose their bright colour and glossy appearance. Chilli and capsicum fruits are cleaned and sometimes graded before sale as a fresh spice. Harvested chillis and capsicum have a moisture content of 65–80%, depending on whether withered on the plant or harvested while still succulent, and moisture content must be reduced to about 10% to prevent fungal infection and loss of pungency. Poor handling of fruits prior to drying can result in bruising, which results in discoloured spots on the pods or splitting, which causes excessive seed loss and thus a considerable reduction in weight.

In addition to other factors previously described, drying and handling methods can significantly influence colour and pungency. Drying may be in the sun or artificially, and sun-drying fruits immediately after harvesting remains the most common. The disadvantages of natural drying have been discussed in the vanilla and pepper chapters, and are relevant to chillis.

Sun-drying. Sun-drying is most common throughout Asia, Africa, Central and South America, and on smallholder crops everywhere. The method varies from spreading on roadside verges, dry paddy fields, on mats around the homestead, on flat roofs in Nepal and Iran, and fruits are thus contaminated to a greater or lesser degree and the quality is generally poor (Fig. 9.12). In periods of inimical weather losses can be 70%. In South America fruits may be hung in string bags weighing 5–12 kg on house walls to dry,

Table 9.3. Annual average costs of producing fresh capsicums and chillis in Queensland, Australia.

Costs	Capsicums		Chillis	
	A$ per carton[a]	A$ ha^{-1}	A$ per carton[b]	A$ ha^{-1}
Growing	2.3	9.0	1.7	10.0
Harvesting and packing	3.7	15.0	10.0	58.0
Marketing (all costs)	2.6	10.5	3.0	18.0
Total	8.6	34.5	14.7	86.0

[a] 8 kg carton.
[b] 3 kg carton.
Based on average wholesale prices (1998–2000) of A$10 per carton for capsicum and A$15 for chillis.
Figures do not include interest, depreciation, local rates and electricity.

Fig. 9.12. Chillis drying on hut roof – Central Africa.

which may take several weeks. In many countries drying is carried out by traders who buy the peppers from smallholders. These semi-dry fruits are allowed to remain undercover in heaps 'to ripen' until all have become a uniform red, usually after 2–3 days. Drying continues in the shade as direct sunlight can produce white spots on fruits.

The common drying method in most countries is basically similar to that used in India, modified to suit local conditions. Fruits are spread in thin layers in the sun on hard dry ground, concrete floors, or the flat roofs of houses, and covering the surface with a tarpaulin or black plastic reduced drying time by 20%. Fruit is stirred frequently to ensure drying is uniform, with no discoloration or mould growth. At night or if showers threaten, fruits are heaped and covered. After several days, the larger types are flattened by trampling or rolling, then packed in bags for storage and transport. Drying requires a minimum of 20 h sunshine, usually takes 5–15 days depending on the weather, and produces 25–35 kg dried spice from 100 kg fruit. Larger traders use permanent single or multi-tier mobile racks, which can be moved in and out of the sun until fruits are dried below 10% whilst retaining colour

and pungency. This system reduces by half the normal drying time and produces a more uniform product. The CFTRI (Central Food Technological Research Institute, Mysore) has developed a chemical emulsion, Dipsol, to be sprayed on fresh fruit to accelerate sun-drying.

Artificial drying. Artificial drying was first introduced in the southern USA for the pungent forms of *C. annuum*, finger peppers, and for the very pungent Tabasco chillis, and tobacco barns were often utilized. Notes regarding solar dryers for peppers and vanilla are relevant for chillis in less developed countries such as Tanzania, and the Pacific Islands where copra dryers can be used. Suitable solar dryers are shown in Fig. I.2 (see Introduction). A solar dryer developed in India has a batch capacity of 30–50 kg, takes 4–7 days to dry one batch, and is suitable for a variety of crops. A locally designed forced-air dryer gave best results when air temperature was 50–52°C with a velocity of 1.5 m sec⁻¹.

A drying method adaptable to almost any producing region is known as dehydration to distinguish it from forced-air dryers. All that is required is a large windowless building, preferably fitted with one or two large interior fans to circulate hot air, and an exterior heater. Fruits are placed on trays then loaded on to shelves of mobile racks, and manually moved into the shed until it is full. Drying time varies with the pepper type, and after a basic period the final drying is by inspection. This system produces good-quality dried fruit virtually free of contamination. A vertical dryer heated by a fire at the base is common in Spain (Lozano and de Espinosa, 1999).

Commercial pepper producers in developed countries may deliver harvested fruits to drying centres, unless they are large enough to be able to own their own drying plant; on arrival fruit is cleaned, washed and checked for diseased pods etc. Fruit may be dried whole or sliced, in a variety of hot-air dryers, at a temperature of 60–75°C to reduce the moisture content to 7–8%, but to 4–6% if fruits are required for grinding. In the USA, temperatures up to 100°C are common, but overlong exposure to the upper limit can result in soft (cooked) fruits. Sliced pods generally reduce drying time by half with no loss of colour or pungency, but raising the temperature to reduce drying time often results in poor quality, especially if the initial moisture content was high. A number of companies produce suitable dryers with detailed operating instructions covering the various pepper types.

Distillation

Fruits of bell and chilli peppers contain a small amount of essential oil, which is obtained by steam distilling mature ripe fruits. Fruits can be crushed, sliced or macerated prior to distilling, and the operation is straightforward with no special equipment required. The oil is seldom produced, generally only on demand, as the solvent-extracted oleoresin is far more important.

Oleoresins are generally obtained by solvent extracting dried and ground fruits and the operation is simple and routine, with suitable plants in a range of sizes available from manufacturers. A constraint on the wider use of solvent extraction plants is a local shortage of an approved solvent. Restrictions on solvent residue levels in foods precludes some of the more common solvents used in extracting oilseeds, for example. Supercritical fluid or gaseous extraction methods are now more common and are suitable for large or small levels of production. Extraction methods vary considerably, but plants are available for a range of operating conditions. The extraction of a particular component may require a special technique; capsaicin from paprika, for example (Zang et al., 1999).

A major requirement for extraction is that fruits must be free of all other plant material including stalks (peduncles), leaves, etc., and dried to a maximum moisture content of 8%. They are then ground to suit the equipment used, and although a fine grind often results in a higher yield of oleoresin, too fine grinding may cause problems of heating in the mill with a consequent loss of quality, produce excessive fines in the oleoresin, or problems with clogging in the solvent-extraction plant. Solvents in descending order of preference are ether, hexane, ethanol and acetone.

Oleoresin solvent extracted from whole paprika or chilli fruits also contains a fixed oil removed from the seeds (up to 25% has been reported), whose retention in the oleoresin rapidly induces rancidity. By removing the seeds prior to extraction the amount of fixed oil in the oleoresin is negligible. Fixed oil remaining after solvent extraction can be removed by cold-washing oleoresin in ethanol with a substantial reduction in the red colour, offset by concentration of the capsaicin content. Techniques exist, some patented, to avoid or remedy these problems during or after processing to produce a branded product of standard colour and pungency. Oleoresins reflect the parent material and thus vary in colour and pungency, and a very wide range is available to suit end-users. Supercritical fluid extraction produces oleoresins whose pungency and organoleptic characteristics are considered to be close to the original fruit (Jaren-Galan et al., 1999). Since the red colour is important in assessing the quality and subsequent value of paprika oleoresin, the method of determining colour intensity is important, as in Spain (Minguez-Mosquera and Perez-Galvez, 1998).

Transport and storage

Many fungal diseases contracted in the field may be transferred on harvested fruit to stores, where they can cause fruits to decay; in India, for example, 67 species of fungi were isolated from chilli fruits received for storage (Prasad *et al.*, 2000). Thus, fruits should be inspected on arrival at stores, and either hand-sorted to remove those diseased, or chemically treated. Allowing diseased fruits to remain in a sack or bin quickly renders the remainder unfit for sale. Methods of decontamination used on black pepper may be suitable.

Fresh peppers can be stored for 2–3 weeks at a temperature of 7–10°C and a relative humidity of 85–90%. Chilling below 4°C causes softening, pitting and induces decay. Fresh peppers are generally transported in waxed corrugated cartons in refrigerated trucks for long distances. Similarly to other spices, capsicum fruits are attractive to rodents despite their pungency, and stores should be protected against their depredations, since far more fruit is destroyed by contamination from their faeces than is eaten by them.

Storage of dried chillis and capsicum fruits appears to have little effect on pungency compared with other effects already discussed; its effect on colour is more important. Colour is a major factor influencing the price of ground pepper (the spice) and the degree of loss is frequently related to time in storage. Deep red fruits tend to retain their colour longer than those of a lighter shade. Retention of the red colour is most important for paprika pepper, as a strong red colour is considered a characteristic of paprika powder at the retail level. Milling can reduce the carotenoid (red colouring) content of powder, the amount depending on the cultivar, but the loss can be partially regained by including seeds (Minguez-Mosquera and Hornero-Mendez, 1997). Colour stability of paprika powder in storage was improved in Hungary when the pericarp was included when grinding (Markus *et al.*, 1999).

When chillis are to be stored for a period, medium-sized fruits are preferred as they remain intact, while long pods tend to break at the distal end. A fairly thin pericarp is better than a thick one, as its moisture is less and drying more easy, and, in contrast to long pods, remains unwrinkled and bright in colour. However, this colour loss is also directly linked to cultivar and origin, with some cultivars losing up to 75% of their original colour and others only 10% over the same storage period. It was noted in Spain that greenhouse-grown fruit was more red than fruit of plants grown outside, but the former lost more of the red colour in storage (Gomez *et al.*, 1998).

Well-dried fruits or spice kept in sealed containers in dark, cool storage lose less colour and pungency, but commercial processors may add an antioxidant to prevent deterioration. Peppers can be transported in sacks, casks or boxes, but most shippers now use corrugated cardboard cartons. Peppers for export are required to meet the stringent regulations regarding microbial contamination of many importing countries, and fresh fruits may also require a phytosanitary certificate. Irradiation can prevent or combat contamination, but is illegal in many countries. A method of detecting irradiation has been developed in India (Satyendra *et al.*, 1998) and Spain (Correcher *et al.*, 1998).

Products and specifications

The large-fruited bell peppers, green, yellow, orange or red, are sold as a fresh vegetable and are not a spice as they have very little pungency, and are frequently marketed as sweet peppers. They are eaten raw or cooked. In most producing countries, sun-dried capsicum or chilli pepper fruits are sold in local markets and buyers select those that suit their taste or purpose. Such is the variety of peppers now available in any particular region, Mexico or India, for example, that dried fruit of almost any shape, size, colour and pungency is on display, and consumers show a remarkable knowledge of their different characteristics.

Such dried fruits are also available in developed countries and many varieties are available in speciality food and spice shops, with the cultivar or origin indicated. The growing populations of peoples from Africa and Asia in the West has broadened the range of peppers available, and fresh green and red chilli peppers are flown in by air to meet demand from this new consumer group.

Commercial traders and spice producers differentiated the dried fruits mainly by variety (cultivar) and origin, and gave them names – East African, West African, Indian, Japanese, Central and North American – and purchased those suitable for their end product. This separation was more for convenience than botanical or regional accuracy, as many are virtually indistinguishable and interchangeable in the finished product. Most pungent types usually available on the market have a capsaicin content of about 1.0%. Today most fruits are internationally traded on their colour and pungency levels, irrespective of origin. Accelerating this trend is that modern cultivars sought by processors are rapidly distributed around the world. However, in the USA, which has a huge domestic market, a range of cultivars are produced for specific end uses. Capsicums and chillis used in food preparations must be of high quality based on the following characteristics: a good pungency level, a bright red colour, a good flavour, a medium-sized fruit with a moderately thin pericarp, a smooth, glossy surface, few seeds and a firm stalk.

Methods of classifying dried chilli and capsicum fruits have been proposed, but such is the variation in size, colour and degree of pungency made more variable by seasonal differences that none are satisfactory, and indeed may not be required. Modern cultivars can be differentiated on one or more characteristics, but these often become less obvious after several seasons in the field. Additionally, as has been noted in the literature, the character of fruit from a particular area may change gradually over time either through controlled varietal selection or random hybridization, with the highly pungent small-fruited types being particularly prone to degeneration through the last named.

The constituents of paprika, capsicums and chillis that affect the products described are basically their pungency, as determined by the level of capsaicin and other vanillyl amindes, and red colour, due to the presence of carotenoid pigments, principally capsanthin. Thus, a considerable amount of research on the chemistry of these and related compounds is available in the literature. The traditional method for evaluating pungency of capsicum products has been the organoleptic method devised by Scoville in 1912, which although it has its drawbacks is still widely used. Pepper powder is classified into five pungency groups using Scoville heat units (SHU): non-pungent paprika 0–700 SHU, mildly pungent 700–3000 SHU, moderately pungent 3000–25,000 SHU, highly pungent 25,000–70,000 SHU, and very highly pungent above 80,000 SHU. Research continues into developing alternative physicochemical methods for assessing the pungency of cultivars using their capsaicin (or total capsaicinoid) content.

Colour is usually determined at two levels and is usually expressed on a scale specified by the American Spice Trade Association (ASTA). Extractable colour, which measures total pigment content, is measured using a spectrometer and designated in ASTA units with the higher the number the brighter the colour; surface colour of pepper powder uses a colour difference meter and is also stated in ASTA units. A third laboratory method analyses and measures the carotenoid profile and is often used to determine colour loss during storage etc.

In addition to the various powders, a range of capsicum products are available as noted, and some require special cultivars or processing techniques. Canned or bottled fruits are a very common retail product and are internationally traded. Two well-known types are pimiento and jalapeno, and both can be processed whole or skinned. In Europe, bottled bell and paprika peppers are popular, whole or stuffed. Pickled peppers, usually in brine, are more common in the Americas. Fermented peppers are not a common retail product, but are used mainly as an ingredient in a range of pepper sauces; a popular brand is Tabasco sauce, described later. Whole or skinned peppers may be either conventionally or flash frozen, and a range of frozen and chilled sliced, diced or crushed products are also available.

Traditionally in the West, the smaller-fruited types are called chillis and valued principally for their high pungency. The somewhat larger, mildly to moderately pungent types are known as capsicums and also valued for their colour. However, in general most fruits are ground and sold as powdered spice, and broadly differentiated by consumers as paprika (mildly hot and spicy), chilli pepper (hot) and cayenne pepper (very hot), to be incorporated in cooked dishes and seldom used directly on meals as is black pepper. Powdered pepper may also be prepared for a particular market or use; in the USA, a particular brand contains chocolate, oregano, cumin and other ingredients, which are a trade secret, and formulated for the very popular TexMex-style dishes.

Red or chilli pepper. This is a spice prepared from moderately pungent varieties mainly for domestic culinary purposes, in curry powder, and by food manufacturers for seasoning processed foods (Fig. 9.13). The spice should be a relatively coarsely ground, uniform red-brown to red, smooth powder; the taste is spicy and hot but not burning. Red

pepper is milder than cayenne and is prepared from the larger-fruited, dark-red, less pungent capsicums. Manufacturers may adjust the final colour and pungency to produce an identifiable brand.

Cayenne pepper. Cayenne pepper is an extremely pungent spice prepared by blending small pungent fruits of any origin. The name is a misnomer as the fruits do not originate in Cayenne (French Guyana)! The chillis are mechanically ground to pass through a 40-mesh sieve. Producing the spice was regarded as a dangerous operation as the dust irritated eyes, nose and lungs. Modern machinery is completely enclosed to solve this problem. The spice should be a uniform fine powder, orange to dark red, the taste very hot and biting, and is normally incorporated in dishes during cooking. Cayenne powder is used by food manufacturers to give a pungency to their products, including a range of specially formulated additives either powdered or liquid for Mexican and similar foods, processed meats, soups and pickles. A widespread use for cayenne pepper was as a sexual stimulant as

Fig. 9.13. Dried chilli as spice.

it causes itching or tingling in the genital area, resulting in sexual stimulation. It has now been replaced by more effective drugs.

Paprika pepper. Paprika was initially obtained from varieties of *C. annuum* grown in Hungary since the 16th century, hence the common name, but modern cultivars are now widely distributed in Europe, North and South America. The spice is a relatively coarse, bright to brilliant red powder, and the taste spicy and mildly pungent, but never bitter. The capsaicin content can vary from 0–0.03%, but always below 0.05%. Many proprietary brands of paprika are available differing mainly in the degree of pungency, which is controlled by the production method and whether the whole fruit or pericarp only is used. These carry the brand name of a specific company or are identified by region, i.e. Hungarian, Spanish, etc. Methods of detecting toxic residues of microbial contamination in paprika powder have been developed, for example, ochratoxin (Bassen and Brunn, 1999).

Paprika is a traditional ingredient of Hungarian goulash, to which cayenne powder may be added to increase the pungency. A major use for paprika is as a food colourant where pungency is not required (Buckenhuskes, 1999). Research also continues into the medicinal properties of chilli powders, and several studies have shown that the addition of chilli powders to food can assist in the prevention of blood clots, while eating fruits can produce a temporary and harmless mental 'high' accompanied by a feeling of well-being.

Essential oil. Analysis of the essential oil of fresh Californian green bell peppers (*C. annuum* var. *grossum*) showed one of the major components, 2-methoxy-3-isobutylpyrazine, possessed an aroma characteristic of the fresh fruit and dominated the organoleptic profile. However, the oil from locally grown fresh red Tabasco peppers (*C. frutescens*) in the USA had a quite different composition, and although 125 compounds were identified in the oil, it contained no pyrazine compounds. Unlike bell pepper oil, no single component possessed an aroma characteristic of the fresh fruit. Analytical results obtained in many of the world's pepper-growing areas indicates the difficulty of drawing general conclusions about oil composition, due to the very wide range of constituents in the oil from the same cultivar in a relatively restricted area or between seasons, and the equally wide range from the same cultivar grown in different geographical regions. This is also apparent in paprika cultivars.

Oleoresins. These are extracted from dried fruits with or without seeds, as described in the section on Distillation, and basically replace powders in the food-processing industry. Oleoresins were originally prepared in consuming countries, but are now as frequently prepared in producing countries; however, very high pungency oleoresins are currently produced mainly in China and India.

A special oleoresin is prepared from fruits with a high capsaicin content and is used almost exclusively in the pharmaceutical industry to prepare counter-irritant creams for external application in the treatment of rheumatism and similar complaints, and also in stomachic, carminative and stimulant formulations. A more recent use is as an ingredient in special sprays used for personal protection or by police to control criminals.

Oleoresin red pepper is obtained from similar fruit used in the production of chilli and cayenne powder, and commercial preparations are usually available at pungency levels of 80,000 and 500,000 SHU (approximately 0.6–4.0% capsaicin, w/w), ranging in colour to a maximum 20,000 units. It is available as a free-flowing or slightly viscous liquid, or dispersed on a dry carrier. Manufacturers claim that 1 kg oleoresin containing 200,000 SHU units can replace 10 kg good-quality pepper powder.

Oleoresin paprika is obtained by solvent-extracting the dried fruits of paprika peppers, and the composition varies according to the method used, as noted in Spain (Guaday *et al.*, 1997). It is normally a free-flowing to slightly viscous fluid, dark to ruby red, with little odour, and spicy but never bitter. A number of preparations are available with proprietary names, differing in their colour and pungency. Major producing countries are, in alphabetical order, India, Israel, Mexico, Morocco, Spain and the USA.

Chilli extracts are prepared by extracting chilli fruits with a water-miscible solvent to produce oleoresins that are a solution of capsaicin, up to 14%, in fatty oil, with little colouring matter. Chilli extracts are used in the manufacture of ginger beer and similar beverages, and in the preparation of pepper vodka. A famous extract to sprinkle on curries to make then even hotter was known in East Africa as Jungle Juice, and prepared by soaking very hot fresh chilli fruits in gin. Paprika extracts are prepared in a similar method to chilli, and produce an oleoresin with a relatively large amount of colouring matter but little pungent material.

Pepper sauces are prepared from pepper extracts or are proprietary mixtures of pepper and other spices. A famous brand is Tabasco, made by the McIhenny Company of Lousiana, USA, established after the Civil War, which claims to sell 150 million of its small bottles annually. The hot red fruits are harvested, crushed and placed in wooden barrels of brine to ferment for 3 years. The mash is then blended with vinegar, stirred for another month, strained and bottled. A more recent entrant is Walkerswood Caribbean Foods of Jamaica, which producers Jerk seasoning sauce based on hot peppers. From a tiny workers cooperative formed 20 years ago, the company now exports sauces worth $2 million annually.

In the USA, capsicum (red, cayenne and paprika) powders and their oloeresins and extractives are given the status of generally regarded as safe (GRAS 182.10 and GRAS 182.20). Paprika powder and oleoresins are also approved as colour additives for food under exempt from certification (73.340 and 73.345). Capsaicin-containing topical products are approved for over the counter (OTC) and prescription drugs in a number of countries.

10

Umbelliferae

The *Umbelliferae* is one of the great families of flowering plants, and contains about 300 genera with 3000 species, and virtual all are perennial, biennial or annual herbs, generally native to the temperate zones. Members are often so similar that chemical or biotechniques may be necessary for differentiation. Many have long been cultivated, and include culinary herbs, spices and important vegetables such as carrot, *Daucus carota* L., and parsnip, *Pastinaca sativa* L.

Umbelliferous plants usually have sturdy, hollow, erect stems bearing large alternate leaves, often much divided and carried on petioles with swollen, clasping bases. The characteristic feature of the family is the inflorescence, which forms a simple or compound umbel of small regular flowers. A simple umbel consists of a number of peduncles each bearing a single flower, all arising from the apex of the stem and subtended by an involucre of one or more bracts. In the compound umbel, the main inflorescence axes each bear small terminal umbels, each secondary umbel being subtended by an involucre of small bracts. The flowers of an umbel and its component umbellets open in sequence from the outer whorl to the centre. Pollen for protandrous outer flowers is supplied by protandrous inner flowers; pollen for inner flowers is supplied by outer flowers of a secondary umbel, continuing in sequence.

The small flowers are regular except in certain compound umbels where peripheral flowers tend to have the outer petals enlarged. Each flower consists of five small petals, inrolled in the bud, with or often without a small external calyx of five pointed lobes. The five stamens are carried on thin filaments surrounding two short styles, which arise from the inferior ovary. Anthesis is normally completed in all flowers of an umbel in 3–4 days in coriander and caraway, 9–12 days in fennel and dill, but with considerable cultivar variation. The ovary is composed of two united carpels, is bilocular and each loculus contains one pendulous ovule, the whole structure often surmounted by a nectar-secreting disc. Most of the species are cross-pollinated, the stigmas only becoming receptive after the petals and stamens have been shed.

The *Umbelliferae* are characterized by their schizocarpic fruit, the cremocarp, separating at maturity into two mericarps each containing a single seed. The outer wall of each mericarp is variously ridged and furrowed by both primary and secondary ridges, and the whole surface may be more or less bristly. In the furrows of the mericarp wall, or below the secondary ridges, longitudinal canals contain resins, gums and volatile (essential) oils, giving the fruits their characteristic flavour and smell. The indehiscent single-seeded mericarps, or fruit, are generally known as seeds when traded or sold as the spice.

Many early records confused fruits of the various *Umbelliferae* and are thus often specifically unreliable, but are included in the text for their valuable or interesting contents. Fruits of the crops discussed are generally known as seeds when traded or sold retail and will be so designated in the text, but fruit and seed where botanical accuracy is required.

The crops discussed in this chapter require basically similar methods of production, pest and disease control, etc., which are described in these introductory sections. Special requirements will be noted as necessary. The commercial production of horticultural crops of all kinds is a major industry in most countries, and a huge volume of easily accessible published material exists, much directly applicable to herbs and spices. New entrants to the growing of umbelliferous crops should seek advice from local specialists before beginning commercial production.

Soils

Sandy loams are the preferred soil type with a neutral reaction, although so adaptable are these crops that they are planted by smallholders on any available land, with pH 5.5–8.0, although saline and alkaline soils should be avoided. Soils must also be free-draining since waterlogging is usually fatal. Soils that produce good vegetable crops locally are the most suitable.

Fertilizers

The fertilizers required are discussed for each crop. However, few data are available on the merits of using one fertilizer type over another and thus in general the cheapest should be chosen. Data on the effect of varying levels of a specific fertilizer on fruit composition are also lacking for most of these crops. In essential oil crops time of harvesting, which also modifies fruit composition, is directly affected by fertilizers, thus further research on spices should be profitable (Weiss, 1997).

Much data from fertilizer trials are concerned with the yield of green material, and are thus of little value for seed producers except where trials demonstrate a specific shortage can adversely affect plant growth. Minor elements are frequently deficient in specific regions, and this should be alleviated. The small amounts required are usually supplied as part of a compound mixture applied to the seedbed. Locally manufactured compound fertilizers can often be obtained containing the minor or trace elements necessary.

Farmyard manure (FYM), rotted plant residues and similar bulky organic fertilizers can be applied where their use is profitable. In general, these materials should be applied to a previous crop, or ploughed in several months before sowing. In drier regions their use is not recommended, since there is often insufficient soil moisture to allow decomposition, and their decay may also require additional nitrogen. FYM and similar material should not be applied shortly before sowing coriander as this produces rank growth and delays ripening. Due to the cost of spreading and ploughing in, these sources of plant nutrients are seldom used on large commercial crops grown in a rotation. When such material is used, an analysis of available plant nutrients is necessary to enable any deficiency to be remedied.

Whilst fertilizers of any kind to supply plant nutrients normally increases seed yield, the profitability of so doing must be accurately established. Large-scale commercial growers have little difficulty, but small-scale producers may have problems in determining the optimum fertilizer level in relation to the increased return obtained from their use.

Cultivation

The most important requirement is as level and weed-free a seedbed as possible, since the seeds are small and poor competitors with most local weeds. This ideal is achieved only on well-managed commercial crops, and excepting small areas grown in the vicinity of homesteads, most smallholder crops once sown have virtually to fend for themselves. A multiplicity of seeding and transplanting machines is available from which to choose that most suitable locally. Since high-quality seed may be difficult to obtain and expensive, optimum seed rates should be accurately determined. Seed should be treated with a fungicide and an insecticide if soil-living insects or their larvae are a local problem. Since seed is small, mixing with an inert carrier will increase the volume and thus allow more accurate calibration of drilling equipment. Modern machines are electronically controlled, and thus automatically monitor seed rate.

Weed control

Weeding is essential while plants are small, but requires care as most crops have spreading or superficial root systems easily damaged by too deep cultivation. Since weeding in broadcast crops is difficult, it is often neglected and yields are thus low, especially in areas of low rainfall, as many weed species are very efficient competitors for soil moisture. Herbicides are seldom used, mainly because they are expensive, but also because of the increasingly stringent legislation regarding their residues on or in food crops. A herbicide specially recommended for any of these crops will be mentioned. It must be emphasized that local trials are essential to test the suitability of a herbicide, since many factors affect their action in a particular region or local crop.

Irrigation

For irrigated crops the land or bed surface should be allowed to dry out following a pre-planting irrigation, to allow even depth of sowing and subsequent even emergence.

Pests

Many pests attacking local horticultural and umbelliferous crops will also usually attack the crops described, and a wealth of information on their control is available. Whether it is profitable to do so depends on the scale of the enterprise and its markets. An important aspect of chemical pest (and disease) control is the residues in foliage and fruits, especially the organophosphates and organochlorines, detected in samples of herbage and dried fruits in many countries,

including Egypt (Abou-Arab and Donia, 2001). Insects pests attacking umbelliferous crops are noted below; specific pests will be in the relevant section.

Locusts, grasshoppers, mirid and stink bugs, white flies, aphids, the caterpillars of various moths and butterflies all attack herbage, and other pests bore maturing fruits. Eelworms can also cause severe reduction in plant populations when grown in infected soils, as can larvae of various cutworms (*Agrotis* spp.) in warmer areas.

Diseases

These are treated as for insect pests. Probably the most important diseases in terms of economic damage are the various root and stem rots and wilts, since these can substantially reduce plant populations and spread rapidly in suitable conditions.

Harvesting

While manual harvesting of herbage and mature plants remains common, this operation is often partially or wholly mechanized. A range of suitable equipment is available, and any special techniques will be mentioned. The green herb should be cut and distilled without undue delay, and harvesting and still operations closely coordinated. Cutting should be related to still capacity, and herbage should not be left in heaps near the still overnight, as it will heat and change oil characteristics and composition. Cutting, chopping and blowing green material into special trailers, which may also be mobile still bodies, as a combined operation has been successful. The system has much to recommended it, since it allows greater control over the type of oil produced. Similar systems are used on essential oil crops and have been described in detail (Weiss, 1997).

Transport and storage

In general, the storage and transport of these spices is virtually identical. Seed from commercial crops should be cleaned and stored in new jute bags, lined with polythene if necessary, in a cool, dry store. Unless seed is to be distilled where grown it should be transported to merchants or processors as soon as possible, since fresh seed has the highest essential oil content. Smallholder crops are generally marketed as and when desirable, and thus the quality is very variable. Most *Umbelliferae* are well visited by bees at flowering, thus honey production can be a profitable sideline and may assist pollination in areas where wild bees are not numerous.

Anise

The genus *Pimpinella* L. comprises about 150 herbaceous species mostly native to Eurasia and northern Africa, and many have aromatic plant parts. The most important spice producer is *Pimpinella anisum* L., or anise. A Malesian species, *Pimpinella pruatjan* Molkenb. has aromatic roots. Star anise, *Illicium verum* Hook, is included here for convenience, as the main essential oil constituent of anise and star anise is *trans*-anethole, and both have virtually the same uses. It is discussed following anise.

Anise probably originated in the eastern Mediterranean region, but as with many other members of the *Umbelliferae* it was carried worldwide by either armies or traders. It was used as a spice in Dynastic Egypt about 1500 BC (Joret, *Les Plantes dans l'Antiquite*, 1897), and later by the Greeks, Romans and Hebrews. Anise was popular with Romans in the 1st century AD, especially to flavour the small cakes baked in bay leaves, known as *mustaceus*. This use survives on the island of Majorca, where cakes of minced figs flavoured with aniseed and wrapped in fig leaves are given to friends at Christmas (Stobart, 1977). Cultivation of anise spread rapidly throughout Europe and by the Middle Ages, the 7th–12th centuries, was cultivated in most suitable environments, although seldom grown in Britain. However, anise seed became very popular in England, and in the reign of King Edward I, taxes and tolls on anise helped to pay for repairs to London Bridge in 1305. His descendant Edward IV scented his clothes with 'lytil bagges of fustian stuffed with ireos and anneys' according to the Royal Wardrobe

Accounts of 1480. Currently in Europe the main use for anise extracts is to flavour drinks and confectionery, especially aniseed balls.

Anise was carried along the Silk Roads to Persia and India, where it is mentioned in many early writings as a component of herbal medicines, and finally China and Japan, although it never displaced star anise. It is recorded that Java shipped fruits to China in AD 1200, although this was apparently a trans-shipment of western produce (Burkill, 1966).

Anise is currently grown commercially in many countries with a suitable climate, including China, India, Iran, Italy, Spain and Turkey, previously in England and France. Although anise was known in India from early times and grown by smallholders for their own use, it was not grown commercially until the 1960s, using a cultivar introduced from France. It is virtually unknown in sub-Saharan Africa although occasionally occurring in West African gardens, and its use was probably introduced by the trans-Saharan traders of antiquity. In the late 17th century, a French doctor, Mr Charles J. Poncet, returning from a visit to Ethiopia, remarked that in trading posts near Dongala, no money was used, only barter. Among the main items traded were anise, fennel and cloves (Foster, 1949).

Anise is seldom grown in South-east Asia, China and Japan, where it is generally replaced by Chinese star anise. The fruits are widely used as spice and in herbal medicine. In Malaysia, they are used to treat stomach pains or colic, in a tonic taken after childbirth, and in the treatment of gonorrhoea in conjunction with leaves of *Phyllanthus* spp.; in Thailand, anise is considered antipyretic, and used to prevent morning sickness.

Botany

P. anisum L. (main syn. *Apium anisum* (L.) Cranz.; *Apium vulgare* Gaert.; *Apium officinarum* Moench.), whose genus name is said to be derived from the Latin *bipinella*, or two-winged, is known in English as anise or sweet cumin, in French, German, Spanish and Russian as *anis*, in Italian as *anice*, in Arabic as *yanisun*, in Indonesian and Malaysian as *jin-tan manis* and in Japanese as *anisu*. There are no botanically accepted subspecies although named cultivars exist. Germplasm collections are maintained at the Plant Genetic Resources Department, Aegean Agricultural Research Institute, Izmir, Turkey; Gene Bank of the Institute for Plant Genetics and Crop Plant Research in Gatersleben, Germany; and Institute of Plant Introduction and Genetic Resources, Sadovo, Bulgaria.

Anise is an erect, annual, aromatic herb. Seedlings quickly develop a long, thin, tap-root, which makes transplanting difficult. The taproot continues to grow and produces lateral roots and extensive minor roots. The sturdy stem is up to 50 cm, sometimes much higher, and is grooved and finely pubescent. The leaves are alternate, entire to pinnately compound and the blade of lower leaves is kidney-shaped, coarsely toothed reniform and 2.5–5.0 cm. The blade of middle leaves is pinnate or trifoliate with incised leaflets, and the blade of the uppermost leaves is tri-partite and subsessile (Fig. 10.1). The petiole is 4–10 cm on the lower leaves, gradually

Fig. 10.1. Flowering shoot – anise.

shortening to become absent in the upper, and is always sheathing at the base. The leaves contain an essential oil similar to that in fruits. Young leaves and thinnings are commonly used as a vegetable, in salads and finely chopped as a dressing.

Cell cultures are assisting in determining the biochemical pathways by which the essential oil and its major constituent, anethole, are produced. Anethole and myristicin have been detected in undifferentiated cultures, but the amounts are variable and both may be absent; conversely epoxy-pseudoisoeugenol-2-methylbutyrate (EPB) and β-bisabolene are always present. Leaf callus cultures can accumulate substantial amounts of EPB, some anethole and pseudoisoeugenol-2-methylbutyrate (P2B), but root organ cultures accumulate EPB, no anethole and only traces of P2B (Chand *et al.*, 1997).

The inflorescence is a terminal compound umbel (Fig. 10.1). The peduncle is 2.5–7 cm, the involucre of bracts absent or short, and one to two foliate. There are 4–15 primary rays, 0.5–2.5 cm long and hairy, and 7–13 secondary rays (pedicels), 1–5 mm long. The flowers are white and bisexual, with the calyx indistinct, and the corolla with five obovate-cordate petals, 1–1.5 mm, the margin ciliate, and the apex inflexed. There are five stamens, the filaments longer than the petals, inflexed at the apex; the pistil has an inferior, bilocular, two-carpelled ovary, with two styles each with swollen stylopodium at the base and globular stigma at the top. The floral biology of anise and related plants has been described (Nemeth and Szekely, 2000).

The main flowering period in Europe and the USA is mid-summer to early autumn, depending on the cultivar and locality, and February–March in India. Anise is basically cross-pollinated, with insects, especially bees, being the main pollinating agents.

The fruit is a short hairy, brownish, ovoid schizocarp, 3–5 mm, splitting at maturity into two five-ribbed mericarps, with numerous oil ducts, containing a single seed; the fruit wall is connate with the seed testa. The fruit, either whole or ground, are the spice

(Fig. 10.2). The detailed morphology and histology of the seeds has been published (Parry, 1965). The weight of 1000 seeds is 1–4 g. Seed can remain viable for up to 3 years. Fruits contain approximately, water 9%, protein 18%, fat 16%, carbohydrates 35%, fibre 15% and ash 7% (including per 100 mg Ca 646 mg, Fe 37 mg, Mg 170 mg, P 440 mg, K 1441 mg, Na 16 mg and Zn 5 mg). Energy value is about 1400 kJ per 100 g. The lipid fraction includes approximately 10% mono-unsaturated and 3% poly-unsaturated fatty acids.

A pale yellow essential oil up to 4% can be distilled from fruits, whose main constituents are *trans*-anethole up to 90% and methyl chavicol up to 5%. The main characteristics of a commercial oil, often called aniseed oil, are: specific gravity at 25°C, 0.978–0.988; optical rotation at 25°C, +1° to −2°; refractive index at 20°C, 1.553–1.560; slightly soluble in water, soluble in 3 vols alcohol, freely soluble in chloroform and ether.

Fig. 10.2. Anise seed.

Ecology

Anise is native to regions with a warm-temperate climate, but widely distributed in antiquity, and under cultivation has developed local cultivars adapted to subtropical climates. When grown for herbage only, its range is greatly extended. It is a naturalized escape in many countries to which it was introduced. Anise was originally considered day neutral, but later research has indicated that long days promote maximum flowering.

Anise flourishes where there are no sudden fluctuations in either rainfall or temperature. A mean annual maximum temperature of 12–18°C is the optimum; plants can withstand 24 °C for short periods, but a temperature of 6°C will depress growth. Once stems have begun to elongate, frost is usually fatal, and plants normally require a frost-free season of about 120 days when grown for their fruit. However, selected north European cultivars can be planted in the autumn where no hard frost occurs. The crop is sown in October–November in northern India, and in late March–early April in Europe.

A rainfall of 1000–1500 mm annually will produce excellent crops provided a substantial amount falls between stem emergence and flowering, and up to 2000 mm is tolerated. A pre-flowering soil moisture shortage will substantially reduce yield, and for this reason sandy soils are unsuitable for rain-grown crops. High winds and hail can cause severe damage to mature plants.

Soils and fertilizers

Anise can be successfully grown on a wide range of soils, but highest yields are obtained on sandy loams or clay loams if well drained with pH 6–8. Heavy clay and very sandy soils are not recommended for commercial crops, although the black cotton soils of India produced good yields. Saline soils are unsuitable.

Few data are available on the nutrient requirements of anise, and in its absence fertilizers recommended for similar crops in the district should be applied until formal trials establish the optimum. Data indicate that the nutrients required to produce 100 kg fruits is 3.5 kg N, 0.65 kg P and 3.3 kg K, and most is taken up in the period from stem emergence to flowering. When grown in England, an NPK mixture containing the equivalent of 80 kg N, 100 kg P and 80 kg K ha^{-1} was generally applied.

Cultivation

Anise is propagated by seed, which remains viable for 1–3 years, but fresh seed should be sown. Plants can also be propagated *in vitro*, and multiple shoot formation achieved from callus cultures derived from shoot apices, root and stem explants and seeds. Somatic embryogenesis has been observed in callus cultures. Plantlets have also been directly regenerated from shoot apices. Germination can be hastened by soaking seed overnight in water, and direct sowing is recommended as seedlings are difficult to transplant and easily damaged. Seed should be drilled 1–3 cm deep, and a row width of 40–50 cm and inrow spacing of 8–12 cm is common. Seed rates range from 6 to 15 kg ha^{-1} depending on cultivar, soil type and region. However, as with other similar crops, spacing to suit the equipment available should be determined by formal trials, which can also establish the optimum plant population. Smallholder crops can be broadcast at 15–20 kg ha^{-1}.

Germination normally takes 17–25 days, and initial growth is slow, 35–40 days from germination to stem emergence. Seedlings are thinned to the required spacing when about 5–8 cm tall, around 30 days after planting. Flowering normally begins 65–75 days after germination, and seed matures after a further 20–25 days. Time of sowing directly affects the time to flowering; in Turkey, this was 88–102 days for plants sown in March, 175–200 days for plants sown in November.

Weed control

A weed-free seedbed is essential as the often extended period to germination and the slow initial growth renders anise seedlings vulnerable to faster growing weeds, which are very difficult to control in the young crop. It is also important to maintain a low weed population, since a high content of their seeds in a harvested crop is not only difficult and expensive to remove, but will downgrade the crop or contaminate the oil during distillation.

Pests and diseases

Pests and diseases of anise have been poorly reported. A major pest in the Mediterranean region and West Asia is the moth *Depressaria pimpinellae*, whose larvae eat young shoots and bore fruit. A grasshopper, *Calliptamus italicus*, attacks a wide range of crops including anise from the Mediterranean to India and Afghanistan. The stink bug *Graphosoma italicum* and the aphis *Hyadaphis coriandri* have been reported as damaging umbelliferous crops including anise from West Asia to Iran. In India, anise may be severely infected at fruiting by *Trichothecium roseum*, a fungus producing toxins, which, if ingested over a prolonged period, may cause neurological disorders in humans.

Harvesting

Plants are ready to harvest when fruit is greyish-green, usually in the European autumn, but as fruits ripen first on the main stem and later on the side branches, general ripening is uneven. Harvesting too early will yield immature fruits of low quality, while late harvesting increases fruit loss from shattering. Thus, anise is generally harvested when 80–90% of the fruits turn greyish-green or when the main umbel fruit is completely ripe.

Few sowings are large enough to be mechanically harvested, and plants are pulled by hand or by cutting off the tops, then stooked 2 m high in the field with the heads toward the centre and left for up to a week for further drying and fruit ripening. The moisture content of fruits should be 10–12% before threshing. Large areas can be combine harvested, or cut and left in windrows for later pick-up combining. Average fruit yield ranges from 0.7 to 1.5 t ha^{-1}, and for anise oil is about 15 kg ha^{-1}. However, trials at experimental stations indicate that better weed control, and a relatively simple selection programme could rapidly increase local average yield, although it is apparently more difficult to increase oil content.

Processing

Following threshing, fruits are cleaned and all unsound fruits and other extraneous material removed; if necessary, sound fruits may be further dried in the shade, either on racks on in sacks with the tops left open. Fruit for use as spice should be packed in new jute bags and kept in a cool, dry store. Cleaning, drying and bagging are fully mechanized in Russia and Poland, where forced cold or warm air dryers are used.

To obtain the essential oil, fruit must be distilled as soon as possible after harvest, as essential oil content rapidly decreases in open storage. If distilling is delayed, seed must be stored in sealed bins in a cool and dark store, as data show an oil loss of 1% monthly in open storage. However, no significant difference occurs in oil composition between fresh or properly stored fruits. Fruit must be crushed or comminuted immediately before steam distilling to reduce oil loss. Distilling is straightforward with no special requirements. The residue remaining after distillation may be used as cattle feed.

Storage and transport

Seed should be stored for the minimum period and transported to market, or to traders for export as soon as possible, to preserve seed oil content and thus its flavour. A carbon dioxide gas fumatorium to control storage pests was successful in Egypt, and unlike chemical control left no residues (Hashem, 2000).

Products and specifications

Anise produces a fruit, which when dried is the spice (anise seed or aniseed), an essential oil used as a flavouring, and the less important fresh and dried leaves used for flavouring and garnishing.

Seed. The seed has a characteristic sweet smell and a pleasant aromatic taste, and crushed seed or powder is used to flavour curries, sweets, confectionery and baked goods, and is an ingredient of various alcoholic drinks or beverages including French *anisette*, Turkish *raki*, South American *aguardiente* and *anisado* of the Philippines. Anise seeds are considered stomachic, carminative, diaphoretic, diuretic, antispasmodic, antiseptic, stimulative and galactagogue. Seed may be adulterated with exhausted fruits (after distillation), and other anise-like fruits and seeds. Ground anise may be adulterated with ground fennel (*Foeniculum vulgare* Mill.).

Essential oil. The oil is a colourless to very pale yellow mobile liquid, with the characteristic odour of anethole, and an intensely sweet, mild, but tenacious taste. Anise oil is placed in the sweet, non-floral, candy flavour group by Arctander (1960). Yield varies with the method of extraction: 1.9–3.1% for hydro-distillation, 2.3% for hydro-diffusion and 4.3–5.1% for carbon dioxide extraction.

The main oil constituents are phenols and derivatives 90–95%, hydrocarbons 2–5%, and less than 1% each of alcohols, carbonyls, acids and esters (Lawrence, 1998c). The oil contains 80–95% (E)-anethole, which is the main flavour compound, and up to 5% methyl chavicol (estragole); (Z)-anethole, which has toxic properties, is usually below 0.2%. Additionally, oxidation products of the anetholes, including anisic alcohol, anisaidehyde, anisic acid, anethole epoxide, *p*-methoxyphenylacetone and *p*-methoxypropiophenone, may indicate oil deterioration. Also reported are coumarins, flavonoid glycosides, β-amyrin, sitosterol and myristicin.

The essential oil replaces fruit in the food processing and confectionery industries, in perfumery, soaps and other toilet articles, and as a sensitizer for bleaching colours in photography. Synthetic anethole isolated from sulphate turpentine has replaced the oil where the true flavour is not important, and along with fennel oil is a reported adulterant.

Anise oil is mainly used as a flavouring and can replace the seed in most applications. The oil is traditionally blended with liquorice (*Glycyrrhiza glabra* L.) extract in sweets and thus the flavour of anise is often confused with that of liquorice and erroneously described as liquorice-like. Highest average maximum use levels are 0.06% (570 p.p.m.) in alcoholic beverages and 0.07% (681 p.p.m.) in

sweets. The scent of anise is not only attractive to humans, it is frequently used to scent the lure substituted for a fox in drag hunting, and as bait for mice, rats and fish.

In human medicine the essential oil is widely used to mask undesirable odours in drugs, as a carminative, stimulant, mild spasmolytic, antibacterial, expectorant, and included in cough mixtures and lozenges. In Germany, a mean daily dose of 3 g seed or 0.3 g essential oil is allowed as a bronchial expectorant and a gastro-intestinal spasmolytic, and preparations with 5–10% essential oil are prescribed as a respiratory inhalant. Anise oil may cause skin irritation and induce nausea, seizures and pulmonary oedema due mainly to the anethole component, but long-term studies showed anethole is not carcinogenic. The reputed oestrogenic activity is probably due to the anethole polymers dianethole and photanethole. A monograph on the physiological properties of anise oil has been published by the Research Institute for Fragrance Materials.

The oil possesses activity against potato virus X, tobacco mosaic virus and tobacco ringspot virus. The spice and oil are antibacterial and antioxidant, the oil and anethole have antifungal activity, and anethole, anisaldehyde and myristicin exhibit mild acaricidal and insecticidal properties.

Anise oil from *P. anisum* is interchangeable in the USA with that of star anise *I. verum* Hook., although the former is superior to the latter which has a harsher note. The US regulatory status generally recognized as safe has been accorded to anise, GRAS 2093, and anise oil, GRAS 2094.

Other Pimpinella *species*

The characteristics of fruit and oils from other species occasionally used as a spice or flavouring have been described: *Pimpinella squamosa* in Azerbaijan (Mekhtieva, 1997), and *Pimpinella anisoides* in South Italy (Hammer *et al.*, 2000).

Unrelated species

Anise pepper is obtained from *Xanthoxylum piperitum* D.C. (*Zanthoxylum piperitum*),

Rutaceae. The dried red berries are the fruit of a small, feathery-leafed, spiny tree. The flavour is hot and aromatic. It is also known as Chinese pepper, and is an ingredient of Chinese Five Spice. Anise root is obtained from *Annesorrhiza capensis* Cham. & Schlect., *Umbelliferae,* native to South Africa where the roots are eaten. Certain cultivars of fennel (*F. vulgare* L.) are known as anise or sweet anise due to their taste. A plant with similar uses to anise in the USA is *Osmorhiza longistylis* D.C., and oil produced from its roots is sold as anise or sweet cicely oil.

Star Anise

The genus *Illicium* L. (*Magnoliaceae*) comprises about 40 species of mainly trees and shrubs, distributed from eastern North America to South-east Asia, China and Japan. The most important is *I. verum* Hook, star anise, which most probably originated in the south-east China–northern Indo-China region, but is now known only from cultivation, although semi-wild populations exist descended from abandoned plantations. Chinese star anise has been used as a spice and medicine for over 3000 years. It is cultivated in Hainan, Taiwan and Japan, and although introduced to other countries has not thrived. Chinese star anise was believed in Europe to originate from the Philippines, as in 1578 the navigator Thomas Cavendish brought the first fruits to Europe via the Philippines, although unaware they originated from southern China.

The common names star anise and star anise oil are generally used for *I. verum* products. Several other *Illicium* species produce similar fruits, which are often also named star anise. Whenever possible the botanical names should be indicated. To avoid confusion in this text, the name star anise will apply to Chinese star anise *I. verum* and its products, and Japanese star anise to *Illicium anisatum* L.

Botany

I. verum Hook f. (syn. *Badianifera officinarum* Kuntze), whose genus name is stated to be derived from the Latin *illicium,* allurement, alluding to the fruits sweet smell, is known in English as star anise or Chinese star anise, in German as *sternanis,* in French as *anise de Chine,* in Italian as *anice stellato,* in Spanish as *badian,* in Indonesia and Malaysia generally as *bunga lawang,* in Thailand as *chinpaetklip,* in the Philippines as *tagalog, sanque,* and in Vietnam as *hoi sao.* Germplasm collections are maintained at unspecified locations in China and Vietnam, but there are no reported breeding programmes.

It is an evergreen tree generally 8–15 m tall but up to 20 m; the DBH is up to 25 cm, and the bark is white and aromatic. The wood is fine-grained and suitable for pulping, but because of its slow growth is not recommended for plantations. The leaves are alternate, simple, coriaceous and glandular-punctate; the petiole is about 1 cm long, the blade elliptical to obovate or lanceolate, 5–15 mm × 1.5–5 cm, the margin entire, the apex acute, and the lower side pubescent (Fig. 10.3). Fresh leaves and twigs contain 0.3–0.4% essential oil similar in composition to fruit oil with which it is often mixed.

The flowers are axillary, solitary, bisexual, regular, 1–1.5 cm in diameter, and white-pink to red or greenish-yellow (Fig. 10.3). The pedicel is 0.5–1 cm. There are 7–12 perianth lobes, spirally arranged, 11–20 stamens, spirally arranged, with short, thick filaments, and usually eight carpels, free, in a single whorl. The fruit is an etaerio, 2.5–4.5 cm in diameter, consisting of 5–13 (usually eight) follicles arranged around a central axis in the shape of a star (hence star anise); each follicle is boat-shaped, 1–2 cm, rough, rigid, reddish-brown, splitting along the ventral edge when ripe, and containing one seed (Fig. 10.4). The dried fruit is the Chinese star anise of commerce.

The fruit contains fixed oil (in the seed), minerals, catechins, pro-anthocyanidin, and 3.0–3.5% essential oil as described in the Products and specifications section. The seed is cylindrical to compressed ovoid, 8–9 mm × 6 mm, smooth, glossy, light brown, containing copious, oily endosperm. Decorticated seeds contain 55% fatty oil, including oleic acid 60%, linoleic acid 20%, myristic acid 10% and stearic acid 8%.

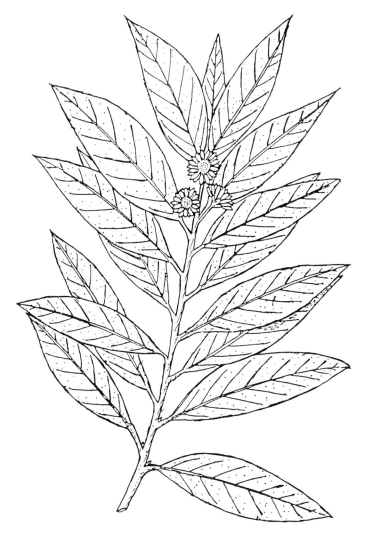

Fig. 10.3. Flowering shoot – star anise.

Ecology

The natural environment of star anise is undefined, but it is mainly cultivated in the cooler tropics and subtropics of the Indo-China region up to 2000 m, with average annual temperatures of 12–18°C, with average annual precipitation of 1000–2000 mm, and on soils with a pH of about 5.8. There is no information on the trees' nutritional requirements or responses to fertilizers. In China and Vietnam the official recommendation is for FYM at 7 kg per tree plus a small amount of ammonium sulphate if available, applied at the beginning of the rains (de Guzman and Siemonsma, 1999).

Cultivation

Star anise is propagated by seed, collected from selected 15–20-year-old trees, and only fully matured, brown seeds are selected. Seed rapidly loses viability and should be sown within 3 days; soaking seeds for 6 h in water at 35–37°C reportedly stimulates germination. After 1–1.5 years in nursery beds, seedlings at the 4th leaf stage are transplanted into growing beds about 25 cm using a hooked knife on a long pole. In some areas of China young twigs and fresh leaves are also harvested to be distilled separately or together with fresh fruits, since the oils are similar. An average of 5–10 kg fresh fruit per tree is obtained from 13–25-year-old trees, 10–20 kg per tree from those older than 25 years, and in high-yield years up to 40–45 kg per tree. Similarly to clove, yield appears to be cyclical with high following low-yield years. Trees can remain productive for many decades.

Processing

The traditional method of drying fruit is in the sun; harvested fruit is placed in flat baskets, exposed for about 10 days and then kept in a cool, dry store until packed and sold. However, this is a time-consuming, labour-intensive operation, and although the latter is of little importance in Vietnam and China with huge reservoirs of labour, quicker and more efficient methods will reduce the time fruit is exposed to contamination and inclement weather. A small heated dryer with the capacity of 120–160 kg per h^{-1} developed in China produced a high-quality dried fruit with excellent aroma. Another dryer used solar energy but required a longer period to dry fruits fully. Both were superior to sun-drying (Liu *et al.*, 1999). On drying, 100 kg fresh fruit yields 25–30 kg dried fruit. Fresh fruit yields 2.5–3.5% essential oil, dried fruit 8–9%.

To obtain the essential oil, fresh or partially dried fruit are steam distilled for 48–60 h, and often no further processing is carried out. Thus, crude oil may contain water, fruit residues and other material, and require filtration and rectification. Crude oil is stored in drums in a dark store, but should be cleaned and refined as soon as possible to reduce containers and storage space, then kept in full, airtight, opaque drums, preferably below 25°C. Distillation methods must be substantially improved to enhance yield and quality of oil; one is supercritical carbon dioxide extraction, which is faster, produces a more even quality oil, and small and large plants are available (Della Porta *et al.*, 1998). Storage and transport are as for anise.

Fig. 10.4. Fruit and seed – star anise.

apart, where they are left for another 3 years before planting out, 5–7 m apart, in pits filled with a mixture of topsoil, compost or animal manure. Regular weeding is necessary, and mulching at the beginning of the rainy season retains soil moisture through the dry season. Trees begin flowering at about 6 years, then twice annually, and fruit is mature about 3–4 months after flowering. Star anise does not apparently suffer serious damage from either pests or diseases, but this lack of data may be due to little interest in, or funds available for, crop protection.

Harvesting

First harvest is possible from 7–10-year-old trees, and steadily increases for the next 10 years. Fruits are harvested before they are fully ripe when the essential oil content is highest, usually August–September, although two harvests per year are possible. Harvesting is carried out by climbing trees and picking fruits, or they are detached by

Products and specifications

The major products from Chinese star anise are the seeds (fruit), which are the spice, and an essential oil distilled from seeds.

Seed. Seeds are used as flavouring in a wide range of foods and drinks including chewing gum, baked goods, gelatine, meat and meat products, liqueurs and brandies. They are also a constituent of Chinese five spice. The dried ripe fruit is often found in pot-pourris. Bulk supplies of star anise fruits from producing countries are often adulterated with *Illicium cambodianum* Hance. or *I. anisatum* fruits, and the latter introduce a bitter taste to end products and may be toxic, as noted. In products with anise as the main flavouring principle in the USA and Europe, the normal use level is 5–10 mg per 100 g, and the minimum perceptible is 0.3–0.6 mg per 100 g. In traditional regional medicine, a powder or decoction of the fruit is used to treat a range of stomach ailments and diarrhoea; the fruit is used to promote menstruation, increase milk secretion and facilitate childbirth.

Essential oil. The oil, star anise oil, is obtained by steam distilling fresh or partly dried fruits, and is mostly present in the fruit wall, not in the seed. The oil is a clear, colourless to pale yellow liquid, with an anise-like odour and sweet, never bitter taste. The oil is placed in the sweet, non-floral, candy flavour group by Arctander (1960). The major constituent is (E)-anethole at 70–95%, with methyl chavicol the only other constituent over 2%, with a very small amount of *cis*-anethole, which is many times more toxic to animals than *trans*-anethole. Oil from China differs from that of Vietnam (Formacek and Kubeczka, 1982; Cu *et al.*, 1990).

The main characteristics are: specific gravity (25°C) 0.978–0.988 and optical rotation +1° to −2°; refractive index (20°C) 1.553–1.560; slightly soluble in water, soluble in 3 vols alcohol, freely soluble in chloroform. Bulk oil is often contaminated with water, fruit residues, etc. and must be cleaned and sometimes refined before use.

Japanese star anise oil also contains mainly anethole, but also safrol and eugenol. The oil is mainly used as a flavouring material similarly to anise oil, with minor usage in soap, perfumery, tobacco and dental cream, with anethole isolated by freezing.

In the USA the regulatory status generally recognized as safe, applies to star anise, GRAS 2095, and star anise oil/oleoresin, GRAS 2096. The maximum permitted level of star anise oil in food products is about 0.07%.

Essential oil is administered therapeutically as a bronchial expectorant for upper respiratory tract congestion and as gastrointestinal spasmolytic, and has a permitted mean daily dose of 0.3 g. The oil is used as starting material for production of synthetic oestrogens, including diethylstilbestrol, diethyistilbestrol di-propionate and for perfumes as *p*-panisaldehyde. It is interchangeable with anise oil in alcoholic drinks. The oil is claimed to have stimulant, antiseptic, stomachic, carminative and mildly expectorant properties, but may cause dermatitis in susceptible persons. In Indo-China, it is used to treat body lice and protect against bed bugs. A monograph on the physiological properties of star anise oil has been published by the Research Institute for Fragrance Materials (RIFM).

Chinese star anise products compete with those of anise, as the major oil component in both is anethole. Star anise oil can be distinguished from anise oil by the presence of small amounts of 1,4-cineole. A major competitor of both is synthetic anethole, but the amount of *cis*-anethole in the latter can be much larger than in natural anethole, which consists mainly of harmless *trans*-anethole.

Other Illicium *species.*

I. anisatum L. (syn. *Illicium religiosum* Sieb. & Zucc.), which occurs wild and is cultivated in Japan, southern China and Taiwan, is the species most commonly confused with *I. verum* (Small, 1996). The fruit is smaller than that of *I. verum*, is toxic, and does not form a regular star due to the abortion of some carpels. The usually eight follicles are not swollen in the middle and are more pointed at the apex; they are more irregular with wrinkled sides, and their odour is balsamic and the taste bitter. An aqueous extract is used as an insecticide in China.

I. cambodianum Hance. is a wild tree native to Myanmar (Burma) and Indo-China; its fruits usually consist of 12–13 follicles, which contain little essential oil or taste. It is occasionally used to adulterate fruit of other species.

Illicium floridanum Ellis. is indigenous to the south-eastern USA from Florida to Louisiana. The evergreen shrub is up to 1.5 m, bearing purple flowers, and is grown there and in Europe as an ornamental, but has recently been evaluated in the USA for its economic potential (Tucker and Maciarello, 1999).

Illicium griffithii is native from India (Agarwal *et al.*, 1999) to Vietnam where its fruit is often exported to China to adulterate star anise; chemotypes exist (Nguyen *et al.*, 1998).

Illicium lanceolatum A.C. Smith (formerly *I. anisatum* L.), Japanese star anise, produces anise-tasting fruits used as food flavouring in the region.

Caraway

The genus *Carum* L. contains about 30 species mainly native to the Mediterranean and West Asian regions; many are aromatic and some are important spice producers (Nemeth, 1998). Probably the best known is caraway, *Carum carvi* L., so long domesticated that its use predates recorded history, as its remains have been found in food debris at Mesolithic sites. It was used in Dynastic Egypt, the ancient Mesopotamian civilizations and by the Greeks and Romans. Cooked roots mixed with milk and moulded into cakes are believed to be the *chara* of Julius Caesar and eaten by Valerian's legionaries. Similar cakes are still made in Scandinavia. Dioscorides, in his *Materia medica* of the 1st century, recommended an extract of the fruit as a tonic for 'pale faced girls', while the *Codex Aniciae Julianae* of AD 512 calls it *karia*. It remains to this day a widely used and important ingredient in cooking, herbal medicines and liqueurs.

Caraway was initially brought to Britain by the Romans, but was not cultivated until much later. Caraway spread up the Nile valley from Egypt to the Sudan, but not apparently to Ethiopia, and it has been found cultivated only as a garden herb in East Africa (E.A. Weiss, personal observation).

Commonly used in many baked products in Central Europe during Medieval times, Austrians and Germans are today probably the world's greatest caraway users. In the Middle Ages in England, chopped caraway leaves were added to soups and salads, but the seed was more highly regarded as a medicine than a spice and described thus in the 12th century *Macers Herbal*: 'The virtue of hym [the seed] is that it destroyeth wycked wyndes and caughs, and heleth men that hath the frenzy, and biting with venemous beestes.' In Elizabethan England it was 'deemed to confer the gift of retention, preventing theft of anything containing the seed, and holding the thief in custody within the violated house'. It was also considered to keep lovers constant, and a few seeds in a husband's pocket prevented him from straying! Continuing the tradition of constancy, modern pigeon fanciers mix caraway seeds in their birds feed to ensure they always return to their lofts. Caraway seeds were used generally in cooking and baking, while a favourite dish was roast apples served with caraway. In Shakespeare's *Henry IV*, Falstaff was invited by Master Shallow to partake of 'a last years pippin [apple] of my own grafting with a dish of caraways'. This dish was served until quite recently at some colleges in Cambridge, England, and at London Livery Company dinners.

Botany

C. carvi L. (main syn. *C. velenovskyi* Rohlena.), whose genus name is said to be derived from the Greek *kapov* (caraway), is known in English as caraway, considered to be from the old Arabic *karauya*, in French and Italian as *carvi*, in German as *kummel*, in Dutch as *karwij*, in Spanish and Portuguese as *alcaravea*, in Swedish as *kummin*, in Russian as *tmin* and in Arabic as *karawiya*.

The basic chromosome number of the genus is x = 10; caraway is a diploid with $2n = 20$, and tetraploids have been induced using growth hormones including CCC and TIBA. Intraspecific hybrids with other members of the *Umbelliferae* have been successful, but with no reports of viable seed produced. A detailed study of flowering and pollination indicated that inbreeding and hybridization may be used for genetic improvement (Nemeth *et al.*, 1999).

Gene pools of spring caraway are maintained at national gene banks in Egypt and Israel, and to a minor extent in The Netherlands. There are no botanically accepted species of caraway, although two types are recognized, one sown in winter (biennial), the other in spring (annual). Selection is carried out in Egypt and Israel to find superior regional strains, and in The Netherlands for annual types adapted to European conditions, which produced the non-shattering spring cultivars Karzo and Springcar, and two major winter cultivars Volhouden (shattering) and Bleija (non-shattering).

Caraway is basically a biennial but usually treated as an annual, and is a glabrous and erect herb. The well developed taproot is strong, tapering and fleshy, with few laterals, and generally resembles a small parsnip. A boron deficiency affects roots, which tend to become hollow, or cracked and brown in severe cases. Similar to many *Carum* species, the roots are cooked as a vegetable. An essential oil can be extracted from the root whose composition varies with age of plant. Fast root growth in biennial cultivars ensures plants flower in the following year, and the degree of flowering is directly related to root size in the spring. The stem (one or more) is up to 1.5 m but usually less, the base up to 2 cm in diameter, striate, terete, hollow and branching in the upper part. Daylength has a significant effect on annual caraway: long days produce taller plants, short days smaller, but the increased height may not greatly increase fruit yield.

The leaves are bright green, alternate, in an approximately 2/5 arrangement, pinnately compound, all sheathed, with the margin membranous and the apex auriculate. The blade is subtriangular, 6–15 × 2–8

cm; the lowest leaf segments are at least twice as long as wide, the ultimate lobes linear-lanceolate to linear, up to 25 mm long. The petiole is up to 13 cm on lower leaves, gradually becoming shorter until absent on the highest (Fig. 10.5). An essential oil can be distilled from green herbage, whose main component is germacrone D up to 80%, and is discussed at the end of this section. Growth regulators including CCC and GA applied to seedlings affected growth, seed yield and seed constituents, but the effects were not constant, and differed with region and cultivar. The biennial type produces a rosette of bipinnate or tripinnate leaves in the first year, then resembles the annual in growth. Young leaves and thinnings are used as a vegetable raw or cooked, with a taste of a parsley and dill mixture. In Germany, caraway plants with immature fruits are a highly regarded component of special meadows producing herbal hay, enhancing milk yield and quality in dairy cattle.

The inflorescence is a terminal compound umbel, 4–8 cm in diameter (Fig. 10.5). The peduncle is up to 11 cm, the bracts and bracteoles few or absent. The primary rays (3–16) are unequal and 0.5–6 cm; the secondary rays (6–16), are unequal, up to 1 cm, umbellets about 1 cm in diameter. The flowers are usually white, sometimes pink tipped, bisexual and protandrous, and the calyx is absent. There are five petals, which are obcordate with a short inflexed apex, about 1.5 × 1 mm. There are five stamens and two styles, which are recurved, the enlarged base forming the stylopodium, and the stigma is capitate. Phyllody occurs rendering flowers sterile, and is considered to be due to mite infestation. Insects are the main pollinating agents.

Flower initiation of annual cultivars in Europe begins 6–12 weeks after sowing, with a further 4–8 weeks to maturity. The sequence of flowering in umbels and umbellets is from the outside to the inside. The inner umbellets and the latest formed umbels tend to produce predominantly male flowers. A detailed description of flowering dynamics and pollination has been published, and eight flowering stages were distinguished (Nemeth *et al.*, 1999).

Fig. 10.5. Flowering stems -- caraway.

Bisexual flowers are protandrous, the pollen being carried by wind but mainly by insects. Dry, sunny weather is optimal for insect activity and also for pollen dissemination. The main flowering period in Europe is June–July, later in cool seasons, May–June in North America. However, caraway may not always flower in the year following autumn sowing, and this is more common when a regional cultivar is grown elsewhere.

The fruit is a schizocarp, ellipsoidal, laterally compressed, 3–5 mm long, splitting into two dark brown, sickle-shaped mericarps, with five prominent ribs and wide, solitary vittae (Fig. 10.6). While fruit size varies considerably, that from a particular cultivar or region is reasonably uniform. The average content of commercial grade dried caraway fruit (per 100 g) is water 10 g, protein 20 g, fat 14 g, carbohydrates 37 g, fibre 13 g and ash 6 g, including (per 100g) Ca 689 mg, Fe 16 mg, Mg 258 mg, P 568 mg, K 1400 mg, Na 17 mg and Zn 6 mg. Vitamin A is 363 IU per 100 g, energy value is about 1395 kJ per 100 g.

Fig. 10.6. Caraway fruit.

Seed weight of biennial type caraway is 3.0–4.5 g per 1000 seeds, while the average seed weight of annual caraway as determined by image analysis was 5.2 g per 1000 seeds (Frank *et al.*, 1996). The seed is dorsally flattened, and smooth or slightly grooved on the inner face. A detailed description of the histology and morphology of the fruit has been published (Parry, 1962).

The herbage and roots contain a volatile oil, detectable at an early stage of seedling development. The content of volatile oil in the vegetative organs increases progressively with plant development and reaches a maximum in the above-ground parts at flowering, about 0.1–0.2% on a fresh-weight basis. The chemical composition of essential oil from various plant parts is shown in Table 10.1. Analyses indicate that the carvone content of herb oil increases from zero at anthesis to a maximum at harvest while limonene is present over the whole period reaching a maximum of around 1%. Growth regulators,

including CCC and GA, applied to limit plant height, can also adversely affect herb oil content, but the effect is not constant and varies with the rate of application, cultivar and region. Oil from mature plants contains 68% aliphatic aldehydes, but from seedling roots the main constituent was germacrone-B up to 50%, which gradually decreased as roots aged.

The essential oil is present in the seed from an early stage, and is usually at a maximum in terms of weight ratio when seed has reached the soft-milling stage. Oil content falls substantially to physiological maturity, rising again when fruit is fully ripe (brown and hard). Near infra-red spectroscopy (NIRS) can be used as a non-destructive method of determining essential oil and monoterpene content of seeds (Fehrmann *et al.*, 1996).

Fruit from annual caraway air-dried to 10–12% moisture usually contains 1.5–5.0% essential oil, from biennial caraway 3.0–7.0%, but up to 10% is recorded from selected lines of either. In both types the oil yield and ratio of major constituents is directly affected by season, soil conditions and time of harvesting: early harvesting gives a slightly smaller fruit and a higher oil content. Oil from the biennial type is considered superior due to its stronger taste and scent. Fruit essential oil consists mainly of two monoterpenes: *d*-carvone at 45–60% and *d*-limonene at 35–55%, and analyses show that either may be the greater. The biosynthesis of the monoterpenes limonene and carvone in fruit has been investigated (Bouwmeester *et al.*, 1998; van der Mheen, 2000; Zawirska-Woitasiak and Wasowicz, 2000).

Oil characteristics also change with maturity; optical activity is reduced, but the refractive index increased. In maturing fruit, an inverse relationship apparently exists between optical activity and refractive index of the oil, and a direct relationship between specific gravity and refractive index. Oil composition also changes from early in seed development but has been less investigated. It appears the initially high limonene content falls with maturity, whilst a low initial carvone level increases to its maximum at maturity.

Table 10.1. Main components (%) of oils obtained from caraway in India.

Component	Fruit	Flower	Leaf	Stem	Root
α-Pinene	+	76.3	–	–	–
Sabinene	+	+	–	(+)	+
α-Phellandrene	0.2	+	–	+	+
Limonene	47.3	+	14.6	3.5	34.0
γ-Terpinene	+	(+)	–	–	5.6
Menthone	+	+	(+)	(+)	1.5
Linalool	+	+	+	+	–
Camphor	+	+	+	+	1.2
Caryophyllene	–	3.5	19.5	26.4	3.7
Dihydrocarvone	+	+	+	+	2.1
β-Cadinene	–	17.2	61.7	39.1	6.3
Dihydrocarveol	+	+	+	+	+
Carvone	–	2.3	+	8.7	4.1
Carveol	+	–	–	+	37.1

Brackets indicate tentative identification.
+, present in minor amounts.
–, absent.

Commercial samples of the oil should not contain less than 50% by volume of ketones calculated as carvone, which is responsible for the characteristic flavour and biological properties. The oil from a particular region is fairly uniform in its constituents and characteristics, but varies between regions, as demonstrated by analysis of fruit from various areas in the Czech Republic (Sedlakova *et al.*, 2001). Dutch cultivars have an oil content of nearly 10%, while Indian types are seldom half this.

Ecology

Caraway is native to the cool temperate regions of Europe and Asia Minor and occurs naturally from southern Europe and northern Africa through to India. Its popularity as a spice resulted in caraway being introduced to most countries where it would grow, and as plants rapidly become naturalized, it can now be found wild around the world in the warm temperate zones. Its range has also been extended by introduction of suitable cultivars to regions where other *Carum* species are indigenous, parts of India, for example, or where the climate is suitable, as in the USA, China and Japan. Caraway, as noted, is divided into two types,

a winter and a spring (annual). The former is native to the mountain ranges of Eurasia and is cultivated in East European countries and The Netherlands; the annual is indigenous to the eastern part of the Mediterranean and cultivated in Egypt, Sudan and Israel.

Winter caraway occurs naturally in meadows, grassland, forest edges, along roads, rivers and as a weed of cultivation, from sea level up to 4000 m. Annual caraway is now a well-established escape in many countries but is generally found at lower altitudes, and is often an early colonizer of waste ground. Annual caraway thrives in the cool short days of the eastern Mediterranean winter and the Indian plains.

Commercial crops of caraway are usually located in moderate to high rainfall areas of the temperate regions, up to 1500 mm annually, with rain normally falling over the life of plants. Thus, the main producing areas are in northern Europe and the USA, where biennial types are grown. Adequate soil moisture is essential and should preferably be 50–70% of soil capacity during the main period of growth. Above and below this level, yield is adversely affected. In low rainfall areas of the Mediterranean and West Asian countries, annual caraway can be grown under irrigation.

Caraway can withstand frost provided it has reached the rosette stage after sowing in the autumn, but very severe frost without a snow cover may cause substantial damage. Selections vary in their cold-hardiness, and some Russian cultivars are very frost-resistant, but this characteristic may be linked to longer vegetative growth. Autumn-sown caraway should have at least five well-developed leaves in the rosette stage for vernalization to occur, or plants may not flower the following year. Northern Eurasian winters are sufficiently protracted to vernalize winter types, provided plants have produced a tap-root 8 mm in diameter. Such plants require 6–8 weeks exposure to temperatures below 10°C. In Saskatchewan, Canada, caraway was tested as an annual crop since winters are too severe for biennial cultivars (Arganosa *et al.*, 1998a). Deep freezing of seeds in Russia strongly and adversely affected subsequent plant growth (Tikhonova and Kruzhalina, 1997).

Temperature affects growth and yield of annual caraway as demonstrated in phytotrons, when 18°C/12°C or 24°C/12°C (day/night) and a 10 or 16 h photoperiod were compared. Time of flowering was reduced most by the long-day treatment, as was seed yield; the latter was highest under a short-day regime. Day temperature had a minor effect compared to day length (Putievsky, 1983). In general, light intensity is more important than day length, and long periods of cloudy weather or shading from other crops at flowering substantially reduces seed yield.

In warmer regions, caraway is grown at higher elevations, for example, above 3000 m in Kashmir, India. In general such regions produced lower seed yield ha^{-1} than in lower altitudes in Europe, as factors other than temperature may be involved. Whilst altitude can affect growth and seed yield of caraway, latitude generally has only a minor effect. When grown in West Germany and at three sites in Finland, for example, the same cultivar showed no significant differences in either seed yield, oil content or carvone content of the oil (Halva *et al.*, 1986).

Warm, sunny conditions promote flowering, extend the flowering period and, provided adequate soil moisture is available, have a positive effect on early fruit growth. Conversely, very hot windy conditions at this time have an adverse effect on flowering and fruit set, as does heavy rain. Hot sunny weather when seed is physiologically mature but not ripe can affect oil composition by increasing the carvone content, but reduces total oil content. Cool, humid conditions have the opposite effect. It is widely believed in The Netherlands that proximity to the sea produces a high carvone content in the seed, but this is probably due to a combination of favourable climate and a locally high-carvone caraway strain. Both caraway types are susceptible to wind damage at maturity, and strong winds can flatten plants when ready to harvest. Hail and heavy rain have similar effects, and can virtually destroy a crop if a storm occurs just prior to harvest.

Soils

Commercial caraway crops are usually grown on free-draining clays or heavy loams, the heavy artificially drained clays of the northern Netherlands for example, but caraway will flourish on a range of soils provided moisture is adequate. Conversely, waterlogging except for short periods is often fatal, especially to seedlings or plants that have begun stem elongation in the spring following autumn planting.

A soil reaction of pH 6.5–7.5 is preferred, and above or below this, yield is progressively reduced, although there may be no major difference in vegetative growth. Below pH 6.0 plants generally make poor growth and many die. Liming can adjust soil acidity, but heavy application may induce manganese and boron deficiencies. Acid or alkaline soils should be avoided.

Fertilizers

Total nutrients removed by biennial caraway crops have apparently not been determined, but it is possible to estimate the amount by comparison with other vegetable crops grown in the same district. It is essential optimum fertilizer rates and application times should be determined by field trials, since considerable variation in crop response due to cultivar and region occurs. Incorporation of FYM and similar organic manures on sandy soils increases water retention, essential for high yields.

When caraway follows crops that have received significant amounts of phosphate (P) and potassium (K), additional amounts are seldom necessary unless a significant local deficiency exists. Both should be applied to the seedbed, immediately before or during sowing. Potassium uptake is generally highest prior to and at flowering. Nitrogen is essential to obtain high yield, and should be applied as a top dressing. Seedbed N may encourage excessive growth in a mild autumn, and thus make plants susceptible to frost damage. No difference in plant response to various types of commonly available N fertilizers or methods of application has been noted. For annual caraway, seedbed N has significantly increased seed yield, and if only one application is possible, then it should be at sowing. A combination with P is recommended, i.e. diammonium phosphate. The level and timing of top dressing should be accurately determined, since too much N depresses seed yield. In Israel, seed yield following 300 kg ammonium sulphate in one application early in growth almost equalled seed yield produced following 600 kg in two applications, and was significantly higher than 600 kg in four applications. The earliest top dressing should be applied only when further frost is unlikely, the last not later than first flowering. When caraway was undersown to spring barley in Denmark, a top dressing of N was applied after the barley was cut in the autumn, followed by a further application in spring.

In general, fertilizers have no significant effect on oil content of fruit or composition, but a specific compound or application time may do so. There are indications that high levels of ammonium nitrate or potassium chloride may affect optical and refractive indices of the oil.

Cultivation

This is generally as for other local vegetable crops, but as biennial caraway tends to be grown on heavier soils, extra care is required to obtain a suitable seedbed and to ensure soil does not cap or crust and prevent seedling emergence. The most important factor governing the choice of cultivar is fruit oil content as this is genetically controlled. In general, if a low-oil-content cultivar is selected, then no amount of fertilizer or agricultural skill will increase oil yield ha^{-1} above that of a high-oil type. Thus, it is important to carry out trials to establish the most locally suitable cultivar. However, as oil yield ha^{-1} is generally most important, it must affect the choice of cultivar when commercial plantings are being considered. Thus, a moderately high-oil-content variety with a very high seed yield ha^{-1} can be the most profitable.

Prepared seed, preferably treated with a fungicide, should be sown in rows at least 30 cm apart, with an in-row spacing of 8–15 cm for biennial but half this for annual caraway. A slightly higher plant density is preferable to a lower one in biennial caraway, since there are usually fatalities over the winter. Seed rate varies from 5 to 12 kg ha^{-1}, sometimes higher in smallholder crops. Method of sowing is important to obtain high seed yield, and broadcasting or cross-drilling to produce dense pure stands usually results in a lower fruit yield than row planting.

Since plant density has a significant effect on yield and oil content, it is vital to establish the local optimum, irrespective of the cultivar grown. It is equally important to determine the germination percentage of the seed. Clean sound seed, which has been correctly stored, normally has a germination rate above 80% and can be nearly 100%. Seed stored for more than 2 years should preferably not be used, as not only will the germination percentage be reduced, but emergence tends to be erratic. Some Mediterranean cultivars can have a dormancy period of 2–3 months after harvest, but this is vary variable and may not be constant. Hence the need for germination tests. Seed should be sown about 1–2 cm deep at approximately the correct spacing, or sown at a higher rate and seedlings mechanically thinned to the required distance. Correct spacing ensures an optimum population as caraway has only a limited ability to compensate for missing plants. Germination is epigeal and when field-sown, caraway normally germinates after 10–15 days, more slowly if the weather is cold and wet following sowing.

Time of sowing is important and should be determined locally. Biennial caraway can either be sown in late spring–early summer in areas with a relatively mild winter, or in autumn where winters are more rigorous. In areas where there are very cold winters, caraway should be sown in late July to ensure vernalization occurs. In The Netherlands it is frequently sown in March–April for harvest in the summer of the following year. Annual caraway should be sown as early as possible in the spring when the ground has warmed after winter. A soil temperature between 10 and 15°C gave the highest germination percentage in Israel, and germination time was halved when seeds were leached with water and dried before sowing. Annual caraway can be sown under cover early in the year and seedlings transplanted when conditions are suitable. This produces an early herb crop or a high seed yield, but is profitable only near a high-value urban market, or for domestic use. Micropropagation using *in vitro* culture of tissues derived from petiole, hypocotyl and seedling shoot tip is reportedly successful.

Weed control

Pre-sowing cultivation should aim to reduce the weed population to the minimum, not only to reduce competition with the crop, but also at harvest, because many weeds are also *Umbelliferae* and their seeds are difficult to separate from caraway fruit and reduce its value. Additionally some also have fruit that contain an essential oil, whose distillation together with caraway reduces oil quality and thus its value.

Caraway can be mechanically cultivated once well established, but herbicides are now routinely used to control weeds. A number of compounds have proved successful, including afalon, diuron, glyphosate, nitrofen, prometryn, propazine, terbutryn and treflan, singly or in combination. However, it is essential to carry out trials to determine the local effect of herbicides on both biennial and annual caraway, since time of application and method used can be critical in avoiding crop damage. Herbicides recommended for use in caraway have no recorded effect on seed oil content or oil characteristics.

Irrigation

When annual caraway is grown as an irrigated crop, or with irrigation to supplement rainfall, it is particularly important to maintain adequate soil moisture in the period prior to flowering. Where water is limited or expensive, one watering should be prior to planting and another prior to flowering.

Intercropping and rotations

Whether caraway should be sown in pure stands or with a cover crop is a local decision based on experience. Either usually significantly lowers fruit yield, but this may be offset by proceeds from the other crop. Shading alone can reduce seed yield as noted in the Ecology section. Peas, beans or clover for seed are common in Europe, and caraway can be undersown in spring barley in Denmark. In England, coriander can be sown with caraway, the former being harvested the same year, with caraway in June–July the following year. However, this practice may not be suitable in other countries, since the pathogen *Ramularia coriandri*, which is a major cause of leafspot of coriander, can also attack caraway when the two are grown in close proximity.

Pests

Many pests of field and vegetable crops grown in the same area as caraway are polyphageous, and will thus cause damage to caraway. These pests, together with their control methods, are well known, and local experience and recommendations should be followed. Some insect pests are partially specific to caraway including *Aceria carvi*, which causes phyllody of caraway in eastern European countries; also in Europe the carrot fly, *Psila rosae*, attacks caraway roots and young plants become stunted and may die. The aphid *H. coriandri* is frequently recorded on caraway from the Middle East to India and is damaging in particular seasons. Birds and rodents can be major pests.

Diseases

Caraway can be infected by a number of common field crop and vegetable diseases, similarly to insect pests, and thus local control measures are applicable. Some caraway cultivars are known to be resistant to one or more locally important diseases, and these should be grown when possible. Some wild or naturalized strains also have a degree of disease resistance and this could prove of value in breeding programmes.

Most commonly recorded are diseases caused by *Fusarium* spp., *Verticillium* spp., *Sclerotinia* spp., especially *S. sclerotiorum*, which has a very wide host and geographical range, and *Phomopsis* spp., especially *Phomopsis diachenii* in Europe; *Ramularia* spp., more often recorded as a serious leaf spot of coriander, may occasionally attack caraway, especially when in proximity. A major disease of spring caraway in The Netherlands is the soil-borne *Sclerotinia* stem rot, which can only be effectively controlled by crop rotation (Fig. 10.7). Anthracnose due to *Mycocentrospora acerina* occurs widely in Europe, and a crop management programme can result in a major reduction in the damage caused (Evenhuis *et al.*, 1999).

Harvesting

Biennial caraway should be ready to harvest from July to September in the year following sowing depending on region and cultivar, annual caraway after August, but both must be harvested before the first frost. Undersown caraway in spring barley in Denmark was left to regenerate and produce seed crops in two successive years. Time of cutting is important and is usually when the first fruits turn dark brown, since loss from shattering can be high in overmature crops. Trials in Germany indicated that colour was a good indicator of fruit ripeness, and thus time of harvesting (Quilitzsch and Pank, 1996).

Caraway is normally reaped with a binder, preferably in the early morning or late evening when dew is on plants to reduce seed loss. Sheaves are stooked to dry, and threshed either in the field or taken to a stationary machine. Caraway can also be cut, windrowed and then either made into sheaves by a binder with the knife removed, or threshed with a pick-up harvester. Both methods require a high degree of local knowledge and skill to be successful. Direct combining is possible but uncommon. Development of truly non-shattering cultivars with a short but profuse flowering period to ensure even ripening would materially assist the use of combine harvesters. Smallholders usually thresh manually, or utilize the now commonly available small, manually operated threshers. Threshed plants are frequently used as livestock feed, either as hay or processed stockfeeds.

Fig. 10.7. *Sclerotinia* stem rot – right-hand side.

Fruit yield is very variable, depending mainly on cultivar and environment, between 0.75 and 3.0 t ha^{-1}. Oil content of fruit is equally variable ranging from 1.5 to 6.5%. Thus, it is essential to select the most suitable cultivar for a particular area to obtain maximum seed or oil yield ha^{-1}. However, important though a high potential seed yield may be, more important is the actual in-bag seed yield ha^{-1}. For this reason a non-shattering variety with a lower potential seed yield can be more profitable. Not only will there be less seed loss in the field, but harvesting operations will be simplified, or the period extended; both invariably reduce costs. The two major cultivars of winter caraway in The Netherlands are Volhouden and Bleija, and well illustrate the characteristics influencing cultivar choice. The former shatters easily at maturity and must be windrowed before pick-up. However, threshing and cleaning is easy, and its fruit is well suited for use as spice. Bieija fruits are non-shattering but usually require recleaning after harvesting. Two commercial non-shattering cultivars of spring caraway available are Karzo and Springear.

Processing

Following threshing, seed can be further dried either naturally, in cold or hot air dryers at 30°C, to a moisture content of about 12% and comments on drying coriander are generally applicable to caraway. To comply with stringent trade standards, recleaning is often necessary to remove stalks and other plant material from fruits.

Provided fruit is properly dried and stored, little loss of the essential oil occurs, although there may be some change in constituents; after 6 months' storage in Israel the carvone content fell from 80% to 60% but limonene was unchanged. However, once ground a steady if small loss occurs. In general, there are no significant changes in characteristics between oil from fresh ripe fruit and similar fruit after 1 year in suitable storage. Fruit is normally milled or crushed prior to distillation, and use of uncrushed fruit is not recommended. Distillation to obtain the oil is straightforward, and no special techniques or equipment are necessary. Fruit can also be solvent-extracted to produce an oleoresin. Distillation residue can be processed into stockfeed and may contain up to 25% crude protein and 15% fat, but is usually burnt.

Storage and transport

Seed must be inspected when received at stores as it frequently contains plant debris, other solid impurities or insect remains, which must be removed by recleaning or decontamination before storage, as in Italy (Locatelli *et al.*, 2000). Correctly dried whole fruit can be stored in sacks or in small bins until required for grading and shipping. Essential oils should be stored in full, sealed, opaque containers until required for use or export. The comments in the coriander section are generally applicable. Caraway seed is attacked by the major stored product pests, but especially *Anthrenus coloratus* in the Sudan and India.

Products and specifications

The main products are the fruit, known generally as the seed, which is the culinary spice, herb and seed oils.

Seed. Caraway seed has a characteristic distinct warm, slightly sweet, very sharp, somewhat acrid but pleasant taste reminiscent of cumin, a pleasant aroma and can be used whole or as powder. Caraway is currently less popular as a domestic spice, although a frequent ingredient in prepared sauces and processed meats. Caraway is quite often and incorrectly included in Western cookery recipes as an ingredient of curries and similar dishes, when cumin is in fact required. Seed is used to flavour a range of breads (it is often the flavouring in rye breads), meat dishes, salads and especially sauerkraut; it can be coated in sugar and rolled into 'sugar plums' but is not generally used in confectionery, and can be used as a flavouring in alcoholic drinks including kummel, schnapps and vodka. For processed foods it has been replaced by the essential oil. In medicine, the seed is considered a carminative and stomachic, and in herbal remedies it is believed to enhance lactation in women.

Essential oil. Caraway fruit oil (caraway oil) is obtained by steam distilling fully ripe fresh, but usually dried fruits, and inclusion of unripe fruit, twigs or other plant debris reduces oil quality. Caraway oil is a mobile liquid, almost colourless to pale yellow, although it may become brownish to dark brown depending on the time it has been exposed to light. The oil has a strong characteristic odour due to the carvone content, somewhat nauseous initially, with a warm spicy, sometimes very hot taste. Rectified oil, also known as double-distilled, is colourless to pale yellow, has a strong odour with no unpleasant overtones, and a more biting taste. The oil is placed in the caraway group by Arctander (1960).

Major components of caraway oil are *d*-carvone up to 65% and *d*-limonene up to 40%, but these proportions are very variable, as is the ratio between them. Carvone is considered the most important constituent and the principal source of the characteristic odour of the oil. Limonene supplies the citrus-like note. Approximately 60 compounds have been identified, including small amounts of *α*- and *β*-pinene, *β*-phellandrene, *p*-cymene, and larger amounts of myrcene and sabinene (Bourrel *et al.*, 1995; Lawrence, 1996b). Minor components have been identified and compared with other *Umbelliferae* (Schulz *et al.*, 2000). Major components are shown in Table 10.1. Commercial samples of the oil should not contain less than 50% by volume of ketones calculated as carvone, but around 60% is normally preferred.

The oil has virtually replaced the seed in processed foods, and is extensively used as a flavour component in processed meats, pickles, sauces, seasonings and similar preparations, in alcoholic and non-alcoholic drinks including the German kummel and a Scandinavian akavit, in cosmetic products such as soap, perfume and toothpaste, and in pharmaceutical preparations. The maximum permitted level for caraway oil in food products is about 0.02%. Caraway oil is frequently adulterated, principally with synthetic *d*-carvone and *d*-limonene, and as

these compounds are now inexpensively produced, they also compete with caraway oil in some products, restricting its usage. Additionally dill fruit oil contains 20–50% *d*-carvone, and could serve as a substitute for caraway oil. In the USA the regulatory status generally recognized as safe has been accorded to caraway, GRAS 2236, and caraway fruit oil, GRAS 2238.

In human medicine, the oil is spasmolytic, antimicrobial, and used to treat dyspeptic complaints such as mild gastrointestinal spasm and bloating. It is effective as an anti-sprouting agent in stored potatoes, and also controls most storage fungi; it is also a disinfectant and insect repellent. These qualities are attributed to the carvone content. A monograph on the physiological properties of caraway oil has been published by the RIFM.

Oleoresin. Caraway oleoresin is prepared either from the oil or directly from the fruit. It is usually a greenish shade of yellow, and normally contains 20–25% volatile and 60–75% fixed oil. Commercial samples in the USA require a minimum of 60% volatile oil with a dispersion rate of 5%. The high fixed oil content usually requires the addition of an antioxidant to the legal limit.

Chaff oil. Chaff oil is obtained by steam distilling material left after threshing fruits. It is much inferior to the seed oil, containing less carvone and more terpenes. It thus has less of the characteristic odour of the seed oil and is harsher, with a somewhat bitter taste. This oil is produced on a very small scale and is also an adulterant of the fruit oil.

Herb and root oil. Herb oil is currently produced to order only, but could quickly expand to commercial production as an alternative to seed, as it is similar in its characteristics and composition and can be a direct substitute. The oil consists mainly of sesquiterpene hydrocarbons up to 90%, including germacrene D up to 75% plus *β*-caryophyllene, *β*-eleme, humulene, germacrene A and B, and two cadinenes. If herb oil can be produced more cheaply than seed oil, this would probably expand or at least maintain caraway oil's present usage. Root oil can be obtained by distilling minced roots, and consists mainly of oxygenated compounds, with aldehydes up to 70%, including octanal, nonanal, *cis*-dec-4-enal and *trans*-dec-2-enal.

Carvone. The essential oil constituent, *d*-carvone, is a nearly colourless to pale yellow liquid, which darkens with age; it has specific gravity (20°C) 0.960 and refractive index (18°C) 1.4999; boiling point 227–230°C; insoluble in water, but soluble in alcohol, ether and chloroform. Its odour is warm, herbaceous, bread-like, spicy and slightly floral; the taste is sweet, spicy and bread-like. Carvone reportedly has certain cancer-preventative properties.

Other Carum species

Carum bulbocastanum W. Koch. (syn. *Carum nigrum* Royle.), or black caraway, is known in Hindi as *shah-zirah*, in Kashmir as *gunyun, siyah-zira*, and in Arabic as *kamene*. It is native to northern India where it grows in profusion, and is a locally common weed of cultivated crops. It is an annual herb up to 75 cm high, its fruits are collected and used as a caraway substitute, and its root as a vegetable. The fruit yields an essential oil up to 2%, containing 18% aldehydes, of no commercial importance. In countries with native *Carum* species, the roots are frequently cooked and eaten as a vegetable, even when no other plant parts are utilized. Other species include *Carum capense* Sund. in South Africa and *Carum gairdneri* (H & A) Gary. in North America.

Unrelated species

Several *Trachyspermum* species may have been confused with *Carum* species; the most common are noted below.

Trachyspermum ammi (L.) Sprague ex Turrill, *ajwan, ajowan*, or bishop's weed, is widely cultivated in India and to a lesser extent in Afghanistan, Iran and Pakistan, and as *netch azmud* in Ethiopia. *Ajwan* is a herbaceous annual up to 90 cm; the stem is profusely branched, the leaves pinnate and the flowers small and white. In India, the seeds are used as a spice and as flavouring in pickles, biscuits, confectionery and beverages. *Ajwan* oil is distilled from seeds, contains 30–60% thymol and is the major local source of this compound; other main constituents are α-pinenc, *p*-cymene, dipentene, γ-terpinene and carvacrol (Husain *et al.*, 1988). The oil is placed in the thyme group by Arctander (1960). Although *ajwan* fruit and oil production can be locally substantial in India, which has exported about 500 t of seed annually for decades mainly to neighbouring countries and the Middle East and Gulf States to satisfy a demand from the many expatriate Asian workers, both seed and oil are of little international importance.

Trachyspermum roxburghianum (DC) H. Wollf, previously *Carum roxburghianum* Benth., is cultivated on a small scale in India, Sri Lanka and some parts of the Far East for its chopped leaves used as a condiment, the fruit as a spice and commonly used in traditional medicines.

Coriander

The genus *Coriandrum* L. contains two species, *Coriandrum tordylium* and *Coriandrum sativum*, the latter being the source of the important spice coriander and one of the earliest spices used by mankind, as a Babylonian recipe on a clay tablet lists coriander, cumin and five other spices as ingredients in a stew (Bottero, 1985). It was used in Dynastic Egypt for culinary and medicinal purposes, mentioned in the Medical Papyrus of Thebes (the Ebers Papyrus) in 1500 BC and seeds have been identified in Pharoanic tombs. Coriander was probably carried up the Nile by traders, as it is wild and cultivated in Egypt, the Sudan and especially Ethiopia. It was known to the ancient Israelites, referred to in the Christian Bible, and is one of the bitter herbs involved in the Jewish Passover ritual.

Hippocrates, in about 400 BC, records its use as a drug, Pliny in his *Historia Naturalis* of AD 77 stated the best-quality came from Egypt, and Marcus Cato (234–149 BC) in the *De Re Rustica* praised it as a food flavouring. The Arab al-Biruni has provided information on the common cultivated herbs and spices of Islamic gardens at the beginning of the 12th century, including coriander, cumin, safflower and saffron (Harvey, 1975).

In the Middle Ages of Europe, the Emperor Charlemagne in AD 812 ordered some 70 herbs, including coriander, to be grown on the Imperial farms, and in 1611 Carmelite monks in Paris included it in their *Eau de Carmes,* used oddly as a toilet water or cordial! Because of its popularity, the plant was widely disseminated in Europe and the Middle East. Coriander seeds were not only used in the kitchens of Europe, but were frequently included in herbal remedies. William Turner in his *A New Herball* (1551) states, 'Coriandre layd to wyth breade or barley mele is good for Saynt Antonyes fyre' (erisipelas). Such small cakes were baked in rural communities for centuries.

Coriander is mentioned in Sanskrit literature as *kustumburu,* and remains a commercial Indian crop and popular food flavouring. A type with ovoid fruits was introduced into South-east Asia initially from India, a globoid-fruited type later from China, and the typical European cultivars much later. However, it is cultivated in the region only as a domestic plant. The first Chinese records of its use, in local medicine, are from the Han dynasty of 207 BC to AD 220. It was also believed that an extract of the whole plant, or fruit, could preserve a persons spirit or essence after death, and for this reason such compounds were an important accessory at funerals. It is still considered to prolong life and is incorporated in current herbal medicines.

Coriander was apparently introduced into Britain by the Romans, but it is interesting to note that it is not specifically mentioned in Gerard's and other early herbals. When the *Thousand and One Nights* as translated by the explorer Sir Richard Burton was published in England in 1885, it mentioned coriander as arousing passion, and the spice subsequently became known and widely used in Victorian England, quite incorrectly, as an aphrodisiac.

It was one of the first herbs grown by the European colonists of North America, and apparently introduced into Massachusetts in the mid-17th century. Small-scale commercial cultivation is mainly in South Carolina. Coriander later spread to South America generally, and is particularly popular in Peru. It is traditionally grown in the Ukraine, Russia, central Europe, North Africa especially Morocco, West Asia, India, Mexico and more recently in Argentina (Bandoni *et al.*, 1998).

Botany

C. sativum L., cultivated coriander, and *C. tordylium* (Fenzl.) Bornm. are native to the eastern Mediterranean region, although up to six species have been included in the genus at various times, hence the large number of synonyms. The generic name is derived from the Greek *koris,* or bedbug, as the unpleasant, fetid, odour of the green unripened fruits is likened to bedbugs, although the stink bug (*Peritato midae*) is considered by some to be more apposite.

C. sativum L. (syn. *Coriandrum majus* Gouan.; *Coriandrum diversifolium* Gilib.; *Coriandrum globosum* Salisb.; *Selinum coriandrum* Krause., etc.) is known in English as coriander, in German as *koriander, schwindelkraut,* in French as *coriandre,* in Italian as *coriandolo,* in Spanish as *cilantro,* in Dutch as *koriander,* in Russian as *koriandr,* in Indian as *dhania, kotmili,* in Arabic generally as *kizbara* or *kasbara,* in Ethiopia as *dembilal, debo* and in Indonesia generally as *ketumbar.* Coriander is a diploid, with $2n = 22$, and tetraploid plants have been artificially produced and interspecific hybrids obtained by growing coriander on various rootstocks, later noted. Although there are no botanically accepted subspecies, many are listed in the literature classified on various plant characteristics; for example, four subspecies are recognized in Russia: *C. sativum* subsp. *sativum, indicum, asiaticum* and *vavilovii.*

One accepted classification divides commercial coriander into three categories by fruit size: the small-fruited, 1.5–3 mm diameter *C. sativum* subsp. *microcarpum* DC., represented by Russian coriander with its high volatile (essential) oil content fruit, the largefruited, 3–5 mm diameter *C. sativum* subsp. *vulgare* Alef, which includes Moroccan and related types, all with very low oil content and the third, *C. sativum* subsp. *indicum* Stolet, with ovoid fruits of low oil content including Indian and related cultivars. Separation into nine regional types – European, North African, Caucasian, Central Asian, Syrian, Ethiopian, Indian, Bhutanic and Omanic – is considered to reflect accurately the species' evolutionary pathway. In this text coriander refers to *C. sativum* and its derivatives.

Large collections of coriander germplasm exist in Russia, Germany, the USA and to a lesser extent in India; however, geneticists consider some geographical regions are poorly represented, notably West and Southeast Asia. Due to neglect or environmental reduction, local cultivars in several regions are considered at risk of genetic erosion. Additionally, other species of *Coriandreae* have potentially useful genetic resources but with the exception of the genus *Bifora* F. Hoffm., they are apparently not represented and neither is the close relative *C. tordylium.*

Initial breeding programmes, especially in East European countries and India, were concerned with increasing fruit yield, essential oil content, disease and pest resistance, and are well reported in the literature; recent research has been noted in the text. Hybridization with other *Coriandreae* spp. including *Bifora* spp. has apparently not been successful. Interspecific hybrids have, however, been used to ascertain the effect on fruit yield, oil content and composition, by grafting coriander on to various rootstocks. Plants from seeds resulting from coriander grafted on to fennel contained more oil; when grafted on caraway, fruit oil had a higher linalool content; coriander on parsley had less fruit loss from shattering but had a lower fruit oil content with a higher than average linalool component; coriander grafted on the related *Libonotis montana* had a longer vegetative period and a lower fruit oil content. Gamma radiation induces variability in coriander seeds, but the dosage must be accurately controlled to avoid a reduction in viability. *In vitro* and tissue culture has successfully produced plantlets for growing on, but in Russia evidence of clone ageing was noted (Kataeva and Popowich, 1993).

Coriander is an erect, herbaceous annual, varying considerably in height up to 1.5 m, with small white flowers, and producing small, brown, aromatic fruits in the autumn. Volatile oil is present in all plant parts. Coriander has a well-developed taproot, up to 20 mm diameter, which is yellow-brown, fleshy and fascicled, with many laterals. Secondary roots are extensive and frequently superficial under cultivation. Root-oil content rapidly diminishes once the stem elongates. *In vitro* culture of the root callus contains geraniol, but no other flavouring principle associated with the spice. The root may be harvested and cooked as a vegetable and is especially relished in Thailand.

There is normally one slender stem, 20–90 cm high, usually 60–80 cm, depending mainly on the cultivar and standard of crop management; the base is up to 2–5 cm in diameter, subterete, solid, and smooth. Older internodes may become hollow and vertically ridged. The stem is light green with darker green ribs, sometimes with a violet tinge or white bloom. The stems are usually unbranched on the lower section, profusely and corymbosely branched on the upper. Stem essential oil content reaches its maximum at flowering.

The leaves are alternate, rather variable in shape, size and number, with a yellow-green, scariously margined sheath surrounding the supporting stem for up to three-quarters of its circumference; the petiole and rachis are subterete and sulcate. The leaves are shiny light green, often with darker green veins, and waxy. The lower one to three leaves are usually simple, carried on long petioles, withering early, with the blade ovate, deeply cleft or parted, 2.5–10.0 × 2.0–7.5 cm; there are usually three incised-dentate lobes. Higher leaves are decompound, with the-blade ovate or elliptical, 30 × 15 cm, usually pinnately divided into 3–11 leaflets, which may be again pinnately divided. All higher leaves are compound, with the petiole restricted to the sheath, the blade divided into three leaflets, the central one largest, each often variously divided into ultimately sublinear, entire, acute lobes (Fig. 10.8).

Freshly cut leaves are one of the world's most widely used culinary herb, the dried

Fig. 10.8. Flowering stem – coriander.

leaves being less popular. Leaves are often cut from plants subsequently harvested for seed, and in India it was found that one cutting generally had no effect, but two reduced seed yield. Some cultivars can be cut more often than others, and still produce a reasonable seed yield. Spraying foliage with N-fixing strains of *Klebsiella* bacteria increased foliar and fruit yield (Mukhopadhyay and Sen, 1997), and with mixtalol (aliphatic alcohols) enhanced foliar growth, chlorophyll content, flower number and subsequently fruit yield (Lal *et al.*, 1996).

Fresh foliage contains 0.1–0.2% essential oil, which reaches its maximum at flowering. Some 40 compounds have been identified in the oil, with decyl and nonyl aldehydes accounting for about 80% and alcohols 16% (Potter and Ferguson, 1990). Herbage also contains (per 100 g fresh foliage) up to 12 mg pro-vitamin A, up to 60 mg vitamin B2, and up to 250 mg vitamin C; the last diminishes rapidly on storage. The seedling acquires the distinctive coriander smell, which is retained by the leaves and stalk until the final phase of fruit ripening, when the vegetative parts wither.

The inflorescence is an indeterminate compound umbel approximately 4 cm in diameter; the peduncle is up to 15 cm. The flowers are small and white or pinkish; there are up to 20 in an individual umbel, and up to 150 per compound umbel, but this varies with the cultivar (Fig. 10.8). Flowers have small linear bracts at the bases of secondary umbels. Outer flowers of the whole inflorescence are zygomorphic, having the two outer sepals and outer two to four petals much enlarged. The five petals are heart-shaped, very small (1 × 1 mm) in male flowers, and in bisexual peripheral flowers there are usually three petals, which are larger. Hermaphrodite and staminate flowers may occur in each umbel; male flowers are usually central. There are five stamens, with spreading white filaments up to 2.5 mm. The pistil is rudimentary in male flowers; in bisexual flowers with an inferior ovary, there is a conical stylopodium bearing two diverging styles up to 2 mm, each ending in a minutely papillate stigma. The number and sexual status of flowers can be affected by soaking seeds in growth regulators such as IAA and ascorbic acid prior to sowing, while the number of umbels per plant can be affected by fertilizers; nitrogen increased umbel numbers in New Zealand (Reddy and Rolston, 1999), and planting after the optimum date reduces umbel numbers, as in Turkey (Kaya *et al.*, 2000).

The flowers are protandrous, the anthers maturing before stigmas becomes receptive. Coriander is cross-pollinated, usually from adjacent flowers and by many insects. The stigma is receptive for 5 days, the first 3 giving the highest set; pollen usually remains viable for a maximum of 24 h. For emasculation and hand-pollination, buds should be at least 4 days old. Flowering begins in the early morning and reaches a maximum between noon and 2 p.m. Flower opening and anthesis within the compound umbel and within the umbellets proceeds from the periphery inwards. Anthers dehiscence over about 10 hours, with individual anthers appearing consecutively and shedding pollen. Anthesis of all umbel flowers is completed in 3–4 days.

Insects, particularly bees, are significant pollinators, although the degree of cross-fertilization is a cultivar characteristic. The importance of bees as pollinators was indicated in Yugoslavia, where an average fruit yield of 800–1200 kg ha^{-1} increased to 1800–2000 kg ha^{-1} when bee hives were placed around coriander fields. Bees apparently chose coriander in preference to other plants flowering at the same time, and fields were a prolific source of honey. In India, yields of plants from which insects were excluded were 50% lower than those exposed to insects, especially bees.

Main flowering is in early or late summer in the northern hemisphere, depending on local cultivar (Fig. 10.9). Duration of flowering may extend to 30 days depending on the number of branches and weather conditions. Thus, umbel maturation proceeds successively, and ripe fruits of the primary umbel may shatter before later umbels are fully mature. Uniformity of flowering is thus an

Fig. 10.9. Coriander flowers.

asset in mechanized crops. Essential oil is present in umbels at flowering and rises steadily until fruits begin to ripen, when it rapidly falls. Research indicates that, in general, late flowering and maturing cultivars have high essential oil content fruits, preferred for distillation. Earlier flowering and maturing cultivars have larger, lower essential oil content fruits more suitable for spice.

The fruit is an ovoid to globose schizocarp up to 3–5 mm in diameter, yellow-brown when ripe, with ten straight longitudinal ribs alternating with ten wavy longitudinal ridges, often crowned by the dry, persistent calyx lobes and stylopodium with styles. The fruit contains two mericarps, each bearing on their concave side two longitudinal, rather wide vittae containing essential oil. A relationship exists between the size (surface area) of fruit and its essential oil content. Fruit does not usually split at maturity (Fig. 10.10).

Fig. 10.10. Coriander fruit.

Immature fruit possesses the characteristic unpleasant aldehydic odour, which gradually lessens and disappears on full ripening and drying, due to changes in essential oil constituents, related to changes in fruit structure. Two types of essential oil canals are present in unripe fruit; those located on the periphery contain oil in which aldehydes predominate; inner canals buried in the mericarp have a high linalool content plus α-pinene, γ-terpinene, geranyl acetate, camphor and geraniol. As fruit ripens, the peripheral canals become flattened, gradually lose their oil, and after drying contain no oil; the inner canals remain intact, and oil develops the characteristic odour and composition. The essential oil is discussed at the end of this section.

Ripe, dried, fruit contains essential oil, fixed oil, proteins, cellulose, pentosans, tannins, calcium oxalate and minerals. The major constituents are fibre 23–36%, carbohydrates 13–20%, fatty oil 16–28%, proteins 11–17% and essential oil 1–1.5% with small fruited cultivars up to 2.0%. These amounts and ratios can vary with cultivar, season and maturity at harvest. Analysis of air-dried fruits gave the following average content: water 11%, crude protein 11%, fatty oil 19%, carbohydrate 23% (including starch 11%, pentosans 10%, sugar 2%), crude fibre 28%, minerals 5%. Fruit contains two small seeds, each enclosed in a mericarp, which is concave on the commissural, convex on the dorsal side; the testa is attached to the fruit wall. The seed is whitish to yellowish with two flat, thin, circular cotyledons and a conical radicle; there is copious grey-white endosperm, and 1000-seed weight is 7–17 g.

Fixed oil. Fruit contains a dark brownish-green fixed oil with a similar pleasant odour to the essential oil, and often solidifies with storage. Fatty oil content and composition in ripe fruit endosperm ranges from 12 to 25%, mainly dependent on environmental conditions. The saponifiable portion of the fatty oil accounts for some 90% of the total, and contains a very high content of octadecenole acids; petroselinic and oleic acid occur at similar levels and together comprise 75-85%, linolenic 7–16% and palmitic 4–8%, but the relative abundance of each is a cultivar characteristic. The main characteristics of an Indian oil were: specific gravity (15°C) 0.9262–0.9284; refractive index (20°C) 1.4704; iodine value 93–100; unsaponifiable matter 2.3%. A detailed analysis of the fatty oil and its constituent monoeconoic acids and their biosynthesis has been published (Suh *et al.*, 1999).

Essential oil. Oil content of fruit is very variable, especially between regional types and cultivars, and ranges from 0.2 to 2.6%, but is usually 0.4–1.0%, although experimental selections with oil contents up to 3.5% have been reported associated with lower yield. Plants with medium or large globose fruits with low or medium essential oil content have developed in the West Asia/Mediterranean region, Europe and the New World. Plants with small, globose fruits occur mainly in the Caucasus and Central Asia, and include forms with the highest essential oil content, always containing camphor, mycene and limonene.

Dehydration occurs as fruits mature owing to collapse of peripheral volatile oil canals, as noted, and approximately one-third of the essential oil present in the immature fruit is lost. Oil content of mature fruit depends basically on its position; in small-fruited varieties it is highest in the central umbels, lower in the second-order, and lowest in the first-order umbels. Major oil component is *d*-linalool (coriandrol) between 55 and 75% mainly depending on cultivar, location and ripeness at harvest. Other constituents are α-pinene and γ-terpinene up to 8%, camphor up to 6% and geraniol up to 2% (Anitescu *et al.*, 1997)

The main characteristics of coriander oil are: specific gravity (20°C) 0.863–0.875; refractive index (20°C) 1.4620–1.4720; optical rotation (25°C) +8° to +15°; almost insoluble in water, soluble in 3 vols 70% alcohol, very soluble in chloroform, ether and glacial acetic acid. A fairly wide variation occurs in these characteristics, since cultivar, time of harvest, storage and method of distillation all affect the type of oil produced. Inclusion of unripe fruits or leaf and stalk material, for example, reduces the specific gravity, optical

rotation and linalool values, while the carbonyl value is enhanced.

Physiochemical properties of European oils are normally within fairly close limits; the optical rotation value is usually in the range +9° to +13°, while ester number is normally well below the upper limit of standard specifications. Data for Moroccan oils show an upper limit for the optical rotation value of +10°, and ester numbers up to 30. Traditional Indian oils have refractive index values at or below the lower limit of standard specifications, but very high ester numbers. Low essential oil content cultivars from India contain little or no camphor, mycene and limonene, but considerable linalool. Despite the relatively low essential oil content, this type may be preferred because of its specific flavour.

Ecology

Coriander is naturally a crop of the warm temperate zones of the northern hemisphere, but has been widely introduced and become locally adapted to a range of habitats, including higher altitudes of Ethiopia, Indonesia and Malaysia above 1500 m. For fruit (spice) production, coriander is grown in tropical highlands, subtropics and temperate regions, but as a green herb in the lowlands. Local selection and breeding programmes have produced cultivars well adapted to specific areas. In general, these cultivars are most productive in their homeland, and when grown elsewhere tend to vary greatly in their growth, yield and fruit characteristics. Similar to other long-domesticated plants, coriander probably no longer exists as a truly wild species in Europe, but escapes from cultivation can frequently be found growing in suitable locations throughout the region.

Warm sunny days with a temperature of 16–22°C are preferable during the period of maximum growth and when fruit is ripening. The optimum temperature range is 17–20°C for small-fruited types and 22–27°C for larger-fruited types. An average range of 16–20°C during the period from germination to harvest is common for coriander over a large area of Russia. Low temperatures or very cloudy periods have an adverse effect on flowering, but in Russia cool, rather wet summers increased fruit oil content of local cultivars. Very hot periods at flowering or early fruit maturity can inhibit both. In tropical regions with cool and hot seasons, India, for example, coriander is grown in the former. When the plant is required to produce a green culinary herb only, the range of coriander is greatly increased and it can frequently be found in gardens in monsoon Asia throughout the year.

Plants are frost-sensitive at all growth stages, but cold resistance is a feature of some Russian cultivars and seed will germinate down to 4°C. After stem elongation, coriander is sensitive to low temperatures but resistant to drought. Temperatures of −8°C to −9°C generally kill roots, and −13°C to −14°C killed leaf rosettes of seedlings in Russia. Thus, coriander can be grown successfully from northern Europe to southern Siberia, and strains have been selected with a high degree of cold hardiness. Time of planting is usually important, as autumn-sown plants in the Caucasus had one-third higher fruit yield with nearly one-fifth more oil than the same cultivars sown in the spring. Specially selected short-season cultivars from Siberia and Norway have high oil content fruit. Planting on 1 November, 1 March or 1 April in Turkey had no effect on yield, although a variation in fruit constituents occurred, particularly linalool (Kaya *et al.*, 2000). However, in countries with long hard winters without snow cover, coriander is best treated as an annual. Selection of an appropriate cultivar is also important, as in Finland small-seeded coriander with high-oil-content seeds could not mature in the time available. It is, however, possible to reduce the period required to produce a seed crop; in China, germinating seed maintained at 3°C to 4°C for 15 days hastened flowering by 10 days, and a similar method was successful in Russia.

Time of sowing is important in regions with a defined rainy season, or hot and cold seasons. For example, in Bengal and Uttar Pradesh, India, coriander is sown in the rabi (cold) season, in Maharashtra during the monsoon, and in Tamil Nadu in the drier autumn. In Karnataka and Madras, corian-der can be grown in two seasons, May–August and October–January. However, in Argentina biomass was greater in early sown plants but fruit yield was little different between early and late sowings, although seasonal weather conditions affected yield (Gil *et al.*, 1999).

Coriander is basically day-neutral, but long days slightly accelerate development, and in Europe and the northern hemisphere generally, fruit essential oil content tended to increase from south to north; southern Indian selections, however, gave higher fruit yield per plant compared with northern selections. Whether this is due to temperature or daylength remains unclear, since there are indications that coriander may be affected by light intensity or duration. Coriander plants flowered and matured earlier under long-day treatment than in short days in a phytotron trial in Israel. Irrespective of daylength, plants flowered earlier at the higher day temperature. Daylength before flowering had a greater effect than temperature on duration of the vegetative cycle in Bulgaria. An interesting effect of location on oil composition was noted in Egypt, with considerable variation in main oil constituents between seed from plants grown in relatively arid Upper Egypt and seed in the wetter Delta, with specific cultivars having a high average at one location.

Coriander is normally grown below 500 m in its native lands, but in more tropical regions is grown at progressively higher altitudes as the mean maximum daily temperature increases. In Malaysia, Thailand and Indonesia, for instance, coriander grows well to flowering in many areas, but will only produce fruit in the cooler hill regions. It appeared to be well adapted to the Ethiopian highlands between 1500 and 2500 m, and pilot plantings flourished at 1500 m in Kenya (E.A. Weiss, personal observation). The effect of altitude on fruit yield and fruit oil content from a specific cultivar has apparently not been investigated.

A rainfall of 400–600 mm, with half falling during the period of main growth and flowering and ceasing after fruit has matured, will produce excellent coriander crops. Higher rainfall will be tolerated if soil is free-draining. Short-season cultivars can be grown where the rainfall is between 200 and 400 mm provided there is ample sunlight during growth. A moisture shortage at the seedling stage stunts plants and reduces yield more than a shortage at later stages of growth. High winds when the crop is full-grown cause extensive damage, and a hot, dry wind when fruit is ripe causes extensive loss through shattering. Hail has a similar effect, stripping foliage and breaking stems on mature plants, and can destroy a field of young plants or seedlings.

Soils

A medium to heavy loam with good drainage is preferred for commercial crops in Europe, and truck crop soils in the USA, but coriander will also grow well on the more sandy soils of North Africa and West Asia. The heavy, black cotton soils of the Indian Deccan and southern India are well suited to coriander, which is also grown on rich silt loams of the Gangetic plain. A major requirement is that the soil should be well-drained or free-draining, since coriander is intolerant of waterlogging for other than relatively short periods. Smallholder crops everywhere are planted on any available land, especially when only green herbage is required. In general, weather conditions during growth are more important than soil type, especially with respect to fruit essential oil content.

A neutral to slightly alkaline soil, pH 6.5–8.0 is acceptable, but acid and saline soils are unsuitable for spice crops, although they can be used to produce green herb. Seed yield and quality were significantly reduced in India at the low salinity level of 2.5 mmhos cm^{-1} (1 mmhos cm^{-1} = 1 $\times$ 10^5 siemens m^{-1}), but vegetative growth was relatively unaffected; germination was zero at 10.0 mmhos cm^{-1}. In a trial in India some recently developed cultivars showed high tolerance to salinity (Yadav et al., 2000).

Fertilizers

Coriander generally responds well to applied fertilizers, the amount usually being determined by the local selling price of the spice. Thus, fertilizer usage is greatest in more developed countries including the USA and Europe. Phosphorus and K are the most frequently required nutrients, the N requirement generally being moderate to low. An indication of the nutrient uptake by coriander in Tamil Nadu, India, showed that a yield of about 650 kg ha^{-1} removed 12 kg N, 2 kg P and 9 kg K, all ha^{-1} (Sadanandan et al., 1998) (Table 3.4). In India generally, only N significantly increases yield of herbage and fruit (Vinay and Bisen, 1999), but the response has been somewhat irregular elsewhere; in New Zealand, levels of N from 50 to 200 kg ha^{-1} had no significant effect on yield (Reddy and Rolston, 1999), while in Argentina up to 75 kg N ha^{-1} significantly increased seed yield, and up to 150 kg N ha^{-1} increased total biomass (Lenardis et al., 2001). Where N has been shown to be beneficial, an optimum rate of application must be established, as above this excess herbage is produced. The tall narrow-stemmed succulent plants resulting are most susceptible to wind damage. It is doubtful whether applications above 50 kg N ha^{-1} are necessary, preferably in two applications, as in Harayana (Thakral et al., 1997). Half can be applied in the seedbed as part of an NPK mixture, the other half as a top dressing when first flowers appear. The type of N fertilizer may also affect yield, but is usually unimportant at the rates normally applied (Hussain, 1995), as is the effect of N on seed-oil content, demonstrated in the Czech Republic (Strasil, 1997).

Data indicate that coriander absorbs most of its P in the early growth stages, and as this nutrient is frequently deficient in tropical soils, applying P to the seedbed at sowing increases fruit yield. In Europe and the USA, this is normally a standard procedure, using an NPK mixture. However, in India the yield increase from P application has varied from moderate to nil, irrespective of the levels of applied or available N and K. An NP mixture applied at planting followed by an N top dressing just prior to flowering proved the most profitable. The local optimum ratio between plant nutrients should be determined by fertilizer trials, since there are indications large- and small-seeded cultivars respond differently to the same level of applied fertilizer. Potassium should be supplied only where a deficiency limits growth, and is best applied in the seedbed. When coriander follows a crop to which K has been applied, no more may be necessary.

Most data from fertilizer trials concern yield of green herbage or mature dry fruit. Only occasionally is the effect on oil content of fruit or oil composition mentioned. Data obtained in India are shown in Table 3.4. An increase in fruit yield and thus oil yield from application of the nutrients compared to a control was common, but was small and apparently unprofitable. The effect on oil linalool content is interesting; data indicated that increasing N may reduce linalool content, whilst P and K apparently had a plateau level above which they also had a depressing effect. In Germany, the petroselenic content of fruit oil was affected by varying P levels (Rohricht *et al.*, 1999), while increasing the P level at the same N rate depressed oil content in Romania.

Laboratory trials have indicted that coriander plants are capable of absorbing heavy metals, including lead and cadmium, which appear in the oil (Zheljazkov and Jekov, 1996; Lima *et al.*, 1999). This has implications for coriander grown in eastern European countries where soil contamination by uncontrolled release of heavy metals from industrial plants is common. Selenium uptake and its effects on coriander growth, oil yield and composition, indicated no adverse effect on humans (Lee *et al.*, 2000). Pot trials with minor nutrients borax, ammonium molybdate, Cu, Zn, Mn and Fe all affected growth of coriander seedlings, but copper sulphate increased resistance to stem gall disease, *Protomyces macrosporus*.

Cultivation

Because of its great popularity, coriander is grown at every level from pot-plants in towns or gardens, to commercial-scale field crops, and frequently occurs as a weed of grain crops in eastern European countries to which herbicides have not been applied. Coriander can be sown in seed boxes or seedbeds under cover and transplanted into the open when conditions are suitable. This system is practi-

cal only on a very small scale by domestic growers for their own use, or near urban centres where returns from selling the green herb are high. Soaking seed in water for a day prior to planting hastens germination. In less-developed areas, seed can be broadcast or small manual seeder units used.

The whole fruit should be mechanically rubbed to free the two seeds, and large sound seed should be selected, since large seed germinates more quickly and usually has a higher germination percentage. Seed should be cleaned and preferably treated with a fungicide. Coriander seed remains viable for several years under reasonable storage conditions, but it is recommended seed less than 2 years old be used whenever possible. Where seed has to be stored, it is often fumigated or treated with a fungicide. Choice of the correct chemical may be important, as it was found in India that paradiclorobenzene reduced germination by 25% after 6 months. Seed normally has no dormancy period, but very fresh seed may germinate erratically. Heat treatment of the fruits after harvest or before sowing, either artificially or by exposure to the sun, will promote physiological ripening and ensure even germination. Plants grown from gamma-irradiated seed in Bulgaria yielded 25% more fruit, which contained up to 28% more essential oil than non-irradiated plants (Jeliazkova *et al.*, 1997).

Seed is usually sown 2–6 cm deep in rows 25–75 cm apart, both depth of planting and row width depending mainly on the soil type, normal rainfall and equipment available. Seed rate will thus vary from 10 to 25 kg ha^{-1}, and this rate will also vary with cultivar; for example, 40 kg ha^{-1} was the local optimum in north-west India, and 30 kg ha^{-1} in New Zealand, with 10 and 50 kg ha^{-1} reducing fruit yield (Reddy and Rolston, 1999). In terms of spatial distribution, a 20 cm spacing in 30 cm rows gave highest seed yield in India (Nehra *et al.*, 1998), but increasing the row width to 60 cm reduced yield in New Zealand; however, in Argentina planting date was more important than spacing (Gil *et al.*, 1999).

For commercial spice production in cooler regions, coriander is direct drilled in spring when the soil has become warm; a temperature between 15 and 20°C ensures rapid germination. Under optimum conditions, sound coriander seed germinates in 10–15 days, but may take up to twice this if cold weather follows sowing, or if soil moisture is deficient; cultivars also vary. Germination is epigeal; the hypocotyl is up to 2.5 cm, the cotyledons pale green, opposite, oblanceolate, and up to 3 cm × 4 mm. Some genotypes of coriander form several basal leaves, others start stem elongation immediately, or after the second leaf (Fig. 10.11).

Time of planting has a significant effect on seed or oil yield, as noted previously, and it is essential to determine the optimum period by field trials. Coriander sown in March in Hungary gave the highest oil yield ha^{-1}, but mid-April sowing gave the highest essential oil content with yield from later sowings reduced by 40%; in Europe, sowing in March–April gives highest yield; in India, middle to end-October gives highest seed yield, but highest essential oil content from early to mid-October sowing. Thus, planting outside the optimum time seriously reduces fruit and/or oil yield. Seed sown in March in Ethiopia produced only small seedlings 2 months later, but seed sown in May following a rainy period produced similar seedlings in less than a month. When grown to produce foliage only, seed can be sown in the open from April to mid-September for harvesting as required, but must be protected in polythene tunnels for winter use. Time of sowing can also affect the incidence of insect pests or the degree of damage sustained, and this effect has been noticed on coriander in Africa and Asia (E.A. Weiss, personal observation).

Weed control

Two or three weedings are normally required during crop growth, but manual weeding is unpopular in many countries, especially eastern Europe, owing to the unpleasant odour from plants. Herbicides are an alternative and the following have been used successfully in coriander: fluchloralin, glyphosate, linuron, oxadiazon, prometryne, propanil, and propazine applied pre-emergence. The importance of field trials to determine coriander's susceptibility to herbicides was demonstrated in the USA, where it was damaged to some extent by the 14 compounds tested (McReynolds and Abraham, 1999).

Fig. 10.11. Young coriander plants.

Irrigation

Coriander is usually grown as a rain-fed crop, and normally irrigated only where water and a distribution system are already available. Saline or brackish water from bores is unsuitable, as data indicate that coriander has low salinity tolerance. Wholly irrigated crops are seldom profitable except when adjacent to high-value urban markets for the fresh herb or fruit. When grown under irrigation, one watering is given prior to sowing, then two to three applications at approximately 30-day intervals depending on local conditions and soil type. The last application should be made when fruit is filling. Irrigation to supplement rainfall generally increases fruit yield and essential oil content, but in Canada it decreased oil linalool content and 1000 seed weight, not noted elsewhere (Arganosa *et al.*, 1998b).

Intercropping and rotations

Coriander is usually grown in pure stands, but can be sown round field margins, or strip-cropped to suit local preference. In India, it is commonly interplanted in millet, cotton or beans, and less often with sugarcane or geranium; in Ethiopia, in barley, sorghum or teff. The main reason given by smallholders for interplanting local crops with coriander is an overall reduction in insect attack, but this is not supported by data. Coriander can also be underplanted in tall-tree plantations, in India neem, *A. indica* (A). Juss. Coriander can also be sown as a mixed crop with caraway, as noted.

Coriander is also grown in rotations – in India with garlic, cowpeas, onions, sorghum, wheat and irrigated rice, in the USA and Canada with small grains or vegetables and in Europe with canola and grains. In general, coriander should follow grain crops since the herbicides used reduce the subsequent weed level in coriander.

Pests

Coriander is frequently reported as being generally free from major insect pests. Whilst this may be correct in respect of a particular crop or district, it is not generally so. For example, coriander grown for domestic use in both Tanzania and Kenya was attacked by the usual range of pests found on other vegetables and spices (E.A. Weiss, personal observation). Inspection of small plots grown by Asians and Arabs produced the same result. Chemical control measures are uncommon, and pesticides seldom used unless coriander is sprayed as part of an overall crop protection programme. It is possible the unpleasant secretions produced by the plant may have some deterrent effect on mammals, but it is certainly not evident in respect of insects. Bees swarm on coriander immediately plants flower, and remain active throughout the flowering period.

In East Africa, lygid bugs, whiteflies, aphids and various caterpillars including the armyworm, *Spodoptera litura*, were commonly found on leaves and stems, and developing fruits were bored by a weevil. Larvae of the chalcid fly, *Systole albipennis*, can cause serious damage to flowering heads and fruit from West Asia to India and in Eastern Africa, but in Europe, *Systole coriandri* is responsible. Damage increases when coriander follows coriander. A range of pests similar to those recorded in East Africa has been reported from India, plus thrips and mites. In screening trials, some Indian cultivars were found to be resistant to aphids and whiteflies, including one Romanian introduction. The ubiquitous and polyphageous *M. persicae*, which heavily and continuously infested coriander in Brazil, was controlled using an easily produced spray made from local tobacco leaves.

Time of sowing can also influence the incidence of and damage caused by insect attacks. In India, the optimum time of sowing in Rajasthan is late October. Delaying sowing by 15-day intervals progressively increased the population of the coriander aphis, *H. coriandri*, and decreased seed yield. Damage was generally less on early rather than late-flowering cultivars. This aphid is a most damaging pest in India generally, together with the tetranycid mite, *Petrobia latens*. Several species of nematodes are reported from coriander roots, causing varying degrees of damage. Trials in India indicated that local cultivars had some degree of resistance to *M. incognita* (Das, 1998).

Plants in the highlands of Kenya were usually raided by birds immediately seeds began to ripen (E.A. Weiss, personal observation), and most local growers reported their unprotected plants were similarly attacked. However, in India an apparent attack by local sparrows on coriander was in fact found to be the reverse; birds were feeding on larvae of *Plusia orichalcea* and efficiently controlled the infestation.

Diseases

More information is available on the various diseases and the degree of damage caused to coriander than for insect pests. Environmental factors have a big influence on the incidence of disease and extent of crop damage. In India, for example, plant mortality and pathogen population were found to be highest when soil had a moisture content of about half field capacity, a pH of 5.8–6.9, and a temperature around 28°C. Fungicides used to control diseases may themselves cause problems; for example, endosulfan residues have been detected in herbage and seeds. However, most chemicals have a withholding period of about 10 days before fresh herbage can be sold for human consumption. Ripe, whole seed is seldom contaminated.

Many diseases of coriander are seedborne and in India 46 were isolated from local fruit. Most damaging generally are *Aspergillus flavus*, *Alternaria alternata*, *Curvularia lunata* and *Cladosporium oxysporum*. *Xanthomonas translucens* has been recorded on seed in Russia and Eastern Europe. Fungicidal or heat treatment of seed before sowing reduces the damage, but crop rotation may also be necessary.

The gall-producing *Protomyces macrosporus* causes damage to coriander plants from West Asia to South-east Asia, but is a major disease in India where yield can be reduced by 20% in a heavily infested crop (Lakra, 1999, 2000). Symptoms are small elongated swellings, 12 × 3–5 mm, on all plant parts,

and infected fruits are often larger than normal. Results of fertilizer trials in India noted that phosphate in the absence of nitrogen increased the damage, but manganese had the opposite effect. Some Indian cultivars are partially resistant. Another leaf blight is caused by *Colletotrichum gloeosporioides*. A bacterial leaf and umbel blight due to *Pseudomonas syringae* pathovar, designated *P. syringae* pv. *coriandricola*, is now a major disease in Europe, especially Germany where it first appeared in 1987 (Toben and Rudolphi, 1996). Whole fields can be affected, and *P. syringae* has also been recorded on coriander in Florida, USA.

A wilt disease due to *Fusarium oxysporum* f. *corianderii* can cause serious damage wherever it occurs; up to 50% of potential yield may be lost in India. An associated *Fusarium* spp. causes similar symptoms and damage in the USA, as does a complex of three *Fusarium* spp. in Argentina (Madia *et al.*, 1999). Plants are attacked at all growth stages and severity of infection increases with age. Typical symptoms are drooping shoots and leaves, progressive yellowing of foliage from the base of the stem upward, and badly affected plants finally wilt and die. A major outbreak of *Fusarium* wilt is often devastating, with a majority of plants in a particular crop killed, or so weakened that they produce no fruit. Some Indian cultivars are reportedly resistant. Another wilt of lesser importance, but which can result in severe damage to individual crops in warmer regions, is due to *Verticillium* spp.

Stem or root rots attack individual or small groups of plants in a field, and seedlings in beds. Most commonly noted is *Sclerotinia sclerotiorum*, which attacks a wide range of crops and occurs in most warmer regions, with *Phytophthora* spp. and *Ramularia* spp. less often recorded. However, in Russia and some eastern European countries, *R. coriandri* is considered one of the most damaging coriander pathogens. These diseases are most severe where soils become locally saturated, or plants remain wet for some time. *Rhizoctonia solani* is a serious disease of seedlings wherever it occurs, and *Curvularia pallescens* causes root rot of seedlings and young plants.

Mildews, especially *Erysiphe polygoni* or powdery mildew, frequently occur in coriander crops when environmental conditions are suitable, and a severe attack causes substantial seed loss through a reduction in leaf size and efficiency. Chemical control is effective (Ali *et al.*, 1999), and varying time of sowing can also influence the degree of damage, as in India (Kalra *et al.*, 2000).

Harvesting

When required as a green herb, young plants may be uprooted about 60 days after sowing, but in commercial crops several cuts are possible if suitable cultivars are grown. The third (at 13 weeks) of four cuts had the highest essential oil content in Egypt. Fresh herb yields of 20–24 t ha^{-1} are common in northern Europe. In smallholder crops grown for spice, entire plants are uprooted or cut at the base when the fruits of the primary umbel are ripe and shatter when touched. Plants are then dried and threshed.

When grown in Europe as an annual for spice, coriander is ready for harvest 90–140 days after emergence depending mainly on the season and cultivar, but in India and Africa some slow-maturing types may require 200–220 days to produce fully ripe seed. In western Europe, coriander is normally sown in April, flowers in June, and is ready to harvest in August. Highest yield was obtained in Virginia, USA, at 52 days from plants sown in summer (June–September), and at 73 days from plants sown in autumn (August–November) (Rangappa *et al.*, 1997). As noted, cultivars with a short growing season (early types) generally have larger fruit with a low oil content whilst long season cultivars (late types) usually have smaller fruit with a high oil content. Some cold-resistant cultivars are sown in the autumn for harvest the following summer.

Since umbels flower sequentially, fruits also ripen in sequence, and plants must be harvested when first fruits are fully mature to avoid loss through shattering. Chemical desiccation of standing plants prior to harvest has proved successful on other crops where shattering is a problem, sesame, for example (Weiss, 2000), but has been little used on coriander. A common dessicant is magnesium chlorate, and when applied to a Russian crop at the waxy-ripe stage, it significantly advanced ripening and made harvesting easier. Fruit quality and essential oil content was unaffected, although the linalool content fell slightly. Fruit ripens unevenly, as noted, but with little difference in oil content and composition if harvested fruit is allowed to dry naturally before threshing.

The optimum time to harvest varies widely according to cultivar and season, but in general plants are cut when at least half the fruits on the plants turn from green to grey-brown, and in Russia when fruits on central and first-order umbels become chestnut-brown. Deciding when to harvest a field of coriander to achieve maximum seed yield without loss of quality or from shattering requires considerable local experience.

Harvesting should commence early in the morning while dew is still on plants to reduce shattering. The crop can be cut with a mower or binder, tied into sheaves, then stooked to dry, or cut, windrowed and then threshed by a pick-up harvester. However, the last requires considerable skill in deciding the correct time to carry out the various operations. Whatever method of harvesting is used it is essential that fruit are fully ripe and dry before threshing to ensure the unpleasant odour has disappeared. Degree of ripeness of fruit at harvest is a major factor determining not only the quality of the subsequent spice, but also that of the essential oil. Immature seeds can contain more oil but its quality as defined by market requirements is low.

Yield of dried fruit varies widely, depending mainly on the level of management and cultivar. In Pakistan and India, the average yield for commercial crops is 750–1000 kg ha^{-1}, but over 2000 kg ha^{-1} has been achieved from crops in Mysore State. In Britain, yield may be up to 2000 kg ha^{-1}, in Yugoslavia 2000–2200 kg ha^{-1}, in Bulgaria and Russia 800–1000 kg ha^{-1}, but individual crops of newer Russian cultivars yield 3000 kg ha^{-1}. In a specific locality where rainfed crops usually produce 400–700 kg ha^{-1}, irrigated crops yield up to 2500 kg ha^{-1}.

In major coriander growing regions, breeding aims to increase fruit yield, or high fruit-oil content and to fix these characteristics in improved cultivars. These programmes have produced strains with yields of 3000 kg ha^{-1}, and fruit-oil content approaching 3.0%, with indications that these levels could be increased. Additionally, plants with specific fruit-oil characteristics have been identified.

Processing

Moisture content of harvested fruit can exceed 20% and must be reduced to 9–10% before storing. The cheapest way to dry fruit is by sun-drying, but care is necessary to avoid overexposure, which will reduce oil content and adversely affect quality. Fruits of a Russian cultivar dried naturally from a moisture content of 35% to 12% lost 4–5% oil, the main loss occurring in the initial stages. Solar-assisted or artificial drying of both fruit and herbage using cold or hot air is becoming more common, as in India (Pande *et al.*, 2000).

In general, supervised hot-air drying has no significant adverse effect on either oil content or quality. The two most important factors are initial moisture content and strict temperature control. The optimum air temperature is 80–90°C; above 100°C the oil acquires a burnt note; at temperatures over 140°C loss of oil by evaporation becomes significant. Fruits should preferably be dried naturally to a maximum 18% moisture before artificial drying, while fruit harvested with a moisture content below 18% and dried at temperatures below 90°C lose little volatile oil. Fruits with moisture content above 18% lost oil during artificial drying due to internal steam distillation, and the higher the drying temperature the greater the reduction. The loss is considered due to evaporation from internal oil reservoirs as walls and tissues are permeable in fruit with a high moisture content. During artificial drying, the volatile-oil content may initially increase slightly reaching a maximum between 80 and 90°C, apparently due to stimulation of the biosynthetic pathways responsible for volatile-oil formation. Dried

fruit may require recleaning and grading if necessary, before being packaged for storage.

Distillation

The small-fruited types of coriander grown in Europe possess the highest essential oil content, but as chemical composition and organoleptic properties of the distilled oil vary, cultivars whose fruit yields oil with specific properties may be selected. Western consumers generally dislike the characteristic odour and thus only fully ripe fruits are distilled for the spice oil.

Essential oil yield is basically dependent on whether fruit is crushed or distilled whole, and where a very high-quality oil is required, freshly dried fruit only should be distilled. Fruit should be crushed prior to distillation, and the degree affects distillation time; light crushing in New Zealand reduced the time and produced more oil (Smallfield *et al.*, 2001). Overheating during crushing and delay in loading the still, results in oil loss by evaporation. Distillation is complicated and requires skill to produce quality oil. Oil cells are located within the mericarp protected by a thick cell wall and high fat content, which tends to occlude the essential oil and reduce vapour pressure. During distillation, changes may occur in oil composition; small quantities of *p*-cymene are produced as an artefact, linalyl acetate and other esters are hydrolysed, and *d*-linalool becomes its optically inactive isomer geraniol. Manipulation of distilling time altered the linalool content of oil in New Zealand (Smallfield *et al.*, 2001).

The USSR was the largest producer of coriander oil, and continuous automatic distillation equipment was introduced to replace conventional batch methods. Throughput was reported at 100–320 t crushed coriander per 24 h, and 1 kg oil required 41–48 kg steam; volatile-oil content of distillation residue was about 0.01%. A large factory at Alexeev was stated to process about 200 t day^{-1} with an average oil yield around 1.0%, about 20% above steam distilling. Distillation waters were continuously extracted by an internal auto-cohobation system, and residues roller-milled and solvent-extracted to yield 17–18% fatty oil. Time of harvesting (maturity), method of drying, period and type of storage are major factors influencing the type of oil produced. Residues remaining after extraction can be used as stockfeed, and may contain the following (approximately): water 20%, protein 13%, fat 4%, nitrogen-free extract 21%, cellulose 34% and ash 7%. The actual content and ratios between constituents depends on the method of extraction and the raw material used.

While distilling remains the most common method of obtaining seed oil, the introduction of supercritical carbon dioxide extraction techniques offered a viable and more efficient alternative. Such oils are considered more accurately to reflect the true aroma and taste of the spice. Extraction plants can be manufactured with a selected capacity, and are easily installed although requiring skill to operate and a local supply of gas.

Storage and transport

Bulk storage of fruit is not recommended unless modern facilities including fumigation are available, as fruit quickly becomes infested with insects or attacked by rodents. Fruit for export may have to meet statutory requirements of importing countries regarding purity, foreign matter, or fungal contamination. Bulk shipments of coriander fruits from most producing countries are routinely fumigated on receipt by importers, and the fumigant used is important; phosphine, for example, reacts with the oil constituents α-pinene and linalool to form monomers and associated compounds (Padmanabha and Rangaswamy, 1996).

Whole, dried, sound fruits stored in new sacks in a cool, dry building, experience only minor loss and changes in oil composition. There is generally a slight increase in linalool and ester constituents up to 1 year, little difference up to 2 years, but storage for longer periods is not recommended. Reports on the extent of oil loss and changes in oil characteristics in stored fruit and type of packaging are conflicting, and it may be necessary to conduct specific trials locally to determine the most efficient to prevent loss and organoleptic deterioration.

In India, for example, coriander seed could be stored for only 6 months in polythene or cotton bags, while aluminium foil bags were superior for ground seed (Sanjeeve and Sharma, 1999).

Fruit damaged or split during harvesting and drying may lose up to 50% of its essential oil during 6 months storage due to oil diffusion, as 40% of oil is in the husk and the remainder dissolved within the fatty oil of the kernel. The greatest loss is in ground coriander, which, if left exposed to the atmosphere for more than a week, rapidly loses volatile oil with marked organoleptic deterioration. During prolonged storage of the spice, the free fatty acid content gradually increases and this value is a good indicator of age.

Coriander is liable to infestation by common storage pests, and as the fruit is attractive to birds, rodents, etc., protection or fumigation is necessary. For prolonged storage or transport, sealed containers are recommended, but movement in bulk should be avoided.

Products and specifications

The coriander plant yields four main products: fresh green herb, fruit as spice, and herb and fruit essential oils as flavourings.

Herb. The herb is widely available fresh in local markets worldwide, and also ground, frozen or preserved fresh in jars in Western supermarkets. Fresh herb packed in 30 μm biaxially orientated polypropylene (BOPP) and stored at 30°C had the longest shelf-life in Australia. However, the herb is not popular and is little used in Europe and North America, although an ingredient in many prepared foods. However, coriander herb or its substitute cilantro, later described, is probably the most commonly used flavouring from West Asia to China, Central and South America. Harvesting at a specific time can influence palatability of the herb, since the (E)-2-dodecenal content falls with seed formation (Potter, 1996).

Herb oil. Herb oil is obtained by steam distilling green herbage and consists almost wholly of aliphatic aldehydes, mainly decylaldehyde. About 65 constituents have been detected, some not fully identified. Oil produced from pre-flowering and flowering plants differs, but not enough to affect either odour or flavour. At the young green herb stage the distilled oil has a specific gravity (20°C) of about 0.850, optical rotation below +1°, a low alcohol and high aldehyde value. Oil distilled from fresh plants bearing fruits at progressive stages of maturity have intermediate values between those of herb and spice oil. Published data on herb oil are conflicting, for the same reasons noted later for seed oil. Herb oil is not available in commercial quantities, as the main component, decylaldehyde, is easily and cheaply synthesized. A full analysis of herb oil is contained in Potter (1996).

Seed. Dried whole, mature fruit is commonly known as coriander seed and will be so designated in this section, and is the spice of commerce and the kitchen. Correctly harvested and dried seed has a warm, distinctive, fragrant odour, and a mild, sweet but slightly pungent taste, reminiscent of lemon and sage. The mature round seed is often reddish-brown to brown if of Asian origin, or creamy to pale green if from Europe, but colour also differs with the cultivar.

Two basic types of seed are recognized, large and small. The large-seeded cultivars are grown mainly in tropical and subtropical countries, have a low essential oil content of 0.1–0.35%, and are used extensively for grinding and blending. Smaller-seeded cultivars from temperate regions including Europe have an oil content above 0.4% and are preferred for spice. The very small-seed cultivars from the former USSR and other European countries contain 1.0–2.0% essential oil. Moroccan and Indian coriander have oil contents usually less than 0.4%. Organoleptic qualities also differ with origin; Indian and Moroccan seed is generally considered commercially inferior to European, and Russian seed has a highly regarded spicy note. The main factor controlling quality is composition of the essential oil, discussed later.

Seed should be stored in an airtight container and freshly ground for domestic use to obtain the best odour and flavour. Whole seed is seldom used either domestically or industrially, and is usually available as a coarse or finely ground powder. The whole seed is used in pickling, chutneys and similar products, to flavour alcoholic beverages, including gin, vermouth and vodka, and crystallized in sugar as a sweetmeat. The major use of the ground spice worldwide is in flavourings, mixed spices and is an important ingredient, between 25 and 40%, of curry powders. It is extensively used in processed foods, meats, sausages, baked goods, including speciality breads, and sauces.

Coriander is credited with a multitude of medical functions, as an aphrodisiac and to treat a wide range of aches and pains. It has been a constituent of traditional medicines from antiquity, and seeds and foliage remain common ingredients of herbal remedies. Seed or extract is used medicinally, and considered a carminative, diuretic, stomachic, anticatarrhal, galactagouge, emmenagogue, and as a flavouring to disguise the taste of unpleasant medicines. A monograph on the physiological properties of coriander fruit oil has been published by the RIFM.

Fixed oil. A fatty oil can be expressed from seeds, and non-volatile, extractable minor constituents, some occurring in the non-saponifiable component, have been identified. The major fatty acid is petroselinic, which gives the oil physico-chemical properties suitable for special technical purposes.

Essential oil. The oil, oil of coriander, is obtained by steam distilling dried fully ripe fruits, and inclusion of unripe fruits or other plant parts produces off-flavours in the oil. The oil is a pale to colourless mobile liquid, with a characteristic odour; the taste is sweet, pleasant, somewhat woody-spicy and the floral-balsamic undertone and peppery-woody topnote are characteristic. The oil is placed in the sweet, delicately floral group by Arctander (1960). The flavour is mild, sweet and spicy-aromatic, somewhat warm and slightly burning. Moroccan and Indian coriander oils are generally regarded as having an inferior odour to European, and Russian oils have a spicy note.

The main oil component is linalool up to 90%, but variation is great; European oils are usually high in linalool and Asian oils low, while oil extraction methods also affect the level of linalool and other components. Over 200 constituents have been determined, some not fully identified (Lawrence, 1997a; Pino and Borges 1999b). For most European oils, the monotorpene hydrocarbon content is 16–30%, linalool 60–75%, and the remainder mainly other oxygenated monoterpenes; Russian oil traditionally exhibited a high linalool content of 69–75%. Indigenous Indian spice oils differ from European in possessing a lower linalool content and comparatively high ester content. The large-fruited Indian type also appears to be unique in that its oil contains thymol up to 7%. The major monoterpene hydrocarbon constituents of European seed oils are γ-terpinene up to 10%, α-pinene, limonene and p-cymene up to 7% each; the most abundant non-linalool oxygenated monoterpenes are borneol up to 7%, geranyl acetate up to 5%, geranial and camphor at 4% each, and geraniol up to 2%. Anethole, which is present up to 0.5% in some Bulgarian oils, may be absent in other European oils. While the main components account for between 95 and 97% of the oil, many minor constituents occurring at 0.1% or below are considered to give the oil its characteristic odour and flavour.

Composition and organoleptic properties of an oil are mainly determined by cultivar characteristics; both can be modified by other factors including maturity at harvest, changes during drying, storage and distillation, and in the oil during storage. The characteristic properties of European seed oils are generally similar: optical rotation between +9° and +13°; ester number around 22. Moroccan oils generally have an optical rotation below +10° and ester numbers up to 30; many Indian oils may have a refractive index below 1.400 and very high ester numbers. Coriander oil is frequently adulterated with lower-quality oil, sweet orange, cedarwood and anise oils, turpentine, anethole or other synthetic compounds, and all reduce oil quality. Synthetic linalool is also often added, and small amounts of other compounds as modifiers or extenders.

The oil is extensively used as a flavouring in a wide range of foods, soft drinks, confectionery, alcoholic drinks, sauces and seasonings of all kinds, and as a tobacco flavour; it blends well with a range of other oils, and is used to a minor extent in perfumery and cosmetics. The oil has antifungal, antimicrobial and antioxidative activity (Baratta *et al.*, 1998). Seed and essential oil in the USA have the regulatory status generally regarded as safe, GRAS 182.10 and GRAS 182.20.

Oleoresin. Coriander oleoresin is produced on a very small scale, most frequently on demand, but also by manufacturers for their own use, and little information is available on its characteristics and composition. The oleoresin is normally obtained by solvent-extracting seed and thus varies considerably in composition, since the proportions of volatile oil, fatty oil and other extractives depend on cultivar, fruit quality, processing method and solvent. Oleoresin commonly contains 90% fatty oil and 5% steam-volatile oil but up to 40%, and small amounts of distilled oil may be added to oleoresin to enhance the aroma and flavour; thus volatile-oil content of commercial oleoresins varies between 5% and 40%. The oleoresin is generally a direct substitute for the oil in the majority of its uses.

Unrelated species

Eryngium foetidum L. (syn. *Eryngium antihystericum* Rott.), also known as cilantro in the USA and the Philippines, is native to Central and South America, but is now widespread in the region and commercially cultivated in Mexico to supply fresh leaves to the North American market. The plant is also a common and popular green herb in the Pacific region where it was reportedly introduced by the Chinese in the mid-19th century.

Cilantro is an erect green herb up to 80 cm in height, the stem unbranched in the lower part, with spreading branches at the top. The leaves are simple, petiolate, with a somewhat fetid scent when bruised, and the flowers are small and greenish-white. Young leaves are harvested and should be marketed fresh as soon as possible since shelf-life is about 4 days. When packaged chilled or frozen for transport, leaves can be stored for 2 weeks at 10°C.

Persicaria odorata (Lour.) Sojak. (syn. *Polygonum odoratum* Lour.), the Vietnamese coriander of South-east Asia, is of no international significance. Fresh leaves have a slight coriander flavour.

Cumin

The genus *Cuminum* L. is now considered to contain a single species, *Cuminum cyminum* L., although a number of forms and a Sudanese species are noted in the literature. Cumin is another spice long associated with man, and its use and cultivation of great antiquity. Thus, the earliest records are from the ancient Mesopotamian civilizations of the Euphrates and Tigris valleys, where its fruits were highly prized as a flavouring, and in Pharoanic Egypt for its medical properties, as mentioned in the Ebers Papyrus of 1550 BC. Similar to coriander, it was well known to the Romans, who used the seeds as an alternative to pepper, or used them ground to a paste and spread on bread and meat. Cumin was associated with cupidity by the Greeks, and Roman emperor Marcus Aurelius (AD 121–180) was nicknamed 'Cumin' because of his avarice. Pliny in his *Historia Naturalis* of AD 77, gives an amusing use for cumin paste – it can apparently whiten the skin, and scholars used it to make their faces appear pale to convince their tutors they were spending long hours in study! However, cumin may have been difficult to cultivate as Theophrastus in the 4th century AD wrote, 'They who grow cumin say it must be cursed and abused while sowing if the crop is to be fair and abundant!'

Cumin is mentioned in the Christian Bible and Jewish Torah, and ground fruits remain an essential ingredient of many traditional dishes of the Mediterranean, especially Egypt, Turkey and the Levant. Cumin was well known in Rome's north African colonies, and traders carried the seeds along the great trans-Saharan caravan routes, although it is now seldom found in West Africa except in gardens. Cumin spread up the Nile valley to the Sudan where it continues to be sown by smallholders in the winter season to provide a flavouring, and on to Ethiopia where small-scale cultivation is widespread and fruits commonly available in local markets. In eastern Africa, it is grown only as a herb in gardens, and specimen plants flourish in the western Kenya highlands at 1500 m, but not at the coast (E.A. Weiss, personal observation).

Cumin has a long but unclear history in India, as it is mentioned in Ayurvedic manuscripts, but its early usage is obscured by translators confusing cumin and caraway, since both are known there as *jeera*, although caraway is little used in the region. This confusion continues in European countries as many modern cookery books wrongly substitute caraway for cumin in Indian recipes. Although cumin was carried to South-east Asia in trade by West Asian merchants it never became popular, and except in Indonesia and China is little cultivated. Introduced into North America by either the Spanish or Portuguese, it is not a generally used domestic spice, although as an ingredient of spice mixtures its use is increasing. It is believed to have been introduced by Roman armies to Europe including Britain, where it was popular in the Middle Ages mainly as an ingredient in traditional medicine. Its use as a spice steadily declined, and in many countries, including Britain, it now has only minor use in veterinary and herbal medicine. Although cumin plants may be confused with caraway, their seed tastes quite different.

Botany

C. cyminum L. (syn. *Cuminum odorum* Salisb.; *Ligusticum cuminum* (L) Crantz.; *Selinum cuminum* L. Krause), whose genus name is derived from the Greek *kuminon*, itself probably derived from the Babylonian *ka-mu-na*, is known in English as cumin, in German as *stachelkummel*, in French as *cumin*, in Italian as *cumino*, in Spanish as *comino*, in Dutch as *komija*, in Russian as *kmin*, in Arabic as *kimum akhdar*, in Indian as *jerra* or *zira*, in Indonesia and Malaysia generally as *jinten* with regional suffixes and in Chinese as *ma-chin*.

The basic chromosome number is x = 7 with cumin being a diploid, 2*n* = 14. Although a number of potential species have been described based on morphological differences, none are botanically acceptable. In the main areas of cultivation, many cultivars exist differing in fruit colour, flavour and essential oil content. Germplasm collections are maintained in India at various Institutes for Spice Research, where breeding programmes have produced high-yielding cultivars and fruit with increased essential oil content. Investigations into plant characteristics showed wide variability in plant height, branch numbers, yield and 1000-fruit weight, and showed that plant height was positively correlated with fruit-oil content but not herb-oil content. Additionally, assessing African and Indian cultivars indicated that the highest contribution to genetic diversity was made by seed yield followed by days to flowering. Seeds subjected to gamma-radiation produced plants that flowered and matured earlier (Koli *et al.*, 2000).

Cumin is an erect or suberect, glabrous, annual herb. The taproot is light brown, variable in length, thin, 3–5 mm diameter, with the number of laterals dependent on soil type. The root can be used as a vegetable although it may have a bitter taste. The stem is generally 20–50 cm but can be up to 80 cm, and is 3–5 mm in diameter; it is well branched throughout, subterete and finely sulcate, and in colour it is grey to dark green, sometimes brownish on lower stem, and grooves may become whitish with age.

The leaves are alternate, compound and blue-green; the blade consists of three slender filiform leaflets, each forked two to three times into filiform lobes up to 7 mm; the petiol is terete, 2–25 mm long, finely sulcate, with sheathing at the base (Fig. 10.12). A high content of phenolics in herbage is considered to inhibit fungal spore germination and mycelial growth (see Diseases). The application of growth regulators on seed or seedlings gave varying results, but generally influenced germination, growth and resistance to foliar disease (Omer *et al.*, 1997). Finely chopped thinnings are used as a food flavouring. In Peninsular Malaysia, leaves are often pounded together with leaves of other species and applied in poultices to treat a variety of diseases; in Ethiopia, pounded leaves are similarly used to treat skin disorders.

The inflorescence is a compound umbel up to 3.5 mm in diameter (Fig. 10.12). The peduncle is dark green, up to 7 cm and finely sulcate. There are often as many bracts as primary rays, with sheathing at the base, linear, and often three-forked into lobes 2–35 mm. There are two to ten primary rays per umbel, which are terete, unequal in length, up to 18 mm and finely sulcate. There are three to five bracteoles per umbellet, which are linear, up to 3–25 mm, with sheathing at the base, sometimes two to three-forked; there are three to eight secondary rays per umbellet, up to 6 mm. The calyx has five narrow, triangular sepals, up to 2.5 mm. The flowers are pinkish, reddish at the tips, with five equal petals up to 1.5 mm, and are bisexual, regular and protandrous. The five stamens have filiform filaments up to 1.5 mm; the pistil has a ribbed ovary, with two styles on a conical, persistent stylopodium, and the stigma is semiglobose, with the ovary inferior. Local cultivars can vary in numbers of primary and secondary rays in the inflorescence, in length of bracts and bracteoles and in colour of flowers and fruits. Flowers are basically wind-pollinated, with insects of secondary importance. Although cumin flowers are protandrous and cross-fertilization normal, self-pollination can be up to 70%.

The fruit is a yellow-brown slightly curved schizocarp, ovoid-oblong, 3.5–6.5 × 0.8–1.5 mm, tapering to both ends, and crowned by persistent sharp stylopodia and sepal bases (Fig. 10.13). There are eight primary ribs, with secondary ribs prominent and alternating with and wider than the primary, and whitish setose. The fruit varies from glabrous to densely setose; it is generally indehiscent, but splits into two mericarps under slight pressure. The mericarp is strongly concave ventrally, convex dorsally, usually bearing one oil duct (vitta) below each secondary rib and two on the commissural ventral side. Dried cumin fruits contain an essential oil discussed at the end of this section containing fatty oil to 22%, protein to 18%, 14 free amino acids, flavonoid glycosides, tannin, resins and gum.

Fig. 10.12. Flowering stems – cumin.

Fruit from Ethiopia had the following approximate composition (per 100 g): water 7 g, protein 18 g, fat 4 g, carbohydrates 29 g, fibre 17 g, ash 6 g (containing Ca 605 mg, P 570 mg, Fe 175 mg), niacin 8 mg, ascorbic acid 3 mg, foreign matter such as sand etc. 17 g; plus tannin, resin and gum (Jansen, 1981). Indian fruits contained the following trace elements (average content in p.p.m.): K 1069, Ca 603, Ti < 40, Cr < 35, Mn < 30, Fe 297, Zn < 25, Br 25, Rb 12 and Sr 57 (Joseph *et al.*, 1999). The seed testa is adnate to the

Fig. 10.13. Cumin fruit.

fruit wall, the embryo white with a conical radicle, and the endosperm copious, grey and fatty; 1000-seed weight is 5–15 g, depending on cultivar.

Fruit contains 2.5–5.0% essential oil consisting of hydrocarbons 30–50%, aldehydes and ketones 50–70%, alcohols 2–5% and ethers less than 1%, and their relative abundance depends mainly on cultivar and maturity at harvest. Cuminaldehyde is the most important constituent, at 20–75%, and also varies among cultivars, location and season. The characteristic flavour of cumin is probably due to dihydrocuminaldehyde and monoterpenes. The average characteristics of commercial oil are: specific gravity (25°C) 0.900–0.9350; optical rotation (20°C) +4° to +8°; refractive index (20°C) 1.4950–1.5090; almost insoluble in water, soluble in 10 vols 80% alcohol, ether and chloroform. The petroleum ether-soluble fraction reportedly has antioxidant activity in lard.

Ecology

The origin of cumin is still in dispute, but it is native to an area from northern Africa, via West Asia to Central Asia. Basically a plant of the warm temperate regions, cumin is so adaptable that a wide range of cultivars have developed and been given local names. Plants grow well in a temperature range of 10–24°C, below 5°C generally slows growth, and frost at any stage is usually fatal. In general, a growing season of 3–4 months of cool dry weather with full sunlight and low humidity, especially at flowering, produces the highest yields. Thus, cumin is suitable for dry areas with a moderate winter or up to 2200 m in the tropics. Although often found in semiarid areas, plants are not tolerant of high temperature, especially at flowering. In Gujarat and Rajasthan, it is usually sown in the middle 2 weeks of November when the daytime temperature has fallen to within the optimum range; in Europe it is sown in mid-March to mid-April depending on locality and cultivar.

Plants grow best on residual soil moisture rather than depending on rain falling during growth, but cumin is generally not drought-resistant. Cumin is usually sown towards the end of a rainy season in Iran, in Ethiopia after the main rains, which occur at different times around the country, in Egypt and the Sudan on a falling Nile flood. Cumin is basically a rabi (dry season) crop in India and planted to flower in February and March when humidity is low. Waterlogging is not tolerated, especially in the seedling stage. Cumin is considered a short-day plant but some local cultivars have extended its northern limits, to Scandinavia, for example, although long days are generally not suitable for spice production. Vernalization has not proved a substitute for selection to extend cumin's northward range.

Soils

Cumin grows well on a range of soils from light to more heavy clay loams provided the latter are well drained. Very sandy soils are less suitable as data indicate that plants on these soils have a greater incidence of wilt diseases. Although the reason is unclear, application of trace elements shown to be lacking reduces both incidence of the disease and its severity. Neutral soils are preferred, but a range of pH 6.8–8.3 is acceptable. Acidity tends to reduce yields more than alkalinity, but both soil types should be avoided.

Fertilizers

An indication of nutrient removal by cumin in India showed that an average yield of 550 kg fruit ha^{-1} removed 18 kg N, 2.2 kg P and 8.5 kg K, all ha^{-1} (Rethinam and Sadanandan, 1994), and additional data are shown in Table 3.4. A general recommendation for commercial crops in India is 100 kg ha^{-1} of a 5:8:6 NPK mixture in the seedbed, followed by a top dressing of 30 kg N ha^{-1} when plants are about 15 cm. However, this is often higher than most growers can afford, and the fertilizer rate should be related to soil analysis. When grown as a winter crop in Rajasthan, N at 30 kg ha^{-1}, half in the seedbed and half as a top dressing 30 days after emergence produces the highest seed yield. The method of application is apparently unimportant provided the coverage is even. Urea at 25–30 kg ha^{-1} is recommended for irrigated crops, applied via the irrigation system 40–50 days after emergence. A recommendation for smallholder crops in India is 15–20 t ha^{-1} FYM ploughed in well before the crop is sown.

Cultivation

Cumin is usually grown from seed, broadcast for smallholder crops and by a seeder for commercial crops. Propagation of cumin by tissue culture is difficult but possible (Hussein and Amla, 1998; Tawfik, 1998). Soaking seed in water for 24 h prior to sowing accelerates germination, but is not recommended for mechanically sown crops as seed tends to clog planter chutes. A seed rate of 15–20 kg ha^{-1} is common for smallholder crops, 10–12 kg ha^{-1} when mechanically sown, and mixing seed with an inert filler to increase bulk assists in achieving regular spacing and saves seed. A slightly high seed rate is preferred to ensure a full stand and

plants thinned to the desired spacing. In-row spacing of 12–15 cm and a row width of 15–20 cm is usual but up to 30 cm with a closer in-row spacing can be used, depending on the equipment available. Seed should be sown not more than 2.0–2.5 cm deep; below this emergence is irregular and erratic.

Emergence is epigeal and is usually completed in 14–28 days depending mainly on location and cultivar. In the Alemaya region of Ethiopia, altitude 2000 m, full emergence varied between 4 and 6 weeks. Once stem elongation begins, plants grow rapidly and flowering commences in 6–12 weeks, again depending on location and cultivar. The main flowering period in the northern hemisphere is late spring to early summer.

Weed control

Regular early weeding of cumin is very important and should begin as soon as plants are about 5 cm in height, with further weeding as necessary to keep the crop weed-free; in India, an additional two to three weedings is common to control *Cynodon dactylon* L. Pers., *Chenopodium* spp. and *Heliotropium* spp. Weeding is often accompanied by thinning to the local in-row width and thinnings eaten as a vegetable or sold in the local market. In Ethiopia, *Plantago psyllium* L. is considered a troublesome weed as it resembles cumin, but has opposite leaves and the inflorescence is not an umbel. Herbicides are seldom used in cumin as it is generally cheaper to control common local weeds in preceding crops. The following compounds have been used as directed sprays or pre-emergence: glyphosate, metalachlor and oxyfluorfen.

Irrigation

Cumin is very susceptible to waterlogging and thus raingrown crops may be severely affected in years of above average precipitation especially in the seedling stage. Crops sown at the end of a rainy season, as in the Sudan, Egypt and India, then irrigated to maintain soil moisture generally produce high yield of good-quality fruits. When grown as an irrigated crop, a pre-sowing watering to wet soil to about 30 cm is preferable to one post-planting to avoid standing water, then allowing the surface to dry for ease of planting. A low water velocity just after seedlings emerge is important to ensure they are not displaced. The watering interval is about 20 days initially increasing to 25 days, depending on soil type and cultivar, and the last when fruit has formed, as excess soil moisture and high humidity when fruit is maturing can reduce oil content and quality and increase disease damage.

Intercropping and rotations

Cumin can be grown as a monocrop or in rotations with wheat and legumes, as in India. It should not be grown for more than 2 years in succession on the same land to avoid disease build-up. Cumin can also be a catch crop, as some short-season cultivars can be sown and harvested in less than 3 months.

Pests

Insect pests in general are less damaging than diseases, and it is surprising that so few are of economic importance. The major pests of umbelliferous crops also attack cumin in the same area and are easily controlled by pesticides. Most important are aphids especially *M. persicae*, leaf-eating caterpillars especially *Prodenia litura* and *Helicoverpa* spp., and larvae of a chalcid fly, *Systole albipennis,* which feed on flower heads. Pesticide residues have been detected in cumin herbage up to 10 days after spraying, but not in seed.

Diseases

Cumin is susceptible to a range of diseases, and the most important are blight, wilt and powdery mildew, described later. Many are seed-borne (Kishor and Jain, 1999; Prabha and Bohra, 1999). Little selection for resistance has occurred, and information on resistance to a specific disease in regional cultivars is generally lacking. Diseases in fact can become so damaging as to preclude cumin cultivation in a particular region, Gujarat, India, for example, and, as noted, cumin crops should not be grown for more

than 2 years in succession for this reason. However, in Ethiopia cumin was reportedly virtually disease-free (Jansen, 1981).

A major disease is leaf blight caused by *Alternaria burnsii,* occurring in most cumin-growing areas. Symptoms are most obvious at flowering, and infected plants have small, whitish, necrotic areas especially on the tips of young leaves initially, which gradually spread to other plant parts. Affected areas enlarge, coalesce, become purple, brown, finally black and leaves wither. The infection spreads rapidly in cloudy periods or when plants remain wet, and a whole crop may become infected, with greatly reduced yield. Another blight due to *A. alternata* damages seedlings and young plants, and was shown to be seed-borne in India (Chand *et al.*, 1999).

Another important disease, which is also very widespread but easily controlled, is powdery mildew, *Erysiphe polygoni.* The initial symptom is a whitish powdery mass on leaves and stem, which if not controlled rapidly spreads to cover the whole plant. Such plants produce little or shrivelled seed. In India, the disease can reduce yield by half, and in some years is so severe that crops are devastated.

Wilt caused by *Fusarium oxysporum* and its pathovars especially f. sp. *cumini,* which is damaging on cumin crops from India to West Asia, is apparently extending its range westwards (Pappas and Elena, 1997). The disease is most damaging on young plants when soil remains saturated following emergence, or on older plants during periods of high humidity. The initial symptom is a general yellowing of plants in isolated patches in a field, which does not usually spread. Affected plants wilt and die, and the roots are very dark brown and rotten. The disease is apparently soil-borne, and soil type had some influence on incidence of the disease in Egypt (Omer *et al.*, 1997). Some Indian cultivars are reportedly tolerant (Mandavia *et al.*, 2000). A root rot due to *Macrophomina phaseolina,* which occurs worldwide and also has a wide host range, can be damaging in drier regions.

Harvesting

Cumin is ready to harvest about 4–5 weeks after main flowering, when first fruits become yellow-brown. Plants can be manually harvested by cutting or uprooting, and stooked to dry for several days but no longer than a week. If rain falls on stooks, these should be carefully turned to allow drying to continue. Plants can be threshed in the field or at the homestead. Mechanized crops can be treated as for coriander.

Average yield from smallholder crops seldom exceeds 100–200 kg dried fruit ha^{-1}, but up to ten times this can be achieved from well-managed crops. A small increase in the standard of field management, especially timely weeding, usually results in a large increase in yield. Yield from commercial mechanized crops appears to be very variable often from successive crops grown in similar seasons, in the range 1–3 t ha^{-1}, occasionally higher.

Processing

Harvested fruits should be cleaned and dried as soon as possible after threshing, as retention of herbage fragments or soil rapidly reduces fruit quality or introduces disease. Cumin can be artificially dried similarly to coriander.

Distillation

Distillation should be as soon after harvest as possible, as stored seed has a lower oil content. Fruits must be fine-milled and distilled immediately, as up to half the essential oil may be lost in the first hour following milling. Particle size, weight of charge and distilling time all influence oil yield and constituents. In Turkey, for example, reducing the particle size from 0.9 to 0.5 mm increased oil yield from 60 to 90%, but increasing batch weight from 35 to 160 kg decrease oil yield from 93 to 64% (Beis et al., 2000). Commercial oil distilled from previously ground cumin seed reportedly contains more cuminaldehyde than other aldehydes and no 1,4-p-methadien-7-al, while oil from freshly ground seed contains primarily the last named with much smaller amounts of cuminaldehyde.

Transport and storage

Sound, whole, dried fruits store well in bulk or in bags, but must be kept cool and dry.

Most smallholder crops are quickly moved to large traders or market as storage is usually inadequate. On local markets, cumin is often sold mixed with other seeds, in Ethiopia with fennel and dill, in Indonesia with coriander. Bulk shipments are often adulterated with other similar seeds, and must be recleaned before packing or processing. Seed may also be irradiated and a rapid detection method has been developed (Satyendra et al., 1998). Common storage pests attack cumin seed, especially the cigarette beetle, L. serricorne.

Products and specifications

The most important products are the dried fruit used as spice, and the essential oil used as flavouring. Cumin fruit is generally known as cumin seed and will be so designated in this section.

Seed. Whole, sound, dry seed has a strong and persistent aromatic odour, with a hot, spicy, somewhat bitter taste. Cuminaldehyde and the major menthadien derivatives have very similar odours and are mainly responsible for the characteristic scent of unheated whole fruit. When heated, the chief odour characteristics are due to 3-p-menthen-7-al in combination with the three other aldehydes. Cumin seed is a major ingredient of curry and chilli powders, speciality bakery products, processed meats and sausages, prepared condiments, pickles, soups, relishes, salad dressings, chutneys and sauerkraut, and is an ingredient in certain teas. Only small quantities are needed to impart the characteristic taste. The maximum level in the USA is about 0.4% (4308 p.p.m.) in soups. Cumin seed can be differentiated from caraway seed by being lighter in colour, bristly rather than smooth, and almost straight instead of curved.

In traditional medicine, seeds are mixed with other ingredients to treat diarrhoea and colic, and ground seed or extracts are considered to be a stimulant, antispasmodic, carminative, diuretic and emmenagogue and incorrectly an aphrodisiac. An interesting finding was that an aqueous seed extract inhibited the haemolytic activity of scorpion and snake venoms in Jordan (Sallal and Alkofahi, 1996). Pigeon breeders consider cumin seed in the diet of birds is a remedy for many diseases, especially scab.

Essential oil. Cumin oil is usually obtained by steam distilling the milled spice, although hydro-diffusion gives a higher yield, and more recently by supercritical gaseous extraction, which is claimed to give an oil closer to the aroma and taste of the spice (Eikani *et al.*, 1999). The oil is a mobile liquid, initially colourless, becoming bright yellow with age, with a strong, warm, pungent and persistent odour. Cumin oil is not categorized by A7ctander (1960). Average characteristics of distilled commercial oil are: specific gravity (25°C) 0.905–0.935; optical rotation (20°C) +4° to +8°; refractive index (20°C) 1.4950–1.5090; almost insoluble in water, soluble in 10 vols 80% alcohol, very soluble in chloroform and ether. The main constituents are cuminaldehyde and three other aldehydes up to 70% but this level can vary widely due to many factors including distilling method, cultivar and season. Dominant monoterpene hydrocarbons (total about 50–55%) are β-pinene, γ-terpinene and *p*-cymene, plus myrcene α- and β-phellandrene and limonene, with minor amounts of sesquiterpene hydrocarbons (Baser *et al.*, 1992; Pino and Borges 1999d). Cumin oil is sometimes adulterated with synthetic cuminaldehyde, which is difficult to detect. The oil is a raw material for the production of thymol.

The oil is a partial substitute for the powdered spice in similar food products, and the highest average maximum use level is about 0.025% (247 p.p.m.) in relishes and condiments. Oil is also is used in cosmetics and toiletries to scent creams and lotions, and in perfumes with a reported maximum use level of about 0.4%. A current use is in veterinary medicine. The cuminaldehyde content of the oil has strong larvicidal, fungicidal and antibacterial activity: however, undiluted oil has a distinct phytotoxic effect on mammals apparently not due to the cuminaldehyde content. An unusual oil use is as a bird repellent to reduce the nuisance of roosting birds on buildings (Clark, 1998). A monograph on the physiological properties of cumin oil has been published by the Research Institute for Fragrance Materials. The regulatory status of cumin and cumin oil in the USA is generally regarded as safe, GRAS 2340 and GRAS 2343.

Oleoresin. Cumin oleoresin or absolute is produced in very small quantities, either by the end-user or to order. Few details are available, but both are used chiefly as a food flavouring where very small and accurately controlled amounts are required.

Unrelated species

Bunium persicum (Boiss.) B. Fedtsch is native to the Middle East and especially numerous in south-eastern Iran where it is known as black cumin to differentiate it from cumin, usually called green cumin. Black cumin seed is harvested from the often abundant wild plants and sold locally. Seed is less bitter than cumin and is roasted and eaten.

Nigella sativa L. seed is also widely known as black cumin in the Middle East and Asia, while *P. anisum* L. is known as sweet cumin, *cumino dulce*, and cumin as *cumino aigro*, hot cumin, in Malta.

Dill

The genus *Anethum* L. currently includes two species and two putative species, the most important spice producer being *Anethum graveolens* L., dill. The generic name is derived from the Greek *anethon* as used by Aristophanes (448–388 BC), and probably derived from *aemi*, meaning I breathe. The original home of dill is unknown but it is probably native to the Mediterranean and West Asia. Similar to other umbelliferous spices, dill was widely distributed in antiquity, and is now grown worldwide. Escapes from cultivation have become naturalized in many countries, it is a common weed of cereals in southern Europe, and is reported to be a major weed of cultivated crops in Uruguay and Paraguay.

Dill was one of the herbs used as a flavouring in dynastic Egypt, the Mesopotamian civilizations, and by the Greeks and Romans as a flavouring and medicine. It was carried by Roman armies northward into Europe generally, and was commonly known as dill as early as AD 1000. By Medieval times, it had become a common culinary herb and pickling spice. When dill was introduced to Britain is unknown, but it was named anise by 14th-century herbalists owing to a translation error from *anethon*, an error that persisted for centuries; it was also called false fennel, and by the 16th century was widely grown and used. Dill was commended by the 10th-century Archbishop of Canterbury Alfric for its medicinal qualities especially to alleviate flatulence. He also recommended it should be 'grown by every household for hindering witches and countering their enchantments'. Dratons *Nymphidia* (AD 1627) contains the verse 'Herewith her Vervain and her Dill/That hindereth Witches of their Will.' Dill cucumbers were apparently a favourite of King Charles I according to his cook Joseph Cooper, in his *Receipt Book* of 1640.

A 15th-century herbal by Rembert Dodoens, *A Niewe Herball* (1578), recommends 'A decoction of the toppes and croppes of dill with the seed boyled in water and drunken, causeth women to have plentie of milke', and was probably the repetition of a similar recommendation by the Greek physician Dioscorides about AD 60. This belief is common today, as is an association with witchcraft. Parkinson in his *Paradisus* of 1629 says, 'It is also put among pickled cowcumbers where it doth very well agree, giving unto the cold fruit a pretty spicie taste or rellish', and pickled dill cucumbers remain popular. Culpeper in *The English Physician* of 1652, noted that, 'Dill added to oils or plasters dissolved impostumes of the fundement.' Dill is a favoured herb in Scandinavia, central and eastern Europe, especially Russia, the Balkans and Romania, and in Turkey and Iran.

Dill and Indian dill, sowa, are grown in India but the latter predominates, although seed of either are sold as dill in local markets. It is unclear whether sowa is native to India is or a well-adapted cultivar from an early and unrecorded introduction of European dill, i.e. via the Silk Road. Indian writers state it is referred to in early records including the Hindu *Ayerveda*. Interestingly, Watt (1908) makes only passing reference to dill or sowa as *Peucedanum graveolens* Benth. In India, the local dill was originally described as *Anthum sowa* (Roxb. ex Flem.) in the literature, currently as *A. graveolens* subsp. *sowa*.

European dill was deliberately introduced to the Jammu and Kashmir regions of India as a potential commercial crop in the early 1950s and later to other suitable areas. It has never achieved the same importance as sowa, a well-established local winter crop grown on a large scale. Sowa fruit oil generally has a lower content of the desirable carvone. Dill seed and herbage extracts are used in local medicines to treat ailments similar to those in European countries, which may support an early introduction. Like many other plants, it was probably introduced to Indonesia by early Indian traders, and is now commonly grown on Java, where it is known as *adas sowa*. Local sowa cultivars occur in China and Japan, and may also be introductions.

Introduced to the USA at the beginning of the 17th century, dill was listed by John Winthrop (1605–1676) as grown by early European settlers. Current cultivars were later introduced and are now commercially cultivated in the North Central and Pacific Northwest States mainly to produce dill oil, while Californian crops are grown mainly for the herb market, fresh or dried, and often sold as dillweed. Dill subsequently spread southward and is now naturalized in the Caribbean region and South America. References to dill in the text refer to European cultivars of *A. graveolens* and their derivatives unless otherwise stated, and sowa to Indian cultivars.

Botany

A. graveolens L. (syn. *Anethum arvense* Salis.; *P. graveolens* (L) Hiern; *Peucedanum sowa* (Roxb. ex. Flem.) Kurz; *Selinum anethum* Roth.) comes from the Greek *anethum* and the Latin *gravedens* for strong smelling. The common name dill is believed to be derived from the Norse *dylle* (to lull) or *tylle* (to sleep), referring to the believed carminative properties of seed extracts. In English, Swedish, Dutch and German it is known as *dill*, in Russian as *ukrop*, in French as *aneth batard*, in Italian as *aneto*, in Spanish as *eneldo*, in Arabic *shibith*, in Ethiopian as *ensilal, kamun* and many others, and in Indonesian and Malaysian generally as *adas* with regional suffixes. The somatic number of the genus is x = 10 and dill is a diploid with $2n = 20$. Germplasm collections are few and small; the largest is held at the North-Central Regional Station of the US Department of Agriculture, and small collections of sowa cultivars at various Spice Research Stations in India. Official dill breeding programmes are unreported, but seed firms in Europe market their own named selections to retain exclusivity.

Dill is a very variable species and several classifications have been attempted based on physical differences among various plant characteristics; none is botanically sustainable. A classification based on regional cultivars would be more useful. Several chemotypes of dill exist founded on essential oil composition: type one contains limonene, carvone, myristicin and dillapiole; in type two myristicin is absent; in type three limonene and carvone are the only major components (Badoc and Lamarti, 1991). Physically dill resembles fennel *F. vulgare* M. (later described), but they are easily distinguished by the shape and smell of bruised fruits – dill smells bitter and slightly pungent while fennel smells of liquorice, and dill fruits are lens-shaped and narrowly winged, while fennel fruits are less flat and wingless.

Dill is an erect, annual herb, with long, fine leaves, compound umbels of small yellow flowers, and small pungent fruit. All plant parts have the characteristic strong smell when crushed. The root resembles a carrot, being yellowish, 1.0–1.5 cm in diameter and 10–15 cm long, frequently with many laterals 10–12 cm long, the number, length and extent usually related to cultivar or soil type, i.e. more and larger laterals develop in light rather than heavy soils. Secondary roots can be restricted or spreading again depending mainly on soil type. Root essential oil may contain around 10% dillapiole and 20% pars-

ley apiole. In many countries roots and fruit, bruised and boiled in water, are applied as a poultice to ease the pain of swollen joints.

Dill usually has only one main stem arising from the root, unlike fennel to which it bears a strong resemblance, although rattooning may occur if the main stem is damaged early in plant growth. The stem is up to 1.5 m but sometimes taller, the base up to 12 mm diameter, blue-green, lighter at nodes and lightly furrowed with fine lighter-coloured vertical lines; it is terete, with the internodes often hollow. Stems are dichotomously branched, and branching can begin at any height, the number of branches varying from few to many depending mainly on plant spacing. Isolated or pot plants can present a bushy appearance and branch from low down on the stem. Environment normally has a greater effect than heredity. A relationship may exist between the length of the main stem and fruit yield, as some data indicate tall plants give a lower yield than shorter sturdier plants; the latter characteristic is also related to larger umbels. Smallholder crops are sometimes topped to increase branching and reduce height, and the croppings used as a flavouring. Mechanical pruning of commercial crops has been suggested to increase branching and thus the ratio of umbels to other non-oil-bearing plant parts, and thus increase herb oil yield ha^{-1}. However, published data indicate similar results can more easily be obtained by using other techniques, i.e. plant population. The effect of fertilizers on plant growth is discussed later.

The leaves are alternate, decompound tripinnate and sheathed, the sheath forming a cone with whitish margins. The petioles on the lower leaves are longer than the upper, which may be almost sessile. The leaf colour is blue-green or dark-green, lighter below, and the blade is triangular to ovate, 15 cm × 25 cm but can be twice this, pinnately divided into two to six pairs of primary pinnae and one top pinna, again pinnately divided two to four times into linear or filiform, acute lobes up to 60 mm × 1 mm (Fig. 10.14). Lobes on the lower leaves are usually broader and shorter than on higher leaves. Leaf size and number is variable especially between regional cultivars. The fresh and

Fig. 10.14. Dill plants.

less frequently dried leaves are widely used as a flavouring and fresh dill leaves have a bitter, pungent taste. Approximate composition of dry dill herb (per 100 g) is water 7 g, protein 20 g, fat 4 g, carbohydrates 44 g, fibre 12 g, ash 12 g, ascorbic acid 60 mg; energy value is about 1060 kJ.

The leaves contain more essential oil than stems, thus leafy types are to be preferred where herb oil is more important than fruit. The essential oil content increases gradually with advancing growth and development, peaking at the young-fruit (milky) stage, and then declining to seed maturity. Herb essential oil (dillweed oil) content varies greatly between 0.1 and 1.5%, and generally contains less carvone but more α-phellandrene than fruit oil. The biochemical pathways leading to oil synthesis and its composition have been partially determined (Faber *et al.*, 1997). Herb essential oil is discussed further at the end of this section.

The inflorescence is a compound umbel, 4–16 cm in diameter (Fig. 10.15). The pedun-cle is up to 30 cm, and the bracts and bracteoles usually absent. The primary rays (5–35 per umbel) 2–7 cm and unequal, with the outer longest; the secondary rays (3–35 per umbellet) are 5–10 mm. The terminal umbel on the main axis is very large in comparison with secondary and subsequent umbels. Umbels of the first order bear 20–40 umbellets, the latter progressively diminishing in number as the umbel ranking is lowered. A very wide variation in the size of primary and secondary umbels exists, and among local selections. A comparison of 17 cultivars in India showed an average primary umbel diameter ranging from 15 cm in a Greek cultivar to 28 cm in one from The Netherlands; secondary umbels ranged from 12–15 cm overall. Whilst umbel size is important in producing high seed yield, too large a diameter umbel may be a disadvantage. With sunflower, very large heads often have a large number of 'pops' at the centre (Weiss, 1966), and a similar trend can be found in very large dill umbels. Thus, a selection or breeding programme would probably achieve a substantial seed yield increase by concentrating initially on strains with medium-sized umbels and perhaps 20–25 umbels per plant. This would also assist mechanized harvesting by ensuring a shorter maturation period. Seed from highest ranking umbels frequently has the highest germination percentage, and plants from these seeds had a generally higher average yield than seeds from lower (later) umbels. The floral biology of dill and related species has been described in detail (Nemeth and Szekely, 2000).

The flowers are yellow, regular, bisexual, pentamerous, dichogamous and open centripetally; flowering is generally completed in 9–12 days, although this period is very variable. The calyx is normally vestigial, and the corolla with five petals is distinct, subovate, up to 1.5 × 1 mm, with the top strongly inflexed and notched. There are five stamens, the filaments are yellow, up to 1.5 mm, the pistil has an inferior, bilocular ovary and a fleshy, conical stylopodium bearing two spreading styles up to 0.5 mm. Anthers dehisce in the early morning and remain receptive until about midday, depending on ambient temperature. Out-crossing is high

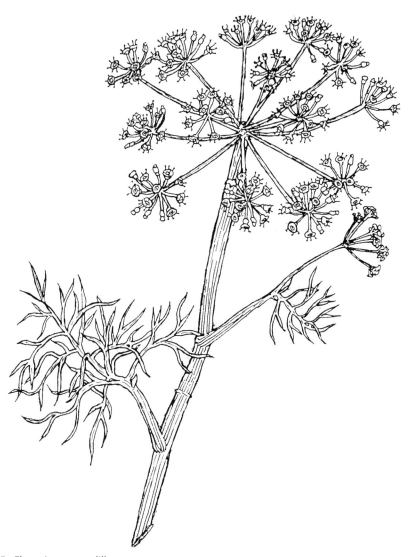

Fig. 10.15. Flowering stem – dill.

and is usually between flowers on the same umbel or adjacent umbels and selfing normally reduces fruit set. Growth regulators, including gibberellic acid, applied to young plants may accelerate flowering, and also affect fruit-oil content and carvone content. Flowers are attractive to insects, especially bees, which are thus the main pollinating agents. The main flowering period is July–August in Europe, February–March in India, and December–January in Australasia. In some African and Asian countries, the dried flowers are used to flavour stews, etc. Fresh and dried inflorescences are used in the ornamental flower industry.

The fruit is a lens-shaped schizocarp, generally 3.5–6.0 × 2–4 mm, light to dark brown with a paler margin (Fig. 10.16). The fruit splits at maturity into two one-seeded mericarps attached at their top to an erect, thin carpophore. The mericarp is flat, usually with three longitudinal prominent ridges and two flat, wing-like commissural ridges; on the commissural side, there are usually

Fig. 10.16. Dill fruit.

two dark brown longitudinal vittae, and on the dorsal side between each two ridges is one vitta. Fruits are crowned by the persistent short stylopodium and styles. The approximate composition of dry dill fruits is: water 7–9%, protein 16–18%, fat 15–20%, carbohydrates 25–35%, pectin 5–7%, fibre 20–30%; energy value 1275 kJ per 100 g. The fruit also contains dillanoside, coumarins, kaempferol, vicenin and other flavonoids, petroselenic and phenolic acids. These constituents can vary regionally and between cultivars, for example, in Poland (Zarwirska-Woitasiak *et al.*, 1998, 2000). The fruit contains an essential oil discussed at the end of this section. The seed is dark brown, very small, with the testa adnate to the mericarp; the hypocotyl is 5–25 mm, the cotyledons opposite, linear, 15–50 × 1–2 mm and entire, and the endosperm copious, grey and fatty. A detailed description of the morphology and histology of the seed has been published (Parry, 1962). The 1000-seed weight is 4–5 g.

Herb essential oil. The essential oil distilled from fresh herbage contains mainly carvone, α-phellandrene and limonene, plus about 50 other components, as noted in the section on Products and specifications. Oil is present very early in crop growth, easily demonstrated by crushing seedlings in the hand when a faint but characteristic odour is released. The oil content of stems and leaves increases very slowly during growth and remains at a very low level until plant senescence, or may decline after flowering, mainly depending on local strain. The greatest increase in oil content occurs during the reproductive cycle, and it is then that the characteristic oil develops. Oil content is greatest in umbels and maturing fruits, but oil composition varies as seeds mature, and composition may vary from primary and subsequent umbels as their fruit is more or less mature (Faber *et al.*, 1997; Mheen, 2000). During umbel development, a major change in oil composition occurs, initially from a majority of phellandrene and very low level of carvone, to a majority of carvone. Thus, an oil will be influenced by the degree of maturity of umbels and fruits, i.e. evenness of flowering. The latter can be influenced by plant density precluding branching, low soil moisture levels, or similar factors directly controlled by growers.

Carvone apparently increases in all plant parts with the onset of senescence, but the rate at which it increases and the time over which this increase occurs varies between strains. In New Zealand, carvone content peaked and then fell with increasing maturity of fruits; conversely, in Hungary, the highest carvone level was maintained through fruit maturation until harvest. Similar observations on carvone development have been made in other countries, and it has been suggested that the carvone development cycle could be used to differentiate regional strains.

The main characteristics of a USA commercial oil were: specific gravity (15°C) 0.880–0.910; refractive index (20°C) 1.4770–1.4840; optical rotation +86° to +105°; soluble in 0.5–1.5 vol. 90% alcohol.

Fruit essential oil. The essential oil content of fruit ranges from 2.5 to 5.0% in most commercial cultivars, the variation due mainly to season and fruit maturity at harvest. A higher oil content up to 8% may be achievable although some trials appear to indicate a negative correlation between seed size, weight and oil content, but this has still to be accurately established. A similar relationship has been noted with sunflower grown in East Africa (Weiss, 1966). The main oil components are carvone and *d*-limonene, and these two often account for 95% of the oil, α-phellandrene, β-phellandrene, limonene and anethofuran, which normally account for almost all the remainder, with major regional differences in abundance and ratios. For example, carvone content of fruit from the North American cultivars is usually 50–60%, from European cultivars 35–60%. The main characteristics of commercial oil from the USA were: specific gravity (15°C) 0.900–0.905; optical rotation (20°C) +70° to +80°; refractive index (20°C) 1.481–1.492; insoluble in water, soluble in 1 vol. 90% alcohol.

While regional herb oils could be identified on carvone content, as noted, the cycle is different in fruit, which has a far higher degree of variation over a given period. This may be relatively long or short, i.e. over the period of seed development or within a few days. In Bulgaria, there was a diurnal variation in carvone content and other major aroma compounds of seeds at the dough stage, with the highest level occurring between 21.00 and 24.00 hours. Similar effects were noted in Finland, India and Tasmania. The metabolic processes affecting other major oil components remain to be determined in detail.

Similar to herb oil, the greatest variation is between European and Asian oils, particularly in respect of dillapiole. However, general comparisons are of little value except to traders. Of more interest to agronomists and plant breeders is a comparison of oil obtained from dill plants from a number of countries grown at one location. Oil composition is also a regional characteristic, and the effects of climate will be discussed in the Ecology section.

Ecology

Dill is basically a cool-temperate plant, native to Europe (but oddly not to Great Britain) and the cooler regions of Asia. However, the plant is extremely adaptable, and within the rather extensive geographical region covered by this description, dill rapidly becomes naturalized as an escape from cultivation. It is also for the same reason a major weed of cultivation in Uruguay. It can also be grown in warmer areas to produce herb for flavouring. An annual rainfall of 1000–1700 mm is desirable mainly falling over the vegetative period. Higher rainfall may be tolerated on well-drained soils, but waterlogging is usually fatal, especially to seedlings.

Dill is primarily a summer crop of temperate climates, and thrives in full sunlight. A monthly average temperature of 14–20°C is suitable with a night temperature approximately half that of the daytime. A temperature above 25°C or below 10°C usually retards growth, above 30°C and below 7°C stops growth and if prolonged may be fatal. Data from phytotron trials in several countries indicate that a day/night temperature of 24/12°C accelerates maturity but reduces plant height and branch numbers compared with 18/12°C, which increases fruit yield. In general, long warm days increase growth and accelerate maturity, but not long cool days. Conversely, short cool days prolong growth, short warm days the reverse.

The presence of dillapiole in dill essential oil was widely believed to differentiate Asian from European oils, and while this remains generally true, it appears that total heat units or hours of sunshine on a crop during growth may also affect the dillapiole content. This may explain the presence of dillapiole in oil from European dill cultivars grown in India and the USA.

Some dill cultivars are cold-resistant, can be treated as perennials, and can overwinter if frost is not too severe, or with a snow covering, but in general frost will usually kill young and severely damage older plants. A frost-free period of 70–80 days is required to produce foliar herb, 120–150 days for fruit. A strong cultivar/region adaptation exists, as cultivars selected to produce fruit under very cold conditions are less successful in warmer areas. This is less obvious when only the herb is required. Since altitude reduces temperature, dill will grow well in the Indian and Ethiopian highland from 1500 to 2000 m, but will grow up to 3000 m; in Russia the highest oil content was in plants growing in the Caucasus mountains.

Dill is usually day-neutral, but the effect of temperature and other climatic components can be modified by daylength, and vice versa. Dill is considered a long-day plant in terms of flower initiation, with a critical period of 11–14 h, but it may well be that long days merely enhance anthesis. Dill's apparent photoperiodism may also be partially explained as the reaction by well-adapted cultivars to being planted outside their natural range. In Scandinavia, southern cultivars produced progressively less herbage the further north they were grown, and were outyielded by the local cultivar. Period to maturity was also very different; in the south it was 3 weeks shorter than in the north – 50 days compared to 74 days. Oil composition also differed; the northern plants had less α- and β-phellandrene but more p-cymene and benzofuran derivatives.

Light intensity and duration directly influence growth, with unshaded plants producing significantly more leaves, a larger leaf area and a higher herb yield than shaded plants receiving 30–70% natural light. Flower bud development in unshaded plants is faster and herb oil content several times higher than in shaded plants. Oil constituents are also affected by light intensity and/or duration, especially the dillapiole content, and the reasons for this remain to be fully determined.

Climate not only has a major effect on plant growth, it also affects the amount and composition of herb and seed oil. Thus, inter-country, and indeed inter-season, comparisons of oils in terms of yield and composition are of little value. Long-term averages are more relevant. The effect of environmental conditions can in fact be so important that individual dill crops must be regularly monitored when maturing to establish oil composition, to ensure harvesting at the correct time to produce the type of oil required by the market, or even a specific buyer. Oil content is less subject to variation, and once synthesized is little affected by weather conditions, which may have a substantial effect on oil composition. It is thus very important to establish the optimum sowing date in a particular area where dill is to be grown on a commercial scale, as it is usually unprofitable to compensate by applying additional nutrients, etc.

High winds when plants are well grown and branched can cause heavy damage, and hot winds during flowering or whilst fruits are maturing also reduce seed yield. In regions where dill is grown for its fruit and high winds are common at harvest, permanent windbreaks are recommended. Hail at any growth stage causes extensive damage.

Soils

The preferred soil for commercial dill production is a fertile rather sandy loam, and although plants will grow well on most soils, heavy loams or clays must be well drained, as dill plants are very susceptible to waterlogging. A soil reaction of pH 6.0–7.5 is the optimum, but a local cultivar may grow on somewhat acid soils of pH 5.6; below this growth and yield are poor. When grown for herbage, a wider pH range is acceptable and local experience is the best guide. In general, heavy clays, alkaline and acid soils should be avoided. The effect of soil salinity levels on dill in India confirmed that saline soils are unsuitable for commercial production (Ravender and Kundu, 2000).

Fertilizers

Fertile soils produce high seed or herb yields, and dill responds well to fertilizer applications, but the amount of individual plant nutrient applied depends on the end product required. A high herb yield requires a higher rate of N, for example, than for high seed yield. In general, N is of major importance in herbage and oil production, whilst P is equally important for seed and seed oil, while the level of K is variable. Data indicate that fertilizers can affect seed and seed oil composition, but not at the levels normally applied.

A wide range of the three main nutrients are applied as compound fertilizers; in Finland an NPK mixture of 60:40:40 is common, but trials indicated the optimum was 40:16:68; in Bulgaria 70:70:70 is used. The levels of K are higher than necessary if dill followed other well-fertilized arable crops. An NPK ratio of 60:30:30 would be a general recommendation until trials proved an alternative was necessary.

Whatever level of fertilizer is used, it must also be related to the system of crop management; for example, a direct relationship exists between time of N and water applications, which also directly influences herb oil character and composition, as is further discussed in the Irrigation section. An interesting finding is that dill, similar to caraway and fennel, absorbs heavy metals including cadmium, lead and zinc, but that residues detected in herbage and oil would allow human consumption (Zheljazkov and Jekov, 1996).

A major effect of N applied in the seedbed or early in plant growth is to prolong the vegetative cycle, promote green material production and thus herb oil yield. The level of P and K necessary is directly related to that required to ensure the optimum result from the N applied. Thus, the most important ratio is the optimum N:P levels to be applied in the seedbed, with an N top dressing if shown to be desirable. Few data comparing various types of N are available, although urea is normally chosen for irrigated crops, and it appears that the level of N application is more important than type. When dill is grown for fruit, seedbed N with or without an additional top dressing early in growth, usually promotes branching and thus the number of umbels per plant. Provided other nutrients and soil moisture are adequate, this results in more fruits per umbel and thus total oil yield ha^{-1}. The carvone level in seeds is then optimized by cutting at the correct time.

From an evaluation of trial results over a number of countries, 40–45 kg N ha^{-1} would appear to be a generally acceptable level until a higher rate is shown to be necessary; in Canada, 120 kg N ha^{-1} gave the highest herbage yield; in India generally 30 and 40

kg N and P ha^{-1} is a common rate, with a further 30 kg N as a top dressing 30 days after sowing; in the Punjab, 90 kg N ha^{-1} gave the highest yield (Randhawa *et al.*, 1996b); in Egypt, urea was superior to ammonium sulphate or calcium nitrate. Elsewhere date of sowing had a major effect on seed yield and seed oil content, and this affected optimum fertilizer levels.

Fruit from plants grown in soil with a low natural level of N or in control plants of fertilizer trials have lower oil content but with a higher carvone content. When grown to produce herb oil, lack of N results in lower oil content, but the carvone content is little affected. In The Netherlands, the level of N that gave the highest seed yield was 60 kg N ha^{-1}, but the highest carvone content at about 45 kg N ha^{-1} (Wander, 1997; Wander and Bouwmeester, 1997). Increasing N from nil to 160 kg N ha^{-1} on dill grown to produce herb oil in Finland had little effect on oil content, but increased total oil yield ha^{-1} as the total green material increased. However, the proportion of the four main constituents did vary; benzofuranoid was highest at moderate N level, 30–40 kg N ha^{-1}, whilst β-phellandrene, limonene and α-phellandrene were little affected, although the highest rate of N tended to reduce the level of these compounds (Halva *et al.*, 1987).

While there may be some variation in these effects in other areas, particularly in warmer regions where climate can affect nutrient uptake, carvone content is most affected by time of harvest, i.e. maturity of herb or fruit. Thus, the effect of fertilizers, N in particular, are of little importance in deciding the carvone content unless they significantly affect rate of growth and thus plant maturity.

Once the basic P requirement to ensure optimum uptake of other nutrients is satisfied, further application is unnecessary. In the absence of adequate P, response to other nutrients is low, and reflected in low-oil-content fruit. In India, increasing levels of N in the absence of P tended to reduce fruit-oil content, as did increasing levels of P in the absence of N. The effect of P on fruit-oil content and composition is extremely variable, but data from fertilizer trials indicate that plants receiving no

P had lowest fruit-oil content, although the effect on oil composition was not constant.

Potassium normally has a high residual effect, and if high levels are available dill plants will often take up luxury amounts with no significant yield increase. However, it has also been noted that high levels of N combined with low levels of P and K are often associated with higher incidence of leaf and stem disease. In India, where K was not applied, highest herbage and oil yield was obtained with N and P at 120 kg ha^{-1}.

Cultivation

The whole fruit may be sown, but should preferably be rubbed to free the two seeds, cleaned and large sound seed selected, since such seed normally germinates more quickly and emerges more evenly, both valuable attributes in mechanized production. Seed should be treated with a fungicide and an insecticide if the latter is normally required locally when sowing similar crops. Dill seed remains viable for several years if suitably stored, but preferably seed less than 2 years old should be used. Dormancy is unusual, but fresh seed may germinate erratically, and soaking seed in water for a day prior to sowing hastens germination. Seed may also be treated with growth promotants but their effect is erratic and usually unnecessary. Irradiation with low levels of gamma rays does not affect germination and may increase yield. In Bulgaria, irradiated plants made heavier growth than untreated plants and thus gave a higher oil yield ha^{-1}, but had no higher oil content. Neither was the relative ratio between major oil components affected.

Dill seedlings may be transplanted, but there can be high mortality, and for commercial crops seed should be sown where plants are to mature. When grown as an annual, seed should be sown in the open as early in the year as possible in Europe and North America, when the danger of frost has passed. A seedbed temperature 10–15°C ensures rapid germination, but cold, wet weather after planting extends this period. In warmer regions, sowing following the cessation of the main rains is usual, but dill seed will germinate at a soil temperature around 30°C provided soil moisture is sufficient to ensure seedling growth.

Seed can be broadcast, sown with small manual seeders, or direct drilled using modern equipment. Since the seed is small and expensive, accurate adjustment of machinery is important to reduce loss or an excessive seed rate, and mixing seed with an inert carrier to increase the volume passing through seeders, or pelleting, will assist even distribution. Seed should normally be sown 2–6 cm deep in rows approximately 25 cm × 15–20 cm in-row spacing for herb oil, and 45–75 cm × 10–15 cm to produce seed or seed oil, the optimum spacing to be determined by local trials since it directly affects profitability, or by equipment available. A seed rate of 5–10 kg ha^{-1} of clean selected seed is usual depending on cultivar, and sowing slightly more seed than the local rate will ensure a high plant population. Smallholder crops grown for seed are commonly sown in 25–30 cm rows with plants thinned to 10–15 cm, the thinnings sold in local markets.

Dill seed normally germinates in 7–14 days, but up to twice this depending on the weather following sowing and the cultivar, as considerable cultivar differences exist. Germination is epigeal, the hypercotyl is 5–25 mm, and the cotyledons are opposite, linear, entire, and 15–50 × 1–2 mm. Time of planting has a significant effect on plant growth and seed yield, but is more critical when the latter is the main product required. In Europe, sowing in March–April is common, but occasionally dill is sown late in the year to be harvested the following year. This is a hazardous choice, and is not to be generally recommended unless known to be successful for local conditions or cultivar. Flowering occurs about 8–12 weeks after emergence, but varies with the cultivar and seasonal conditions; in Russia, this variation ranged between 50 and 80 days. Flowering and fruit maturation periods are extended as umbels flower and ripen sequentially, as noted in the Botany section. Because dill flowers are strongly protandrous, cross-pollination and fertilization is high, and bees are the predominant insect pollinators.

Time of harvesting creates a major difference in oil composition between fruits and pedicels. At full fruit maturity, phellandrene is present in seeds at a very low level, but at a high level in pedicels, limonene is similar in both, and carvone is very high in seed but negligible in pedicels; the reverse applies to anethofuran. Benzofuranoid (anethofuran) is usually highest in umbels at flowering, but declines as the seed matures. The specific period in plant growth at which the required peak occurs is extremely variable, and appears to be directly related to the existing environmental conditions.

These basic characteristics have significant practical implications for growers. The ideal dill plant should be short, sturdy, with few branches, having few large umbels, with a non-shattering characteristic. Plants could then be sown at a high density to encourage level growth, would mature more evenly, and thus a specific oil type could more easily be obtained by harvesting at a particular stage in plant development. Sowing at intervals would also allow more efficient use of stills and harvesting machinery, as harvesting bottle-necks would be avoided. The increasing consumer demand for organically grown produce has stimulated the hydroponic production of green herbs, including dill, near large towns in many European countries, including Poland (Rumpel and Kaniszewski, 1998) and organic field production in Germany (Hoppe, 1996).

Weed control

Two or three manual or mechanical weedings are normally required, or herbicides may be used. The following have been successfully used pre-emergence: entholfluralin, fenuron, glyphosate, linuron, prometryne and trifluralin.

Irrigation

Where the type of oil produced for local consumption is not usually critical, as in India, plants are irrigated as required to maintain growth or when water is available. When grown primarily under irrigation, a pre-planting application to wet the seedbed to about 15–20 cm is preferable to watering shortly after emergence since seedlings are susceptible to root and stem diseases. The use of saline water to irrigate dill is of general importance, but especially in India where the area of salt-affected soil is increasing, and results are encouraging (Nayak *et al.*, 2000).

For commercial dill crops, the amount of water applied and timing of applications is critical. Since stems and older leaves have a very low oil content compared with umbels and maturing seeds, control of plant growth by judicious N application and timing of irrigation will produce the highest yield of high carvone content herb oil. In New Zealand, seedbed N and early watering, with a second watering following main flowering, gave highest oil yield and highest carvone content. In contrast, N top-dressing and irrigation prior to flowering reduced oil yield and carvone content in the green herb. Thus, where a high carvone content herb oil is required, timing of N application is important.

Intercropping and rotations

Dill may be intercropped with longer-season crops, for example, sage in Hungary (*Rauvolfia* spp.) and mint in India, but in general dill should be grown in pure stands to achieve the best results. In domestic gardens, dill is usually intercropped with onions, parsley or carrots. Dill is normally grown as a pure crop in Europe and North America, and an interval of 4 years is recommended in any rotation.

Little information is available on the comparative profitability of dill oil or seed production in comparison with other local crops. In New Zealand in the mid-1980s, there was a gross return to dill oil producers of NZ\$1420–1720 ha^{-1} compared with NZ\$760 for barley grown on the same land in the following year. In Canada, an oil price of C\$4.50 kg^{-1} and an oil yield of around 100 kg ha^{-1} was considered the break-even level in the mid-1970s.

Pests

Dill is frequently attacked by insects locally important on other local horticultural crops or indigenous *Umbelliferae*, but it has been remarked that dill appears to generally suffer less damage at the same level of infestation. The reason for this has still to be determined, and specific trials may prove the observation incorrect. When grown in proximity to carrots, parsley or celery in Europe, dill roots can be mined by larvae of the carrot root fly, *Psila rosae*, a major pest of many vegetable crops. Pentatomid bugs cause damage in warmer regions from southern Europe, North Africa through to Asia. Most often mentioned are *Graphosoma* spp., especially *Graphosoma semipunctatum* in North Africa and *G. italicum* in Iran, *Tholagmus flavolineatus* also in North Africa, and less often *Calidea* and *Nezara* spp.

White flies occur on almost all dill crops, those frequently mentioned being *Trialeurodes* spp. Aphids are commonly reported from dill, including the widespread *Cavariella aegopodii* with *Dysaphis* spp. causing heavy damage in the Mediterranean region, and *Dysaphis foeniculus* in East Africa and Ethiopia. The chalcid fly, *S. albipennis*, is an important pest in those countries where it occurs. Larvae feed on and destroy seed heads, and can reduce seed yield in India on sowa by 25% when uncontrolled. Nematodes have been reported from dill roots, but frequently the species is not identified, nor the degree of damage.

Diseases

Climate and soil conditions have a major effect on the incidence of, and the damage caused by, disease in a dill crop. Both are higher in cool wet or warm moist conditions, with climate being more influential in Europe and North America, and soils in West Asia through to China. Dill is susceptible to a range of pathogenic fungi, but damage is frequently limited in respect of root or stem rots to specific areas within a particular crop, and only leaf diseases are generally more widespread. The relative importance of the various diseases in Germany has been investigated (Kusterer and Gabler, 2000). The most common or damaging diseases of dill are described below.

Most commonly noted of the stem and root rots are those caused by *Fusarium* spp.

generally, especially *Fusarium culmorum* in the USA. Prolonged wet weather in temperate climates encourages spread of the disease. The main symptom is a general yellowing and wilting of plants, which often die (see cumin diseases). Another damaging root rot is due to *Phoma* spp., usually *Phoma anethi*. Symptoms vary, but lower leaves generally show interveinal and chlorotic areas, and dead plants have severe rotting of stem bases and roots. Damping-off caused by *Pythium* spp. is one of the commonest diseases of vegetable crops, with *Pythium mastophorum* widespread in Europe and Russia. *Rhizoctonia* spp. (see coriander) cause substantial damage to seedlings and young plants generally, and a severe infection can destroy a crop. Both *Pythium* and *Rhizoctonia* spores are soil-borne and seed dressings and rotations are necessary where the disease is well established. *Alternaria radicina* can cause severe damage to dill seedlings in Germany and Eastern Europe.

Scab caused by *Fusicladium* spp. can cause severe damage if uncontrolled, particularly in Eastern Europe and especially Hungary and Russia. Leaves become covered with scab-like areas, wither and die, and severely infected plants are defoliated. The most common leafspot of dill in Europe is *Ascochyta anethicola*, with the rust *Puccinia petroselini* occurring more widely. Wet weather at flowering favours infection by *Botrytis* spp. and *Alternaria* spp., head rots, which prevent fruit set or destroy developing fruits. Powdery mildew *Erysiphe polygoni* can occur in almost any dill crop at any growth stage when circumstances favour its spread. Another mildew, *E. anethi*, attacks dill between the early seedling stage and flowering. Both are easily controlled by spraying. A virus obtained from stunted bushy plants with narrow, yellow to bronze leaf lamina in Italy was identified as the celery mosaic polyvirus (CeMV) (Bellardi *et al.*, 1998).

Harvesting

When required for fresh herb, plants are cut either by hand or a specially adapted mechanical harvester. Cultivars chosen for large-scale production remain a uniform green with no yellowing, and are late flowering, as in Germany (Bomme, 1997). Dill is normally cut for herb oil at or during bud appearance on primary and secondary umbels, about 40–60 days after sowing. In general, this is when herb-oil content and oil-carvone content are greatest. Since this period is very short, successive plantings should allow the optimum harvesting period to be extended. Significant differences exist among regional cultivars in rate of vegetative growth, oil content, retention of maximum oil content and level of carvone content during maturity; for example, Hungarian cultivars retain their maximum oil and carvone content longer than New Zealand types. Management techniques affect plant growth, as previously noted, and thus a specific crop can deviate widely from the local average. Continuous sampling over the budding and pre-flowering period is necessary if a high carvone oil is required, since it is a most important factor in determining crop profitability.

If oil with a high phellandrene and limonene content is required, plants must be cut to include leaves and immature umbels, generally just after flowering commences. Since stems and older leaves have a low oil content, the height of cut should be adjusted to leave as much of this material as possible in the field. This will reduce the amount of material passing through stills and significantly lower cost of oil production. In Finland, in a crop yielding 8000–14,000 kg fresh herb ha^{-1} the proportion of stems fell from 45% when plants were cut 7.5 cm from the ground to 32% at 15 cm. However, in other countries little difference occurred in total oil yield between various cutting heights. Local trials are thus essential to optimize still usage. Yield of green herb and oil content at harvest is so variable, even when the same cultivar is grown in succession, that inter-regional comparisons are of little value. In Bulgaria, average fresh herb yield is 8000–10,000 kg ha^{-1}, with an oil yield of 1.0–1.5% (9000 × 1.25 = 112 kg oil ha^{-1}).

When grown to produce fruit for spice, plants should ideally be harvested when fruits are fully developed but still green, and before they become dark grey, as oil content over this period can fall from 2.5–4.0% to 1.5–2.0%. Since this is difficult as fruits mature over a period, local growers rely on their experience and cut when the fruit on primary and secondary umbels is turning brown or red (depending on local cultivar), and over the years this usually produces a high average fruit and oil yield. The optimum time to cut Indian dill, sowa, is slightly earlier when fruit are mature and greenish-yellow and before they begin to turn grey. At this period dillapiole is usually low in relation to more desirable compounds.

Plants are usually cut, bound in sheaves and stooked to dry before being threshed. It is also possible to cut plants rather higher than usual to leave a stubble, windrow plants and allow them to dry on top of the stubble, as is often done with oilseed rape, canola. Plants are then picked up by binder or may occasionally be threshed directly from the windrows by a combine fitted with a pick-up reel. Both the latter techniques require skilled judgement and the right weather conditions. It is usually safer to stook and thresh. To avoid shattering loss, plants should be cut in the early morning or evening when wet with dew. Shattering is also reduced by harvesting before fruits are fully ripe, but when the desired oil composition has been reached.

Threshing is often in the field, with stooks carried to the machine. It is advisable to stand the thresher on, and cover the ground in its immediate vicinity with, tarpaulins or heavy-duty plastic to collect seed lost during this operation. If sheaves are to be transported, trucks or trailers should be similarly lined to prevent seed loss. Smallholder crops are often threshed by hand in the field or at the homestead, and a suitable hard, clean surface, which can be swept, or a tarpaulin should be used.

Yield of fruit from smallholder crops when grown in pure stands is low and seldom exceeds 750 kg ha^{-1}, usually half this in India. Commercial crops yield 0.2–1.5 t ha^{-1}, and at least 1.25–1.5 t ha^{-1} is necessary for mechanized crops to be profitable. Oil yield obtained varies from a low of 25 kg to a high of 200 kg ha^{-1} with 125 kg ha^{-1} considered acceptable, and a carvone yield of 40 kg ha^{-1} obtained in The Netherlands (Mheen, 1996). Similar to herb oil, good management is the main factor producing high seed or oil yield.

Processing

Dill herb is marketed fresh for culinary use, and to maintain its keeping quality should be sealed in polythene bags and stored at 6–12°C. For domestic use, fresh herbage can be chopped and frozen with water in ice-cube trays. To produce dried herb for commercial sale, the drying temperature should be 80°C initially, lowered to 40°C halfway through the drying process, and the product have a minimum essential-oil content of 0.15% and maximum water and ash contents of 8% and 6%, respectively. The effect of freeze- or air-drying on aroma content is considerable; fresh dill averages ten times and freeze-dried herb four times the aroma compounds in dried herbage.

Following harvest, smallholders' plants are dried under shade, threshed and the fruits spread in a thin layer and turned frequently until dry. This operation is fully mechanized by commercial seed producers who either hot- or cold-air dry fruits, and the remarks relating to coriander are applicable. Fruit should be cleaned, dried and graded as necessary to produce a uniform sample for retail sale, unless it is sold in bulk for distilling, when appearance is less important.

Distillation

Distilling green herb or fruit is uncomplicated, and no special techniques or equipment are necessary other than those noted under distilling in the coriander section. Time of harvesting directly affects the type of oil produced, and once plants are cut it is difficult to alter the basic character of the oil by varying distillation procedures. This applies equally to the green herb or mature plants grown for seed. It is, however, possible to select different oil fractions by segregation during distilling; for example, the majority of carvone in green herb distils over early in the cycle.

For herb-oil production it is recommended that only as much fresh plant material should be cut as can be distilled during 1 day. Prolonged drying in the field results in considerable loss of oil by evaporation, especially of the more volatile terpenes; too long exposure allows fruits attached to herbage to continue to ripen and the oil thus obtained approaches the composition of dill seed oil, undesirable in dill herb oil. Thus, herbage should be distilled as fresh as possible. The effect on oil content and oil composition of wilting herbage prior to distillation has been little investigated although many growers normally allow plants to wilt for up to 24 h prior to distillation. In Bulgaria, wilting for up to 72 h reportedly had no significant effect on oil composition and characteristics; in Canada, field-wilting for 14 h had no effect on oil yield but carvone content was slightly increased. After oil extraction, the residue contains approximately 15% protein and 16% fat, and can be used as cattle fodder, wet or dried. Fruit is distilled in a similar manner to coriander, and when high-quality oil is required, fresh fruit should be distilled.

Storage and transport

Fruit (seeds) once threshed should be cleaned, dried and stored in new sacks or ventilated bins, and provided seed is correctly stored, little oil loss or change in composition occurs. Herb or fruit oil to be stored for extended periods must be kept in full, sealed, airtight and light-proof containers; the last two factors have a significant adverse effect on physico-chemical characteristics, while temperature has a minor effect.

Products and specifications

The dill plant produces fresh or dried herb and fruit (seed) as spice, and a herb and seed oil used as flavouring. Dried seed is the most important in international trade as spice and as raw material for seed oil produced in consuming countries.

The flavour of fresh herb is likened to a spicy parsley, and quite different to that of the rather bitter fruit. Dried herb is available in retail packets but does not compare with the fresh herb. The herb is mainly sold fresh in markets, but frozen or freeze-dried for the supermarket trade. Regular users should produce their own, as dill is easily grown in pots and preserved when necessary in vacuum packs rather than glass jars. Chopped green herbage is sprinkled on foods, especially fish dishes, used in sauces and may also be used to flavour pickles, vinegar and vodka.

Herb essential oil. Herb oil, also known as dill herb oil, is obtained by steam distilling fresh or wilted herbage, with oil yield and characteristics varying with plant age and the dominant plant parts in the still charge. The oil is a mobile, almost colourless to pale yellow liquid, with a strong but not pungent, fresh sweet-aromatic, pepperish odour; the taste is warm, pleasant, sharp spicy but not pungent, reminiscent of anise. The oil is placed in the fresh-peppery, warm, light group by Arctander (1960). Some 50 constituents of herb oil have been determined, not all fully identified. The main components are α-phellandrene, β-phellandrene, limonene, anethofuran and carvone, and these five normally constitute about 90% of the oil. Other constituents include terpinene, α-pinene, dillapiole, myristicin and two coumarans (Lawrence, 1996b). Commercial herb oils usually contain less than 45% carvone, frequently between 25 and 35%. The ratio between the main aroma compounds varies with the country of origin, but is often 'adjusted' to an accepted standard. The main characteristics of a commercial herb oil from the USA are: specific gravity (15°C) 0.880–0.911; optical rotation +86° to +105°; refractive index (20°C) 1.4770–1.4840; soluble in 0.5 to 1.5 vol. 90% alcohol.

Herb oil is frequently adulterated with *d*-limonene, a by-product of sweet-orange oil concentrate, and with synthetic carvone. Considerable variation exists in regional oils, but in general the greatest difference is between European and Asian oils. European herb oils are often low in carvone, below 30%, while North American oils are usually above, and the US-EOA (Essential Oil Association) specification stipulates a carvone content between 28 and 45%. Herb oil is mainly used in the food industry for flavouring and seasoning, especially for pickling (virtually all American oil is so used) and has largely replaced the whole herb. The oil is also used in perfumes (maximum reported use 0.4%), and as a fragrance component in cosmetics, soaps, creams, lotions and in detergents. A herb oleoresin is also available, usually on demand, with a volatile oil content of 30–40%. Herb oil is mainly produced in the USA, and seed oil in the EU.

Fruit. The dried fruit is the spice dill seed, and will be so designated in this section. The seed has a warm, sharp, slightly bitter taste, and an aromatic smell, somewhat similar to caraway. Dill seed is a minor culinary spice, but in many countries is less popular as a flavouring than the green herb. It is used whole as a pickling spice for a range of products especially dill cucumber and dill vinegar. The seed is usually sold whole in retail packs, and should be kept in an airtight opaque container and freshly ground for domestic use in meat dishes, bread and pastries. The seed, either whole or ground, has a major use as a flavouring in the food-processing industry, and is added to a wide range of products; highest average maximum use level in the USA is about 3%. Dill is only a minor constituent of curry and similar mixtures, and can generally be ignored in favour of cumin.

The seed is considered a panacea in Asia and China, although with little medical basis for the belief, and is used to treat a range of illnesses including haemorrhoids, bronchial asthma, neuralgia, colic, genital and other ulcers, and dysmenorrhea. In Europe, dill was used in herbal remedies from very early times; the Greek physician Dioscorides recommended a decoction of the seed to increase milk flow in women, for stomach pain, vomiting and flatulence, all probably reflecting long use to alleviate these complaints. In Western pharmacopoeia the seed is known as *Fructus anethi*, and an extract is stated to have carminative, stomachic, stimulant and diuretic activity. In Ethiopia, mature dry fruits are occasionally roasted then brewed in hot water as a drink, while whole plants or fruit are used to flavour the local alcoholic drink, *katikala* (Jansen, 1981).

However, recent research indicates that eating foods containing dill seeds can cause allergic reactions in susceptible persons (Chiu and Zacharisen, 2000).

Seed essential oil. The oil, commonly known as dill oil, is generally steam distilled from the crushed, dried fruit. The oil is a very mobile liquid, almost colourless to very pale yellow when fresh, becoming darker with age. The odour is light, pleasant, fresh but hot, strong spicy, resembling caraway, with a warm, sweet, aromatic but more burning, camphorish taste than herb oil, and is placed in the caraway group by Arctander (1960). The oil normally contains at least 50%, sometimes 60%, carvone (Lawrence, 1996b; Pino, 1999a). Carvone content in dill oil ranges from 50 to 60% in the USA and 35 to 60% in Europe. Wide differences in published data of essential oil composition can be partially explained by varying extraction or distillation methods, and by the cultivar providing the raw material.

While seed oil can be used in similar applications to herb oil with highest application rates of 0.075% in snack foods, it is less popular. Overall usage is small in proportion and could decline further in favour of herb oil, as apparently an oil virtually indistinguishable from seed oil is obtained by adding additional carvone to herb oil. Should this become general practice and the product acceptable to food processors, then herb oil production could increase at the expense of seed oil, as it is easily distilled in producing countries. Another competitor is caraway oil, as there is little flavour difference between small amounts of dill and caraway oils. In the USA, 'rye bread' is flavoured with carvone or caraway, and thus carvone is often popularly said to smell like rye bread. Dill oil is an important ingredient of certain medicines, particularly those used to treat gastro-intestinal disorders. It remains a common constituent of dill or gripe water, diluted 1:20 oil to water, to relieve flatulence especially in children. The oil is considered strongly antiseptic, and exhibits anticarcinogenic activity. It is interesting to note that *in vivo* animation of the ring-substituted compounds in dill and fennel oils can result in a series of three narcotic amphetamines.

Indian seed oil, sowa oil, is not considered a direct substitute for European or North American seed oils, since it contains dillapiole. Whether this constituent has any dietary disadvantage at the level at which it would normally be included in human food is doubtful. Sowa oil is a pale yellow to yellowish-brown mobile liquid, with a similar odour to other dill oils, but more sharp and less like caraway; the taste is more harsh and less pleasant than other dill oils. The oil contains less carvone, usually below 20% sometimes down to 5%, but contains dillapiole in varying amounts from 15 to 35%. Egyptian seed oil differs from dill and sowa oil in that it may also contain camphor and linaly acetate.

Monographs on the physiological properties of dill oils have been published by the RIFM. In the USA, the regulatory status generally recognized as safe has been accorded to dill, GRAS 2382, dill herb oil, GRAS 2383, and dill seed oil, GRAS 2384. Herb and seed oil have also been widely investigated in respect of their insecticidal and fungicidal properties. That both oils have the ability to control certain insects and plant pathogens is well known, and is mainly due to the constituents *d*-carvone, apiol, *d*-apiol and myristicin. These compounds are probably more important as synergists of synthetic insecticides. Carvone is considered a germination suppresser in root crops, including potatoes.

Oleoresin. Dill oleoresin is usually obtained by solvent-extracting dill seed, normally in the country of use, as it quickly deteriorates with storage. The oleoresin is a pale amber to greenish fluid, depending on its composition, since the addition of a permitted antioxidant is essential for storage. It generally contains 65% fixed and 20% volatile oils. Commercially offered oleoresins have a volatile oil content of 10–70%. Sowa seed oleoresin is prepared in very small quantities in India for local sale. The main use for oleoresins is as substitutes for the oil, but this usage remains very small and is apparently unlikely to expand. Dill oleoresin may also be a concentrated aromatic powder, obtained by extracting the fruits with alcohol and then drying the extract; it is especially recommended for low-salt or salt-free diets.

Other Anethum *species*

A. graveolens L. subsp. *sowa* (Roxb.) ex. Flem. (syn. *A. sowa* (Roxb). ex Fleming; *P. graveolens* (L.) Hiern.; *P. sowa* Kurz.) is known as Indian dill, and is now described in India as *A. sowa* (Roxb. ex. Flem) Gupta. It is also called East Indian dill but is more widely known as sowa, which designation has been used in the text. It behaves almost similarly to dill when cultivated, and thus the previous sections generally apply. The substantial variability in dill cultivars make visual differentiation between the two difficult where both are cultivated in proximity, as in India.

Sowa is generally similar to dill. The leaves have an ultimate segment that is longer and more fusiform. The inflorescence has fewer primary rays, with less variability between primary and subsequent umbels and the number of compound umbels is variable, up to 50, with less variability between primary and subsequent umbels. The fruit is less dorsally compressed than in dill, 3–5 × 1.5–2.5 mm, with three longitudinal, pale, angular ridges on both sides, which are less prominent than in dill. The vittae are similar, but have a more compact cell formation per segment, with irregular marginal walls.

Fruit contains 2–6% essential oil, kaempferol, isorhamnetin, quercetin, petroselinic acid triglyceride, β-sitosterol, fats and protein. The essential oil is mainly composed of 13 monoterpinoides, four phenyl derivatives, two methylene dioxy-phenyl derivatives, two sesquiterpene hydrocarbons, and contains less carvone than dill, but dillapiole is absent in European dill oil. The oil is placed in the same group as dill by Arctander (1960). The green plant also contains 0.06–0.10% essential oil with a high proportion of phellandrene and carvone. The composition and characteristics of sowa herb and fruit oils are described in the Products section. The main characteristics of Indian dill oils vary widely; an Indian commercial oil had the following characteristics: specific gravity (15°C) 0.900–0.9150; refractive index (25°C) 1.4810–1.4920; optical rotation (25°C) +70° to +80°; should dissolve in 10 vols 80% alcohol.

Fennel

The genus *Foeniculum* Mill. contains either five species, or two species and several varieties, according to the authority consulted, and all are biennial or perennial aromatic herbs. A most important member is *F. vulgare* Mill., the spice fennel, which most probably originated in the Mediterranean region, and like many of the long-domesticated *Umbelliferae* has been carried and cultivated worldwide. It is now a naturalized escape in many countries from Europe to the Far East, in sub-Saharan Africa and especially Ethiopia.

The Greeks believed Prometheus stole the spark of fire from Olympus and brought it to earth inside a stalk of giant fennel. The Greeks named the plant *marathron*, to grow thin, a reputation supported by later herbalists. Fennel was a favourite green herb of the Romans, whose armies distributed it throughout Europe including Britain, where it was well established in local cookery and medicine by the end of the 1st century AD, especially as a complement to fresh and salted fish dishes. So popular did fennel and its seed become that the household accounts of the English King Edward I in the 13th century recorded using the equivalent of 4 kg seed in 1 month. Fennel root was also used to flavour the English sack, a drink based on mead made from honey. Fennel became so common in European gardens that in Gerrards *Herbal* of 1597 he summarily dismisses it so: 'The Fennel is so common amongst us that it were but lost labour to describe the same.' Parkinson in his *Theatricum Botanicum* (1640) states that fennel's culinary usage at that time derived mainly from Italy, where it was especially favoured in the preparation and cooking of fish dishes.

Fennel's association with witchcraft and the occult persisted for centuries, and together with St John's Wort (*Hypericum perforatum* L.), fennel was hung over doors and stables on Midsummer's Eve to ward off evil spirits. A tradition the author remembers was maintained by Suffolk villagers when he was a boy, while wagoners on the farm nailed it over the stable entrance. The spread of fennel in Europe was stimulated by the

European Emperor Charlemagne, who required its cultivation on the imperial farms. However, it was soon found that fennel tended to suppress growth of other herbs and vegetables. A 12th-century English gardening manual urged, 'Plant fennel near your kennels', since not only would this keep fennel out of the main garden, but its dried powdered leaves and seed sprinkled in the kennels reduced flea numbers on the inmates. William Cole in his *Nature's Paradise* of 1650, stated, 'Seeds, leaves and root of garden fennel are much used in drinks and broths for those that are grown fat, to abate their unweildiness and cause them to grow more gaunt and lank.'

Traders probably transported fennel to Asia, South-east Asia, China and Japan, and now it is cultivated everywhere as a flavouring. The plant was apparently taken to North America by Spanish priests and planted round their missions. It was first grown commercially by a religious sect known as The Shakers around the end of the 18th century. Fennel is now naturalized wherever conditions are suitable, especially in California.

Fennel was also highly regarded for its medicinal qualities. The Greek physician Dioscorides who lived in the 1st century, wrote in his *De Materia Medica* that eating fresh fennel plants and seeds encouraged the flow of mothers' milk, the finely ground root mixed with milk cured rabies, and a decoction of the flower stalks alleviated bladder and kidney disorders. Pliny (AD 23–79) ascribed a score of remedies to the plant and its seed. He also quoted the very ancient, but quite inaccurate belief that, 'Serpents eat it when they cast their old skins, and they sharpen their sight with juice by rubbing against the plant.' This association of fennel juice with improved sight or as a cure for eye diseases persisted in Europe until very recently. The American poet Longfellow (1807–1882) alludes to this virtue thus: 'Above the lower plant it towers/The fennel with its yellow flowers/And in an earlier age than ours/Was gifted with the wondrous powers/Lost vision to restore'. When the seed oil was first obtained is unrecorded, but the crushed seeds were used in European cooking and herbal remedies by the 10th cen-

tury. In Morocco, oil from seeds of wild plants was used to treat local illnesses, and it was known that the dosage given could induce an epileptiform fit of madness and hallucination.

The taste of fennel varies greatly; wild fennel and closely related local types grown in Central Europe and Russia are slightly bitter and have no anise flavour. Florence, sweet, or Roman fennel, cultivated in southern Europe, the Americas and elsewhere for its bulbous leaf bases used as a vegetable, lacks the bitter principle and tastes strongly of anise due to the high anethole content of the essential oil. Other fennels are cultivated for their small edible stalks (carosella).

Botany

Generally known as *Foeniculum vulgare* Miller. (syn. *Foeniculum capillaceum* Gilib.; *Foeniculum officinale* Allioni), some taxonomists consider *Anethum foeniculum* L., currently its main synonym, is more correct and that the species should be included in the genus *Anethum* L. The genus name *Foeniculum* is derived from the Latin *foenum*, hay, thus *foeniculum* means little hay because the dry leaves were apparently thought to resemble fine dried hay. The word was corrupted in Medieval Europe, when it was commonly known as fenkel or fennill. It is known in English as fennel, in French as *fenouil*, in German as *fenchel*, in Spanish as *hinojo*, in Italian as *finocchio*, in Swedish as *fankal*, in Dutch as *venkel*, in Russian as *fyenkhel*, in Arabic as *shamar*, in Ethiopian as *insilat, kamun*, and many others, in India generally as *saunif*, in the Philippines generally as *anis*, in Indonesia generally as *adas*, in Malaysian as *das pedas* and in Thai generally as *thian-kiaep*.

Foeniculum is frequently stated to have three species, *F. vulgare* (fennel), *Foeniculum azoricum* Mill. (Florence fennel, a vegetable) and *Foeniculum dulce* Mill. (sweet fennel, also a vegetable). *F. vulgare* has no subspecies although subdivision into subsp. *piperitum* (wild taxa) and subsp. *vulgare* (cultivated taxa) has been proposed, as has division into *F. vulgare* var. *azoricum*, *F. vulgare* var. *dulce* and *F. vulgare* var. *vulgare*. However, the botanical and anatomical differences used are not considered consistent (Rapisarda *et al.*, 1998). *F. vulgare* is better classified into regional cultivar groups, which have also been grouped by essential oil composition (Bernath *et al.*, 1996; Muckensturm *et al.*, 1996). The basic chromosome number of the genus is x = 11, thus fennel is a diploid with *2n = 22*.

Disagreement also exists as to whether the present genus designation should continue or that of Linnaeus be reinstated and fennel placed with dill, which it closely resembles, in the genus *Anethum*. Additionally, as noted, the number of *Foeniculum* spp. is in dispute and whether there is but one, *F. vulgare*, all others being cultivars. However, as commercial fennel seed is obtained only from *F. vulgare* (or *F. vulgare* var. *vulgare*), the taxonomic argument is unimportant to this text. All references to fennel in the text refer to *F. vulgare*, its seed and derivatives unless specifically noted.

Substantial fennel germplasm collections are available in Russia at the St Petersburg N.I. Vavilov Research Institute of Plant Industry, in Germany at the Braunschweig Institute of Crop Science, and in France at the Groupe d'Etude et de Controle des Varietes et des Semences, Brion. Since fennel is cultivated and naturalized worldwide there is no danger of genetic erosion. Fennel is predominantly cross-fertilized, thus most cultivars are highly heterogeneous. Current research has developed cultivars with high fruit yield, fruits with high anethole content, resistance to certain pests and diseases, and for small fruits (Pank *et al.*, 2000). Other objectives include reduced height and earlier maturity. Some of the factors involved have been determined in Indian cultivars (Agnihotrei *et al.*, 1997).

Fennel is often confused with the closely related dill, *A. graveolens*, previously discussed, which easily crosses with fennel. Fennel and dill can be distinguished by their scent – fennel smells of anise or liquorice, dill is bitter and slightly pungent – and by their ripe fruit as fennel fruit is wingless, while dill fruit has a wide wing. Immature fennel plants may be recognized by their finely dotted stems, longer and broader leaf-sheaths and the usually shorter secondary rays in the umbel.

Fennel is a tall erect, glabrous, glaucous biennial or perennial herb, with yellow flowers and aromatic seeds. All plant parts contain an essential oil in secretory channels, which varies in composition according to plant part, development stage and cultivar. In callus cultures and cell suspensions, little or no aromatic compounds were found, while in tissue colonies derived from plantlet apices, anethole was produced in very small amounts.

The taproot is yellowish to light brown, up to 2.5 cm in diameter, which in suitable soils can become deeply penetrating, with few large laterals but many side rootlets. Root oil contains (on average) α-pinene 1.0%, *p*-cymen 0.3%, β-fenchyl-acetate 1.0%, *trans*-anethole 1.6%, eugenol 0.2%, myristicin 3% and dillapiole 87%. By comparison, the root and bulbous stem bases of Florence fennel contains less than 1% of dillapiole but 70% of *trans*-anethole, giving a very different taste. Fennel root may be used as a vegetable either raw or cooked, and when pounded into a paste and mixed with water or milk is believed to alleviate stomach pains.

There is normally one (sometimes more) erect, stout, main stem up to 1.5 m, occasionally to 2 m, with the base 2.0–3.0 cm diameter, usually profusely branched on the higher portion (Fig. 10.17). The stem is terete, striate, smooth, shiny, green to blue-green with light green ribs (frequently described as glaucous), and yellowish at the nodes, which may become hollow with age; it is usually profusely branched on the higher stem. For the effect of growth hormones on young plants, see the Cultivation section. Stems prior to flowering can be scraped or peeled and eaten raw or cooked.

The bright green to blue-green leaves are alternate, decompound, with a leaf sheath, 2–15 cm, with scarious margins, forming an open cylinder at the base embracing the stem. The leaf blade is triangular, the lower ones up to 30 × 50 cm, the upper much smaller, and pinnately divided as to be almost filiform; the primary pinnae are odd numbered, from 3 to 19 (Fig. 10.18). The petiole is up to 13 cm, broad, exceeding the sheath, and longitudinally striate; the lower leaves are usually long petiolate, the higher leaves with blades often sessile on the sheath. Herbage contains

Fig. 10.17. Fennel plants at flowering.

1.0–2.55% essential oil, generally about 2.0%, whose main constituent is *trans*-anethole up to 75%. Data indicate that anethole and fenchone concentrations generally increase from the bud stage until fruit ripening, the α-pinene and limonene concentrations decrease, and the estragole concentration remains almost constant. Leaves picked before bud formation are used in salads, and fresh or dried as a culinary herb.

The inflorescence is a large, flat compound umbel up to 20 cm in diameter but usually not more than 16 cm, borne terminally on branches (Fig. 10.18). The peduncle is 5–16 cm but up to 30 cm, terate to subterate, finely sulcate, light to blue green, with bracts and bracteoles usually absent. There are 5–30 primary rays per umbel, 0.5–12 cm long, unequal in length, with the shortest in the centre; there are 10–30 secondary rays (pedicels) per umbellet, up to 1 cm but unequal; the involucre and involucels are absent, and the calyx is vestigial at the top of the ovary. The five golden yellow petals are distinct, up to 1.5 × 1 mm, with a strongly inflexed notched apex, the margins entire, and a thin membranous outgrowth on the ventral side of the midrib. There are five stamens, up to 1.5 mm, dehiscing via longitudinal slits. The pistil has an inferior, bilocular ovary, with two styles, each with a basal stylopodium and superior stigma. The floral biology of fennel and related plants has been described in detail (Nemeth and Szekely, 2000). A relationship may exist between the number of rays in the main umbel and factors affecting seed yield, including plant height and duration of the period to flowering. The main flowering period is July–August in the northern hemisphere, December–January in the southern.

Flowers are bisexual, actinomorphic, and open centripetally. Anthers dehisce in the early morning, and out-crossing is high, since some cultivars may have high self-incompatibility. The flowering dynamics and pollination of umbels indicate that inbreeding and hybridization could be successful in genetic improvement (Nemeth *et al.*, 1999). Flowers are attractive to insects, especially bees, which are often the main pollinating agents, although wind pollination also occurs. In India, for example, seed yield from plants from which insects were excluded was reduced to 1300 kg ha^{-1} compared with non-netted plants of 2800 kg ha^{-1}.

The fruit is a light green to dark brown (depending on cultivar) lens-shaped schizocarp, oblong-oval to elliptical, 3–6 × 2–3 mm usually with a long pedicel and

Fig. 10.18. Flowering stem – fennel.

short stylopodium (Fig. 10.19). The fruit splits at maturity into two yellow to grey-brown mericarps. The mericarps are flat or slightly concave at the commissural side, slightly concave at the dorsal side, usually with three to five longitudinal prominent ridges, with oil-vittae between. In India, small seeds generally had higher oil content than large. Fruit composition varies, and a recent analysis showed it contained (per 100 g edible portion) water 8.8 g, protein 15.8 g, fat 14.9 g, carbohydrates 36.6 g, fibre 15.7 g and ash 8.2 g (containing Ca 1.2 g, Fe 19 mg, K 1.7 g, Mg 385 mg, Na 88 mg, P 487 mg and Zn 4 mg). The vitamin content was A 135 IU,

niacin 6 mg, thiamine 0.41 mg and riboflavin 0.35 mg; energy value was about 1440 kJ per 100 g (Bernath *et al.*, 1996).

The fruit contains a fixed oil up to 30%, but usually 15–20%, and a volatile (essential) oil up to 12%, but usually around half this; both are discussed at the end of this section, and in the Products and specifications section. The fruit also contains flavonoids, iodine, kaempferols, umbelliferone and stigmasterol and ascorbic acid. Traces of aluminium, barium, lithium, copper, manganese, silicon and titanium have been reported. The morphology and histology of the fruit and seed have been

Fig. 10.19. Fennel fruits.

described in detail (Parry, 1962). The whole or split dried fruits are commonly known and traded as fennel seed. A non-destructive method of determining oil constituents has been developed (Fehrmann *et al.*, 1996). The pericarp normally encloses a single seed, which is adnate to the pericarp, with the embryo at the apex of the mericarp. The seed is white, 2×0.2 mm, with a conical radical, and the endosperm greyish and fatty; the weight of 1000 seeds is about 4–8 g. Seed from regional cultivars varies greatly, especially the anise characteristic.

Fatty oil. Most of the fixed oil is contained in polygonal cells in the seed endosperm, and contains about 10% total monounsaturated and 2% total polyunsaturated fatty acids; main components of an expressed oil are petroselinic acid up to 75%, oleic acid up to 25%, linoleic acid up to 15%, and palmitic up to 5%, depending on the sample tested. The main characteristics of an Indian oil were:

specific gravity (15°C) 0.9304; refractive index (15°C) 1.4795; optical rotation +35°; saponification value 181.2; iodine value (Wijs) 99; unsaponification material 3.7%. The expressed oil is classed as semi-drying, and is a source of lauric and adipic acids.

Essential oil. Analyses of oil from various plant parts tend to be very variable, especially between regional cultivars, as in Turkey (Karaca and Kevseroglu, 1999) or between wild types as in Italy (Piccaglia and Marotti, 2001). In European and Argentinean types of *F. vulgare*, limonene concentration in the whole plant may not exceed 10%, but α-phellandrene in leaves is between 23 and 25% and in stems between 22 and 28%. In contrast, the limonene content in young leaves and stems of *F. dulce* in European and Indian types is between 37 and 40% and 28 and 34%, respectively, and decreases with age, whilst the α-phellandrene concentration at 1–4% is low and remains constant. While these findings are generally true in respect of the basic difference between fennel and sweet fennel herb oils, the actual level of limonene in a particular cultivar can vary significantly. Similar variation in a specific constituent has been noted in many countries, and recorded in the literature.

In mature fennel, about 95% of the essential oil is located in the fruit, with highest content in fully ripe fruit; hydro-distillation yields 1.5–3.5% depending mainly on cultivar. In general, anethole and fenchone are higher in the waxy and ripe fruits than in the stems and leaves, whereas α-pinene was higher in the latter. Many published analyses of fennel oils fail to state the cultivar or geographical origin, and thus comparisons are difficult and may be misleading. Very wide local variation in seed oil constituents can occur in regional strains; in Turkey, *trans*-anethol ranged from 76 to 87%, limonene from 4.3 to 9.2, estragole from 3.3 to 5.2, fenchone from 1.0 to 2.8, γ-terpinene from 0.9 to 1.6 and α-pinene from 0.5 to 1.2 (Akgul, 1986); in Israel, significant differences were noted in indigenous regional fennels (Barazani *et al.*, 1999); in Germany, four chemotypes were distinguished in 46 cultivars tested (Kruger and Hammer, 1999); and in Italy, major differences were noted in trials with ten cultivars and clear phytochemical differences were notable, for example, the *trans*-anethole and estragole occurred in inverse proportions (Miraldi, 1999), which suggests a common precursor.

Few data are available on factors affecting fruit oil content or composition, other than the gross effect of environment and similar undifferentiated phenomena, and probably the most significant factor is genetic. Heredity has a direct effect on fennel characteristics; in Japan, progeny from clonally propagated plants had a more uniform anethole content than those sexually propagated, 0.7–3.0% compared with 0–13%; in Russia, selection within a local cultivar, Pomrie, increased seed yield by nearly 60% over the local average for the cultivar, and the anethole content of the seed oil by 11%. However, the extent to which this variation is inherited, or due to the local environment is difficult to judge; what is apparent is that there is a genotype–environment reaction, since it is widely reported that the same cultivar varied greatly in seed yield but less widely in seed-oil composition in succeeding, climatically different seasons.

Ecology

Fennel is basically a plant of the warm temperate regions, typically with a Mediterranean climate, is native to southern Europe, North Africa and the Middle East, and naturalized in many other countries with similar climates to which it has been introduced. Selection and local adaptation has considerably extended fennel's range, and it is now cultivated or can be found growing wild, in both colder and warmer regions. However, climatic variation is frequently reflected in differences in seed yield, seed-oil content and composition, or viability, as in Israel (Fait *et al.*, 2000). Cold, wet conditions adversely affect growth in temperate regions, and warm humid in semitropical areas.

Fennel thrives on long, sunny days, and although considered day-neutral, trials in Tasmania (Australia) showed that a local cultivar was a long-day plant, with umbel initiation and stem elongation occurring at a minimum of 13.5 h day^{-1} (Peterson *et al.*,

1993). A temperature of 15–20°C is the optimum, and above 25°C for extended periods usually retards development and in early growth may result in premature flowering and very low seed yield, or death of young plants especially if associated with low soil moisture. Below 5°C, growth is generally retarded and long cold periods suppress growth and adversely affect flowering. Fennel plants, once established, can withstand moderate frost without damage, and overwintering plants well covered by snow or straw suffer little harm from quite severe frosts. However, once new growth has commenced, frost will cause severe damage and may be fatal. Annual fennel seedlings must not be planted out until all danger of frost has passed. High-altitude cultivars in Ethiopia can withstand light frost prior to flowering.

Fennel is thus an annual crop in temperate regions; in northern India it is grown in the cold season, at higher elevations up to 2000 m in the south and Pakistan, and up to 2500 m in Ethiopia. Pilot plantings in west Kenya at 1500 m flourished during the dry (cool) season (E.A. Weiss, personal observation). When only herbage is require, fennel is successfully grown in the lowland tropics.

A rainfall in the range 400–750 mm falling over the main growing period will produce good seed crops, but irrigation will be necessary if an extended dry period occurs before flower development. Although fennel is intolerant of high levels of soil moisture, a rainfall of 1000 mm or even 2000 mm causes little damage if the soil is free-draining, or artificially drained. In Israel, a correlation was noted between rainfall and oil constituents – in high-rainfall areas oil had higher estragole and lower *trans*-anethole content and vice versa (Barazani *et al.*, 1999) – and between seed weight and viability (Fait *et al.*, 2000).

High winds when seed is mature can cause shattering, and very hot winds at flowering reduce seed set, and when seed is maturing may also cause premature drying off, with a consequent loss of seed oil. Hail is damaging mainly early in growth, but a severe hailstorm just prior to harvest will severely reduce yield by stripping seed heads from plants. Such storms are more common in the semi-tropics.

Similar to many other crops grown as annuals, the seed yield of fennel is frequently directly related to time of sowing or planting out. The optimum sowing time in some countries is often very restricted, especially where there are definite seasons, i.e. wet and dry. In Rajasthan, India, for example, sowing on 30 October gave a higher seed yield than sowing on 15 October or 14 November; in Harayana around 15 October was the optimum (Yadav *et al.*, 1997), and in Italy, highest yield was from sowing on 20 November rather than 1, 2 or 3 months later (Leto *et al.*, 1996).

Soils

A soil that produces good crops of local vegetables is suitable for fennel, but in southern Europe areas suitable for grapes are considered the best for commercial cultivation. However, fennel is a popular smallholder crop and grown on a very wide range of soils, from fairly heavy clays to sandy loams, stoney uplands to fertile river valleys. Wild fennel is often found on limestone and chalk soils in Europe. The main requirement is that the soil should be well drained, since waterlogging even for short periods is undesirable. It is widely believed by growers in India that too fertile a soil promotes vegetative growth at the expense of seed production, and thus more-sandy loams or the local black sands are preferred.

A neutral to slightly alkaline soil in the range pH 6.5–8.0 is preferable, but saline soils should be avoided, as even low levels of salinity significantly reduce seed yield, although vegetative growth may be less affected. Seed germination in India was only 12% at 10.0 mmhos cm^{-1} (1 mmhos cm^{-1} = 1 $\times 10^5$ siemens m^{-1}) compared to 58% in control, and seed yield was severely affected by a salinity level of 2.5 mmhos cm^{-1}.

Fertilizers

Similar to other *Umbelliferae*, fennel responds to the application of fertilizers, and where a certain level has been found necessary for other similar local crops, this is suitable for fennel until fertilizer trials establish the optimum levels. When fennel follows crops to which substantial amounts of farmyard or similar manure, P and K have been applied, little more is usually required. This is especially so in soils with a high residual effect. Heavy metal content in soils can be absorbed by fennel, but in Poland the residues in seed and oil were not considered a danger to health (Zheljazkov and Jekov, 1996).

A fennel crop can develop a large vegetative mass (40–60 t ha^{-1}) for which ample nutrient supply is required. An indication of nutrients removed by fennel to produce a seed yield of 1300 kg ha^{-1} is shown in Table 3.4. In India, 20–30 t FYM ha^{-1} is recommended applied prior to planting and 45 kg N ha^{-1} as a top dressing. While commercial fennel crops normally receive the necessary fertilizers, smallholders who have to purchase fertilizer from scarce cash resources are advised to apply the minimum, and increase seed yield by greater attention to weeding and harvesting, as these normally require additional labour only.

Nitrogen is necessary to promote plant growth, but the level must be accurately determined in relation to seed production. An excess encourages very tall, leafy plants without a proportional increase in flowering heads. The ratio of N to P and K must also be established. An indication of the nutrient requirement of fennel in India showed an irrigated crop yielding 113 kg fruit removed about 27 kg N, 5 kg P and 17.5 kg K, all ha^{-1}. The amount of N applied must also be related to plant population, since a variation in one reacts directly on the other. Thus, fennel closely spaced to produce uniform growth and flowering to assist mechanical harvesting may be etiolated by the level of N suitable for more widely spaced plants.

Fennel for seed is usually grown as an annual and thus a proportion of the N should be applied to the seedbed, either at sowing or when planting out seedlings. The amount is normally less than given as a top dressing, and is usually applied as a compound mixture with P, and with K where necessary. Since fennel prefers a slightly alkaline soil, the more acidic nitrogenous fertilizers should be avoided where possible. Diammonium phosphate (DAP) can be used in the seedbed, whilst ammonium nitrate or calcium ammonium nitrate (CAN) are to be preferred to ammonium sulphate as a top dressing. In various states of India, trials indicated that 80–100 kg N ha^{-1} and 40 kg P ha^{-1} is optimal. In Italy, a common NPK mixture is 2:1:1 at 50–100 kg ha^{-1} depending on district, applied in the seedbed.

It is doubtful if more than 40 kg N ha^{-1} will normally be necessary, with up to half in the seedbed, the remainder as one or two top dressings depending on the management system used. The method of application is immaterial, provided the fertilizer is evenly spread or applied. The first (and normally only) top dressing should be applied when plants are about 30 cm, or some 60 days after sowing, and if a second is applied this should not be later than when first flower shoots appear. Foliar application of N increased monoterpene content of fruit oil (Khan *et al.*, 1999), while spraying herbage with the N-fixing *Klebsiella* bacteria increased growth and fruit yield, but the highest levels reduced seed and oil yield (Mukhopadhyay and Sen, 1997).

Phosphate is normally applied to the seedbed, usually as superphosphate, but rock phosphate and similar less soluble types can be broadcast and ploughed in during pre-planting cultivation. The amount required is basically governed by two factors: that required to satisfy plant needs, and that required to ensure optimum use of other nutrients, N in particular. Thus, fertilizer trials are essential to determine the optimum amount of P. Superphosphate is probably the most easily obtained worldwide, but where seedbed nitrogen is desirable, the compounds previously mentioned should be used. Placement alongside the seed or seedlings and just below the root zone is preferable.

It is essential to establish the optimum level of K required to produce a seed crop, and fertilizer trials combining the three main plant nutrients will establish the rate required. Potassium can be taken up in luxury amounts by plants with no increase in seed yield, and unnecessary amounts merely waste time and money. Excess K may also exacerbate minor element deficiencies, especially magnesium.

Similarly to N, the less acidic types of K fertilizers should be used when possible. Potassium can be applied in the seedbed as a compound mixture with N and P, but some types can be broadcast and ploughed in.

Fennel has also responded to application of zinc and boron, but in general minor nutrients should only be applied when shown to be essential by soil analysis or field trials; an interaction may occur with major nutrients.

Cultivation

These remarks apply to cultivated fennel, but a large amount of fruit is collected from wild or naturalized plants, especially in eastern Africa. The cultivations necessary are described at the beginning of this chapter, with a level, weed-free seedbed the main objective.

Fennel is normally propagated by seed, but division of roots or crown is possible, and plantlets can be grown in *in vitro* culture from portions of hypocotyl, stem or petiole (Schiff *et al.*, 1996). Fennel is usually treated as an annual and direct-seeded, although some smallholder crops are transplanted, a technique not recommended for larger plantings. Splitting existing plants to provide planting material is also not recommended, except by smallholders who can afford the considerable increase in labour required.

Whole fruits are normally sown, and little is gained by selecting large seed, since seed size is not directly related to final seed yield. Fresh seed should be sown, as viability falls after 2–3 years; germination is 70–75% for fresh sound seed, but can be down to 30% in seed saved from season to season by smallholders.

Commercial fennel crops are usually sown by a drill adapted to small seeds, 2–3 cm deep, not more than 5 cm, at 8–12 kg ha^{-1} mainly depending on cultivar. More rather than less seed should be sown, since thinning is easier than in-filling blanks. Row width varies between 40 and 60 cm and in-row spacing between 25 and 30 cm according to cultivar and local preference; in Bulgaria, 80 cm rows with a seed rate of 15–20 kg ha^{-1} gave the highest yield, while in India, it was 40 kg ha^{-1} in 30 cm rows. It is essential to carry out trials to determine the local optimum plant population ha^{-1} since this directly affects profitability. For mechanical harvesting, many plants with few seed heads ripening at approximately the same time is preferable to fewer and larger plants flowering over a prolonged period.

A moist, warm seedbed with temperature between 15 and 20°C hastens emergence and encourages early seedling growth; germination is delayed below 7°C and above 25°C, while temperature variation during germination generally extends the period to emergence. Soaking seed in water or a solution containing growth promotants prior to sowing hastens germination but is usually unnecessary and unprofitable. Germination is epigeal, and fennel seed normally germinates in 10–14 days depending mainly on the weather following sowing. Cold, wet conditions retard germination and if they persist cause extensive mortality in emerging seedlings. The effect of climate on time to germination may also be related to other factors, including latitude, as was noted in Israel (Fait *et al.*, 2000).

Initial development is slow, with stem emergence 2.0–2.5 months after sowing. Hormone treatment of seedlings can increase the rate of growth and plant height, but may adversely affect seed production and/or seed oil content; application rate is usually an important factor. In Egypt, application of gibberellic acid increased plant height, whilst cycocel had the opposite effect – the reaction was reversed in respect of seed production and seed oil content. Hormonal treatment of fennel grown for fruit production is unnecessary and unprofitable, since greater increases are usually more easily obtained from improved agricultural methods. Flowering occurs 3–4 months after sowing on annual crops, but well-established plants in Ethiopia left for several years may flower year-round with a peak after the beginning of the rainy season (Jansen, 1981).

Weed control

Two or three weedings are usually necessary, although herbicides are applied to commercial crops. It is the substantially increased cost of manual weed control in perennial fennel crops that has made herbicide use more common. The following have been successfully used in fennel, with a minimum of crop damage: chlorthal-dimethyl, flauzifop, glyphosate, linuron, prometryn, propazine and trifluralin. It must be emphasized that local trials are essential to test the suitability of any herbicide.

Irrigation

Fennel should preferably be grown in those regions with a regular rainfall, as irrigation of commercial crops is normally unprofitable. Where irrigation is possible in Europe during dry spells, seed yield and oil content are generally higher than in rain-grown crops. A severe moisture stress causes the basal stalk to split and plants later collapse. Timing of irrigation is important, and under dry summer conditions in Tasmania, Australia (the southern climatic equivalent of Europe), the largest increase in total biomass yield occurred when plants were watered during stem elongation, whereas highest increase in umbel dry weight and oil yield occurred when plants were irrigated at late flowering. The anethole content of the umbel oil is not significantly affected by irrigation.

Smallholder crops, especially in India, are usually irrigated where water is available, since the cost of so doing is minimal. The last application should be when first flowers appear, as irrigation after this period will prolong vegetative growth at the expense of seed production. The amount of water is equally as important as timing; on a clay soil watering to a depth of 60 mm gave a higher average seed yield than to a depth of 80 mm (Yadav *et al.*, 1997).

Intercropping and rotations

European cultivars are usually treated as annuals and seldom grown 2 years in succession, but elsewhere plants can be treated as perennials, cut back annually, and left growing for up to 4 years. Fennel can be interplanted between rows of grape vines or similar low-growing permanent crops, but is usually unprofitable when under-sown in tree crops with a fairly dense canopy. Many smallholders (and garden-ers) believe that interplanted fennel sup-presses growth of other crops, but protects against insect attack. Thus, fennel is usually planted in small separate plots, or in lines on field boundaries.

Pests

Fennel is considered to be generally more resistant to most common insect pests attack-ing local herb and vegetable crops, and it has been remarked that in the same area, wild fennel appears less damaged than cultivated fennel by similar pests (and diseases).

Aphids are the most common and often the most important pest worldwide, espe-cially *Cavariella* spp., particularly *C. aegopodii* in the eastern Mediterranean through Egypt to East Africa; *Dysaphis* spp., in particular *D. foeniculus* in north and east Africa; and the local *Hyadaphis* spp. generally. Thrips, usu-ally *Thrips flavus* are occasionally reported as damaging a specific crop but *Hercothrips indi-cus* can be troublesome in India. The chalcid fly, *S. albipennis*, is one of the most damaging pests in those regions where it occurs, and may preclude fennel cultivation in a particu-lar area. Larvae bore and eat shoots and flower heads, and cause up to 40% loss in yield.

Larvae of *Polyphylla* spp. and *Schizonycha* spp. (scarabs) attack roots of a wide range of crops, but the degree of damage has not been accurately assessed; in Bulgaria, *Polyphylla fullo* can become a major pest, and it also occurs generally in the Mediterranean region and North-east Africa. *Schizonycha* spp. are most important in the Nile valley and eastern Africa.

The following have been reported as causing some degree of defoliation: larvae of the cabbage moth *Mamestra brassicae* gener-ally but *Mamestra persicariae* in warm regions; *Depressaria* spp., particularly *D. foeniculus*, but *D. depressella* in Eastern Europe; *Papilio* spp., especially *Papilio machaon* from West Asia to Iran; and the webworms *Loxostege* spp., in particular *Loxostege sticticalis*.

Diseases

Fennel is apparently less susceptible to a number of frequently damaging diseases of vegetables and herbs; most frequently recorded are stem and root rots, which may attack seedlings or mature plants, and leaf spots. Soil type has a significant effect on the degree of damage caused by root and stem rots, and plants growing in lighter or more sandy soils are generally less affected. On other soils, increased susceptibility or degree of damage may be due to minor element deficiency, as noted in the Fertilizers section in the introduction to this chapter. Most often recorded are the widespread *Phoma* spp., *Sclerotinia* spp., *Rhizoctonia* spp. and the locally important *Fusarium* spp. Many dis-eases are seed-borne, and in India nearly 30 species of fungi were isolated from local seeds, and methods of treatment were tested (Bharat *et al.*, 1997; Prabha and Bohra, 1999).

Leaf spots occur on almost all cultivated fennel, but the degree of damage is extremely variable. One of the most damag-ing is cercospora leaf spot or leaf blight, caused by *Cercospora foeniculi*, which can occasionally cause very severe damage in specific years or crops from Europe to Asia. Symptoms are brown to black lesions on above-ground plant parts, and severely affected plants wilt and often die. Less-affected plants have reduced yield directly related to the degree of damage. Another damaging leaf disease is the widespread *Phytophthora syringae*, which has a range of hosts. The main symptom is brownish areas on petioles and leaves, which become water-soaked lesions, and leaves wither and fall. Humid conditions favour spread of the disease.

Alternaria leaf spot or leaf blight is due to *Alternaria* spp. generally but in India is caused by *Alternaria umbellifericola*, but angu-lar leaf spot is due to *Passolra kichneri*. Other leaf spots caused by *Puccinia* spp. and *Septoria* spp. have been reported from Europe and the Middle East. The black leaf spot caused by *Phoma* spp. is less damaging generally, but in more humid regions *Phoma foeniculina* may cause severe damage in spe-cific crops or seasons.

A root rot due to infection by *Rhizoctonia violacea* and bacterial soft rot of lower stems caused by *Erwinia carotovora* occur in Europe and West Asia, and may be more widespread than currently noted. Another root rot is due to infection by *Phomopsis* spp., especially *Phomopsis foeniculi* in Europe where some cultivars may be resistant (Anzidei *et al.*, 1996). Damping-off or dieback of seedlings and young plants is often due to infection by *Sclerotinia* spp., particularly *Sclerotinia libertiana*. The main symptom is a general wilting, and plants often wither and die. Another stem rot is caused by *Fusarium* spp., especially the widespread *F. oxysporum*, which has a wide host range.

Downy mildew, *Plasmopara* spp., commonly occurs when conditions are suitable, and powdery mildew *Erysiphe umbelliferarum* less commonly. Since the latter can severely damage umbels and developing seeds, it must be controlled where crops are grown to produce certified seed for sale. In India, the most important disease is powdery mildew caused by *Leveillula taurica*, which produces white, powdery patches on the plant and shrivelling and shedding of fruits. This pathogen also occurs on a range of local cash crops, and thus cross-infection could increase the damage to fennel grown in their proximity. The false mildew *Plasmopara nivea* is occasionally reported. A disease reported from fennel in India is *Protomyces macrosporus*, which produces galls mainly on stems. It is a major disease of coriander, and fennel is apparently infected only when it is grown following coriander or in close proximity.

Harvesting

Annual fennel ripens unevenly, the first umbels in about 5 months, the last in 7 months, and first umbel fruits produce the highest quality fruit and oil. Fruit yield in the second and third years from perennial crops is generally higher than the first, but then usually deteriorates with less numerous and smaller fruit.

Plants are cut when fruit on primary umbels is hard and turning greenish-yellow or greenish-grey depending on cultivar, approximately 5–6 months after sowing, and trials indicate that fruit colour is acceptable in determining ripeness (Quilitzsch and Pank, 1996). Plants should preferably be harvested slightly earlier rather than later, since physiologically mature seed will ripen on plants, while overripe seed can be lost through shattering. Highest seed yield in India was obtained when plants were harvested when the seed on the main umbels was mature but green, compared to seed turning yellow. Harvested plants should be dried carefully in the shade as sun-drying causes loss of essential oil. Plants can be cut with a reaper–binder or by hand, and sheaves stooked in the field to dry, but an overlong period in the field should be avoided. Direct combining of fennel is possible, but the same remarks apply as for dill, as do the methods of threshing seed from plants. In Italy, a combine harvester covered 0.08 ha h^{-1} with header and threshing losses of 12% and 16%, respectively (Martini and Garbati Pegna, 1996).

Yield of fruit from smallholder crops is extremely variable. In India, where fennel is a major cash crop in northern regions, average yield is 450–650 kg ha^{-1}, but 1500 kg ha^{-1} can easily be obtained by good management, and 2000 kg ha^{-1} at local experimental farms. A yield of at least 2000 kg ha^{-1} is necessary from large-scale commercial crops, widely achieved in Europe and North America. However, much higher yields are possible; in a trial with 29 cultivars in Azerbaijan, yield reportedly ranged from 3000 to 9500 kg ha^{-1}. In Ethiopia, fruit is collected from wild plants, which in some regions are so numerous the yield is 200 kg ha^{-1}. Fennel grown as a perennial usually gives a seed yield 30–50% higher in the second year. The very high yields reported from experimental farms in a number of countries indicate the potential substantially to increase the local average yield by a relatively modest breeding or selection programme.

In a specific region using the local cultivar, oil content is most directly affected by time of harvesting, since this determines the degree of fruit maturity. Oil composition changes as fruit matures, and although major differences in the ratio of main constituents can occur, there is apparently little difference between them over the period of normal harvesting, i.e. between physiological maturity and fully ripe seed. Thus, for all practical purposes these changes can be ignored.

Processing

Bulk fruit delivered by smallholders may contain sand, stem tissue, immature or mouldy fruits or other umbelliferous fruits, and must be rigorously cleaned prior to shipment. Fennel fruits and those of anise, *P. anisum* (previously described), are often confused or substituted for the other.

Fruit can be artificially dried, and the methods used for coriander are acceptable. After cleaning and drying, fruit should be stored in new sacks or ventilated bins, with provision for fumigation. Provided fruit is stored under suitable conditions, little loss of oil occurs or change in its composition. The most general classification for grading is into 'shorts' and 'longs', with the latter preferred in international trade.

Distillation

Herbage essential oil is obtained by hydro-distilling fresh or slightly wilted foliage, and no special techniques are necessary. For highest oil content, herbage should be distilled just before flowering, as in Italy (Bellomaria *et al.*, 1999). Fruit may be distilled at any time following harvest, but is mainly exported to consuming countries for oil extraction. Fruit must be milled or crushed and then distilled immediately to avoid oil loss through evaporation. Distillation is straightforward, basically as for coriander, but the condenser temperature should be maintained at a sufficiently high level to prevent the oil congealing. The dried distillation residue contains about 14–22% protein and 12–18% fat and is suitable for stockfeed. Supercritical carbon dioxide extraction is becoming more common as plants of varying sizes are now available to suit any production levels. The resulting oil is considered nearer the original spice in odour and taste (Simandi *et al.*, 1999).

Storage and transport

Ripe seed that has been correctly harvested, cleaned and dried shows no significant change in oil content or oil composition when kept for an extended period in suitable containers. Polythene or cotton bags were superior in India (Sanjeev and Sharma, 1999). Insect pests can be controlled by fumigation with carbon dioxide (Hashem, 2000). A method has been developed to determine the extent seed has been irradiated to control microbial contamination (Satyendra *et al.*, 1998). Seed should be transported in new sacks or sealed bulk containers.

Essential oil should be stored in full, sealed containers in the dark. Composition of the oil changes when exposed to air at normal temperature, the main effects being an increase in anisaldehyde and fenchone, and a decrease in *trans*-anethole and α-pinene. Other changes may also occur, for example an increase in the *p*-cymene and camphor content.

Products and specifications

The main products from fennel are the green or dried herb, dried fruit known in the trade as fennel seed, herb and seed oils. Since the essential oil is the main flavouring component of herb and seed, it has been investigated in detail.

Herb. The green herb is generally used as a flavouring, either incorporated in dishes or sprinkled on food prior to serving. Bunches of leaves are widely available in markets in all producing countries or from smallholders in those where fennel is not a commercial crop, i.e. South-east Asia. It is also easily grown as a culinary herb in gardens or in pots for personal use. The dried herb is of inferior quality, but freeze-dried or frozen leaves are superior. The major flavouring component is anethole, as noted in the Botany section, which gives the herb the odour and flavour of anise.

Herb oil. Steam-distilled herb oil from whole, often wild plants was formerly of considerable importance but has now declined to negligible amounts produced only to special order, since it has no specific attributes or interest, and few recent data are available. Oil from fresh or wilted herbage is a nearly colourless to pale yellow mobile liquid, which may darken with age; the odour is lacking in anise and the taste is bitter. The main constituents are described in the Botany section. The main characteristics are: specific gravity (15°C) 0.893–0.925; refractive index (20°C) 1.484–1.508; optical rotation +40° to +68°; soluble in 0.5–1.0 vols 90% alcohol (Gunther, 1950b).

Oil composition changes significantly with age of herbage, with basic differences between herb oils from *F. vulgare* and *F. dulce*. Green leaves of the wild *Feronia limonia* Swing. gave an oil containing anethole used as a fennel oil substitute in India. Sweet fennel herb oil has been produced on a small scale, or experimentally, in several countries; a Bulgarian oil had anethole at 62% and limonene plus phellandrene at 28% as major constituents, with an odour similar to the seed oil.

Seed. Fennel seed is a major culinary and processing spice, used whole or ground to flavour a wide range of foods, including baked goods, meat and meat products, snack foods, fats and oils, gravies and similar products; it is also used crushed or ground in herbal tea. The highest average maximum level in the USA is about 0.12% (1190 p.p.m.) in meat and meat products. Main constituents are described in the Botany section. Good-quality sound seed has a bitter rather camphoraceous taste and a pungent odour. Fennel seed is widely known as a stimulant, stomachic, expectorant and carminative, and is official in many pharmacopoeias.

The fruits are used in phytomedicine in Germany against dyspeptic disorders, as a gastro-intestinal antispasmodic, as an expectorant and in syrups to treat children's coughs. Fennel became a major ingredient in Arab and Ayurvedic medicine, and roots are traditionally applied as a diuretic and purgative. In India, leaves are considered diuretic, a fruit extract is used to treat eye infections, and hot infusions of the fruits are applied to increase milk secretion and to stimulate sweating. In Indonesia, the fruit is combined with the bark of *Alyxia* spp. to give an agreeable flavour to medicines, but is also believed to alleviate sprue. In Chinese herbal medicine, fennel is used to treat gastro-enteritis, hernia, indigestion and abdominal pain, to resolve phlegm and to stimulate milk production.

Seed oil. Fennel seed oil, usually traded as fennel oil, is mainly obtained by steam distilling whole or crushed fruit with a yield of 1.5–6.5%, and more recently by supercritical carbon dioxide extraction (Ehlers *et al.*, 2000). In general, oil content is greatest in European and lowest in Asian varieties, but the yield obtained from a particular crop depends almost entirely on the skill of the grower. The oil is an almost colourless to very pale yellow liquid, which can crystallize on standing, and may require warming before use. The congealing temperature should not be below 3°C. Approximately 45 constituents have been determined from fennel oil, some not fully identified; the main constituents are *trans*-anethole (60–65%, but up to 90%), fenchone (2–20%) estragol (methyl chavicol), limonene, camphene, α-pinene and other monoterpenes, fenchyl alcohol and anisaldehyde (Lawrence, 1998b; Pino, 1999b). The major characteristics of commercial grade oil are: specific gravity (25°C) 0.953–0.973; optical rotation (23°C) +12° to +24°; refractive index (20°C) 1.5280–1.5380; slightly soluble in water, soluble in 1 vol. 90% or 8 vols 80% alcohol, very soluble in chloroform and ether. Interestingly the *in vivo* animation of ring-substituted compounds in fennel and dill oils can result in three narcotic amphetamines.

Oil obtained from an indigenous type of Pakistan fennel contained α-pinene 3.0%, camphene 0.65%, α-phellandrene 0.44%, limonene 4.6%, fenchone 10.2%, methyl chavicol 3.5%, anethole 74.9%, anisaldehyde 1.8% and *p*-anisic acid 1% (Ashraf and Bhatty, 1975). An oil produced in Nigeria from fennel of Indian origin had an 80% anethole content but no fenchone, and a Moroccan oil from the Atlas mountains contained piperitone oxide at 2%, identified for the first time from an *Umbelliferae*. A comparison of fennel with other spice oils has been published (Piccaglia, 1998).

The oil has a pleasant, aromatic, anise odour, and a characteristic strong initial camphor-like taste, spicy but somewhat bitter. The oil is placed in the warm-phenolic, fresh-herbaceous group by Arctander (1960). Indian fennel oil is less strong in both odour and taste, sweeter and less bitter. *Trans*-anethole contributes to the warm, sweet herbaceous odour and sweet taste of fennel oil, but other compounds are involved (Bartschat *et al.*, 1997). An important constituent is the ketone fenchone, up to 24% in bitter, seldom above 5% in sweet fennel oil, and the amount of this compound can help to identify the botanical origin of an oil. Some end-users prefer sweet fennel oil because of the high fenchone content in fennel oil, as the camphoraceous odour of fenchone can be a disadvantage in their products.

The main use for fennel oil is as a food flavouring, in liqueurs where a liquorice or anise taste is important, and in industrial perfumery to mask the odour of aerosols, insecticides, disinfectants, etc. It has a minor use in confectionery, dairy products, soft drinks, as a tobacco flavour, and in pharmaceutical products such as cough sweets, carminative and stimulant preparations. The maximum permitted level in food products is about 0.3%, but the actual content is usually less than 0.1%. The oil is also used in perfumery and cosmetics, with a highest maximum use level of 0.4%.

In current Western medicine, fennel oil is administered as a stimulant, carminative, antispasmodic, expectorant, flavouring in laxatives and a weak diuretic. Monographs on the physiological properties of fennel oil and sweet fennel oil have been published by the RIFM. The oil has antioxidant, antifungal and antibacterial activity, and possesses antiviral activity against potato virus X, tobacco mosaic virus and tobacco ringspot virus.

Oleoresin. Fennel oleoresin is prepared by solvent-extracting whole seeds and is normally offered with a volatile oil content of 50% or a guaranteed content in the range 52–58%. Only very small quantities of the oleoresin are produced for specialized use, as it is not a general substitute for fennel oil.

Sweet fennel oil. Sweet fennel oil is distilled from the fruit of *F. dulce*, but production has dramatically declined, a trend forecast to continue in favour of fennel oil. The main constituents of Hungarian sweet fennel oil are limonene 20–25%, fenchone 7–10%, and *trans*-anethole 4–6%. The oil is placed in the sweet, non-floral, candy-flavoured group by Arctander (1960). In the USA the regulatory status generally recognized as safe has been accorded to fennel oil, GRAS 2481, and sweet fennel oil, GRAS 2483.

Anethole. Fennel oil is a source of anethole, and although a synthetic substitute is readily available, some users prefer anethole of natural origin, while in many countries the use of synthetic anethole in food products is illegal. Major competitive sources of natural anethole are star anise, *I. verum* and anise, *P. anisum*, earlier described. Anethole can also be produced from estragole extracted from *Pinus* oil.

Anethole is effective against *Staphylococcus aureus*, *Escherichia coli*, *Candida albicans* and *Corynebacterium* spp., has stimulant and carminative properties, and is allergenic and weakly insecticidal. Long-term studies showed it is not carcinogenic. The oestrogenic activity (increasing milk secretion and promoting menstruation) of fennel is probably due to the polymers of anethole, including dianethole and photanetholes. Estragole is a hepatic carcinogen in mice; limonene limited mammary tumour growth in rats.

Unrelated species

Plants popularly but incorrectly named fennel include the fennel flower (*N. sativa* L), hogs fennel (*Peucedanum officinale* L.), closely related to dill, water fennel (*Oenanthe aquatica* (L) Poir.) and dog fennel (*Anthemis cotula* L.).

11

Zingiberaceae

The *Zingiberaceae* includes some 47 genera, containing approximately 1500 species of perennial tropical and subtropical herbs mainly as ground flora of lowland forests, divided into two subfamilies, *Costoideae* and *Zingiberoideae*. The *Zingiberoideae* includes several important spice producers: cardamom (*Elettaria cardamomum* L. Mat.), ginger (*Zingiber officinale* Rosc.) and turmeric (*Curcuma domestica* Val.).

Cardamom

The *Elettaria* Maton contains seven species in Indo-Malesia including *E. cardamomum* (L), *Elettaria major* Sm. and *Elettaria speciosa* Blume. The genus name is said to be derived from the Sanskrit *elat-eri*. In the *Susruta Samhita* (AD 600) it is named in Sanskrit *ela*, the seed *ela-tari*, and the plant or rhizome *ela-kai*, from whence is derived the modern Hindi *elaichi*.

The earliest reference to cardamom is on a clay tablet from the ancient city of Nippur, Sumaria, dated to 2000 BC, which indicates that ground cardamom was mixed with bread and added to soups. Cardamoms are reportedly described in the Ayervedic literature of India from the 3rd century BC, and recommended for treatment of stomach and urinary disorders. The fruits have been traded in India for at least 1000 years, and known as the Queen of Spices, with pepper the King. Exports from the Malabar coast were described by the Portuguese traveller Barbosa in 1514, although pepper was the main commodity and cardamoms only occasionally mentioned. However, by the time of Garcia da Orta in 1563 the trade in cardamoms was well developed and he differentiated between the smaller, aromatic form (var. *cardamomum*) from India and the larger fruited form (var. *major*) from Sri Lanka. The Arab geographer Idrisi (1100–1166) in his *Kitab Rujar*, completed in 1154, describes the spice as a product of Sri Lanka, although Marco Polo does not mention it. Subsequent writers described in detail the local Sri Lankan cinnamon industry (Chapter 3), but make no mention of the established but small, local trade in wild cardamoms. In India cardamoms are used as a masticatory and may be included in the chewing quid *pan* with areca nuts and leaves of betel pepper (Chapter 8).

Very early Chinese writings mention a spice translated as cardamom, but this was most probably an indigenous unrelated species *Alpinia globosa* (Lour.) Horan (*Amomum globosum* Lour.), still used as a cardamom substitute and in the treatment of urinary infections. Later the true cardamom was imported via Javanese entrepôts, carried there by Hindu merchants. The spice became and remains a common component of many South-east Asian dishes.

The early history of cardamom in Europe is confused, and it is now considered unlikely that either the Greeks or the Romans had access to the true cardamom in any quantity, although Theophrastus, Dioscorides and Pliny used the name *amomum* and *kardamomun* for several unrelated spices, which may have included cardamom. When knowledge of and the true cardamom of the *Elettaria* became freely available in Europe is unknown, but no substitutes were subsequently accepted and the name cardamom became general in European languages. Despite the medicinal use of cardamom in its native habitat and by the Arabs, it is not mentioned in early European herbals.

Cardamom cultivation in India is concentrated in the natural habitat of the species, except for a small area in Maharashtra where it is grown as a subsidiary crop in areca nut gardens. The planted area in India since 1950 has varied from 40,000 ha to 80,000 ha in 2000, with a low in the 1960s due to an outbreak of mosaic disease. Cultivation is scattered throughout the hill forest zone of the Western Ghats, with about half in the Cardamom Hills, Travancore–Cochin. To improve the quality of cardamom for export and provide a marketing structure and facilities for buyers and protection from exploitation for growers, in 2000 the Indian Spices Board opened a cardamom trading centre at Bodinayakannur, Cochin.

In Sri Lanka, about 4000 ha of cardamom are grown in the hills of Kanda Matale and Nuwara Eliya, with two-thirds in the Kandy district. However, yields are low and production does not reflect the area under the crop. Cardamoms have been introduced widely in tropical regions, Malaysia and Tanzania, for example, but are not produced anywhere on a commercial scale, although small local surpluses may be exported to neighbouring countries.

Botany

E. cardamomum (L.) Maton (syn. *Amomum cardamomum* L.; *Amomum repens* Sonn.; *Alpinia cardamomum* (L.) Roxb.) is known in English as cardamom but *cardamomo* in many European languages, in modern Arabic as *hal*, in East Africa, KiSwahili as *iliki*, in India, Hindi as *choti elaichi*, in Malaysia as *biah pelaga*, in Myanmar as *bala* or *pala* and in Guatemala as *cardamomo*.

Collections of cardamom germplasms are maintained at the Research Institute for Spice and Medicinal Crops, Bogor, Indonesia, and at the National Repository of Plant Genetic Material, Idukki, and National Research Centre for Spices, Calicut, India (Ravindran and Peter, 1995). Current breeding programmes include selection for higher fruit yield, resistance to mosaic disease and thrips (Venugopal and Padmini, 1999; Padmini *et al.*, 2000). The basic chromosome number of *Elettaria* is x = 12 and the somatic number of *E. cardamomum* is $2n = 48$ or 52. The systemics of *E. cardamomum* remain confused, as many local races and cultivars cross easily and a range of botanical types exist. A basic division is recognized on fruit size: *E. cardamomum* var. *cardamomum* (the var. *minuscula* of Burkill) including most of the cultivated races of India with small fruit, and *E. cardamomum* var. *major* Thwaites, native to Sri Lanka with a large fruit.

E. cardamomum var. *cardamomum* includes all the cultivated types. Plants are generally smaller, up to 3 m, the inflorescences longer, prostrate and bearing more flowers; the fruits are subglobose up to 1–2 cm in diameter containing numerous aromatic seeds. *E. cardamomum* var. *major* comprises the wild cardamoms common in Sri Lanka and southern India, later described. While this division is botanically acceptable, cultivated cardamoms are also separated in international trade into two major groups, Malabar and Mysore, although in India a third intermediate between the two and believed to be a natural hybrid, Vazhukka, is recognized. It is cultivated mainly in Kerala State.

The Malabar cultivar group comprises plants less than 3 m, with leaves up to 30–45 cm, which are hairy on the lower surface; the panicles are 60–90 cm and prostrate, and the fruit is small, globose or ovoid and lightly ribbed. They are reportedly susceptible to mosaic (katte) virus (see Disease section). The wild cardamom of South India is the Malabar cardamom, including the large-fruited Malai and Manjarabad types, and cultivated mainly in Mysore and Coorg districts of Karnataka State, and also Travancore. It is grown from 600 to 1200 m. Coorg Green is the best-known type.

The Mysore cultivar group comprises large plants with leafy stems up to 5 m, with large, coarse leaves, glabrous below. The panicles are erect and arching, and the fruits small, fusiform, three-angled and lightly ribbed. Plants are resistant to katte virus, and are cultivated on a large scale in Kerela and Tamil Nadu States of southern India, as they thrive under a wide range of conditions, are hardier and require less water. Grown generally from 900 to 1200 m. The Alleppey Green is a Malabar type.

References in the text refer to the true cardamom, *E. cardamomum* var. *cardamomum* and its derivatives unless otherwise stated. Local cultivars differ substantially in appearance, fruits, etc. and the following is a general description. Figures quoted indicate the range occurring over cultivars, not the range of a single cultivar.

Cardamom is a robust perennial herb, with a branched, underground, horizontal, rather woody rhizome with numerous fibrous roots in the surface layer (Fig. 11.1). High carbon dioxide levels showed little effect on root development or ability of roots to absorb additional nutrients (Arnone, 1997). Mycorrhiza may also be associated with cardamom roots, and in India these included *Glomus macrocarpum*, *Glomus fasciculatum* and *Gigaspora coralloidea*. High levels of applied fertilizer tended to reduce the mycorrhiza present.

Up to 20 erect, leafy shoots, to 2–5.5 m, arising from the rhizome produce a thick clump (Fig. 11.2). The leaves are distichous, the ligule entire to 1 cm, the lamina lanceolate acuminate, and 25–90 cm × 5–15 cm. In colour leaves are dark green, glabrous above, and paler beneath. The lower surface may be either glabrous or pubescent depending on the cultivar. The free portion of the petiole is up to 2.5 cm, with sheathing at base, and together with other sheaths forms the pseudostem. Studies in India indicated that leaf photosynthesis was higher under relatively low light intensities.

The inflorescences, 60–120 cm, arise from the rhizome at the base of the leafy shoots, and are generally recumbent or decumbent, seldom erect, slender panicles. The bracts are alternate, lanceolate, 3 × 1 cm, with axillary cincinni, and usually two- to three-flowered. The flowers are hermaphrodite, zygomorphic, 4 × 1.7 cm; the bracteole is tubular, up to 2.5 cm, and the calyx green, tubular to 2.5 cm, shortly three-toothed and persistent (Fig. 11.1). The corolla tube is pale green, similar in length to the calyx, with three narrow, strap-shaped, spreading, lobes up to 1 cm. The most conspicuous part of the flower is the obovate labellum composed of three modified stamens, 1.5–2.0 cm, with an undulating edge, and is white with violet streaks radiating from the centre.

The lateral staminodes are inconspicuous and subulate; there is a single sessile anther, with a short, broad filament, and a connective with a short crest at the apex. The inferior ovary, 2.3 mm, consists of three united carpels with numerous ovules in axile placentation and a slender style with a small capitate stigma. Flowers on the majority of cultivars are self-sterile, and to ensure complete pollination a mixture of clones is necessary. Pollination is by insects, mainly bees, which are most active in mid-morning, and a large bee population is necessary for high capsule yield (Chaudhary and Rakash, 2000); a relationship was established in India between style length and ability of bees to extract nectar (Belavadi *et al.*, 1997). Buds required about 30–32 days from initiation to full bloom, and capsule development an additional 110 days.

Flowers open from the base of the panicle upwards over a long period, and flowering may continue throughout the year on panicles of the current year and those of the previous year. Most flowers open early in the morning, and anthesis is between 6 and 8 a.m. when pollen is shed. Pollen viability is variable, generally around 70%, and quickly loses its viability, to 6.5% after 2 h storage and nil after 8 h. The stigma is most receptive between 8 and 10 a.m. giving about 72% fruit set. Receptivity gradually decreases, with a fruit set of 24% at 4 p.m. and nil at 6 p.m. After the panicles have flowered,

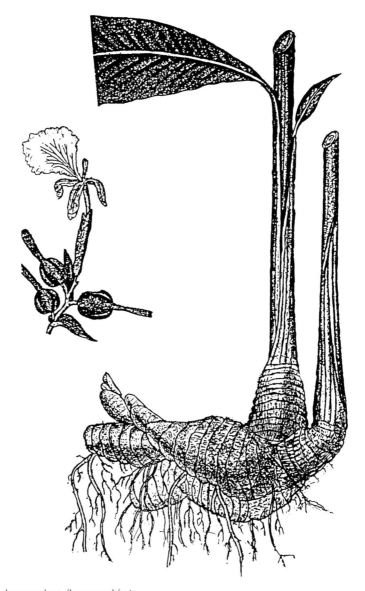

Fig. 11.1. Cardamom plant, flower and fruit.

fruited and died, the vegetative shoots bearing them also die.

The fruit is a trilocular capsule, fusiform globose, varying in size according to the cultivar, 1–5 cm, pale green to yellow becoming brown when fully ripe, and containing 15–20 seeds. The fruit contains an essential oil (mainly in the seeds), fixed (fatty) oil, also mainly in the seeds, and both are described at the end of this section. The fruit also con-

tains pigments, protein, pentosans, sugars, starch, silica, calcium oxalate and minerals. The approximate composition of Indian dried fruit was carbohydrates 42%, fibre 20%, water 20%, protein 10%, fat 2% and ash 6%. In the hulled capsule, crude fibre is about 30%.

The seeds are about 3 mm long, dark brown, angled, rugose and aromatic, with a thin mucilaginous aril. The seeds contain

Fig. 11.2. Cardamom plant.

white perisperm and a small embryo; the major seed constituent is starch up to 50%; 1000-seed weight is 20–25 g. Indian cardamom seeds averaged the following: moisture 8%, volatile oil 8%, total ash 5%, non-volatile ether extract 3%, crude fibre 9%, crude protein 10%, starch 46%, Ca 0.3%, P 0.2%, K 1.2% and Fe 0.012%, plus vitamins (per 100g) thiamine (B1) 0.18 mg, riboflavin (B2) 0.23 mg, niacin 2.3 mg, ascorbic acid (C) 12 mg and vitamin A at 175 IU.

Fatty oil. The fixed, fatty oil content of fresh fruit is up to 10% depending on cultivar. Composition varies for the same reason; constituents include ten fatty acids, the most abundant being palmitic at 25–40%, oleic at 40–45% and linoleic from a trace to 20%; however, some cultivars contain between 15 and 35% stearic acid.

Essential oil. An essential oil can be distilled from all plant parts, but differs in composition. The essential oil content of freshly harvested fruits is about 10–12%, mainly in the seeds; the major constituents are 1,8-cineole and α-terpinyl acetate, up to 50% each. A commercial sample of a steam-distilled oil in the USA had the following characteristics: specific gravity (25°C) 0.917–0.947; refractive

index (20°C) 1.4630–1.4460; optical rotation (25°C) +22° to +44°; phenol coefficient 7.5; dissolves in 7 vols 70% alcohol, soluble in ether, miscible with absolute ethanol, insoluble in water. Indian oils had very similar characteristics, plus an acid value of 0.36–1.3, and a saponification value of 96.5–156.4. The non-saponifiable lipid fraction of Indian oil consisted mainly of *n*-alkene waxes and sterols.

Ecology

Cardamom occurs naturally in the evergreen monsoon forests of the Western Ghats in southern India, Sri Lanka and may also occur in the northern monsoon forests of Myanmar and Indo-China, generally between 750 and 1500 m. Until the beginning of the 19th century, the world's supply of cardamom came mainly from wild plants in these forests.

Cultivated cardamoms are grown at similar altitudes in India: in Malabar at 600–1200 m, in Mysore at 900–1200 m, and in Vazukka at 900–1200 m, with the most productive range between 1000 and 1200 m, although cardamom yields well up to 1500 m. Annual rainfall is usually 2500–4000 mm, with the heaviest falls during the south-western monsoon in

June–September, followed by the north-eastern monsoon to December. The length of the dry season, 3–5 months, is the main factor determining cardamom's natural range; prolonged dry periods or successive poor monsoons endanger the crop. However, cardamoms can be grown with a rainfall of 1000–1500 mm if planted in shaded valleys, for example, with run-off conserved or by heavy mulching. A temperature range of 10–35°C occurs over the main producing regions, with an annual mean of 22–24°C being the optimum, and about 17°C the lower limit.

The crop grows naturally under permanent natural shade with only intermittent sunshine, but will produce good crops with only partial shade if well watered (Korikanthimath *et al.*, 1998b). In general, where coffee flourishes so will cardamom, and this is a major reason why Guatemala has become the world's largest producer and exporter. Cardamom can also tolerate a loftier canopy and a heavier shade than coffee, and maintaining the correct balance is important in cardamom cultivation. In natural forest, thinning of the tree canopy or encouraging tree seedlings is common; in some regions *Macaranga peltata* Muell., which appears spontaneously in forest clearings, is encouraged to provide temporary shade. In cleared areas and under coffee, shade trees such as *Erythrina* spp. may be planted; in the areca nut plantations, shade is provided by the trees themselves. Malabar cardamom is less sensitive to sunlight than Mysore, provided soil moisture is sufficient, but is liable to suffer more damage than Mysore cardamom at lower light intensities if soil moisture is low. Cardamoms are very susceptible to wind damage and in India and Sri Lanka an eastern or south-eastern aspect is preferred.

Soils

The most suitable soil is a fairly deep forest loam, with a good natural mulch in which the shallow mass of fibrous roots and rhizome can develop without constraint. Such soils are often moderately acid, although in Karnataka, India, cardamom soils were pH 5.3–5.8. Details of soil characteristics from the two main Indian cardamom growing regions are contained in Korikanthimath *et al.* (1998b). Soils that produce good yields of coffee are generally suitable, Brazil excepted. However, cardamoms can be grown on a range of soils from deep chocolate-coloured forest loam, red soils derived from laterite and white quartz gravel with only a shallow humic layer, to soil pockets on hillsides and in ravines, provided plants are well mulched. Cardamoms require good drainage and cannot tolerate waterlogging at any stage of growth.

Fertilizers

Cardamoms naturally grow in soils high in organic matter, and cultivated plants require a similar environment. Thus, the usual fertilizer is a heavy mulch provided by slashing surrounding undergrowth or obtained from nearby grassland, plus leaf fall from shade trees, which can amount to 5–8 t year^{-1}. Application of plant residues or animal wastes is recommended, but generally applied only when not required for other crops. Distance from the homestead to the cardamom plantation, often by small paths, is a common reason for the low application rates. The beneficial effect of such applications is both obvious and well documented, with a doubling of fruit yield in some instances. Organic materials include castor poonac, coffee and rice husks, bone and fish meal, and other processing wastes. Cardamoms underplanted in plantation crops may be mulched with leaves of *Emblica officinalis* Gaertn. plus cattle manure; when planted in areca nut gardens and coffee plantations, cardamom benefits from the fertilizers applied to the main crop. An interaction between shade and fertilizers in India showed the former resulted in a higher optimum fertilizer level of NPK at 100:20;100 kg ha^{-1} (Korikanthimath *et al.*, 1998a). When a cardamom plantation is established or replanted in India, pits for the seedlings are filled with topsoil mixed with 50–100 g of the local compound fertilizer, commonly NPK at 14:28:14 or 30:60:30. The few data available indicate that urea is more profitable than other N sources.

It was determined in India that the ratio of major nutrients in a cardamom clump was 6:1:12:3:0.8 for N:P:K:Ca:Mg; another assessment showed the total uptake in kg ha^{-1} was N 26, P 4.5, K 52, Ca 14 and Mg 3.5. The nutrients removed by a cardamom crop to produce 4500 kg ha^{-1} fruit, are shown in Table 3.4. Plant nutrient status at various growth stages showed the rhizome had an increasing K content from panicle initiation to capsule maturity, while the NPK content of leaves and pseudostems declined as rhizomes matured (Table 11.1).

Indian cardamom fruits also contained the following trace elements (average content in p.p.m.): K 1268, Ca < 45, Ti < 40, Cr < 35, Mn < 30, Fe 187, Zn 51, Br < 15, Rb 43 and Sr < 15 (Joseph *et al.*, 1999), and the Ca, Mg, Fe, Zn, Mn and Cu content in all plant parts increased from panicle initiation to capsule maturity (Vasanthakumar and Mohanakurmaran, 1989). Thus, any minor nutrient deficiency as determined by soil analysis must be corrected. An assessment of Indian data indicates that cardamom has a very high K requirement, and soils low in K may require up to 200 kg K ha^{-1}, and up to 220 kg K ha^{-1} if irrigated (Nair *et al.*, 1997). The Indian Department of Agriculture recommendations for annual application to raingrown crops varies regionally, depending on the local soil nutrient status, between 45 and 70 kg N ha^{-1}, 30 and 40 kg P ha^{-1}, and 85 and 200 kg K ha^{-1}; or for an average yield of 100 kg dry capsules, a mixture of NPK at 75:75:150 kg ha^{-1}, plus NPK at 0.65:0.65:1.3 kg ha^{-1} for every additional 2.5 kg capsule yield. In general, half the recommended amount is applied in the year of establishment at the beginning of the rains.

Cultivation

Cardamoms can be propagated vegetatively by division of rhizomes, from seedlings raised in nurseries, but clonal multiplication is a proven technique for raising yields and disease resistance, and a rapid method has been developed in India (Korikanthimath, 1999). Where facilities exist the various micropropagation and genetic manipulation techniques produce large numbers of improved plantlets.

For vegetative propagation, rhizomes are dug, divided into pieces consisting of at least one old and one young shoot, and planted into prepared pits. The method is simple, reliable, less costly and plants begin bearing a year earlier than from seedlings, but division may not produce enough material for large areas. The effect of growth regulators on shoots did not affect the production of either stems or additional suckers (Shanthaveerabhadraiah *et al.*, 1997). Introducing material from elsewhere can be dangerous, as it may be infected with mosaic virus or thrips. However, local Departments of Agriculture have selected and bulked up high-yielding or disease-resistant clones for distribution to local growers (Korikanthimath and Ravindra, 1998; Korikanthimath, 1999).

Table 11.1. Nutrient extraction by cardamom at various growth stages in India.

	Nutrient (kg ha^{-1})				
	N	P	K	Ca	Mg
Young suckers	1.07	0.32	2.85	0.23	0.07
Mature suckers	9.27	2.01	17.97	2.45	0.64
Bearing suckers	12.17	1.40	20.01	8.84	2.32
Rhizome	1.54	0.24	6.50	1.49	0.31
Roots	1.92	0.38	4.78	0.95	0.14
Total	25.97	4.35	52.11	13.96	3.48
Ratio	6	1	12	3	0.8
Nutrient removal kg^{-1} of capsules harvested	0.122	0.041	0.200	0.088	0.023

When large numbers of plants are required, selected seed is sown in special nurseries, established in a sheltered position with a reliable water supply. Raised seedbeds, usually about 1–2 m wide and shaded, are made from soil mixed with equal quantities of well-rotted cattle manure and wood ash. About 1 kg seed (45,000 to 50,000 seeds) should provide enough plants for 1 ha, but as germination is often erratic for as yet undetermined reasons, up to 1.5 kg may be used.

Seed should be collected from sound, ripe capsules on vigorous, high-yielding, disease-free plants, preferably at least 5 years old. However, as cardamom is cross-pollinated there will be considerable variability unless obtained from clonal material. Seed should be washed to remove any mucilage, dried and sown as soon as possible, or if carefully dried and mixed with ash may be stored for up to a week; longer periods quickly reduce germination. Time of collection also affects germination; in Mysore seeds collected in September gave a 72% germination rate compared with 7% from January seeds. September seedlings also developed faster, had greater resistance to leaf spot and leaf rot, and reached about 45 cm for transplanting in August, whereas seedlings from seed collected in other months required up to 12–13 months to reach this size.

Seedbeds are constructed of soil, the surface sterilized by burning straw or grass, and a mixture of well-rotted manure or compost, ash and soil is added. Seed is broadcast on beds at about 60 g per 60 m^{-2} of bed (about 8000 seeds m^{-2}), lightly covered with more soil, sand, grass or chopped straw, kept moist but not wet, and roofed to prevent rain falling on the beds. Seeds germinate in 5–7 weeks, but germination is irregular and may continue for many months. Sowing time is usually August–October in Coorg and Mysore, and February–March in Kerala and Coimbatore. In higher, cooler districts, seedbeds are covered with black plastic, or plastic tunnels, to increase the temperature and improve germination.

Seedlings are carefully transplanted into growing beds when 3–6 months old (about 15 cm high), 15–30 cm apart, kept moist and sprayed with an insecticide and fungicide. Seedlings may remain in these beds for up to 2 years, and experienced growers believe the more mature seedlings have a greater survival rate after planting out. A spacing of 2 × 1.5 m requires 3400 and 2 × 2 m requires 2500 seedlings ha^{-1}, plus 5% for infills.

Planting pits for shoots and seedlings are usually about 0.5 m in diameter and 0.5 m deep, or 60 × 60 cm × 0.5 m deep, filled with a mixture of topsoil and compost or well-rotted cattle manure, and as plants are shallow-rooted, not more than 8–10 cm deep. The preferred local spacing varies from 1.5 × 1.5 m to 3.0 × 3.0 m depending on cultivar and soil fertility. In India, Malabar cardamoms are usually spaced about 2 × 2 m, Mysore and Vazhukka about 3 × 3 m. Seedlings are preferably planted at the beginning of the monsoon and fastened to stakes to prevent damage. Two seedlings are planted in some pits to provide plants to fill gaps. Shallow weeding is carried out as necessary and mulch provided once seedlings are established.

Although cardamoms have been collected from the wild for centuries, it is only comparatively recently that plantations have been established. Initially cardamom plants were encouraged by selective thinning of the forest in which they occurred naturally, the volunteer seedlings thinned, others planted to increase the stand, and periodically weeded. In more remote parts of India, this system persisted for centuries; small patches of forest were cleared and planted with self-sown cardamoms or nursery seedlings and occasionally weeded. Shade was provided by the surrounding trees. Harvesting began in the third year and continued for another 8–10 years. The plot was then abandoned to the jungle and new plots cleared.

With the destruction of forests, this system is no longer possible and plantations are now common. To establish a new plantation, the land is cleared, leaving only the tallest trees to provide shade. However, since jungle is becoming scarce, new plantations may have to establish shade trees in advance, while seedlings are in seedbeds. In India, shade trees include *Erythrina* spp., *Albizia* spp., *Artocarpus integrifolius* (jackfruit) and *M. malabarica* (wild nutmeg), and when mature these should be pruned just before the monsoon to ensure sufficient light reaches plants during the rainy season. Size of a plantation varies with the majority between 2 and 20 ha, although newer commercial estates are up to 500 ha. Following establishment plantations must be weeded, dead plants replaced, old and dying stems removed, and the level of shade adjusted. Mulch must be replenished and organic manures applied as available. The economics of replanting and managing a cardamom plantation in India over the period 1993–1999 have been reviewed and the cost of the various operations tabulated; wages accounted for nearly 60% of the total (Korikanthimath, 2000).

Flowering may commence in the second year after planting out, but in general the first saleable crop is obtained in the third year. Subsequently plants flower throughout the year, but most flowers occur during the rains in April–August. To ensure adequate pollination, four bee colonies ha^{-1} are recommended in India, which also produce very saleable honey.

Weed control

Weed control is usually combined with general crop care, and includes slashing of surrounding vegetation, which provides a mulch. Frequent but shallow weeding is essential in the first year after planting, as the natural vegetation is composed of vigorous grasses, herbs and tree seedlings, which quickly regenerate, three times in the second year, and then once or twice annually as necessary, as once established plants must not be disturbed. Keeping weeds and grasses under control also reduces the level of pest and disease infestation. Since most cardamoms are grown by smallholders, weeding is usually manual, with the surrounding vegetation slashed by hand. This can be dangerous as snakes are common in the undergrowth. On large well-established plantations, ducks-foot weeders set as shallow as possible or rotary slashers control most weeds. Herbicides are seldom used.

Irrigation

The majority of cardamom plantations are rain-grown, with supplementary irrigation only where water is easily available, or in plantations where a drip or overhead irrigation system is integrated in the layout. Irrigation is important mainly to offset the effects of a prolonged dry period, desiccation by strong dry winds, to ensure survival of young seedlings and prevent premature capsule drop. Data from India indicate that maintaining 25–30% available soil moisture produced the optimum yields, up to 2.5 times that of non-irrigated crops. Sprinkler irrigation trials also increased the essential oil content from 8.2% to 10.2% (Vasnanthkumar and Sheela, 1990).

Intercropping and rotations

Cardamoms are often interplanted in coffee or areca nut plantations in India, as a secondary crop in pepper gardens and less often in tea (Korikanthimath *et al.*, 1999; Srinivasan *et al.*, 1999). However in Karnataka, India, coffee and cardamom competed for inputs and resources, but the combination was more profitable than monocrops (Korikanthimath *et al.*, 1998a). Undercropping cardamom in Indonesian rubber plantations was successful, and shade from mature trees had little effect on cardamom yield, nor the cardamoms on rubber yield.

Cardamom is seldom intercropped in Sri Lanka. In Guatemala and other Central American countries, cardamoms are frequently interplanted in coffee. With the diminishing supply of forest, an increasing number of exhausted plantations are being replanted as soon as possible, as no other crops are suitable for a rotation, desirable though this may be in terms of pest and disease control. Some reduction in damage is obtained by clean-weeding and burning off undergrowth prior to replanting. A major advantage is the higher yielding clonal material available from the Cardamom Board's replanting schemes, plus financial assistance in re-establishment.

Pests

A number of pests of cardamom are noted in the literature, often with little information on the degree of damage and whether it occurs to plant parts or fruit. Additionally, identification of the pest is confused by the use of the same, or similar, common name for several insects. The cardamom thrip, *Sciothrips cardamomi*, is a major pest in all Indian and Sri Lankan cardamom-growing districts. Control is difficult as it also infests a range of crops and wild plants. The thrips are small greyish-brown or yellowish insects, 1–1.5 mm, with hair-fringed wings. Eggs are laid in the softer plant tissues and the nymphs hatch in 7–9 days, moult three times in 3 weeks and then pupate for 10–15 days. Thrips attack leaf bases and floral bracts, and flowers may fail to set; attacked fruit develops rough corky scabs known as cardamom itch, resulting in unsaleable split or discoloured fruits. A secondary effect is that thrip damage increases the 1,8-cineole content of the essential oil. All current cultivars are susceptible to some extent (Jasvir *et al.*, 1999).

Whitefly nymphs, *Kanakarajiella cardamomi* (*Dialeurodes cardamomi*), suck the sap of shoots and leaves, and a large infestation can fatally desiccate plants. Eggs are laid on the undersurface of leaves and hatch in 7–10 days. The green nymphs pupate in a waxy web on the undersides of leaves in another 21–28 days. In India, the pest is most damaging in April–May and October–November, has a range of alternative hosts and also natural enemies, and may also be the vector of a virus (Selvarkumaran *et al.*, 1998b). Also causing significant damage in a specific area are *Conogethes* spp., especially *Conogethes punctiferalis* in India, and *Basilepta fulvicornis* also in India. Several polyphagous coccids have been recorded on cardamom, whose identity is suspect; one was probably *S. coffeae* and another *Pseudococcus adonidum*, both having a wide host range. The aphid *Pentalonia nigronervosa* is a vector of the mosaic disease known as *katte* in India (see Disease section). A large moth with a 10 cm wing span, *Eupterote mollifera*, can occur in swarms, and deposits its eggs in large clusters on the undersurface of leaves of neighbouring and shade trees. After the second moult, the caterpillars descend from the trees by threads and can completely defoliate cardamoms. The caterpillars are 7–10 cm, greyish-black with tufts of stinging hairs, and pupate in the soil around plants.

Several borers damage cardamoms, and one of the most important is a brown weevil, *Prodioctes haematicus*, which bores rhizomes in Sri Lanka and India. The adult weevil has three dark lines on the pronotum and six dark spots on the wings. The female lays eggs on the rhizome and larvae tunnel into stems and rhizomes, which also allows entry of root-rot fungus. *Stephanoderes* spp. larvae may also damage roots and lower stems. Larvae of the yellow peach moth, *Dichocrocis punctiferalis*, one of the commonest tropical shoot and capsule borers, is usually a minor pest of cardamom but occasionally becomes more damaging. It also attacks a range of crops including ginger and turmeric. Symptoms are desiccation of the shoot tip and holes about 3 mm in diameter in the stems. Larvae of a pod borer, *Lampides elpis*, enter tissues and pupate in the young fruits. Unidentified borers are reported as attacking roots but usually cause minor damage.

A tingid bug, *Stephanitis typicus*, lays eggs on cardamom leaves, which are subsequently food for its larvae. Damage is usually minor, but may be more damaging when concentrated on adjacent cardamom clumps. Nematodes attack germinating seed in seedbeds and seedlings in nursery beds, which should be fumigated before planting; species recorded from cardamom with the degree of damage and suggestions for their control has been reviewed (Ramana and Eapen, 1999). A most important pest of stored cardamom is the red flour beetle, *Tribolium castaneum*.

A different category of pests that can be most damaging in a specific district are birds, monkeys, porcupines, rats and wild pigs, while theft of plants or fruit, which was previously almost unknown, is becoming more common as the number of landless and unemployed people continues to increase.

Diseases

Disease can be a limiting factor in cardamom production in some regions, and the most important are described below; a general review of cardamom diseases has been published (Sarma and Anandaraj, 2000). The most serious and economically damaging disease of cardamom in southern India, known locally as *katte* or marble disease, is caused by a virus transmitted by the banana aphid, *Pentalonia nigronervosa*, which is the vector of a bunchy top virus of bananas. Cardamom plants may be attacked at any stage of growth, but symptoms usually appear on young leaves. The main symptom is a general chlorosis of the entire leaf with slender, interrupted parallel streaks of pale green following the veins from midrib to leaf margins. Later the streaks become evenly distributed giving the characteristic mosaic pattern, and such plants can be easily recognized in the field by their pale appearance. Affected cardamom clumps rapidly deteriorate, newly formed shoots become smaller, yield is greatly reduced, and during the 10–12 months following appearance of the disease, rhizomes gradually shrivel and the clump eventually dies. If young plants are attacked, they generally die before flowering. The only effective control is careful removal of affected plants, replanting with disease-free seedlings, and spraying to control the aphid vector. Losses can reach 70% of potential yield in the year following infection, and in some areas all clumps become infected in the third or fourth year. Resistance to *katte* disease is being investigated (Venugopal, 1999), and symptoms that may be caused by other viruses have been noted (Saravanakumar *et al.*, 1998).

Causal agents of leaf diseases occurring in nurseries include *Coniothyrium* spp. and *Phyllosticta* spp., produce spotting and eventual necrosis of seedling leaves. Symptoms are small water-soaked spots, which gradually become larger; affected tissues become flaccid, die and shrivel along the midrib. Both are most damaging in the wet season and can decimate seriously infected nursery beds. Various other leaf diseases that cause minor damage include a rust due to *Uredo*

elettariae, a leaf spot caused by *Chlamydomyces palmarum*, and another due to *Cercospora zingiberis*. A capsule rot that has become more widespread and damaging in the last decade is due mainly to *Phytophthora nicotianae,* although other organisms may be present, and can destroy up to one-third of capsules on affected plants. Capsule canker due to *Xanthomonas* spp. is becoming more widespread, and all current cultivars are reportedly susceptible.

Rhizome or root rots due to various pathogens cause a gradual decline in vigour of plants, and are often confused with mosaic, but infected plants can be distinguished by yellowing of the leaves without a mosaic pattern; older leaves often die back, and new shoots are weak and unhealthy. Subsequently rhizomes and stem bases decay, shoots become very brittle, and the plants are easily uprooted. Affected rhizomes are often covered with fungal mycelium. Pathogens isolated from such rhizomes include *Cephalosporium* spp., *F. oxysporum, Pylhium aphanidermatum, P. vexans* and *R. solani*, while the rhizome borer *Prodioctes haematicus* can be present in affected plants. An important aspect of chemical disease control is the residues that may remain on or in capsules. Washing has proved successful with some chemicals (Mather *et al.,* 1999).

Harvesting

Fruits ripen over an extended period and are harvested every 3–5 weeks as convenient. Not only do fruits ripen over an extended period, but fruits on individual stems ripen in sequence from the lowest to the highest. The period to fruit maturity is a cultivar characteristic, but is about 3 months for Malabar and 4 months for Mysore cardamom. Malabar fruits begin to ripen in August and harvesting is completed in December–January; Mysore harvest begins in September and can last until April. Seasonal conditions directly affect the numbers and quality of fruit, but the most important quality factor is time of harvest, i.e. ripeness.

Fruits are picked individually with their pedicels when fully developed but unripe, as fully ripe fruits tend to split and lose colour on drying. Malabar cardamom is harvested when the fruit begins to lose colour and before it assumes a yellow tinge; Mysore fruits remain green and considerable experience is necessary to judge the correct time. For both types, fruit is ready for picking when it breaks cleanly when bent back from the stem. Unripe green fruits have a low essential oil content and tend to shrivel during drying, while small immature fruits produce a shrivelled, badly coloured sample. A most important factor influencing cardamom quality at harvest is that seeds should have changed colour from white to brown or black. Fruits should be handled with care to avoid bruising since this results in lower quality spice. Although modern plantations are partially mechanized, harvesting remains almost wholly manual as noted, and in 1989, harvesting required an average 800 working days ha^{-1} (Korikanthimanth *et al.*, 1998b).

Yield varies directly with the standard of management, size of holding and number of plants ha^{-1}, and is thus of little value in comparisons. Most Indian smallholders obtain a fruit yield of 50 kg ha^{-1}, but 500–700 kg ha^{-1} from modern plantations, especially those using clonal material; the average smallholder yield in Guatemala is much higher at 300 kg ha^{-1}, higher from plantations. However, the plant multiplication methods currently used in India could rapidly increase the average yield, although unless drying methods improve, as mentioned in the Processing section, final yield will continue to be reduced.

The economic life of an Indian plantation depends mainly on the cultivar and standard of management, from 7 to 10 years for Malabar and 10 to 15, exceptionally 20, years for Mysore cardamoms. A method of extending plantation life without replanting the whole area in one operation is to interplant rows with young, preferable clonal plants and then remove the original rows when young plants begin bearing. This ensures continuity of total yield and cash flow. The annual cost of maintaining a well-managed cardamom plantation in India in 1988 was calculated as Rs 30,000 ha^{-1}, or Rs 40 kg^{-1} at an average yield of 750 kg ha^{-1}, to produce a gross profit of Rs 60,000 after deducting cost of production. However, this was using clonal material and the yield was many times the national average. Another estimate gave the average cost as Rs 102 kg^{-1} for a yield of 200 kg ha^{-1}.

Processing

Freshly harvested fruits have a moisture content of about 75%, which must be reduced to below 12% for safe storage. Drying must be efficient to avoid mould formation, which causes blemishes and detracts from the appearance of the spice. To prepare green cardamom, fruits should be exposed to light for minimum periods during drying and storage, and to prevent volatile oil loss temperatures must be strictly controlled. Drying is a major factor influencing final yield, and in India the curing percentage as reported in *The Planters Chronicle*, July 2000 issue, was less than 20% for all smallholders (about 95% of the crop) but 25–28% for plantations. To improve this ratio and to improve the overall quality of cardamoms, the Indian Cardamom Board offers technical advice and financial assistance to introduce new, or upgrade, drying facilities.

Sun-drying. Freshly picked cardamoms are laid out in the sun on sacks or rush matting for 3–5 days, turned frequently to allow even drying, and covered at night or when rain threatens. Although this method is cheap, it takes longer than flue-drying and may result in surface blemishes from mould formation and other contamination. Following sun-drying in India, some fruits are bleached, generally by exposure to burning sulphur to suit particular customers, especially Iran. Bleached fruit is usually lower in volatile oil content, but resistant to weevil infestation. Sun-drying includes solar dryers, which are becoming more common.

Flue curing. Despite advances in artificial drying methods, most cardamom fruits are still flue-cured (Fig. 11.3), with about 4000 chambers in use in India; similar methods are used in the main growing regions of Guatemala and Sri Lanka. Freshly picked green cardamoms are washed and the stalks clipped off, although some growers favour

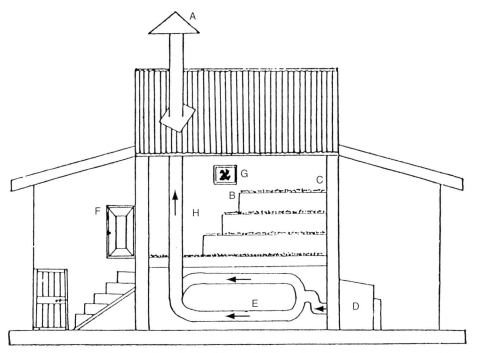

Fig. 11.3. Cardamom drying shed – India. A, chimney; B, wiremesh shelves; C, cardamom spread for curing; D, fireplace; e, flue-pipe; F, curing chamber door; G, exhaust fan; H, curing chamber.

removal of stalks after curing when they are easily detached by rubbing. Curing takes place in a special chamber, a *bhatti*, whose internal measurements are usually 4.5 × 4.5 m and 2.5 m high, sufficient to dry about 2000 kg per annum (Fig. 11.4). The walls are lined with shelves bearing trays, which are switched from top to bottom several times during curing to ensure even drying. The chamber is fitted with an outside furnace and a flue pipe fixed around the inside of the shed just above floor level, exhausting via a chimney stack. The washed, clipped fruits are spread evenly on the 26 × 20 cm trays and laid on the shelves; the curing room is sealed and the temperature raised rapidly to about 50–55°C and maintained for 3 h. This arrests further vegetative development and fixes the colour of the chloroplasts. Ventilators are then opened to allow moisture to escape and to reduce the temperature to around 40–45°C, and it is maintained at this temperature for some 30 h. The temperature is then again raised to 50–55°C for a further 3 h to complete drying. Using an alternative method, cardamoms are dried for about 30 h at 80°C. Research continues to improve the design and use of drying chambers (Sanjay *et al.*, 1999). In a direct-firing method now seldom used, the flue was replaced by a central charcoal fire, whose main disadvantages were the difficulty of maintaining a constant temperature, and the risk of smoke contaminating the fruits.

After drying, cardamoms should be hard and a uniform green; overheating turns fruit yellow and less saleable. Cardamoms are then removed from trays, cooled, winnowed, graded and stored in vermin- and lightproof containers until required for packing. Various minor modifications to this system have been introduced, but it remains basically unchanged since it is cheap to construct, simple to use and requires no skilled maintenance. However, about 0.5 m^3 of wood fuel is required to dry 500 kg of green fruits, and wood is now more expensive to obtain. A combination of solar and wood heating reduces the wood required by 33–50%.

Fig. 11.4. Interior of drying chamber – India.

Artificial drying. Although the methods described continue to be used, modern tray and tunnel dryers introduced by larger traders reduce fruit drying time, retain a good green colour and virtually remove the risk of fungal contamination. Most are still wood-fired although the fuel/dried fruit ratio is greatly reduced; a combined solar/wood fuel is even more efficient. Ongoing research by the Indian Cardamom Planters Association and Cardamom Board is aimed at reducing the solid fuel required, drying time and drying costs, while producing a sound, uniform green product without fungal contamination.

Distillation

Cardamom oil is usually produced by steam distilling crushed, preferably fresh fruits, but fruits that have been stored for a short period are acceptable. The process is straightforward, and no special stills or methods are required. In general, the fresher the fruits, the greater the oil yield and the higher the quality. Data indicate that a distilling period of not less than 4 h is necessary to obtain oil with the full flavour characteristics and organoleptic properties of the spice. Supercritical carbon dioxide extraction is gradually displacing distilling to produce cardamom oil, as the product is considered superior to distilled oils, and extraction plants are generally available. Seed from decorticated fruits can also be distilled, and although the oil yield is higher, oil characteristics and organoleptic qualities are similar to that of the whole fruit. Since several extra operations are required and distilling requires more skill, seed oil is more expensive for little obvious gain.

Storage and transport

Following drying, cardamoms are usually stored for some period prior to export and fruits should be handled with care to prevent bruising. On arriving at storage sheds, fruit is inspected and graded according to customer requirements by colour, size, or export or import specifications. In India, cardamoms must comply with the Agmark grades, as follows: Alleppey Green, five grades with Extra Bold the highest; Coorg Green, five grades with Extra Bold the highest; Bleached white cardamoms, four grades with Mysore/Mangalore Clipped the highest; Mixed, with the highest grade Mixed Extra Bold; Seeds, with Prime the highest grade. Cardamoms for export must also be free from visible mould and insects. Alleppy Green grades account for about 85% of export cardamoms.

The most important storage factors influencing retention of cardamom quality and colour are exclusion of light, moisture and temperature. A direct relationship exists between colour, volatile oil content and organoleptic qualities; as the green colour fades so do these characteristics. Changes are less apparent in bleached fruits. Thus, high-quality cardamoms are packed in lined, corrugated cardboard cartons, which are replacing wooden and other boxes, and kept in cool, shady storage. Lower quality fruits may be packed in new jute bags with a waterproof lining. Cardamom seeds lose their volatile oil, and thus their flavour faster than whole fruit, and should be packed in airtight containers as soon as possible after decortication.

Following distilling, and refining if necessary, oil should be placed in full, opaque, airtight containers for storage. Cardamom oil quickly loses its flavour if exposed to light and air and should remain in sealed containers in dark storage until exported. Data indicate that changes which may take place in stored oil in India were more pronounced in extracted than in distilled oil, and the terpene hydrocarbons content was most affected (Gopalkrishnan, 1994).

Products and specifications

The most important cardamom product is the fruit, which is the spice of commerce and obtained by drying fully developed fruit capsules as described. In international trade, the spice is frequently known as true cardamom to distinguish it from large cardamom and substitutes obtained by drying fruit of various species of *Amomum* in Asia, *Aframomum* in Africa, and wild *Elettaria* known as false cardamoms.

Fruit. True cardamom spice as a flavouring material is most commonly used as whole fruit, occasionally as decorticated seeds, and to a small extent as powder. The major use worldwide is for domestic culinary purposes as whole dried fruit, which may be ground before use (Fig. 11.5). Cardamoms are

divided into shorts, usually broad and plump, and longs, which are finer-ribbed and lighter, and three colour categories, green (flue-dried), bleached (sun-dried and bleached), and straw (sun-dried). Greens and sun-dried fruit account for about 85% of world trade.

Cardamom is the third most expensive spice, and therefore colour and aroma are very important to consumers; thus, cardamoms sold as spice should be a uniform green colour, and lack of colour usually indicates the spice is of poor quality or aged. Malabar cardamom is usually light green to greenish-yellow, while Malabar has a deep green colour. Quality is judged by opening a fruit and examining the seed, which should be brown-black, slightly sticky, with a strong aromatic smell and taste.

The Malabar (Coorg) cardamom has the most pleasant, mellow aroma and flavour, but as the seeds develop their full flavour only when capsules have begun to turn yellow, their appeal to consumers is less than Mysore cardamoms. Cured fruits are generally round to compressed oval (compared to the longer Mysore type) and about 18–19 mm. The Mysore (Alleppey) cardamom has a somewhat harsher aroma and

Fig. 11.5. Dried small cardamom.

flavour, but as seeds mature while capsules are still green, the fruits have a greater 'eye appeal' to consumers. Fruits are three-cornered, ribbed and around 20–22 mm long.

International trade in Indian cardamoms is supported by demand created in two distinct markets as noted; the Middle East cardamoms have been traditionally used for flavouring coffee, and in Scandinavia to flavour a range of baked goods, cakes, buns, pastries and bread. In other European countries and North America, the ground spice is included by food processors in sausage and similar products popular locally (in Germany it may be added to *sauerbraten*), soups, canned and pickled fish, in curry and similar spicy powders and beverages. However, there is little usage of cardamom as a domestic spice in the USA, which is supplied from Central and South American sources. From India to China, cardamom is an important ingredient in meat, vegetable and rice dishes.

Freshly decorticated seeds have a pleasant aroma and a characteristic warm, slightly pungent taste, but rapidly lose their essential oil content, and thus their flavour in storage. Seeds are not popular domestically as many consumers recognize and accept only whole fruit as the spice. Ground seeds have limited use in food processing.

Essential oil. An essential oil is obtained by steam distilling fruits, and according to Burkill (1966) was first obtained by Valerius Cordus in 1540. Supercritical carbon dioxide extraction is now more common, and the oil is obtained mainly in producing countries, with a yield of up to 8% in whole fruits and up to 12% in seeds; the two oils are virtually identical. The oil is a colourless to pale yellow, free-flowing liquid, which darkens on exposure to light. The odour is aromatic, warm spicy, initially with a camphoraceous/cineole note, which later disappears and the oil becomes sweet, woody and balsamic. The taste is strongly aromatic, warm, spicy, and pungent when concentrated. The oil is placed in the sweet, spicy powerful and warm group by Arctander (1960). Oil from Malabar and Mysore cardamoms differs; the latter initially has a prominent camphoraceous note, which is absent in the more mellow Malabar oil.

About 50 compounds have been identified in the oil, mainly oxygenated monoterpenes, monoterpene hydrocarbons and sesquiterpenes, but the amount and ratios between them vary considerably due mainly to the cultivar supplying the raw material. The seven main constituents of cardamom oil, whose ratios may vary according to region, are (approximately): α-terpinyl acetate 50%, 1,8-cineole 25%, linalool 6%, linalyl acetate 5%, α-terpineol 3%, geraniol 3% and terpinen-4-ol 2% (Gopalkrishnan, 1994; Lawrence, 1998b). Commercial oils may be a blend, and some are branded products. The organoleptic qualities of an oil depend on the ratios of its major components, and in general oils with a low cineole but high terpinyl acetate content are preferred by processors. Steam-distilled oil differs from oil obtained by supercritical fluid extraction, which is considered to reflect more accurately the natural odour and flavour of the spice. Cardamom oil is used mainly as an alternative to the spice, for flavouring a wide range of processed and frozen foods, condiments, gelatines and gravies, and liqueurs, cordials, bitters and other beverages; it is also used to a minor extent as a tobacco flavouring, and in cosmetics, soaps, lotions and perfumes. In the USA, the maximum permitted level in food products is about 0.01% and perfumes 0.4%.

Oleoresin. Cardamom oleoresin is obtained by solvent-extracting fresh fruits with a yield of around 10–12%, an essential oil content up to 70% and, depending on the solvent used, a fatty content up to 15%. The oleoresin is produced in small amounts, often to order. Commercial oleoresins, and some are branded products, have a volatile oil content in the range 50–60%, and are normally dispersed on salt, flour, rusk or dextrose. Makers claim 1 kg is equivalent to 20–25 kg of ground spice. Its main use is as an oil substitute in similar products. Both oil and oleoresin can develop off-flavours when exposed to air, and thus their use is confined to processed meats such as sausages, meatloaves and similar products with a short shelf-life. Detrimental chemical changes involving hydrolysis, isomeration and oxidation also take place during food processing when the oil may be raised to high temperatures or exposed to acid conditions.

Cardamoms are retained in British, US and other national pharmacopoeias for use as an aromatic stimulant, carminative, laxative and medicinal flavouring as *Compound Cardamom Spirit*. A monograph on the physiological properties of cardamom oil has been published by the IFM. In the USA, cardamom oil and oleoresin are generally recognized as safe, GRAS 2240 and GRAS 2241.

Other Elettaria species

E. cardamomum var. *major* Thwaites is the wild cardamom of Sri Lanka and southern India. Its major use is as an adulterant or substitute for true cardamom. Plants are robust, up to 5 m, with stems, petioles and leaf sheaths a light pink colouring; the leaves are broad and long, the inflorescence erect, and the fruits large, triangular, elongated up to 5 cm and containing many large, mildly aromatic seeds. The fruits contain an essential oil up to 10%. The composition of the fruit and the essential oil is similar to that of true cardamom but lacks its strength. Cold-pressed oil lacks *p*-cymene, while steam-distilled oil contains a high percentage. The average characteristics of the oil (all at 30°C) are: specific gravity 9.5; refractive index 0.9298; optical rotation +23.3°; acid value 5.5; saponification value 131.

Unrelated species

Amomum aromaticum Roxb. (syn. *Geocallis fasciculata* Horan), the Bengal cardamom, is native to the tropical eastern Himalayas of Bengal, Assam, Nepal and Bangladesh, and is occasionally cultivated. It is a robust herb with a rhizome from which leafy shoots and separate inflorescences appear. The stem is up to 1 m, and the leaves are oblong-lanceolate, 15–30 × 5–10 cm and glabrous. The inflorescence is a short-peduncled, small globose spike, with pale yellow flowers with the labellum twice the length of the corolla segments; the anthers have a large, petaloid, three-lobed connectivum. The fruit is an oblong, trigonal capsule, 2.5 cm, and the seed contains cineol. Plants flower in the dry season and fruits ripen in September in India and Nepal. The fruit is used as spice, the seed and oil in herbal remedies.

Amomum subulatum Roxb. (syn. *Cardamomum subulatum* Roxb.), the Nepal or large cardamom, is known in Hindi as *bara elaichi*, and native to the eastern Himalayas. It is a very minor international spice, although locally important, and is not discussed in detail; for further information see Prithi (1998). Nepal cardamom is a robust, rhizomatous herb, with stems 1–2.5 m. The leaves are narrow-lanceolate, 30–60 × 7.5–10 cm, and sessile on sheaths. The inflorescence is an erect, dense, globose spike, 5–7.5 cm in diameter, with a short peduncle; the flowers are yellow-white, with an emarginate labellum, longer than corolla segments. The stamens have very short filaments and entire anther appendage. Flowering is in March–April, with fruiting in the rainy season. The fruit is red-brown, dark brown when dried (Fig. 11.6), densely covered with soft, irregularly toothed wing-like outgrowths; it is generally globose, but may be triangular or sickle-shaped, 2.5–3.0 cm, containing 30–50 large, aromatic seeds, in a viscid, sugary pulp. The fruit and seed contain an essential oil containing cineole, which differs from true cardamom oil in being highly camphoraceous. Nepal cardamom is cultivated by smallholders in Bengal, Bhutan and Nepal; in Sikkim about 22,000 ha were under the crop in the early 1990s producing 3000–3500 t. Nepal cardamom is grown mainly along river banks and swampy areas. A significant proportion is consumed where produced or in India, and 150–200 t exported to neighbouring countries. The fruits may be used to adulterate true cardamom, and the fruit, seed, and oil are used in herbal remedies.

Additional regional substitutes in the genus *Amomum* L. include *Amomum compactum* Sol. ex Mat., the round cardamom of Java; *Amomum dealbatum* Roxb. (syn. *Amomum maximum* Roxb.), also from Java; *Amomum globosum* Lour., a large round cardamom of southern China; *Amomum*

Fig, 11.6. Dried large cardamom.

krervanh Pierre, Cambodian cardamom; *Amomum ochreum* Rid. and *Amomum testaceum* Rid. of Malaysia; and *Amomum xanthioides* Wall., the bastard cardamom of Thailand. These species are all inferior to the true cardamom, but popular where produced as true cardamom is expensive. Small amounts may become available to world trade, or exported to neighbouring countries.

African cardamom substitutes in the genus *Aframomum* K. Schum. include *Aframomum angustifolium* Schum. of Madagascar and other islands; *Aframomum corrorima* (Braun) Jen. of Ethiopia; *Aframomum daniellii* K. Schum. of West Africa and *Aframomum mala* K. Schum. of East Africa.

Ginger

The genus *Zingiber* Boeh. contains approximately 80 species indigenous to tropical India and South-east Asia, Australia and Japan, with the main centre of diversity in Indo-Malaysia. The genus name is probably derived from the Sanskrit *singabera* (horn-shaped), via the Arabic *zanzabil*, and Greek *zingiberi*. The most important spice producer is *Z. officinale*, ginger, and any reference to ginger in the text without qualification refers to *Z. officinale* and its derivatives.

The original use of ginger was as a spice and so ancient is this use that it pre-dates historical records, as the number of unrelated local names attest. In India, it is mentioned in the earliest Sanskrit literature but apparently not in the oldest Vedic works. The first known Chinese record is in the *Analects* of Confucius (*c.* 500 BC), 'who was never without ginger when he ate' (Bretschneider, 1870), and it was also used to treat rheumatism, toothache and malaria. The related *Languas officinarum* (Hance) Farw. is known as ginger root or Chinese ginger. Ginger was introduced to Japan much later and cultivation is relatively recent. Dried ginger was traded early on from India via Arabia to the Middle East, but whether it was known in dynastic Egypt before 500 BC is unclear; it was certainly in use by the time of Alexander and later became common in Greece and Rome. Ginger is mentioned in Dioscorides' *De Materia Medica*, and later by Pliny.

Potted ginger plants were carried on local vessels travelling the maritime trade routes of the Indian Ocean and South China Sea in the 5th century, and thus ginger was quickly introduced to countries along the way (Svendsen and Scweffer, 1985). In the 13th century, Arabs carried rhizomes on their voyages to East Africa to plant at coastal settlements and on Zanzibar, but not apparently to Ethiopia, where it was uncommon at the beginning of the 20th century (Jansen, 1981). Ginger is now a major cash crop in the Gojam and Begemdir provinces, and for sale in markets generally.

Ginger was one of the plants in the garden of Isidore, Bishop of Seville, in the 6th century, together with saffron and sugarcane. Initially used mainly as a medicine in Europe to treat stomach disorders, it gained popularity as a spice and was well enough known to be included in most herbals from the 9th century; in the 13th to 14th centuries, it was in general use and in England one pound cost 1s and 6d, the value of a sheep. Gingerbread became popular in the reign of the English Queen Elizabeth I, and together with biscuits remains so to this day. It was also an ingredient of mulled ale.

Probably the first European to see ginger growing in its natural habitat was Marco Polo in the late 13th century, and Chinese records show ginger was cultivated in the Malacca region (Malaysia) in 1416. When Vasco da Gama returned to Portugal from his epic voyage to India at the end of the 15th century, he reported that ginger was among the major spices exported from the Malabar coast, and was shipped by Moors to Red Sea ports for carriage overland to Cairo and thence to Europe, a trade virtually destroyed by the Portuguese sea route via South Africa (see also cinnamon, Chapter 3). Subsequently the Portuguese introduced ginger to their West African colonies in the 16th century, and quickly established a thriving export trade in rhizomes sufficient to supply their domestic market, becoming independent of other sources. The Spaniard, Francesco de Mendoza, carried rhizomes to Mexico shortly after Columbus, and ginger subsequently spread throughout the Caribbean islands and Central America. Spanish colonists exported 1000 t of Jamaican rhizomes in 1547, smaller amounts from Barbados in 1654, and Jamaica continues to be a leading producer and exporter of high-quality ginger.

Botany

Z. officinale Roscoe. (syn. *Amomum zingiber* L.; *Zingiber zingiber* (L) Karst.) is known in English as ginger, in German as *ingwer*, in French as *gingembre*, in Italian as *zenzaro*, in Spanish as *jengibre*, in Portuguese as *gengibre*, in East Africa, Kiswahili the plant as *mtangawizi*, the dried rhizome as *tangwizi*, in India, in Hindi/Urdu as *adrak*, in Bengali as *ada*, in Tamil as *inji*, in Malaysian as *haliya*, in Thailand as *khing*, in Indonesian as *jahe*, in China generally as *kiang* and in Japan as *shoga* or *myoga*. The basic chromosome number of the genus is x = 11, and ginger is a diploid with $2n = 22$, but tetraploids with $2n = 44$ have been chemically induced.

Colchicine (CCC)-induced tetraploids have recently given several times the yield of diploids in Australia (Smith and Hamill, 1997; M.K. Smith, personal communication, 2001), while CCC tetraploids in Japan had varying levels of pungency and flavouring compounds (Nakasone *et al.*, 1999).

Germplasm collections are maintained at the Indian Institute for Spice Research, Calicut, Kerala, India, and at the Research Institute for Spices and Medicinal Crops, Bogor, Indonesia. Breeding programmes concentrated on increasing rhizome yield or essential oil, or reducing fibre content. There are many locally important cultivars, at least 400 in India where genetic variability has been evaluated (Rai *et al.*, 1997; Yadav, 1999a), but as ginger is vegetatively propagated, local cultivars often tend to be uniform and develop specific characteristics. Malaysia and neighbouring countries recognized two divisions: *Z. officinale* cv. group *officinale* (ginger) and *Z. officinale* cv. group *rubrum* with small, pungent reddish rhizomes (Theilade, 1996); Indonesian ginger is separated into three main cultivar groups based on rhizome characteristics of aroma, colour, shape and composition. The highly mechanized ginger industry in Queensland, Australia, was founded on a cultivar imported in the 1920s (Whiley, 1981).

Z. officinale is an erect, leafy perennial, usually cultivated as an annual, with purple flowers and a robust branched rhizome growing horizontally near the soil surface (Fig. 11.7). The following description indicates the range occurring over cultivars, not the range of a single cultivar.

The root is a sympodial rhizome, 6–20 cm × 2–5 cm, laterally compressed, often palmately branched, and the lobes are usually 1.5–2.5 cm diameter but in some cultivars much larger, with widely branching, somewhat shallow, thin, white to light brown, fibrous roots (Fig. 11.8). The rhizome is firm, the skin corky and small-scaled, the skin thickness and appearance a cultivar characteristic; scars of leafy stems are visible on dried rhizomes. Skin colour varies from buff to very dark brown almost black, flesh colour from pale yellow to deep orange red, and both are cultivar characteristics.

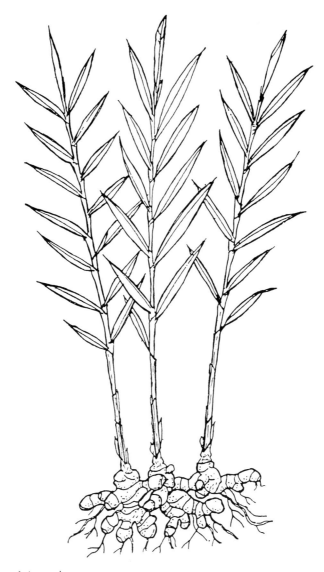

Fig. 11.7. Diagram of ginger plants.

Rhizome morphology and histology have been described in detail (Parry 1962; Nybe *et al.*, 1979), as has the development anatomy of rhizomes (Remashree *et al.*, 1997, 1998). Rhizome shape, size, degree of branching and extent of rooting are directly affected by soil type and soil reaction, while rhizome size was significantly increased in Australia using colchicine (Smith and Hamill, 1996). Details of rhizome composition are given at the end of this section.

Shoots arise from the rhizome, and in China, Korea and Japan the small reddish shoots are especially relished and red-shooted ginger *Zingiber mioga* Rosc. is cultivated specifically for this purpose; it is currently being grown in New Zealand and Australia for export to Japan. The ontogeny of shoot apices has been studied in detail (Shah and Raju, 1975). The effect of growth regulators on ginger has been investigated but their use is of little practical value and may leave residues in plant parts.

Fig. 11.8. Fresh ginger rhizomes.

Stems (pseudostems) are slender, erect, up to 100 cm but generally 50 cm × 5–10 mm in diameter, unbranched, pale green but often reddish at base, with prominent parallel veins, glabrous except for short hairs near the base of the leaf blades, and bearing 8–12 distichous leaves. The leaves are mid-green above, paler below, 5–25 × 1–3 cm; the lamina is thin, subsessile, linear-lanceolate, with the base obtuse or rounded, narrowing evenly to a slender tip, with a pronounced midrib and finely parallel veins; the ligule is up to 5 mm, broad, thin, glabrous and slightly bilobed. Volatile oil cells are present in leaves, shoot and root apex, in addition to the rhizome (Remashree *et al.*, 1999). A direct relationship apparently exists between rhizome size and number of leaves (Premanand *et al.*, 1999), leaf composition (Rai *et al.*, 1999), and canopy photosynthesis (Yu *et al.*, 1996). Straw mulching in China reportedly increased leaf thickness but reduced chlorophyll content (Xu and Li, 2000). In Malaysia, pounded leaves are used as a poultice to

relieve muscular pains, and are boiled and eaten to treat stomach-ache.

The inflorescence generally arises directly from the rootstock and is carried on a slim, leafless stem, 15–30 cm; the scape is slender, 10–20 cm, the upper sheaths with or without short leafy tips. The inflorescence is cylindrical, cone-shaped, 4–7 cm × 1.5–2.5 cm in diameter; the bracts are appressed, ovate or elliptic, 2–3 cm × 1.5–2.0 cm, green with a pale submarginal band, the lower with slender white tips. The flowers are single, axillary in each bract and short-lived; the calyx is 1–1.2 cm, thin, tubular, spathaceous and three-lobed. The corolla tube is 2–2.5 cm long with three yellowish lobes, the dorsal 1.5–2.5 cm × 8 mm curved over the anther. The labellum is roughly circular, 1.0–1.5 cm and dull-purple with a cream-blotched base; the side lobes are 6 × 4 mm wide, ovate-oblong, free almost to the base, and coloured as for the midlobe. The stamen filament is short and broad, the anther cream-coloured and 9 mm, and the dark purple 7 mm connective prolonged to a slender beak-like appendage, containing the upper section of the style, and the stigma protrudes below the apex. The ovary is inferior and trilocular with several ovules per loculus. The pollen morphology and structure of a number of *Zingiber* spp. have been determined (Theilade *et al.*, 1993).

Flowering is a cultivar characteristic; some flower rarely, others regularly, especially when grown undisturbed as perennials. A detailed description of flowering in an Indian cultivar has been published (Pillai *et al.*, 1978), and of pollination (Das *et al.*, 1999). The fruit, rarely produced because of pollination barriers, is a thin-walled, three-valve capsule containing several small black, angled, arillate seeds whose viability is usually very low. However, seeds have been successfully raised *in vitro* and plantlets obtained (Nazeem *et al.*, 1996).

Rhizome. The rhizome composition varies according to cultivar, region, maturity at harvest, etc., and thus the following information is a general guide only, except where specified. Rhizomes contain an essential oil, fats, starch and related compounds, crude protein, crude fibre and a small amount of minerals. Within a local cultivar, composition and characteristics vary little, since vegetative propagation limits major variation. Analyses of Indian rhizomes gave the following percentages by dry weight (average in brackets): moisture 8.5–16.5% (10.9%), crude protein 10.3–15.0% (12.4%), crude fibre 4.8–9.8% (7.2%), starch 40.4–59.0% (45.5%), water extract 14.4–25.8% (19.6%), cold alcohol extract 3.6–9.3% (6.0%), acetone extract 3.9–9.3% (6.5%), volatile oil 1.0–2.7% (1.8%), while analyses of 30 cultivars gave a fibre content of 3.5–7.6% (Nybe *et al.*, 1979). Starch occurs as ovate to subrectangular granules, up to 45 × 24 microns, and is normally absent from young rhizomes but occurs in pith and to a lesser extent in the inner cortex of mature rhizomes.

The essential oil cells are distributed in the cortex and pith, as are the gingerol-containing cells, but the two are independent. The major constituents of the essential oil are: zingiberenes up to 40%, α-circumene up to 20%, farnesene up to 9%; sesquiterpene alcohols up to 18%, with monoterpenes up to 1% and important in determining final odour. Oil from 27 Indian cultivars gave the following characteristics (all at 24°C): yield 0.9–2.50%; refractive index 1.4898–1.4988; optical rotation −34.4° to −49°; specific gravity 0.8474–0.8720. A commercial oil in the USA had the following characteristics: specific gravity (15°C) 0.875–0.885; refractive index (20°C) 1.4880–1.4950; optical rotation −25° to −45°; almost insoluble in water, sparingly soluble in alcohol, soluble in ether and carbon disulphide. The constituents responsible for pungency are non-volatile phenols, gingerols, shogaols, paradols and zingerone, with 6-gingerol the most important. The odour and much of the flavour is determined by the composition of the volatile oil. The chemistry of the pungent principals, the volatile oil and colourants of ginger rhizomes is outside the scope of this text, but has been extensively studied and reported in detail in the literature.

Ecology

Ginger is native to the moist tropical regions of either South-east Asia or India, as it no longer occurs as a truly wild plant. So popular is the spice that ginger is cultivated as an annual much more widely than originally considered possible, but the bulk of rhizomes entering world trade still come from regions lying mainly between the Tropics of Capricorn and Cancer. The proportion of national production by main Indian producing regions are the State of Kerala about 50%, Orissa 14% and Meghalaya about 10%; Australia, Queensland, 95%; in China, Guandong Province 50%; and Nigeria northern provinces 80%.

A rainfall of 2500–3000 mm well distributed over the year is the optimum, but above this or in areas of short, high-intensity storms, soils must be free-draining as ginger is adversely affected by waterlogging even for short periods. A rainfall of 1500–2000 mm will produce good crops with supplementary irrigation. Ginger seldom succeeds as a mainly irrigated crop since the necessary humidity and soil moisture cannot usually be profitably maintained.

Ginger flourishes under warm, sunny conditions, but in hot, dry periods especially at temperatures above 30°C, young unshaded plants may be scorched. Above 37°C without high humidity can cause severe foliar damage at all growth stages, and if the soil is also dry, rhizomes will be damaged and may die. Temperature and light intensity may also affect leaf photosynthesis, and in China the optimum was 25°C (Ai *et al.*, 1998).

When grown as a smallholder crop and thus frequently interplanted among taller plants, ginger benefits from their shade during very hot periods, but in general shading *per se* is unnecessary. The short-term effect of high air or soil temperature is best ameliorated using other methods, including sprinkler irrigation or heavy mulching. Australian growers apply water at a low 2.5–3.0 mm h⁻¹ via overhead rotary sprinklers from 10 a.m. to 3 p.m. during October–December to protect young plants.

A ground temperature of 25–30°C is the optimum for initial rhizome growth, and although this occurs at the correct time of planting in tropical regions, may not in other ginger-growing areas. In Australia, soil temperature at 10 cm frequently does not reach the optimum until several months after the normal planting season and delays full emergence by up to 6 weeks. Thus, low soil temperatures at or following planting could be an important constraint on commercial ginger production where the time available to produce a crop is limited. Irrigating immediately following planting may also have a similar effect on soil temperature, but could be an advantage where soil temperature is above the optimum. Frost at any stage of growth will kill foliage and rhizomes at or near the soil surface. In areas where frost is common, as in the hill regions of India and China, annual ginger is harvested before frosts occur.

Ginger is normally grown at low altitudes, but altitude is apparently of little importance provided other conditions are favourable, since ginger grows well in Assam and Jamaica up to 1000 m. High winds can damage mature plants but are normally of little importance. However, in areas where there are strong prevailing winds, shelter belts or rows of tall plants between ginger plantations will provide protection.

Soils

A friable fertile loam of moderate depth is the most suitable for annual ginger since rhizomes and roots proliferate in the top 25 cm. Soils must be permeable or artificially drained, since ginger is very susceptible to waterlogging. Ginger is, however, very adaptable, and almost any soil will produce an acceptable rhizome yield with good management, as in India, where ginger is grown on sandy and clay loams, tropical red earths derived from laterite and drained paddy fields following the rice harvest. Volcanic soils are favoured on Mauritius and other Indian Ocean islands, but silty-clay loams on Madagascar. Ginger is often grown on hillsides in Jamaica where the soils are mainly clay loams over limestone or conglomerates; moderate to heavy clays are utilized in Australia, and alluvial soils or drained paddy fields in China, Taiwan and Japan, or specially drained marshy areas to produce ginger shoots and young green ginger.

Soil of pH 6.0–7.0 is preferred but ginger will grow with reduced vigour at higher or lower pH levels. While the quoted pH range is usually considered the optimum, ginger can tolerate much lower levels; in a nutrient solution of pH 3.3, yield was half that at pH 6.5, which field trials confirmed (Islam *et al.*, 1980). Acid soils in Queensland, Australia's main ginger-growing region, are normally limed annually to adjust soil reaction to pH 6.5; in India, lime is normally applied only on larger plantations, and in Malaysia, 4–5 t ha^{-1} of magnesium limestone plus an NPK mixture and 17 kg copper sulphate applied to an acid peat produced a green rhizome yield of 18 t ha^{-1}. However, such soils, plus saline and alkaline soils, should be avoided, as they seldom produce a profitable rhizome crop.

Fertilizers

Total nutrient removal ha^{-1} by a ginger crop must be substantial since in addition to the dense foliage, average rhizome yields of 20–40 t ha^{-1} are common and up to 120 t ha^{-1} at experimental stations; in Mauritius, ginger extracted 80.5 kg N to produce 30 t fresh rhizomes, and in India the nutrients removed by 2300 kg ha^{-1} rhizomes are shown in Table 3.4 (Sadanandan *et al.*, 1998). The micronutrients accumulated in rhizomes in Brazil were, in decreasing ranking, Fe, Mn, Zn, B and Cu. Total nutrients removed from the plantation by rhizomes equalled 15% of the nutrients available.

Nutrients should be easily available since absorption is most rapid during early growth, thus P should be in a water-soluble form, particularly on slightly acid soils; subsequently the rate of nutrient uptake is closely correlated with increase in rhizome size. In less developed regions, beds or substantial ridges concentrate nutrients available in the topsoil, supplemented in Africa by soil from termite mounds.

Commercial ginger growers in India apply up to 30 t ha^{-1} of animal manure annually prior to planting and a top dressing of 10 t well-rotted manure ha^{-1} 2–3 months later. In West Africa where animal manure is less available in large quantities, 3–5 t ha^{-1} is common. Mulches of green leaves are an alternative. Night soil in large (but unspecified) amounts is applied in China before planting and as a top dressing. Where castor pomace is easily available in India, up to 5 t ha^{-1} is applied, half before planting and half as a top dressing. The value of castor pomace as a top dressing depends mainly on its N content, up to 6%; pomace also contains significant amounts of P but less potash (Weiss, 2000). Where oilseed cakes are available, up to 400 kg ha^{-1} is applied in three equal dressings, with groundnut cake generally producing the greatest increase in yield (Sadanandan and Hamza, 1998). In Australia, up to 200 t ha^{-1} of mill mud, a phosphate-rich sugarcane processing residue, or aged poultry manure up to 10 t ha^{-1} is applied. Such bulky materials should preferably be ploughed in well before planting, and where ginger is grown on ridges or mounds, incorporated during their construction.

Green manure crops to increase soil fertility or water-holding capacity in drier regions are of value, but their profitability must be established compared with other methods. Green manure or cover crops are usual in Australia, to which mill mud or poultry manure are applied to enable decomposition to be well advanced before planting ginger. Fertilizers as NPK mixtures are commonly applied to the seedbed, and when manufactured locally are known as ginger mixtures to include any minor nutrients necessary; common types are, in India 8:8:8 and 12:8:8, in Fiji 13:13:21 and in Australia 12:14:10.

The effect of fertilizers on rhizome composition remains unclear, and data are often contradictory. High N rates occasionally depressed oil content in India, but generally reduced fibre and ash content in Mauritius. The moisture content of Mauritius rhizomes at harvest was 70–80%; on a dry-weight basis fibre was 3.6–8.1%, ash 2.9–5.8% and ether extract 2.9–5.2% (Owadally *et al.*, 1981).

Nitrogen placed in the seedbed at planting generally increases rhizome yield, and at least one-third of the total N requirement should be so applied, the remainder in one, two, sometimes three top dressings. A total of 100 kg N ha^{-1} will normally be required. In India, the first top dressing is usually 40–60 days after planting, the second around 90–100 days. In Australia, three similar applications are made at almost equal intervals over crop growth but up to ten applications when supplied via irrigation equipment. In Taiwan, 600 kg N ha^{-1} was the optimum in split applications (Xu and Xu, 1999), but in China 40 kg N ha^{-1} was the optimum (Ai *et al.*, 1997).

The end product required directly influences the amount of N used; in Australia where rhizomes are harvested at several levels of maturity for different end uses, up to 350 kg N ha^{-1} in multiple applications is profitable. The effect of varying the number of applications on N recovery and rhizome yield is important to ensure optimum usage. Placing all N in the seedbed in Queensland increased rhizome yield from 35 to 49 t ha^{-1}; from split applications, yield at early and late harvest increased from 32 to 49 t ha^{-1} and 50 to 76 t ha^{-1}, respectively. Recovery of N in shoots plus rhizomes can be low when all N is applied in the seedbed, but rises progressively when the same amount is supplied in several applications (Lee and Asher, 1981), or delayed until plants are making vigorous growth (Xu *et al.*, 1993). The type of N fertilizer is of little importance, but large amounts of ammonium sulphate are best avoided on more acidic soils. In Australia, urea, ammonium nitrate and ammonium sulphate in that order are considered most profitable; in India, calcium ammonium nitrate is used for rain-grown, and urea for irrigated crops.

Phosphate, usually as superphosphate, should be applied in the seedbed and generally at a similar level to N. On some tropical soils there can be a high residual effect and if this is proved to be so locally, the annual amount of P can be reduced. The high rate of P in relation to N that is frequently applied is notable, and indicates the necessity for trials to determine the optimum level and ratio between the three main nutrients.

It is essential to determine the optimum level of K required, since ginger's reaction to this nutrient is often unpredictable. Too little retards growth, especially with higher rates of N and P; too much frequently has no beneficial effect and wastes money. The optimum amount must be determined in relation to levels of N and P required, since both affect the uptake of K. High rates of K are applied especially in the compound mixtures previously quoted, but annual application at these rates may not be justified since there can be a substantial residual effect. Application is directly related to the level of available soil K in Australia. Combinations of 150 kg N + 50 kg K ha^{-1} and 25 kg P + 50 kg K ha^{-1} were the optimum in India (Gowda *et al.*, 1999), but in Shandong province, China, 40 kg N + 7 kg P + 40 kg K ha^{-1} (Ai *et al.*, 1997).

Potassium is normally applied in the seedbed or spread and worked in prior to planting, but can be in two equal applications, at planting and approximately 3 months later. In Australia, three almost equal applications are made, in the seedbed, at 3 and 5 months. Split applications should be judged entirely on the profitability of so doing. The K fertilizer chosen is generally of little importance, but on some soils chloride-free types are advantageous.

Few data on the effect of minor and trace elements on plant growth or rhizome yield are available, but where a response has been shown to occur on other local root crops, it will probably also affect ginger. This is especially so with high rates of applied P, when an Mg deficiency can be induced. Very large lime applications to correct soil acidity may result in Fe, Mn and B deficiencies late in growth; the very small amounts required can be easily incorporated in the compound fertilizer chosen at very little additional cost.

Cultivation

Ginger is grown from smallholder level to large-scale fully mechanized operations, to service a range of markets. When grown by smallholders, the land is cleared of major vegetation, which is dried and burnt, then hoed and weeds heaped round ginger hills. On a larger scale, cleared land is first dug over or ploughed to control weed growth then ridged, either manually or mechanically. These ridges normally include a layer of manure or weeds at the base, and vary in size to suit the locality. For the following season's crop, weeds and manure are placed in the channel and ridges on each side split to build the new ridge. Several rows of rhizomes are planted on each ridge, and

mulched to prevent overheating and suppress weeds. Additional mulch or fertilizers are applied according to local experience and the cash resources of the farmer.

Where ginger is a major cash crop as in the Caribbean, China, India, Sri Lanka and West Africa, cultivation is by draught animals or small tractors, but Australian growers are fully mechanized and machinery is becoming larger, more specialized and efficient. Whatever system is used, the basic operations are similar; land should be thoroughly ploughed, and pre-planting operations aim at removal of roots and previous crop residue and maximum weed reduction. Subsequent cultivation should produce the fine tilth required to produce well-shaped, clean rhizomes. Fields may be ridged or beds constructed prior to planting, or the two operations combined. When ginger is sown on hillsides, ridging on the contour, use of terraces and mulching is necessary, and failure to do so results in the gullies so common on unprotected hillsides in Jamaica and West Africa.

Ginger is usually vegetatively propagated from small portions of rhizome (setts or seed pieces) from the previous year's harvest, but can also be micropropagated using meristems, rhizome sections or tissue culture (Smith and Hamill, 1996; Pandev *et al.*, 1997; Devi, 1999; Arimura *et al.*, 2000). Since 4–10 t of setts are required to plant 1 ha, adequate and suitable storage is very important, and there are many methods. Probably the easiest is to leave a section of field unharvested, cut the foliage to ground level and cover plants with a thick mulch; rhizomes are dug when required. However, in northern India this was discouraged by the local Agricultural Department as it encouraged rhizome rot and soil pests (Rai and Hossain, 1998).

A widely used system is pit storage. In India, fully mature healthy rhizomes are soaked in a fungicide solution, shade-dried, then placed in pits to within 10–15 cm of the top. Each pit is then covered with planks, the centre having a small hole; planks are then plastered with mud to seal the pit, leaving only the hole open. Pits vary from 3.0 to 5.0 m deep, are rendered smooth inside and have recently also been lined with plastic. Smoking rhizomes once or twice before placing in the pits reduces the incidence of disease. A comparison of storage methods in northern India showed pits to be superior for smallholders (Rai and Hossain, 1998). A common practice in Kerala is to preserve seed ginger in smoke houses; rhizomes are first dipped in a thick solution of cow dung, dried and spread on bamboo mats raised on supports to allow the smoke to circulate freely. The room is sealed and smoked periodically until rhizomes are removed about a fortnight before planting. This technique causes some desiccation and loss of viability. In other areas, small heaps of rhizomes are placed on sand beds in the corner of rooms and covered with more sand. Provided the heaps are kept dry and examined regularly to remove those diseased, rhizomes remain sound until planting.

To inhibit sprouting or weight loss, or to control diseases during pre-sowing storage, rhizomes can be waxed, chemically treated, or irradiated prior to storage (Yusof, 1990; Wu and Yang, 1994; Variyar et al., 1997). Cold storage is not recommended as rhizome viability is gradually reduced and may be zero after very short periods below 0°C. In Australia, good-quality rhizomes are kept in controlled storage to provide material to sow a nursery in advance of main planting, and also to store superior clonal material for multiplication. In general, setts 3–6 cm long or 30–60 g have proved most successful, and many trials have shown a significantly higher yield is obtained from larger setts as shown in Table 11.2, which also shows the effect of plant population on yield. Distinct cultivar differences exist, and in some regions of India setts up to 150 g gave highest yield, but Jamaican growers prefer setts with six buds, irrespective of size.

Smallholders usually plant manually but this operation is easily mechanized, and a variety of small planters available. On large estates, lands are ridged or bedded, fertilizer applied and setts placed by fully automatic equipment in one operation. Depth of planting is not critical and 5–12 cm common. However, previous remarks on soil temperature are apposite, and fungicidal treatment

is recommended. A row width of 25–30 cm, but up to 50 cm depending on bed size, is recommended, with in-row spacing of 15–35 cm. Row spacing is also governed by harvesting machinery and thus three rows are normally sown per bed and dug together (Fig. 11.9). Accurate in-row spacing is not necessary provided the optimum number of setts is sown, since ginger tends to expand to fill the area available. About 8–10 t ha^{-1} of setts are required on fully mechanized Queensland estates, 1.5–4 t ha^{-1} under less intensive conditions in India and Sri Lanka, as the raised bed system of cultivation reduces the area sown. On Mauritius, the optimum was 38 × 38 cm square spacing, approximately 60,000 ha^{-1}. Five different in-row spacings of 55,000–166,000 ha^{-1} showed the superiority of close over wide spacing on Fiji, with 60 × 10 cm producing the highest green rhizome yield; 60 × 30 cm gave the highest yield per plant but could not compensate for the low plant density.

Time of planting is important since the soil must be moist and not dry out once setts are sown, and in general the earlier in the season the crop is planted, the higher the yield, especially with rain-grown crops. Recommended sowing times are: in India, in the east March, in the south April, in the north March–April; Sri Lanka March–April; April–May in Jamaica and other Caribbean islands; September in Queensland, Fiji and neighbouring islands; December–January in Mauritius; March–April in Taiwan and April–May in China; April–May in West Africa. Soil temperature at planting should be 20°C, preferably 25°C, but not above 30°C. Emergence depends on depth of planting and cultivar, but the majority of plants should emerge in 14–30 days, average 20–22 days. If emergence is delayed or erratic, better selection or storage of seed material is indicated, or more attention to soil conditions. There are local exceptions to this general observation; in Australia, full emergence can be 6 weeks after planting for reasons stated previously.

When fresh green rhizomes are used as planting material, other factors may affect viability and emergence; in Ghana, fresh rhizomes had a variable dormancy period that

Table 11.2. Comparison of cultivar, population and sett size on yield and mean knob size in Australia.

Cultivar	Yield (kg per plot)	Mean knob size (g)
Queensland	52.9	23.1
Fijian	56.2	29.2
Spacing (cm)[a]		
11.2	63.8	26.7
17.0	55.3	26.2
22.4	44.6	25.6
Sett size (g)		
42.5	48.3	24.6
85.5	60.8	27.7

[a]In 60 cm rows.
Knob, part of rhizome.

Fig. 11.9. Preparing raised beds for irrigated ginger.

affected germination and subsequent yield. Use of growth promotors to hasten emergence is effective but normally unnecessary; sets treated with Ethrel increased root and shoot growth with more regular emergence in Australia, but in India gamma-irradiation significantly reduced shoot growth and rhizome yield (Giridharan and Balnakrishnan,

1991). Where the time available to produce a ginger crop is limited, a method of hastening germination is necessary, and in Taiwan, China and Japan this is by pre-sprouting setts; these may be spread on sacks, straw or similar insulating material and enclosed from flowering to harvest wholly or partially by polythene sheeting to raise the internal temperature. Taiwanese and Chinese growers also construct hot-beds of well-rotted manure and cover these with plastic or straw on which the setts are placed to encourage sprouting, and also grow ginger in plastic tunnels in colder regions (Cui *et al.*, 2000).

The sequence of development of annual ginger from planting to harvest, as determined on ginger in South Africa, can be divided into three phases. Phase 1 covers 35–45 days from planting to shoot emergence, mainly root and shoot development; phase 2, about 150 days, covers the period from shoot emergence to flowering; phase 3, 90 days, covers the period from flowering to harvest. During the last phase shoot and leaf growth virtually ceased, replaced by rapid rhizome development. A field of healthy, well-grown ginger plants in Queensland, Australia, is shown in Fig. 11.10.

Weed control

Ginger plants must be kept free of weeds until foliage is sufficiently dense to suppress weed growth. Hand-weeding is common,

Fig. 11.10. Ginger field – Queensland, Australia.

usually three times is sufficient, and exposed rhizomes should be covered at the same time. Manual or mechanical weeding should be as shallow as possible to avoid rhizome damage. The following herbicides correctly used caused little or no crop damage: pre-emergence, alachlor plus chloramben, atrazine alone and plus metolachlor, fluometuron, simazine and 2,4-D; post-emergence, diuron, monuron and directed sprays of glyphosate. Influencing the greater use of herbicides in smallholder crops is the absence or increasing cost of non-family labour. Since weeding is essential to obtain a high yield of good-quality rhizomes, inability to control weeds or the cost of employing casual labour is becoming a significant factor in ginger production, as in Nigeria (Aliyu and Lagoke, 2001).

Irrigation

Most smallholder ginger is rain-grown, but larger growers and mechanized crops are partially or fully irrigated. Rain-fed crops are generally more profitable; in Himachal Pradesh, India, rain-grown and dry-season irrigated crops had similar rhizome yields, although cultivars differed in both environments (Korla *et al.*, 1999). Irrigation of rain-grown crops is advantageous if dry periods occur, as a soil moisture shortage during growth results in a lower yield of smaller, more fibrous rhizomes. Irrigation can also ensure high humidity in the crop and prevent sunburn. Water is most commonly applied via channels or furrows, which separate the ginger beds, with overhead sprinklers or rainguns equally effective but more expensive. The method of application is usually unimportant provided an adequate soil-moisture level is maintained over the growing period; as a guide this must equal 1500–2000 mm annual rainfall.

When irrigation water is not available or topography precludes its use, mulching to conserve soil moisture can substantially increase rhizome yield or quality; in Assam, mulching with green leaves two or three times doubled rhizome yield to 15,700 kg ha^{-1}, while straw mulch in China retained soil moisture and increased yield over non-mulched (Xu, 1999). Mulching is also popular to reduce soil temperature and a variety of materials is used, including sugarcane trash, rice husks or straw, sawdust and plastic film. Decomposition of plant material provides some plant nutrients, but when sawdust is used the amount of N applied as a top dressing should be increased; in Australia, urea is favoured at 600–750 kg ha^{-1}.

Intercropping and rotations

Ginger is often underplanted in giant castor, coconut, orange and young coffee, and high, light shade can be beneficial; in India, ginger planted under 6-year-old areca nut trees gives a higher rhizome yield than under 2-year-old trees and in the open. However, when ginger was underplanted in various species of fodder trees in Uttar Pradesh State, ginger yield was generally reduced (Bisht *et al.*, 2000).

Ginger is also frequently intercropped but this is not to be recommended where it is a main cash crop. Benefits that may result from the partial shading and soil cover are generally outweighed by a reduction in rhizome yield. Interplanting ginger with maize and dwarf beans in South Africa reduced rhizome yield by 80% and 60%, respectively, compared to mulched ginger. Intercropping with *Sesbania* spp. to provide light shade for young plants was beneficial in southern India, the *Sesbania* later cut to provide mulch. Growers also alternate several rows of ginger and other crops, in West Africa maize or yams, in Taiwan and China soybeans or vegetables, in India maize, pulses or chillis, sometimes pigeon peas or turmeric, which also provide light shade, and in Caribbean countries yams, chillis and vegetables.

A rotation in which ginger is grown only once in 3–4 years will reduce the incidence of pests and soil-borne disease; in India and Sri Lanka, alternative crops are cassava, sweet potatoes, yams or chillis. For irrigated ginger, a rotation of 6 years is recommended to include betel vine, bananas, sugarcane, rice and various vegetable crops. In the Caribbean 2–3 years of ginger is followed by 2 years of soybean, chillis, vegetables or similar crops; in Australia where ginger is continuously cropped, soil and rhizomes are fumigated to reduce nematode and disease build-up.

Pests

The most common pests of ginger are shown in Table 11.3, and the most important described later in this section. Many can be controlled by pesticides, but their use is limited by increasingly stringent regulations covering residues in rhizomes and in their oil, DDT and BHC, for example. Foliage pests may become damaging in a specific crop or season especially on interplanted or intercropped ginger, when a polyphagous pest can cause indiscriminate damage, *Helicoverpa* spp. (*Heliothis*), for example. The most important foliage pest whenever ginger is grown in Asia is the shoot-borer *D. punctiferalis* (yellow peach moth), whose larvae bore and feed in pseudostems and shoots and cause desiccation and dieback. Up to 40% of plants in an area can be severely damaged in India. The moth is polyphagous,

widespread, and with its near relatives probably causes more damage in other countries than is currently reported. It occurs on ginger from China, Japan, Taiwan, though India to Pakistan, Sri Lanka and the Middle East, and south to Australia. Frequently reported are scale insects, especially *Aspidiella hartii* in India, thrips particularly *Thrips tabaci*, leaf-rolling caterpillars including *Udaspes folus*, a borer beetle *Lyctus africanus*, and lacewing bugs *Stephanitis* spp. Leaf miners, especially the widely distributed *Lema* spp., can be damaging in specific years. Chrysomelid beetles are frequently reported on ginger, but are of minor importance; *Monolepta australis* is confined to Australia where it is also an important pest of groundnuts.

Insects attacking the rhizome cause varying degrees of damage, from superficial scoring of the skin to boring the flesh. Damage to the skin is of economic importance where appearance of rhizomes is a major factor; those boring the flesh are more damaging. Pests attacking rhizomes include the polyphagous and widely distributed Fuller's rose beetle, *Pantomorus cervinus*; adults and larvae eat and bore into rhizomes. The related *Adoretus sinicus*, Chinese rose beetle, behaves similarly. A fly whose larvae burrow rhizomes in Asia is *Mimegralla coeruleifrons*, but damage is usually confined to individual rhizomes. Damage to rhizomes in India is generally minor, but subsequent bacterial and fungal infection of the wounds is more serious. It was also noted that the newer, low-fibre cultivars were more susceptible to *M. coeruleifrons* attack than traditional cultivars (Sontakke, 2000). Females of small scolytid beetles bore rhizomes to lay their eggs and larvae damage the flesh; commonly reported are the widely distributed *Stephanoderes* spp. Damage to rhizomes in the field also allows egress to pathogenic fungi, and this secondary infection is often more serious.

Nematodes can limit continuous ginger growing, most commonly *M. javanica*, *M. incognita*, *Pratylenchus* spp. and *R. similis*. Symptoms are galls on feeder roots, cracks in rhizome skin and small, light brown water-soaked lesions in the flesh. Nematode populations increase very rapidly and are usually

Table 11.3. Important insect pests of ginger.

Pest	Common name	Region
Acrocercops irradians	–	Asia
Aedidiotus subterraneus	Coccid	Mauritius
Aspidiella spp.	Scale insect	Asia/Africa
Aspidiotus destructor	Coconut scale	a, b
Aspidiotus sinicus	Chinese rose beetle	b
Calabota spp.	Ginger maggot	Asia
Celphus spp.	–	Asia
Chalcidomyia atricornis	–	Asia
Dichocrocis punctiferalis	Shoot borer	a, b
Drosophelia spp.	Boring larvae	Asia
Formosina flavipes	–	Asia
Hedychrous rufofasciatus	–	Asia
Lema spp.	Criocerine beetles	b
Mimegralla coeruleifrons	Rhizome fly	Asia
Monolepta australis	–	Australia
Panchaetothrips indicus	Thrips	India
Pantomorus cervinus	Fuller's rose beetle	a, b
Stephanitis typicus	Lace-wing bug	b

a, polyphagous.
b, widespread.

controlled by soil fumigation; however, in India various mulches either controlled or reduced nematode numbers (Das, 1999). Immersing planting material in water at 45°C for 3 h also killed nematodes in India (Vadhera *et al.*, 1998), while nematode-free material can be cultured *in vitro* to establish new plantings or for research. For pests of stored ginger, see the Transport and storage section.

Diseases

Diseases are generally more damaging than insects, as the widespread transfer of ginger rhizomes and plants on ancient and modern trade routes helped to spread the pathogens. Thus, disease prevention or control is now an essential component of commercial ginger growing and its neglect is disastrous; soft rot, for example, may destroy one-third of a particular crop. The common diseases of ginger and causal organisms occurring in India have been reviewed (Sarma and Anandaraj, 2000). The most important generally are discussed below. Resistance exists to some of these diseases, but whether it is general resistance or merely to the local biotype has yet to be determined.

Control of rhizome infection with antibiotics may be possible (Singh, D.K. *et al.*, 2000).

Probably the most destructive are rhizome rots, and affected rhizomes are usually destroyed. The most common causal organisms are *Rosellinia* spp., *Pythium* spp. and *Fusarium* spp., with a specific species being responsible in some countries; *P. aphanidermatum* in India, *Pythium myriotylum* in South Korea; with *F. oxysporum* f.sp. *zingiberi* of major importance in South Africa, India, Hawaii, Fiji and Australia (Fig. 11.11). The main symptom is degeneration of the rhizome body into a soft, usually almost black, putrefying mass. The main above-ground symptom is an initial yellowing of the leaf tip, sheath and margins, which gradually spreads over the whole leaf; desiccation and death quickly follow. Waterlogging or badly drained soils exacerbate the damage. The only satisfactory control method is selection of healthy setts and fungicidal treatment prior to planting. Ginger should not be grown on land infected with soft rot fungi except as part of a long rotation or with regular fumigation. In India, various oilseed cakes incorporated in soils infested with *P. aphanidermatum* or *F.*

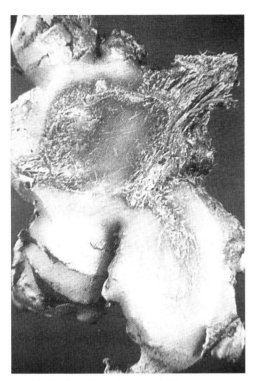

Fig. 11.11. Fusarium yellows and rhizome rot.

crops often grown in rotation with ginger. The bacterial wilt *Ralstonia solanacearum* is ranked the sixth most damaging disease of medicinal plants including ginger in Indonesia.

Leaf spots caused by *Colletotrichum* spp., *Cercospora* spp., *Glomerella* spp., *Phaeoseptoria* spp. and *Septoria* spp. are common and may require chemical control. Specific leaf spots include *Coniothyrium zingiberis* and *Helminthosporium maydis*, *Leptosphaeria zingiberis* and *Phyllosticta zingiberis* in Asia, Japan and China, and *Xanthomonas zingiberi-cola* in the Far East. All are controlled by the chemicals used on outbreaks of more damaging pathogens. Resistance to *Phyllosticta zingiberis* may exist in India (Singh, A.K. *et al.*, 2000). A thread blight due to *Pellicularia fila-mentosa* could become of more importance in India. A virus, similar to the wheat streak mosaic virus, has been reported from India, but with no information on its occurrence or the damage inflicted. Various techniques for controlling local virus infections in propagating material have been developed in China (Gao *et al.*, 1999).

oxysporum considerably reduced the number of infected rhizomes and increased yield. A bacterial root rot due to *Pseudomonas solanacearum* was controlled by hot-air treatment in Hawaii (Tsang and Shintaku, 1998), and a resistant ginger cultivar, Varada, has recently been released by the Indian Institute of Spices Research, Calcutta. Dry rots due to *Diplodia* spp. and *Macrophomina* spp. occur in Asia, *Trichurus spiralis*, a grey rot occurs in parts of India, and *Pyricularia zingiberis*, blast, can be important in Japan and controlled by soil fumigation. Rhizomes can also be infected in store, and once this has occurred it is extremely difficult to control (see Transport and storage section).

A serious bacterial wilt of lower stems and rhizomes is caused by *P. solanacearum* in Asia, but *Pseudomonas zingiberi* in Japan and China (Fig. 11.12). Symptoms are progressive yellowing and wilting from lower leaves to the whole plant, and badly affected stems or rhizomes yield a milky exudate when cut. The disease is widespread, exists in a number of regional biotypes, and attacks many tropical

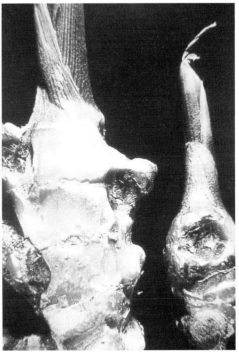

Fig. 11.12. Stem and rhizome rot.

Harvesting

The time of harvesting ginger depends on the cultivar and varies from 7 to 9 months in annual crops. Perennial crops are harvested at the grower's discretion. Time of planting usually determines the final rhizome yield, and planting outside this often very short period reduces the total. In Shandong province, China, for example, planting in the first week of May yielded 45.6 t ha^{-1}, and fell steadily to 27 t ha^{-1} for planting in the last week of May (Xu and Kang, 1999). Harvesting depends mainly on rhizome end use, for general sale whole in local or overseas markets or to be processed for spice, oil and oleoresin. Soil type can also affect rate of growth and thus time to maturity of rhizomes; for instance, on calcareous soils in north-east India, oil content was highest in cultivar Rio de Janeiro at 180–210 days, compared to 240–260 days on the more clayey soils of southern India (Maurya *et al.*, 1984).

Harvesting is usually manual, plants being carefully lifted from the soil to prevent rhizome damage. At maturity, much of the foliage has withered and presents few problems. In India, this operation is only partially mechanized, with rhizomes loosened mechanically to assist hand-lifting, and similarly in Australia to produce high-quality rhizomes for preserving (Fig. 11.13), although general harvesting in Australia is fully mechanized using special equipment. Mechanization affects the method of cultivation, since beds or rows must be spaced to suit the number of rows harvesting equipment can cover; three-row machines are the most common. Whatever method of lifting is used, rhizomes must be handled with care to prevent damage, since breaks in the skin not only allow infection by pathogenic fungi but also loss of volatile oil. Freshly dug rhizomes are generally known as hands and the branches as fingers.

Rhizome yield is basically determined by the cultivar grown, and thus agronomic and climatic factors can only assist in obtaining the highest yield. No genetically governed low-yielding cultivar can produce a higher rhizome yield than a high-yielding cultivar grown under the same conditions. This may sound obvious, but a high degree of manage-

Fig. 11.13. Harvesting high-quality preserving rhizomes – Queensland, Australia.

ment skill is wasted on low-yielding types when much higher yields could have been obtained using another cultivar at lower total inputs (E.A. Weiss, personal observation). The reason usually stated is tradition or local consumer preference. For example, in Kerala, India, dry rhizome yield from common local cultivars is 2–6 t ha^{-1}, average 2–3 t ha^{-1}, but nearly 10 t ha^{-1} from high-yielding selections available from the local agricultural experimental farm. The substantial differences in rhizome yield and oil content between Indian cultivars is well documented, and indicates the considerable potential to increase both.

Yield from smallholder ginger is generally below 3 t dried rhizomes ha^{-1}, indeed in Africa and the Caribbean it seldom exceeds 2 t ha^{-1}. Similar to India, yields obtained at local experimental stations and by more-skilled growers are frequently several times the local average, emphasizing the potential for greatly increased production at little or no additional cost. Australian specialist growers regularly obtain 10–15 t dry rhizomes ha^{-1} and up to 150 t green ha^{-1} from selected strains; the ratio of green to dry ginger is roughly 5:1. Higher yields are expected following the introduction of new cultivars developed in Queensland from tetraploid gingers (M. Smith, personal communication, 2001). Similar to yield, the oil or oleoresin content of rhizomes and their characteristic odour and pungency is genetically controlled. Increases can be achieved only by planting selected cultivars possessing the optimum combination of these factors, but only in Australia has this been achieved on a commercial scale.

Differences in composition of freshly harvested, whole rhizomes have already been noted, and the range in commercial dried rhizomes is well documented. As rhizomes mature, fibre, pungent constituents and volatile oil increase, and these are the most important criteria in determining suitability for a particular product. Rhizomes are left to mature and develop their full aroma, flavour and pungency when required for oil. Time to harvest thus varies with locality and cultivar, and is based mainly on local experience. Leaving rhizomes too long in the ground may affect pungency and reduces oil content; thus, the optimum period for harvesting in a particular locality must be accurately determined. For example, in Australia ginger is harvested three times, the majority, harvested in February–March, consists of young, tender rhizomes with little fibre for consumption as sweet or preserved gingers. The second is in May when rhizomes are about 85% final size and oil and oleoresin content is highest, and are destined for extraction. The third harvest is in June–July when rhizomes are mature. A proportion is retained for seed, the remainder cleaned, sliced, dried and ground for sale as spice and to food processors.

The rate of increase and ratios of various rhizome constituents over time requires more local research, especially the important fibre content since this affects spice palatability. Fibre development is initially slow, but once rhizomes approach physiological maturity, it rapidly increases. In Australia, when green rhizomes reach approximately 50% commercial fibre content by weight, daily fibre deposition became 1% by weight. So important is the fibre level in rhizomes to be used as preserved or green ginger that Australian crops are monitored daily to ensure compliance with the commercial limit of 40–45% fibre-free flesh. Young tender rhizomes about 5–7 months after planting are preferred for preserved ginger, as the fibre content is negligible and the pungency mild. In Australia, the mid-season crop is preferred for oil or oleoresin, the late season for dried spice.

Oil content of mature rhizomes is generally in an inverse ratio to fibre content, and the proportion of oil often tends to fall with age of rhizome, and not merely in relation to total fibre. Both cultivar and season may have a direct influence, with evidence that these relationships can be very localized. In Australia, the volatile oil content normally increases from 2 to 4.5% on a dry-weight basis over the 4 month harvesting period, but remains fairly constant at about 0.4% on a green-weight basis. Fresh weight of rhizomes gradually increases up to 170 days after planting in Fiji, but rises quickly thereafter to 224 days, with fibre content also increasing faster over the latter period.

In southern India and Sri Lanka, maximum oil and oleoresin content occurred 245–265 days after planting, in northern India 180–220 days, in Taiwan and China 230–255 days, and in Australia 270–290 days. In Hawaii, oleoresin was constant at 1% (fresh-weight basis) over 34 weeks of rhizome growth, but the important 6-gingerol principal was highest on a dry-weight basis at 16 weeks. Data on cost of production are scarce; for smallholders in India it was stated as Rs 8.0/9.5 per kg of rhizome in 1996/97.

Processing

The traditional methods of processing rhizomes remain current in many producing countries, mainly because they are well known and usually simple and cheap. Following lifting, rhizomes are washed and roots removed, then they are killed by immersion in boiling water for about 10 min, dried and stored. It is sufficient to state that these operations should be performed as soon as practical after lifting, and any method that ensures clean, whole, dried rhizomes are quickly delivered to store is acceptable. Where sun-drying is not possible, wood-fuelled or solar-heated dryers should be used. Weight loss during drying under controlled conditions is 60–70% to reduce moisture content to 7–12%. Artificial drying minimizes loss of pungency, volatile oil, and can virtually eliminate the incidence of fungal and other contamination. In Australia, gas-heated dehydrators are used and produce high-quality ginger. However, it appears that drying temperature and duration can directly affect the flavour compounds. A major effect was a reduction in gingerol and an increase in the terpene hydrocarbons content (Bartley and Jacobs, 2000). In Brazil, the rate of air flow in a fixed-bed dryer had a greater influence on the rate of drying than temperature (Gouveia *et al.*, 1999).

Many smallholders rely on traditional curing methods, as described (Purseglove, 1981c), but these are falling into disuse for internationally traded rhizomes. Mechanical washers and hot-air dryers with throughputs ranging from hundreds to thousands of kilogrammes are now available to satisfy growing demands from consumers for more hygienically prepared products. For example, in the Philippines, researchers at the Tarlac College of Agriculture have developed pedal-operated machines, which clean 10 kg rhizomes in 10 min and a slicer with a capacity of 20 kg h^{-1}, and have built a solar-operated cabinet dryer, which fully dries fresh rhizomes in 3 days.

The extent rhizomes are treated prior to drying directly affects fibre and volatile-oil content. Removal of the cork skin reduces fibre content but also enhances oil loss through rupture of the surface oil cells; thus, cleanly peeled Jamaican ginger generally has a lower oil and fibre content. Peeling rhizomes also influences pungency, as the constituents concerned are mainly located in the outer skin layers. Fibre content of unpeeled dry rhizomes can be 10%, but in commercial dried ginger is usually 1.5–6.0%, with the pungent constituents, gingerols, in freshly dried rhizomes being 1–2%. Peeling, scraping, slicing, bleaching and similar treatments, which produce edible or preserving ginger, should not be used on rhizomes destined for oil or oleoresin production. Rhizomes for preserved ginger in Australia are washed, dried, chopped and treated in large tanks under strict control to ensure high-quality products (Fig. 11.14).

Fig. 11.14. Processing preserved ginger.

Distillation

Many countries produce dried ginger, as noted, but only Jamaica, Sierra Leone, Nigeria, India, Australia and China in quantities large enough to be internationally traded. Oil is produced mainly from their cultivars, and thus considerable variation in oil yield, characteristics and composition occurs (Table 11.4). The principal difference between oils distilled from dried and fresh (green) rhizomes is that the latter usually contain more lower-boiling-point components. All rhizome types may be used for oil distillation and oleoresin extraction, but coated rhizomes are the most extensively used; preferred material is coated Nigerian splits, Cochin, then Jamaican.

A major factor influencing oil yield is whether rhizomes are peeled or unpeeled, since oil-containing cells are especially numerous in the epidermal tissue, and thus peeled or scraped rhizomes have less oil. A comparison of whole dried rhizomes and scrapings in India gave an oil yield of 2.0 and 0.9%, respectively, illustrating not only the substantial oil loss that occurs but the viability of distilling peelings, whose oil has different characteristics to rhizome oil. For steam distilling, dried rhizomes are comminuted to a coarse powder immediately before being loaded into the still. The charge must be packed evenly, or the powder may be spread on several trays to ensure complete recovery of the oil. Distilling is by live steam, and depending on the weight of charge and steam pressure may take up to 20 h. Cohobation is normally necessary to obtain maximum oil yield. In Sri Lanka and India, approximately 70–75% of oil was recovered from dried ground rhizomes, a further 5–6% by cohobation, and 50–60% from green ginger. Oil yield generally is 1.5–3.0%, averaging 2.0%, with above 3.0 exceptional, but up to 6.0% has been obtained from selected strains at experimental stations. Chinese oil distilled from dried, coated rhizomes yields 1.5–2.5%, but reportedly 4–5% from rhizomes produced in Hunan Province.

Ginger distillation residue in India contained 50–55% starch, which could be recovered by processing. Distilled material, ginger marc, can also be solvent-extracted to produce an inferior oleoresin, sometimes used to dilute or adulterate other oleoresins or ginger oil. Marc is also dried, powdered and used as an inferior spice and can be added to stockfeed, and although fibrous is readily eaten. Rhizome peelings left after preparing edible ginger should be distilled as soon as possible, since delays of only a few hours can result in major oil loss. A yellow dye can be solvent-extracted from ground rhizomes (Popoola *et al.*, 1994).

The characteristics of ginger oil from various origins differ considerably, notably the variation in optical rotation; characteristics are also related to age of rhizome and period of exposure to air and light during storage. The ratios of the main oil constituents vary, in the range zingiberenes 30–40%, α-circumene 8–18%, β-sesquiphellandrene 7–10%, camphene 4–8%, β-bisabolene 4–6%, β-fellandrene, 5%; monoterpenes represent about 1% and are important in determining final aroma. Further details are given in the

Table 11.4. Comparison of ginger rhizomes and oil yield from selected countries.

Country	Rhizome	Flavour and odour	Oil yield (% v/w)	Extracts[a] (%)
India (Cochin)	Light brown, partly peeled	Lemonlike	2.2	4.25
Jamaica	Buff, peeled	Delicate	1.0	4.4
Sierra Leone	Dark, partly peeled	Pungent, slightly camphoraceous	1.6	7.2
Nigeria	Light, partly peeled, fibrous	Very pungent, camphoraceous	2.5	6.5
Japan	Dark, unpeeled	Flat, lemony	2.0	4.6
China	Pale brown, unpeeled	Flat, lemony	2.5	ng
Australia	Light brown, unpeeled	Strong lemony	2.5	ng

[a]Ethylene dichloride.
ng, not given.

Products section. The method and type of drying affects not only the yield of oil, but also its odour; oil with the finest aroma is obtained from fresh green ginger or from freeze-dried ginger. Up to 20% volatile oil can be lost during sun-drying of Indian rhizomes, and the lemony aroma becomes progressively weaker since the major reduction is in the lower-boiling components, including citrals. Australian, Cochin and Calicut fresh rhizomes have a pronounced fresh, lemon-like aroma, and retention of this characteristic in Australian oils is probably due more to careful drying than to inherent rhizome properties.

Ginger oleoresin is obtained by solvent-extracting powdered, dried rhizomes, with significant differences in yield, aroma, flavour and pungency due to rhizome origin, age at harvest, solvent and extraction method. It is important that rhizomes for extraction be mould-free, since mycotoxins may be coextracted with flavour principles. Well-dried rhizomes are ground or pulverized to a coarse powder and extracted with the selected solvent. Using the batch countercurrent method, the powder is packed in stainless columns and extracted by cold percolation; compared with other spices, percolation is notably slow. Fractions of extract containing not less than 10% soluble solids are drawn off and distilled under reduced pressure, to lower potential heat damage to the gingerols, since excessive heat during extraction can degrade them to less pungent shogaols or weakly pungent zingerone and aliphatic aldehydes.

Common commercial solvents are ethanol, acetone, trichloroethane or dichloroethane, and the oleoresins produced differ in their characteristics, odour and flavour. The moisture content of rhizomes may dilute water-miscible solvents, thus the powder is effectively extracted with a solvent–water mixture, which extracts non-flavour components including water-soluble gums. This causes problems during solvent removal and oleoresin concentration, and sludge may separate out during storage. Conversely, a water-miscible solvent is preferred to extract oleoresin for use in soft drinks, to ensure certain water-soluble

flavouring extractives are present. Supercritical fluid extraction dissolves the aromatic and pungent principles from powdered rhizomes in liquid carbon dioxide under high pressure, and these are then recovered. This technique will probably become more common, since it produces extracts considered to be nearer to the true ginger flavour and pungency, and self-contained plants with varying throughputs are now readily available from manufacturers.

Storage and transport

Indian smallholders in some states partially clean harvested rhizomes, whereas in others they leave soil adhering to rhizomes. They are then packed in sacks of 60–160 kg for transport to wholesalers. In Guatemala, most rhizomes are cleaned and some may be washed before packing in 100 kg sacks. Following delivery at traders' or exporters' stores, rhizomes are checked for extraneous and diseased material. A major problem with smallholder ginger is that the traditional methods of washing and sun-drying allow rhizomes to become heavily contaminated with microorganisms, and they must be treated with a sterilent such as ethylene oxide, or gamma-irradiated; a cobalt-60 irradiation treatment at 10 kGy was sufficient in India to render rhizomes free of microbial contamination. Gas treatment may not be sufficient, and consignments can be rejected as not meeting the stringent health requirements of importing countries, which are generally based on the specifications of the American Spice Trade Association. Australian 'first harvest' rhizomes destined for eating fresh or preserving are stored in brine-filled concrete vats while awaiting processing, while controlled-atmosphere storage and modified packaging extends the shelf-life of fresh rhizomes in the USA.

Rhizomes are attacked by the common pests of stored food products and none are specific to ginger; if uncontrolled the damage caused can be extensive. Control should aim at ensuring that no infected rhizomes enter stores by treating sacks or containers with insecticide, while storage areas and buildings should be regularly fumigated. To reduce the potential damage from disease, only healthy, dry rhizomes treated with a fungicide should be admitted to stores, and regular inspection is essential during prolonged storage, especially for seed rhizomes. Bacterial soft rot caused by *Erwinia carotovora* is responsible for serious loss of stored rhizomes in Australia, and a wide range of fungi are commonly recorded on stored rhizomes in India, with *Fusarium* spp. causing most damage (Rana, 1997).

Rhizomes are graded according to local preference or end-user requirements, and for export must conform to Indian Standards Institute (ISI) specification, IS 1908–1961. Rhizomes are then packed in new jute sacks, bales, wooden boxes or increasingly in lined corrugated cardboard cartons for shipment. Processed Australian ginger, including purees, shredded green and pulps, is packed in 200 and 250 kg drums. The following terms are commonly used to differentiate the various forms of dried rhizomes, and are also well-understood grades:

- Peeled, scraped, uncoated. Whole rhizomes with the corky skin removed without damaging the underlying tissue.
- Rough scraped. Whole rhizomes with the skin partially removed, usually on the 'flat' sides.
- Unpeeled or coated. Whole rhizomes with skin intact.
- Black ginger. Whole rhizomes scalded before being scraped and dried, which darkens the colour.
- Bleached. Whole rhizomes treated with lime or dilute sulphuric acid to lighten the colour.
- Splits and slices. Unpeeled rhizomes split or sliced to accelerate drying.
- Ratoons. Second growth rhizomes; small, dark and very fibrous.

Clean, whole, correctly dried rhizomes can be kept in controlled storage for up to 6 months, but in general rhizomes are shipped as soon as possible, since storage is expensive compared to the rhizomes' value. Microbial and fungal contamination of rhizomes in storage is a major danger, and in Sikkim, India, 39 fungal contaminants were found that were either contracted in the field or in storage (Srivastava *et*

al., 1998). When fresh ginger was held in cold storage in India, a temperature of 2–3°C with 90% humidity gave a shelf-life of 120 days, but 46°C and 40% humidity only 4 days. Storage of fresh ginger at 10°C for 30 days and dipping in thinned wax or a proprietary preservative was superior for retail sale in South Africa.

Prolonged storage of rhizomes or oil causes changes in oil characteristics; light and air increases viscosity, the formation of non-volatile (polymeric) residues and decreases optical-rotation value, while detrimental changes in composition, aroma and flavour can occur at temperatures above 90°C. Irradiation of stored rhizomes reduced the loss of oil and subsequently of extracted oleoresin and gingerol content (Onyenekwe, 2000).

Products and specifications

The most important ginger products are the dried rhizome, whole or ground as spice, ginger oil and oleoresin used as flavourings. Rhizomes are also a source of a range of preserved and crystallized products. Green ginger is normally used where produced.

Rhizome. Dried rhizomes are the usual type of ginger offered for retail sale, and its origin determines its characteristics. Jamaican ginger is generally considered superior due to its clean appearance, delicate aroma and flavour. Nigerian ginger resembles Jamaican but is of lower quality and aroma, with a coarser flavour and a pronounced camphoraceous note, which is very pungent, while Sierra Leone ginger has a rather harsh, slightly camphoraceous, very pungent odour. Australian ginger has a strong lemony aroma and flavour; the quality is generally considered intermediate between Jamaica and Sierra Leone. Cochin and Calicut are two Indian gingers with a lemony aroma. Indian rhizomes destined for the Middle East are soaked in water for a day, next in a slaked lime solution, then dried, rubbed to remove the skin and polished.

It is recommended that rhizomes be bruised by a meat tenderizer or hammer before domestic use to release the aroma. Meat cooked with slices of fresh rhizome is often more tender due to the action of a proteolytic enzyme, zingibain, and 1 g obtained from 40 g fresh rhizome is sufficient to tenderize 10 kg meat; rhizome extracts also have an antioxidant effect on meat. Fresh, peeled and pulped ginger is widely used in Asia and the Far East as a major ingredient of local dishes, and is preferred for its milder, less spicy flavour compared with the dried rhizome. The growing Asian population in Western countries has widened the market for green ginger and it is generally available in speciality stores, either fresh or frozen.

Preserved coarsely ground ginger in its own oil is widely available and in general culinary use. Powdered spice is usually prepared in the UK from lower grades of clean-peeled ginger, but other European countries and the USA frequently use unpeeled ginger. Unpeeled dried rhizomes are ground for use in powdered mixed spice. Packeted powdered ginger is a poor substitute for freshly ground ginger, as the volatile oil content is substantially reduced during manufacture. Freshly ground ginger is used in the processed food industry, bakery products especially gingerbread and biscuits, preserves, desserts, mixed spices and is an ingredient in some curry powders, pickles, sauces and chutneys, ginger beer, wine and cordials. A wide range of gingers preserved in sugar syrup, or crystallized, are available in the world's supermarkets, and Australian products are recognized as superior due to the strict national standards governing manufacture. It also has medicinal properties similar to the oil, later noted. A product almost confined to India is ginger paste, made from 50% sliced and macerated ginger rhizomes, 35% similarly treated garlic cloves and 10% salt. The paste is packed in polythene bags containing 200 g for retail sale. Rhizomes of a number of unrelated species are used to adulterate ginger or as substitutes for ginger products. Common is galanga, later described, whose young preserved rhizomes taste of poor-quality, rather spicy ginger, and are sometimes incorrectly but deliberately labelled 'Ginger in Syrup'.

Essential oil. Ginger oil is a clear, pale yellow to orangey-yellow mobile liquid becoming more viscous with age or exposure to air. Its odour is rich, warm, spicy, somewhat lemony, with a floral note, and the taste warm, spicy, pleasantly aromatic, not pungent but occasionally slightly bitter at high concentrations. The oil is placed in the ginger group by Arctander (1960). Jamaican rhizomes generally have a volatile oil content between 1.0 and 1.3% and a non-volatile ether extract of 4.4%; the distilled oil is used extensively in soft drinks. The oil is usually paler, very mobile, and more sweet and lemony than African. Nigerian ginger oil is darker, richer with a heavier odour and much more tenacious; the non-volatile ether extract content is generally between 2.0 and 2.5% and is in demand for oil and oleoresin extraction (Onyenekwe and Hashimoto, 1999). Sierra Leone volatile-oil content averages 1.5–2.0%, the ether extract around 7% and is in demand for oil, oleoresin and flavouring stock feeds. The Australian mid-season crop possesses the greatest pungency and volatile-oil content, up to 4.4% and is in demand for oil and oleoresin extraction. Cochin and Calicut ginger has a volatile-oil content of 1.9–2.2%, a non-volatile ether extract of 4.0–4.5% and is preferred by ginger beer manufacturers.

Some 100 constituents have been detected in ginger oil (Varmin and Paukauyi, 1994; Lawrence, 2000a), the main components being sesquiterpene hydrocarbons at 50–66%, oxygenated sesquiterpenes up to 17% and monoterpene hydrocarbons and oxygenated monoterpenes comprising most of the remainder. With the significant exception of the citrals, the relative proportion of low-boiling-point monoterpenes is generally low, up to 2%. Australian oils have a high citral content of 8–27% (averaging 19.3%) compared with 0.5–4% in most oils; the ratio of citral (as geranial) to citral (neral) in most Australian samples averages 2:1. A low content of zingiberene and a high content of β-bisabolene in Sri Lankan cultivars is notable.

The pungent constituents determining taste and flavour of ginger oil have been widely investigated and results published in the literature. Analysis by electrospray mass spectrometry has confirmed the leading role of 6-gingerol, and that supercritical carbon dioxide extraction gave an oil that contained the pungent and volatile compounds of the original rhizome (Bartley, 1995). The important pungent compounds present in well-prepared rhizomes are substantially reduced by poor post-harvest handling or improper distilling. A pungency comparison of rhizomes from various origins based on these constituents is shown in Table 11.4. The substantial differences in yield, aroma, flavour and pungency between the various types of ginger is reflected in the character of the resultant oleoresin (McHale *et al.*, 1989). The oil is mainly used to flavour a multiplicity of food products, in beverages, soft drinks and special liqueurs, with small amounts used in cosmetics, perfumes and certain pharmaceuticals. A regional ginger oil may be preferred by individual end-users, but regional oils are often blended to incorporate the special characteristic of each. Oils may also be adulterated with monoterpene hydrocarbons.

A range of medicinal properties are credited to ginger and its extracts in both Eastern and Western societies as a stimulant, carminative, sialogogue and anti-emetic, in the treatment of flatulent colic, dyspepsia and atonic dyspepsia, and as a rubefacient when applied fresh externally (Langer *et al.*, 1998). Recent research has indicated that ginger can be effective in the treatment of nausea and vomiting (Ernst and Pittler, 2000). Ginger is also recommended for the treatment of arthritis, high blood pressure, to reduce cholesterol levels and to prevent travel sickness. A monograph on the physiological properties of ginger oil has been published by the RIFM, and by the European Scientific Co-operative on Phytotherapy (EASCOM). The oil has considerable antibacterial and antifungal activity, and has been used as a seed dressing in India.

Oleoresin. The oleoresin is a dark amber to dark brown viscous liquid, frequently depositing a grainy mass on standing. Its odour is very aromatic, warm, spicy and sweet, and the taste pungent, warm and biting, but never bitter. Oleoresins produced from Jamaican, African and Indian rhizomes differ in odour, taste and pungency partly due to the solvent used, since total removal of some solvents is very difficult. Carbon dioxide extraction of rhizomes produces an extract high in 6-gingerol, as heat decomposition does not occur (McHale *et al.*, 1989).

Commercial oleoresins have a volatile-oil content of 25–30%, and 1 kg is usually equal to 28 kg good-quality ground spice. Commercial oleoresins are labelled according to rhizome origin or as a blend, and offered in liquid form or dispersed on sugar or salt. The main characteristics as defined by the US Essential Oil Association (EOA) Standard are: volatile-oil content 18–35 ml per 100 g; refractive index of oil 1.488–1.498 (20°C); optical rotation –30° to –60° (20°C); meets Federal Food, Drug and Cosmetic Regulations in respect of residual solvent; soluble with sediment in alcohol. The main uses are similar to that of the oil, but the oleoresin is considered to have an aroma and flavour closer to that of the original spice. The oleoresin is seldom used in perfumes or cosmetics because of its poor solubility in alcohol, but more often in pharmaceuticals, especially throat lozenges and similar formulations. Jamaican oleoresin is preferred by carbonated soft drink manufacturers, the African heavier-bodied, more pungent oleoresins being favoured by the processed meat industry.

Australian mid-crop ginger is the only commercially available material specifically harvested for extractors when volatile oil content and pungency are highest. Jamaican, Sierra Leone, Nigerian and Indian gingers are generally harvested at full maturity, when volatile-oil and other desirable components have declined from peak values. Rhizomes thus vary considerably in oleoresin yield, and as commercial extractors' operations are confidential, most data are from laboratories and probably higher than in commercial plants. Using acetone, the yield from Australian dry rhizomes over a growing season was 5–11%, using ethanol up to 20%, and oleoresins from major ginger-growing areas had an average volatile oil content of 7–28%. Using trichloroethane, yield was 2.6–4% from Indian up to 7.2% from African rhizomes. Using dichloroethane, a typical extraction from 209 kg material gave solvent at 36 kg and oleoresin at 12.4 kg (a yield of 5.9%), with oil content in the finished product at 34% w/w, and residual solvent at 30 p.p.m. Solvent-extraction residues have similar uses to ginger marc, and dried meal usually contains 10–12% protein and 40–50% starch.

Absolute. Ginger absolute is usually produced from solvent-extracted dried rhizomes; the extract (basically an oleoresin) has the solvent removed, and the residue is further extracted with alcohol. Absolute can also be produced by extracting commercially available oleoresin, but may not reflect the original material. Absolute is normally produced by the end-user, and is not commercially available as it can be up to ten times more expensive than the parent oleoresin. The absolute can replace oleoresin, but is generally restricted to applications where an alcohol-soluble extract is essential. In the USA, the regulatory status generally regarded as safe has been accorded to ginger, GRAS 2520, to ginger oil, GRAS 2522 and ginger extract/oleoresin, GRAS 2521/2523.

Other Zingiber *species*

The following locally cultivated *Zingiber* spp. produce spice or essential oil used as substitutes for, and sometimes adulterants of, ginger and its products: *Zingiber amada* Roxb. (India); *Zingiber cassumunar* Roxb. (India, South-east Asia); *Zingiber elatum* Roxb. (India); *Zingiber montanum* (K) Diet. (India); *Zingiber mioga* Rosc. (Japan, Korea); *Zingiber spectabile* Griff. (South-east Asia); *Zingiber zerumbet* (L.) Smith (India, Far East). The rhizomes of *Z. cassumunar* are widely used in traditional medicine in the South-east Asian region, and their anti-inflammatory activity is apparently due to their curcuminoid content. Rhizomes of *Z. montanum* are used in traditional medicine and for essential oil, those of *Z. spectabile* and *Z. zerumbet* as a spice and in traditional medicine. *Z. mioga* is also cultivated commercially in the State of Tasmania, Australia, mainly for export to Japan. The genus *Asarum* L. contains a number of species used as medicines or food flavourings, including *A. canadense* L., the wild ginger or Canadian snakeroot, with an aromatic root that resembles ginger when dried, but is not a ginger alternative or substitute.

A related member of the *Zingiberaceae* is *Alpinia galanga* (L.) Willd. (syn. *Maranta galanga* L.; *Amomum galanga* (L.) Lour.). In English it is known as galangal, in Italian and Spanish as *galanga*, in French as *souchet long*, in German as *galangawurzel*, in Indonesian, the Philippines and Malaysian commonly as *langkuas* and in Thailand as *kha*. *A. galanga* is a diploid, $2n = 48$. Galanga is a robust tillering herb up to 3.5 m (Fig. 11.15), with a 2–4 cm diameter, many branched, hard, shiny, light red to pale yellow, fibrous, fragrant rhizome (Fig. 11.16). Details of the plant, cultivation, processing and uses have been described (de Guzman and Siemonsma, 1999).

Galanga has been known as a spice in Europe for at least a thousand years, although now seldom used, but is a locally important flavouring ingredient of food and widely cultivated from South-east Asia to China (Fig. 11.17). Production in the eastern Pacific region is substantial but most is consumed domestically with only a minor international trade; The Netherlands imports about 100 t fresh and 25 t dried rhizomes annually. An essential oil can be obtained from the roots whose main constituents are (E)-β-farnesan, β-bisabolene, α-bergamotone and α-pinene, together accounting for at least half of the total. The leaf oil contains myrcene, (Z)-β-ocimene, α-pinene and borneol, together accounting for about 85% of the total.

Turmeric

The number of species in the genus *Curcuma* L. continues to be in dispute, varying from 30 to 70, with many members providing a food or flavouring, the most important being *Curcuma longa* L., the spice turmeric. Genus members occur naturally from South-east Asia to China, and south to Australia. The genus name is apparently derived from the Arabic and Hebrew name *kurkum*. It is believed that the wild *Curcuma* species from which *C. longa* evolved may initially have been a food, but later the fast-growing plant with a strikingly coloured rhizome acquired magical properties, some apparently associ-

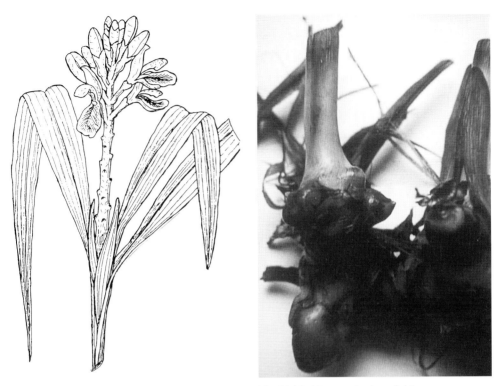

Fig. 11.15. Galangal plant in flower.

Fig. 11.16. Harvested galangal rhizomes.

Fig. 11.17. Dried galangal chips as spice.

ated with human and crop fertility. It is stated the Sumarians and Assyrians used turmeric together with many other spices in their cooking, and if so, the spice must have reached them overland from India. The early Sanskrit name of *haridra* means yellow wood. Garcia da Orta noted in 1563 that it was abundant 'in Cananor and Calicut', and the *Ain-i-Akbari* quoted prices for the rhizome in 1590 indicating a well-established national trade.

The use of turmeric in magical fertility rites became an important part of Hindu ceremonies, and was later adopted by kingdoms in what is now Indonesia and Polynesia, already strongly influenced by Indian Hindu traders responsible for spreading the use of a number of spices, as noted in previous chapters. In India and South-east Asia, turmeric powder is used as a cosmetic, and rice coloured yellow with turmeric is used on social and ceremonial occasions such as weddings. Although the dye is not fast, Watt (1908) stated that Calcutta dyers obtained a brilliant yellow by mixing turmeric with carbonate of soda, and turmeric dye is still used to colour cloth in India. Turmeric is also used as a protective charm and in traditional medicine (see Products and specifications section).

The ancient dhow trade between India and East Africa and the sea connection between the Malagasy Republic and Indonesia are well documented, and it is probable that turmeric reached East Africa by sea in the 8th century and was later distributed throughout sub-Saharan Africa, reaching West Africa in the 13th century. Its major use everywhere is as a dye and ingredient in traditional medicines, derived from its uses in Indonesia. Sea traders probably also took turmeric to China as they did other spices. Early records suggest this occurred around the 7th century AD, and Marco Polo noted in 1280 that it grew at Koncha 'as a vegetable which has all the properties of true saffron, as well the smell and the colour, and yet it is not really saffron'. The rhizome is currently used in the treatment of epilepsy and coma due to febrile diseases.

Turmeric is considered to have reached Europe via the Silk Road from India and was initially used as a dye, although the colour is not fast. Later it became known as Indian saffron and was used to colour foods, particularly some special cheeses, sausages and confectionery. Turmeric was mentioned in early herbals as an ingredient in medicines, but with no specific activity. It was reportedly introduced into Jamaica by a Mr Edwards in 1783, where it became naturalized.

Botany

C. longa L. (syn. *Curcuma domestica* Vale.; *Amomum curcuma* Jacq.) is known in English as turmeric, in Spanish, French and Italian as *curcuma*, in Portuguese as *acafrao da India*, in Dutch as *geelwortel*, in Arabic as *kurkum*, in East Africa, in KiSwahili as *manjano*, in India, in Hindi as *haldi*, in Tamil as *manjal* and many other vernacular names, in Indonesian as *kunyit*, in Malaysian as *temu kunyit*, and in Chinese as *iyu-chin*. In this text, turmeric refers to *C. longa* and its derivatives unless otherwise stated.

Turmeric is probably a sterile triploid, developed over time by selection and vegetative propagation between the wild diploid species *Curcuma aromatica* Salisb. ($2n = 42$), native to India, Sri Lanka and the eastern Himalayas, and other closely related tetraploid species. Turmeric is vegetatively propagated, which is probably why long-established local cultivars of *C. longa* differ, as any variation is perpetuated and has resulted in the proliferation of so-called species. A review of nearly 800 accessions in India determined that the *Curcuma* genus could be regrouped into two sub-genera, *Paracurcuma* and *Eucurcuma* Vale. (Velayodhan *et al.*, 1996). However, the genus is still undefined, with many suspect species, and a reliable cultivar grouping would be an asset. *C. longa* is not now known as a truly wild plant, but has become naturalized in such widely differing locations as Jamaica in the West Indies and in east Java, Indonesia. India is arguably the centre of origin of turmeric but it is certainly the most important centre of diversity.

Substantial germplasm collections are held at the National Repository of Plant Genetic Resources, Idukki, the National

Bureau of Plant Genetic Resources, Thrissur, and also at the various Institutes for Spice Research in India (Ravindran and Peter, 1995). Breeding programmes were hampered by the plants' inherent sterility, but clonal selection (Sunitibala *et al.*, 2001) and more recent biotechniques have partially overcome this problem and allowed development of strains with increased resistance to rhizome rot and with higher rhizome yield. Genetic studies in India showed highest heritabilty for rhizome yield and shape, including number of fingers, while X-ray mutagenesis has produced superior cultivars (Chezhiyan and Shanmugasundaram, 2000). Cultivars differ substantially in appearance and characteristics (Lynrah *et al.*, 1998) sufficiently in India to be given names. The following is a general description; figures indicate the range occurring over cultivars, not the range of a single cultivar.

Turmeric is a stout, erect perennial herb, usually grown as an annual (Fig. 11.18). The fleshy tuber (rhizome) at the base of each aerial stem is ellipsoidal, fleshy, 5–8 cm × 1.5–2.5 cm, ringed with the bases of old scale leaves. At maturity the (mother) rhizome bears many straight or slightly curved lateral rhizomes (fingers), 5–10 cm × 1–2 cm, which again branch at right angles, finally forming a dense clump with the roots. Rhizomes have a brownish-yellow, somewhat scaly outer skin and a bright orange-yellow flesh, with white young tips, and a spicy smell when bruised (Fig. 11.19). The roots are filiform, fleshy, tough, penetrating to depth, and often ending in a small oblong tuber, 2–4 cm × 1–2 cm. Rhizome development has been studied in detail in India (Sherlija *et al.*, 1998), as has the effect of gamma irradiation on growth and quality (Rao, 1999). Details of rhizome composition are given at the end of this section.

The erect, leafy shoots rarely exceed 1 m, bearing six to ten alternate, distichous leaves, surrounded by bladeless sheaths, forming a short pseudostem. The thin petiole, 0.5–10 cm, with narrow erect wings, is rather abruptly broadened to the sheath. The ligule is small, about 1 mm, semi-annular, reflexed, ciliate, membranous, and withers quickly. The leaf blade is oblong-lanceolate, 7–70 cm × 3–20 cm, with the base cunate, and the apex acute-caudate. Leaf colour is dark green above, the mid-rib green, and below very light green and covered with pellucid dots. A relationship apparently exists between the number of leaves per clump and rhizome yield, in India (Hazra *et al.*, 2000), and distribution of metabolites and curcumin content

Fig. 11.18. Turmeric plants.

Fig. 11.19. Turmeric rhizome.

of rhizomes (Deeksha and Srivastava, 2000). Growth regulators, including CCC, adversely affect growth of plants but have no effect on rhizomes. Fish are cooked wrapped in leaves in India to impart the turmeric flavour.

The inflorescence is a cylindrical spike, terminal on the central leafy shoot, erect, with the scape partly enclosed by the leaf sheaths (Fig. 11.20). The peduncle is terete, 3–20 cm, densely hairy and covered by pubescent, bladeless sheaths (scales). The flower spike is cylindrical, 5–20 cm × 3–8 cm, bearing many spirally arranged densely hairy bracts. The bracts are elliptic-lanceolate and acute, 5–7 × 2–3 cm. The upper sterile bracts are white or white streaked with green, pink-tipped in some cultivars; the lower are light green with white longitudinal stripes or white margins. The bracteoles are thin, elliptic, and up to 3.5 cm, surrounding the flowers.

The yellow flowers occur in cincinni of two in axils of bracts, thin-textured, long, narrow, up to 5–6 cm, opening one at a time, and fugacious. The calyx is tubular, short, with three unequal teeth, and split nearly halfway down one side. The corolla is tubular at the base with the upper half cup-shaped with three unequal lobes inserted on the edge of the cuplip; it is thin, whitish and translucent with the dorsal lobe hooded. The lip or labellum is obovate, with a broad thickened yellow band down the centre, and thinner creamy white side-lobes upcurved and overlapping the staminodes. The filament of the stamen is short, broad and united to a versatile anther about the middle of the parallel pollen sacs, with a large, broad, curved spur at the base. The trilocular ovary is inferior with a slender style passing between, and held by, the anther lobes; the stigma is expanded. Fruits are rarely produced, although seeds have been reported from India and treatment with concentrated sulphuric acid speeded germination.

Rhizome. The composition of individual rhizomes and among cultivars differs, and the following information is obtained from Indian rhizomes; ground fresh rhizome may contain (per 100 g) water 11–13 g, protein 6–8 g, fat 5–10 g, carbohydrate 60–70 g mainly starch, fibre 2–7 g; resin: ash 3–6 g,

Fig. 11.20. Turmeric flower.

containing K 2.5 mg, Ca 180 mg, Fe 40 mg, Mg 190 mg, P 270 mg, plus ascorbic acid 25 mg and vitamin C. In addition, it contains sugar, including glucose (28%), fructose (12%) and arabinose (1%). Energy value is 1500 kJ per 100 g.

The essential oil imparts the typical aroma and flavour, and rhizomes contain between 2 and 10%, usually 4–5%, depending on cultivar, as in India (Garg *et al.*, 1999; Rumi *et al.*, 1999). The oil is composed mainly of oxygenated monoterpenes with smaller amounts of sesquiterpene hydrocarbons and monoterpene hydrocarbons. A commercial sample of Indian turmeric oil had the following characteristics (all at 24°C): refractive index 1.5130; specific gravity 0.9423; optical rotation −14°; soluble in 1.8 vols 90% alcohol. An oil produced in the USA had very similar characteristics, plus a boiling point of 220–250°C (decomposed

above 250°C) and soluble in 0.5 vol. of 90% alcohol.

The characteristic yellow colour is due to curcumin, a steam-volatile diferuloyl methane derivative, which quickly degrades on exposure to light. It solidifies as orange/yellow needles, has a melting point of 183°C, is insoluble in water and ether, and soluble in alcohol. A rapid method of determining the curcuminoids in rhizomes has been developed in India (Gupta *et al.*, 1999). Rhizomes from the Alleppey region of Kerala State have a high curcumin content of 5–7%, but it has been noted elsewhere in India that a cultivar with a low rhizome yield often has rhizomes with a high curcumin content (Chandra *et al.*, 1999). For additional information on rhizome and oil, see the Products and specifications section.

Ecology

The natural range of turmeric is in the high rainfall areas of Asia and South-east Asia, but so popular is the spice that it has been widely introduced into many tropical countries in both the eastern and western hemispheres with a similar hot, humid climate, Brazil, for example (Cecilio *et al.*, 2000). Turmeric, as noted, is not known in a truly wild state but has become naturalized in many countries to which it was introduced. A majority of the crop is rain-grown and an annual rainfall of 1000–2000 mm is necessary, falling mainly in the important 120 days of growth to flowering, with 1500 mm the optimum. Less than 1000 mm or in areas where the rainfall can be erratic, supplementary irrigation is necessary. Adequate soil moisture is probably the most significant factor affecting rhizome yield, but plants are very susceptible to waterlogging at any growth stage, even for relatively short periods. Thus, turmeric production in areas where the rainfall exceeds 2000 mm must be on permeable or artificially drained soils, or hillsides.

Commercial crops of turmeric are usually grown between 500 and 1000 m but 450–1500 m is acceptable. In India, turmeric cultivated in the hills is reportedly of a better quality than that grown on the plains, and the same cultivar when grown at higher altitudes is considered to produce more and better quality rhizomes than at lower altitudes, although neither of these beliefs is supported by data. Temperature is important, as the optimum varies with crop growth: 30–35°C to encourage sprouting, 25–30°C during tillering, 20–25°C as rhizomes appear and 18–20°C during enlargement. Since turmeric is naturally a plant of open forests, partial or intermittent shade is desirable for high yields, but this is often sacrificed when turmeric is grown as an annual in open fields or as an intercrop. Higher plant populations are required to offset the lower single plant yield. However, giant castor grown around the field margins of smallholder crops provides additional income and intermittent shade.

Soils

Turmeric thrives naturally on a forest-type loam with a high humic content, which encourages rhizome development, but is grown commercially on a range of slightly acid soils, pH 5.0–7.5, including alluvial or clay loams, tropical red earths, sandy loams and rice paddies. Similar to other rhizome-tous spices, management is generally more important than soil type. Stony and heavy clay soils, however, are unsuitable for rhizome development, while saline and alkaline soils should be avoided.

Fertilizers

Turmeric has a high nutrient requirement and commercial crops require substantial amounts of chemical fertilizer, or animal or crop wastes to produce high rhizome yield. Thus, yields are directly related to soil nutrient status, either natural or enhanced. It is for this reason that smallholder yields are generally low, since the fertilizers necessary are not usually available in the amounts required. An alternative is mulching with any green material available, including sal, *Shorea robusta*, sunnhemp, *Crotalaria juncea, Sesbania* spp., *Dalbergia* spp. or local grasses. In the absence of other fertilizers in India, mulching with leaves after planting at 15 t

ha^{-1} and 60 days later increased yields up to 50%. Should mulch plants be used as a green manure, they must be ploughed in well before planting turmeric, as their decomposition can severely reduce soil moisture availability. In India, locally available plant residues are applied, including castor poonac, silt from local tanks (reservoirs), and distilling and oilcake residues, and occasionally domestic stock are penned on the field prior to ploughing.

In general, the N and phosphorus (P) requirement is high, and potash (K) moderate, with P and K applied to the seedbed, and N as a top-dressing. In India, the nutrients removed to produce 4700 kg rhizomes ha^{-1} are shown in Table 3.4 (Sadanandan *et al.*, 1998). Indian soils vary widely in their nutrient status, and thus most recommendations are only locally applicable. For example, the recommendation in Kerala is 15 kg each N and P ha^{-1} and 30 kg K ha^{-1}, with all the P and half the K in the seedbed, the N and remainder of the K as a top dressing in two applications 30 and 60 days after planting. In Andra and Madhya Pradesh, the very high rate of 190 kg N, 30 kg P and 55 kg K ha^{-1} is recommended, with all the P and K in the seedbed and the N in split applications; the high rate of 150 kg N, 125 kg P, 250 kg K ha^{-1} is also recommended in Bangalore (Gupta and Sengar 1998; Venkatesha *et al.*, 1998). In Tamil Nadu, 25 t FYM or compost plus 50–55 kg N, 25 kg P and 45 kg K ha^{-1} is recommended, all the P and K at planting and half the N in the seedbed and half as top dressing. However, in the Punjab the addition of N to FYM had no significant effect on rhizome yield (Gill *et al.*, 1999). The only minor nutrients noted as increasing rhizome yield are iron and zinc sulphates.

Cultivation

Turmeric is propagated vegetatively using rhizomes or fingers, but *in vitro* culture of excised buds from sprouts or tissue culture results in large numbers of plantlets, and is an excellent method of rapidly increasing clonal material. As 2000–2500 kg of planting material ha^{-1} is required annually in India, any reduction in this large amount, which needs to be protected in storage with the loss of income incurred, is to be welcomed. Similar storage methods to those used for ginger are common. A number of storage methods tested in Bangalore, India, showed that storage in 100-gauge polythene bags with 3% ventilation was the most effective (Venkatesha *et al.*, 1997).

Whole or split rhizomes or fingers with one or two buds are used according to local choice. Whole (mother) rhizomes gave the highest yield in India and Thailand (Yothasiri *et al.*, 1997), but in general fingers are to be preferred as they are more easily protected in storage, are more tolerant of a wet seedbed and less liable to contract soil-borne diseases than cut rhizomes, although fungicide treatment of fingers is recommended. Rhizomes or fingers must be stored for several months from harvest to planting and a number of traditional methods remain common. Fingers are thinly spread under a covering of turmeric leaves, dried grass or straw, or in small heaps covered with leaves, straw and then dry soil or sand. Black plastic is sometimes used, but a major disadvantage is the condensation accumulating on the inner face. Planting material may be mixed with fine ash to resist fungal infection, and must be protected against vermin (see also ginger section). Rhizomes or fingers may be sprouted prior to planting by smallholders, but is not usually necessary as there is apparently little difference in final yield to offset the additional care required at planting.

Land preparation depends on whether turmeric is to be grown in a pure stand or intercropped; if the latter, land is prepared to suit the most important component. For a pure crop of turmeric, lands are generally prepared as for ginger, then beds or ridges constructed to suit the local soil conditions. Ridges are preferred where the crop is to be irrigated or where heavy storms require rapid drainage. In India, bed size varies as does plant spacing, since regional cultivars differ in their growth characteristics, but is seldom less than 1 m wide with two rows per bed and 25 cm between plants. After planting, beds are covered with animal manure or mulch.

Ridges in India are often substantial, 45–60 cm apart and 25–30 cm between plants; in Sri Lanka, smaller ridges and closer spacings of 15 × 25 cm or 25 × 30 cm are considered to give higher yields. Rhizomes can be planted from 5 to 10 cm deep with little difference in yield at the various depths in this range. Planting is usually manual although if turmeric fingers are used semi-automatic small planters are suitable, as for ginger. The fully automatic ginger planters used in Australia, which can also construct beds or ridges in one pass, could easily be adapted to plant large commercial turmeric plantations. The optimum time of planting in India is from April to June depending on the district, with later dates steadily reducing yield. Good-quality fingers germinate in 14–28 days, and treatment with an insecticide and fungicide generally increases percentage emergence. Plants flower in 4–6 months when rhizomes begin to form, the variation being a cultivar characteristic, and rhizome development continues until the crop is ready for lifting. Research in India showed that rhizome colour increases to full maturity but can decline thereafter.

A wide range of local cultivars are recognized and named in India, each popular in a specific locality, although the new numbered clonal types are now grown more widely as they generally give higher yields. Some cultivars are grown for a particular reason; in Maharastra one, *Lokhandi*, has a hard, bright yellow/orange rhizome favoured for dyeing; others are quick-growing, have greater resistance to rhizome rot, large rhizomes, or a more spicy flavour.

Weed control

Turmeric is a robust plant, but young seedlings suffer from weed competition; thus beds and rows should be regularly cleaned and at least four weedings are usually necessary. This operation remains manual except in larger plantations. Herbicides are seldom used, but those recommended for ginger should be suitable for the same reasons.

Irrigation

No accurate data are available on the total water use of turmeric, nor on changes in usage during plant growth; growers rely on their experience. Where turmeric is grown as a partially or fully irrigated crop, control of available soil moisture is important as overwatering can be fatal. A substantial watering at or just after planting is essential if rain has not fallen, and further applications until full emergence. Subsequent waterings should be adjusted to take account of any rainfall. Fully irrigated crops should be monitored to maintain adequate soil moisture levels, the intervals becoming longer as rhizomes mature. Turmeric is susceptible to salinity, and thus the origin and composition of a potential water source should be carefully evaluated.

Intercropping and rotations

Turmeric is generally grown as an intercrop but may be the sole crop in recently established plantations, since the newer clonal types with their potentially greater yield allow a higher level of inputs, including partial mechanization. Turmeric is intercropped in India with maize, finger millet, chillis, onions, aubergine and other quick-growing vegetables as subordinate crops, and may be grown alongside ginger. The dynamics of various combinations have been determined (Sivaraman and Palaniappan, 1996; Bisht *et al.*, 2000). Giant castor is often planted around field margins to provide intermittent shade and additional income.

In Guatemala, turmeric is usually interplanted with chillies, local beans and tomatoes. The fall in international coffee prices and continuing surplus has encouraged growers with small areas under coffee in other Central and South American countries to plant alternative crops including turmeric. One method of combining the two requires uprooting alternate rows of coffee trees and planting several rows of turmeric. The remaining trees providing intermittent shade and windbreaks.

Turmeric should preferably not be planted on the same land for more than two successive crops with a 4-year interval before replanting. When strip-cropping is practised it can be planted annually on different sections of the same field without significant yield reduction. Suitable crops for a rotation

in India and Sri Lanka include rice, sugarcane and bananas on wetter soils, maize, finger millet, chillis, garlic, legumes and vegetable crops on drier soils. Turmeric can also be successfully interplanted/underplanted in coffee, litchi, mango, jackfruit, coconut and woodfuel plantations (Dyal *et al.*, 1996; Sairam *et al.*, 1997).

Pests

In general, those pests of importance on ginger also cause most damage to turmeric, both in the field and in storage, as in India (Sarma and Anandaraj, 2000). These include *Dichocrocis punctiferalis*, whose larvae bore into the central shoot, which wilts and dessicates, the leaf roller *Udaspes folus*, thrips *Panchaetothrips indicus*, the scale insect *Aspidiotus hartii*, and the tingid *Stephanitus typicus*. A serious pest is *Mimegrella coeruleifrons*, the rhizome fly, whose larvae bore rhizomes. Control with various mulches was successful in Karnataka State, India (Kotikal and Kulkarni, 1999). Nematodes of a number of species are usually present in turmeric crops, and in India are most prolific in young, and less numerous in older crops (Poornima and Sivagami, 1999; Ramana and Eapen, 1999).

Diseases

The diseases attacking turmeric are generally more important and damaging than pests, but although a number of leaf diseases have been recorded, few are of major importance. The leaf spot caused by *Taphrina maculans* occurs wherever turmeric is grown in India, and may also occur in Sri Lanka. The main symptom is a large number of irregular yellow spots on both leaf surfaces, 1–2 mm in diameter, which later coalesce. Leaves are not killed, but their contribution to plant growth is greatly reduced and this is reflected in smaller rhizomes. High humidity and temperature favour outbreaks and spread of the disease.

Another widespread leaf spot is due to *Colletrichum capsici* but is most damaging in southern India, where it has reportedly reduced yield in a seriously infected crop by two-thirds. The initial symptom is small elliptic to oblong spots on both leaf surfaces, which may expand to 4–5 × 2–3 mm. Spots have a greyish-white centre and a yellow halo and can coalesce to form large necrotic patches; in severe infections, leaves die and fall. The disease usually appears in August–September in periods of continuous high humidity. Resistance to the disease may exist, but has yet to be fully investigated. A widespread and often damaging leaf blight is caused by *Rhizoctonia solani*, which has a wide host range and is soil-borne. The main symptom is the appearance of water-soaked spots varying in shape and size on lower leaves, which gradually expand to cover the entire leaf. Warm, humid weather encourages spread and virulence of the disease. Other less damaging leaf spots are caused by *Phyllosticta zingiberis*, which also attacks ginger, *Myrothecium roridum* and *Phaeorobillarda curcumae.*

A rhizome and root rot, caused by *Pythium graminicola*, is reportedly becoming of greater importance in India, and incidence of the disease was stated to be higher when turmeric was intercropped with maize in Andra Pradesh State. In Sri Lanka, *P. aphanidermatum* is more important. Rhizome brown rot is apparently due to a combination of eelworm infestation, usually *Meloidogyne* spp., and infection by *Fusarium* spp. Rhizomes are often initially attacked in the field and the rot develops during storage.

Harvesting

Turmeric is ready for digging when the lower leaves turn yellow, and time from planting depends on the cultivar. In India, three major cultivar divisions are recognized: long (9 months), medium (8 months) and short (7 months); foliage begins to wither in December–February and plants are dug in January–April. Manual digging of rhizomes remains the most common method of harvesting, although many small, animal-drawn diggers/lifters are suitable. Remarks on harvesting in the ginger section are applicable to turmeric.

Leaves must be cut prior to mechanical lifting or after manual digging, and whatever system is used, care is necessary to ensure rhizomes are not split or bruised, as such injuries invariable result in fungal infection and rejection. Clumps should be lifted whole, rhizomes cut from the remaining plant material, and earth and long roots removed. Rhizomes are then well washed, quickly sun-dried, fingers separated from the main (mother) rhizome, and stored until processed (Fig. 11.21). Similar to ginger, part of a plantation may be left for later lifting to provide next year's planting material.

Rhizome yields directly reflect the standard of crop management and range from 5000 to 10,000 kg ha^{-1} for rain-grown crops to 27,000 kg ha^{-1} for well-managed, irrigated commercial plantations in India using clonal material. However, even higher yields up to 35,000 kg ha^{-1} have been obtained at experimental stations and demonstration farms using selected clonal material, indicating the potential available to increase yield.

Fig. 11.21. Fresh turmeric rhizomes.

Processing

Curing. The traditional method of curing rhizomes adopted in India and elsewhere was to steam or boil fresh rhizomes in lime or sodium carbonated water. This 'kills' the rhizome, removes the earthy, raw odour, gelatinizes the starch, and generally enhances rhizome appearance. An improved and more hygienic solution for treating rhizomes composed of 20 g sodium bisulphite and 20 g hydrochloric acid per 45 kg rhizomes produces a clean, yellow-tinted rhizome. Rhizomes were then sun-dried. Rhizomes were cured in a variety of containers including the ubiquitous fuel drum, and quality was very variable. To enhance quality and produce a more uniform product, an improved turmeric boiler was developed.

The unit basically consists of two rectangular perforated containers inside a larger container with a lid. The outer container is a rectangular box 1.2 × 0.9 × 0.75 m of 3 mm mild steel sheet with a tight-fitting lid. Two inner containers with a capacity of 75 kg rhizomes each, made of mild steel, 0.5 × 0.75 × 0.5 m, perforated with 12 mm holes, are carried on a stand and provided with lifting handles. Water is heated from below the outer casing by any suitable waste material. The outer container is three-quarters filled with water and sodium bicarbonate added at 100 g per 100 l. The perforated containers loaded with cleaned rhizomes are then placed in the water, heated, and as the outer container is provided with a tight lid, the water boils and the rhizomes are cooked, after which they have a pleasant smell and are soft when pressed between the fingers. The perforated containers are lifted out of the water, the rhizomes unloaded for drying, and the containers refilled with the next batch of rhizomes. Since the perforations allow all water to drain back into the outer container during lifting little is lost; this is topped up and extra salt added for the next batch. Thus, water and heat are conserved. The unit requires only two workers and throughput is between 1000 and 1600 kg per day of 8 h (Sreenarayanan and Viswanathan, 1989). The unit could usually be constructed locally at a 1998 cost of Rs1000.

Similar to other spices previously mentioned, the traditional methods resulted in considerable fungal and other contamination, and with increasingly stringent regulations in importing countries regarding microbial and other content of food additives, more hygienic curing methods became necessary. Thus, large traders or cooperatives have introduced the boiler described or similar cookers, and buy cleaned raw rhizomes from smallholders for curing.

Drying. The majority of turmeric rhizomes destined for international marketing are now artificially dried, and traders or exporters have installed a variety of hot-air, drum, tray or continuous tunnel dryers, which control speed of operation and temperature. An important factor governing throughput and maximum temperature is whether the rhizomes are sliced or whole; slicing also generally produces a more uniform and brighter-coloured powder. Whole rhizomes are usually polished in a manually operated drum following drying to improve appearance, and a special emulsion is added to the polisher, composed of alum 0.04 kg, turmeric powder 2.0 kg, ground castor seed 0.14 kg, sodium bisulphite 30 g, and concentrated hydrochloric acid 30 ml per 100 kg rhizomes. Reported yield of polished rhizomes from 100 kg fresh material varies among cultivars at 15–25 kg, with mother rhizomes giving a higher percentage than fingers. Dried rhizomes are usually chemically treated or fumigated before storage to prevent insect and fungal infestation.

Distillation

Steam distilling fresh or dried, sliced or macerated rhizomes including fingers, yields 2–10% of an essential oil depending mainly on the cultivar. The operation is straight forward and no special equipment is required. However, as the oil contains around 85% high-boiling point sesquiterpenes, distillation time is at least 4 h. Supercritical carbon dioxide extraction of ground tubers is now more common, and in India produced an oil composed of 60% turmerone and ar-turmerone (Gopalan *et al.*, 2000). Oleoresin is obtained by extraction from coarsely ground rhizomes, as for ginger.

Storage and transport

Rhizomes are graded into fingers and whole rhizomes, to national grades including the Indian Agmark, or to suit customer specifications. Provided rhizomes are mature, sound, hard and of good appearance, origin and cultivar are less important compared with a bright, uniform colour, good aroma and freedom from microbial contamination. The pressure to meet international or national standards of importing countries will inevitably force producers and exporters to adopt modern methods of processing or lose their market share.

Well-dried rhizomes can be stored in sacks for some months if kept in cool, dry sheds protected from vermin, and fumigated. Residues from gas or chemical fumigation are negligible after the period required for shipment to an overseas destination. Irradiation can destroy harmful micro-organisms on rhizomes and extend storage life, usually with little affect on aroma or flavour compounds (Suchandra *et al.*, 2000). When finally graded, rhizomes should be packed in lined, corrugated cardboard cartons, heat-sealed polythene bags, or new, double jute sacks, weighing 50 or 100 kg, and exported as soon as possible. Fumigated turmeric powder can safely be stored in laminated polyester/paper/foil bags for a year, and in polyethylene packets for about 3 months. Regulations regarding fumigation of retail products are in force in most consuming countries, but vary among countries regarding the method and permissible residues. In India, for example, a tolerance of 50 p.p.m. bromide ion in cereals and flours applies, and 400 p.p.m. for ground spices sold retail in the USA.

Products and specifications

The most important product is the dried rhizome, the turmeric spice of the international and retail trade, where acceptance is basically determined by its general appearance and colour, and its quality on the organoleptic characteristics of the essential oil.

Rhizome. Traditionally turmeric rhizomes were traded or exported under a multiplicity of names and origins (Purseglove *et al.*, 1981d), and while this continues for specific types or customers, they have generally been replaced by national standards or brands such as the Indian Agmark (Pruthi, 1998). A considerable volume of rhizomes are now traded in bulk, especially from South America, with overall quality guaranteed by exporters. Without this guarantee, turmeric rhizomes are commonly adulterated by inclusion of the many related but inferior species. Rhizomes and branded powders available retail in Western countries are generally unadulterated, due to competition between spice processors who would lose market share by selling an inferior product. Rhizomes from the Caribbean, Central and South America are considered of inferior quality to Indian.

Good-quality rhizomes should be clean, of a uniform colour, with a rough although polished skin, snap cleanly when broken, have a firm, bright orange/yellow flesh, and a musky, spicy, pungent somewhat peppery odour. The taste should be spicy and only slightly bitter. Since rhizomes are less easy to grind into powder than other spices and are seldom used whole, most turmeric for domestic use is sold as a fine, orange/yellow powder. Since the storage life of powder is seldom more than 3 months before it begins to deteriorate, retail packs should always be checked for date of packing or use-by date. Whole rhizomes or powder should be stored in sealed containers in the dark as they rapidly lose their colour.

Turmeric is essentially a colourant, and thus a major ingredient of curry powders and pastes, of mustard pickles and especially the type known as piccalilli. It is generally used as a flavouring and to colour a wide range of processed meat products, confectionery, baked products, cheeses, spreads, soups and similar products, and to a lesser extent in drinks and liqueurs and pharmaceuticals. Turmeric is also known as Indian saffron, and this association with saffron is disastrous in cooking, as turmeric is not a saffron substitute, having the same colour but not the flavour. In the USA, the highest maximum use level in foods is about 20–22%. Turmeric powder and the essential oil are retained in European phytomedicine, while the multitude of uses for turmeric and its derivatives in traditional Indian medicine have been reviewed (Shah, 1997; Khanna, 1999), and its uses in Western medicine reviewed (Grant and Schneider, 2000).

Essential oil. The steam-distilled essential oil, curcuma oil, is a free-flowing liquid, yellow to yellow/orange colour, which can be slightly fluorescent; the odour is fresh, peppery/spicy, with sweet orange and ginger notes, and the taste slightly pungent and bitter but not biting. The oil is placed in the ginger group by Arctander (1960). Oils differ in taste and odour depending mainly on the cultivar, but also on maturity at digging. Aroma of the oil is considered inferior to that of the spice, and thus it is not a direct substitute. The main constituents are tumerones (sesquiterpene ketones) up to 60%, considered mainly responsible for the aroma, and zingiberene up to 25%, although these amounts can vary substantially when obtained from rhizomes from widely separated geographical origins. For example, Indonesian rhizome oil can have little or no zingiberene (Zwaving and Bos, 1992). About 60 compounds have been identified in the oil including 1,8-cineole, α-phellandrene, *d*-sabinene and borneol. The contribution and importance of individual constituents to the aroma and flavour has still to be fully determined. Major adulterants are oils of *Curcuma aromatica* and *C. zedoaria*, detected by the presence of camphor and camphene in both.

The oil has no commercial advantages over the powdered spice or oleoresin and is seldom available commercially. It can be used as raw material to produce a number of turmerone esters, including turmeryl acetate. A range of pharmacological and biological activities have been ascribed to turmeric and its oil, especially the curcumin component, and are fully described in the literature. A monograph on the physiological properties of turmeric oil has been published by the RIFM.

Oleoresin. Oleoresin is obtained by solvent-extracting chopped or ground rhizomes, and the method is similar to that used for ginger as described in the preceding section, with a yield between 6–10%. The oleoresin is an orange/red, relatively insoluble viscous liquid and seldom used in this form, but diluted with an approved non-volatile edible solvent to increase solubility and ease of handling. Commercial products can be liquids, semi-liquids or solids with a guaranteed curcumin content ranging between 7.5 and 45%. The oleoresin is the preferred alternative to the ground spice in many industrial applications; maximum usage in seasonings and flavourings in the USA is 0.883% (8834 p.p.m.). Recent oleoresin-derived products include emulsions/aqua resins, encapsulated water-dispersible powders, and a stabilized, spray-dried curcumin colourant. The regulatory status generally regarded as safe is applied to turmeric oil, GRAS 3085, and oleoresin, GRAS 3087, in the USA.

Other Curcuma *species*

C. aromatica Rosc., yellow zedory or wild turmeric, is native to India, accounts for about 3% of the total area planted to turmeric, and is cultivated mainly in the States of Andra Pradesh and Tamil Nadu. The rhizomes are used as a turmeric substitute and extracted for their yellow dye. The essential oil is an adulterant of turmeric oil. The rhizomes are used in local medicines.

Curcuma xanthorriza Roxb., *temu lawak*, is native to Indonesia. The rhizome is a flavouring and turmeric substitute, and the starch extracted to make a porridge. The essential oil is occasionally used to adulterate turmeric oil.

C. zedoaria Rosc., zedory, is native to India but was later widely introduced to, and cultivated in, neighbouring countries. It has long been cultivated in India for its rhizomes, which were exported early on throughout the region and to Europe in the 6th century. The yellow-fleshed rhizome has an agreeable musky aroma, slightly camphoraceous, and a pungent, bitter taste. It is used as a flavouring, a turmeric substitute, and for its starch known as shoti, used as a medicinal gruel. The essential oil contains α-cineole and an aromatic resin, extracted for use in local cosmetics. The oil is also an adulterant of turmeric oil. The ground rhizome is a component of herbal medicines, especially to treat stomach disorders.

Other members of the *Curcuma* genus produce a rhizome that provides a food or flavouring, including *Curcuma amada* Roxb. (India), mango ginger, *Curcuma leucorhiza* Roxb. (India), East Indian arrowroot and *Curcuma pierreana* Gagnep. (Malaysia), false arrowroot, while others may not belong to the *Curcuma* but to closely related genera. They are too numerous to be described here, but full details are available in the literature.

12

Minor Spices

Capers

The *Rhoeadales* comprises a small number of mainly herbaceous families, many indigenous to Europe. The most important are the *Capparidaceae*, the caper family, the *Cruciferae*, the mustard family (see Chapter 2), *Papaveraceae*, the poppy family and the *Resedaceae*, the mignonette family. The *Capparidaceae* comprises 45 genera containing about 1000 mainly thorny species ranging from small trees, shrubs, and climbers to herbs, occurring in the semi-tropics to tropics worldwide.

Capparis L. is a large genus containing about 250 species, mainly in the European/West Asia region; some 30 are native to the Indian subcontinent, 40 to mainland South-east Asia and about 20 each to Australia and Melanesia (Jacobs, 1965). A caper relative is *Cleome spinosa* Jacq., the spider plant, a popular red-flowered ornamental. The most important spice producer is *Capparis spinosa* var. *spinosa*, the caper bush, while a closely related species *C. spinosa* var. *mariana*, the Mariana caper, is of local importance in the Pacific region. The sharp taste of capers is similar to that of the mustards (*Brassica*) and horseradish (*Cochlearia*), since the *Capparidaceae* and *Cruciferae* are closely allied. Capers have long been a valued condiment, and this accounts for its very wide distribution as it was carried by traders

and travellers along the great trade routes. Capers have been identified at archaeological sites in Europe and southern Russia to India, but not further east, where it is replaced by var. *mariana*, as noted later. Wild caper plants can be found growing in the most remote and often inhospitable regions, whose present inhabitants are often unaware of their uses.

Caper was successfully introduced as a commercial crop into the southern USA, particularly California, but the capers traded internationally are mainly produced in France, Spain, Italy and Algeria. It was apparently introduced into Britain at the end of the 16th century, but was seldom grown outside monastery and similar gardens in the south, and remains uncommon as it is frost-sensitive and usually grown by individuals for personal use. Caper was introduced into Australia after European settlement to provide the spice. Caper has long been used in traditional medicine, and was mentioned by the Greek doctor Pedanius Dioscorides in his *De Materia Medica* of the 1st century AD. The buds or fruit remain a common ingredient of herbal remedies.

Botany

C. spinosa L. is a polymorphic, complex species with several distinct forms that have been incorrectly described as species, and a comprehensive taxonomic study would be welcome. The typical spiny type *C. spinosa* var. *spinosa* L. occurs wild from Europe to India, the less spiny *C. spinosa* var. *mariana* (Jacq.) K. Schum. from India to the Southeast Asian region. This latter population is very uniform and considered taxonomically and geographically distinct, due most probably to being descended from one or two original imports and developing in isolation from other *Capparis* species. The basic chromosome number of the genus is 12, with *C. spinosa* being diploid with $2n = 24$ and others with $2n = 38$.

C. spinosa var. *spinosa* (syn. *C. spinosa* L.) is known in English as caper, in French as *capre*, in German as *kaper*, in Italian as *cappero*, in Spanish as *alcaparra* and in Indian (Hindi) as *kabarra*. *C. spinosa* var. *spinosa* will be designated *C. spinosa* in the text for brevity, caper to indicate the plant, and capers the spice. Germplasm collections of *C. spinosa* are maintained in Italy and Spain, and the wide variability of the species indicates the potential available to plant breeders.

Caper is a rather straggling naturally spiny shrub, but European cultivars are spineless, spreading to 2.0–3.5 m, with the main stem up to 1 m, the branches prostrate and the twigs terete. The root system is extensive and deeply penetrating. The leaves are alternate, simple, dull green, leathery, round to oval, 2–5 cm and veined, with the base truncate, the apex rounded/obtuse, and the petiole short with two spines at the base. The flower buds are conical initially, later spherical, and 2.0–2.5 cm diameter. Fresh buds are bitter but this is lost on drying, and the taste becomes spicy and pleasant. Small groups of rutin crystals occur on buds. Composition of the buds is discussed at the end of this section. An essential oil distilled in India from whole plants with buds was stated to have the properties of garlic oil (Watt, 1908). Flowers occur on year-old branches, are fragrant, white or pinkish, 2.5 cm, solitary, axillary and short-lived, with the pedicel 5.0–7.5 cm, and with four petals, four sepals, and 100–190 stamens, which are reddish–purple, 4.5–6.0 cm and extend beyond the petals (tassel-like); the pistil on the long gynophore is 6–7 cm, with the base hairy (Fig. 12.1). Flowering is June–August in the northern hemisphere. The fruit is an ellipsoidal, olive-green berry, 2.5–5.0 × 1.5 cm, with the pericarp thin and corky, and the ribs distinct. The fruit may be pickled. The seeds are numerous, globose, 4 mm in diameter and imbedded in pulp.

Fresh buds contain (per 100 g): water 79 g, protein 5.8 g, fat 1.6 g, carbohydrates 6.5 g, fibre 5.4 g, ash 1.6 g, plus rutin and a number of kaempferol and quercetin-derived flavonoids, with the flavour mainly due to capric acid. Buds, when pickled as described in the section Processing, become the spice capers.

Ecology

Caper is described as xerophytic, with a complex response to drought, involving osmotic adjustment, regulated stomatal opening, cell wall plasticity and increased root density (Rhizopoulous, 1990). Caper is

Fig. 12.1. *C. spinosa* flower.

thus a plant of warm sunny climates and adapted to Mediterranean and semiarid environments, a good colonizer, and frequently found on waste land, abandoned construction sites and ruins. Caper is distributed from the Mediterranean through the Near East to India, but seldom cultivated outside Europe and North Africa.

Plants flourish under seasonal rainfall conditions, and although cultivated plants require an annual rainfall of 400–500 mm, caper can grow on far less. If soil moisture is sufficient, plants are independent of rainfall. Cultivated caper is usually grown at low altitudes, but wild plants can be found up to the frost line on North African and Middle Eastern mountains. Few data on the effect of temperature are available, but caper is known to grow well up to 40°C, and survive 45°C for short periods. In the northern Sahara, it was noted that leaves remained edge-on to the sun.

Soils and fertilizers

Caper occurs naturally on rather sandy soils derived from lava or limestone, but is cultivated on sandy loams and well-drained clay loams, neutral to slightly alkaline. A valuable asset is that a local caper type may be salt-tolerant; plants have been found growing in Israel near the Dead Sea, and on the edges of the Iranian salt desert south of the railway line from Tehran to Mashad (E.A. Weiss, personal observation). No information is available on the plant's nutrient requirements, and in general no fertilizer is applied, although young plants may be mulched to assist establishment.

Cultivation

Caper is generally propagated from seed, and fresh seed should be used as dried seeds have a variable dormancy period, and germination is erratic (Sozzi and Chiesa, 1995). Dormancy can be broken by soaking in dilute sulphuric acid, as in Turkey (Tansi, 1999). Caper can also be grown from cuttings, taken from vigorous branches at least 1.5 cm in diameter, and these must be kept moist but not wet until well rooted; mist

propagation is excellent. Cuttings 8 cm long with two leaves taken in March and April from the central shoots of 8-year-old plants gave the highest rooting percentage in Italy (Pilone, 1990). Rooted cuttings should be planted out in late summer/early autumn in Europe, and about 2500 are required ha^{-1}. *In vitro* propagation is possible but seldom utilized.

Seeds are sown in early to mid-summer in Europe, 3–5 cm deep, preferably into moist soil, and spaced to suit the equipment available as plants quickly spread to fill available space. Cuttings are manually or mechanically planted, and a spacing of approximately 2 × 2 m is common. Flowering begins about 1 year after sowing and may continue year-round in suitable conditions. Pollination is by insects, and caper is predominantly cross-pollinated.

Plants can be harvested for buds after 1 year, and reach full production after 3–4 years. Pruning is necessary annually as flowers occur only on year-old branches. Plants are cut back in winter to remove dead and disease branches and water-shoots. In southern Europe, the commercial life span is about 15 years.

Pests and diseases

Little is recorded concerning pests on caper, which is usually mentioned as one of the many hosts of polyphagous insects such as locusts, mole crickets, etc., without details of the damage caused. A similar lack of information exists concerning diseases, and usually caper is mentioned only as a host.

A major pest wherever it occurs is the fruit fly *Capparimyia savastani*, especially in the Mediterranean region, but also *Pardalaspis* spp. in eastern Africa. Eggs are laid on developing fruit, which the larvae bore and destroy. In Australia, larvae of the caper white butterfly *Anaphaeis java teutonia* feed on all local *Capparis* species including caper, but its importance or degree of damage has not been noted. Interestingly, it appears that caper may be toxic to human disease pathogens carried in some insects, *Leishmania*, for example (Jacobson and Schlein, 1999).

Harvesting

Buds are ready for harvesting when plump and dark green, and are picked manually about every 10 days as early in the morning as possible. Thus, the operation is labour-intensive and, since mechanical harvesting is virtually impossible, limits the areas where caper can be grown commercially to regions with a plentiful labour supply. Mature plants yield 1–3 kg dried buds, and total yields of 3–4 t ha^{-1} are reported from southern Europe. Time of harvesting (maturity) of buds is important as this is a major factor influencing bud composition and thus flavour (Rodrigo *et al.*, 1992).

Processing and storage

Fresh buds are bitter with an unpleasant taste and aroma, and require processing to acquire the characteristic taste. This flavour is primarily due to capric acid, which develops after pickling. Pure capric acid has a very strong, unpleasant smell reminiscent of goats, but the minor amount in capers is modified by the vinegar and considered to introduce a typical Mediterranean flavour to dishes.

Harvested buds may be kept in the dark for 3–4 h, shade-dried for about 1 day, then placed in casks of vinegar, to which salt may be added. After soaking for 7–8 days, buds are removed, drained, pressed and transferred to fresh vinegar, and the process repeated after another 7–8 days. Capers are then graded by size, the smallest considered the highest quality (*nonpareils*). Capers are then packed in various sized bottles containing vinegar or brine. Buds are also dried, salted and sold loose in local markets in producing areas, as are buds too large for bottling. The fresh and dried fruit of an allied species, *Capparis aphylla* Roth., is also sold in North African markets (E.A. Weiss, personal observation).

Products and specifications

The most important product is the spice capers (pickled flower buds) (Fig. 12.2), with fresh leaves of minor and very local usage.

Fig. 12.2. Preserved capers.

Capers. Capers are mainly used as a domestic spice, generally available in small bottles, and seldom used in food processing except to satisfy a local or specialist demand; for example, pounded capers are used to flavour Montpelier butter and Liptauer cheese. Prepared caper sauce in bottles is also available, and is a capers substitute, or can be added directly to dishes at serving. Capers are frequently added to, or used to garnish, fish dishes and meats, especially mutton. For domestic use, capers should be removed from the bottle when required, the bottle resealed and kept in a dark cupboard. The pickling liquid remaining after all capers are used can be added to sauces or directly to dishes during cooking. Capers have been approved for food use by the Food and Drug Administration (FDA) of the USA.

Capers are expensive, and substitution with flower buds of other *Capparis* species occurs, and in Europe capers are sometimes adulterated with young flower buds of *Caltha palustris* L., *Ficaria verna* Huds., *Sarothamnus scoparius* (L.) Wimm. ex Koch, *Tropaeolum majus* L. and young fruits of *Euphorbia lathyris* L.

Medicinal uses

Fresh flower buds are used to treat lesions of the vascular system, and extracts of root bark to treat dropsy, anaemia and arthritis. In India, root bark is considered a purgative, diuretic, emmenagogue, analgesic, anthelmintic, and is a common ingredient of herbal medicines. The activity of minor components isolated from buds have been investigated (Chhaya and Mishra, 1999). Root and bark extracts from local *Capparis* species, mainly *Capparis tomentosa*, were used in East Africa to treat backache and inflamed joints. Early settlers in Australia and Aborigines reportedly used a similar extract from indigenous species to treat wounds, and it is probable that such usage was introduced by Afghan camel drivers or Asian traders.

Other Capparis *species*

C. spinosa var. *mariana* is a cultigen apparently introduced by the Spanish into the Mariana Islands, which spread and became naturalized in Polynesia, regions of Indonesia, Papua New Guinea and the Philippines. It was commercially cultivated on the island of Guam in the mid-18th century, when the flower buds were harvested, pickled and exported as capers. It is still grown by smallholders in the region but is now only a minor cash crop, although total production is sufficient to satisfy local demand.

C. spinosa var. *mariana* (Jaeq.) K. Schumann (syn. *Capparis cordifolia* Lamk; *Capparis mariana* Jacq.; *Capparis sandwichiana* DC.) is known in English as Marianna caper, in Malaysian as *melada* and in the Philippines as *alcaparras*. It is a shrub, up to 2 m, with branches usually prostrate, and twigs terete and basically spineless, with short, floccose, greyish indumentum, which quickly disappears. The leaves are alternate, simple and glaucous, with the petiole 0.7–1.3 cm; the lamina is ovate-elliptical, 1.5–7.5 × 2.5–5.5 cm, the base truncate to rounded, the apex rounded to obtuse, with five to seven pairs of veins. The flowers are white, turning red-purple, fragrant, solitary and axillary; the pedicel is 4.5–7.5 cm, glabresent and the buds conical when young, later spherical, and 2–2.5 cm. The calyx is strongly zygomorphic, glabrescent, the sepals in two pairs, the posterior sepal deeply saccate, 2.5–4 × 1.7–2.5 cm, the other sepals subovate, 2–2.5 × 0.7–2.5 cm.

There are four petals in two pairs, the upper pair rhombic, 3–5.5 × 2–4 cm, the base thick and fleshy, and the lower pair 3–3.5 × 1.5 cm. The stamens are numerous, white/pinkish, with the pistil on a 5–7 cm gynophore. The fruit is an ellipsoidal, 2.5–5 × 1.3–1.5 cm, olive-green berry with a thin, ribbed, corky pericarp. The seeds are numerous, subglobose, 4 mm in diameter, in yellow fruit pulp (de Guzman and Siemonsma, 1999).

Some South-east Asian *Capparis* species have edible fruit including *Capparis micracantha* DC., and others used medicinally including *Capparis horrida* L., with *Capparis flexuosa* L. similarly in the Caribbean region. In Africa, *Capparis coriacea* Burch. (simulo caper) of South Africa is used in traditional medicine, and the plant has also been introduced to Peru and Chile; *C. tomentosa* Lam. provides grazing and fodder for camels in northern Saharan countries, but in the Sudan, large amounts were considered toxic and *C. decidua* (Forssk) Edgew. (*tundub*) preferred. In India, the latter species is grown for its medicinal properties (Shekhawat, 1999), and is also a grazing plant for camels in the Thar desert. The thorny *Capparis zeylanica* L. is sometimes used as a hedge against children and small domestic stock in Sri Lanka and India.

Unrelated species

Caper is also used as a prefix for unrelated plants, including caper spurge, *E. lathyris*, whose seeds are toxic; while the pickled buds of other plants are used as a capers substitute, including marsh marigold *C. palustris* L., which can cause stomach disorders. However, the pickled buds of nasturtium, *T. majus* L., are an acceptable but inferior substitute.

Nigella

The *Ranunculaceae*, order *Ranales*, is a large and somewhat disorganized family containing about 70 genera and at least 3000 species. The genus *Nigella* L. contains about 20 species of annual herbs indigenous to the Mediterranean and West Asia. The most important spice producer is *N. sativa* L., with *Nigella arvensis* L. of minor importance, while *Nigella damascena* L. is the well-known blue-flowered ornamental 'love-in-the-mist'.

N. sativa is native to the Mediterranean region through West Asia to northern India, and has long been domesticated. Thus, it can be found wild or as an escape over most of this area, frequently as a weed in cultivated crops especially cereals. As black cumin it is mentioned in ancient Greek, Roman and Hebrew texts as a condiment and component of herbal medicines and was reportedly introduced to Britain in 1548. *Nigella* is a minor cultivated crop from Morocco to northern India; in sub-Saharan Africa particularly Niger, and eastern Africa especially Ethiopia, where it is also reportedly used as a fish poison (Jansen, 1981), and in Russia, Europe, and North America. In South-east Asia, nigella seed is used mainly for medicinal purposes.

Botany

N. sativa L. (syn. *Nigella cretica* Miller; *Nigella indica* Roxb. ex Fleming; *Nigella arvensis* Auct. Terr.) is known in English as nigella seed, and incorrectly as small fennel or black cumin, a name also common in many other languages; in French as *nigelle,* or *cumin noir,* in German as *schwarzkummel,* in Italian as *nigella,* in Spanish as *neguilla, pasionara,* in Turkish as *kolonji,* in Indonesian and Malay as *jinten hitam,* in Indian as *kala zira* and many vernacular names, and in Ethiopian as *turkur azmut.* The genus name is said to be derived from the Latin meaning blackish, and *sativa* from 'to sow'.

Confusion exists regarding the common names for nigella as noted above, and it is recommended that in the literature *N. sativa* be referred to as nigella, its seeds as nigella seeds, and these designations will be used in this text. A small germplasm collection is maintained at the Western Regional Plant Introduction Station, USA. Since *N. sativa* is very widely cultivated it is not considered at risk of genetic erosion. A comparison of the chemical and morphological characteristics of different regional types of *N. sativa* is being compiled. The chromatic number of the

genus is x = 6 and *N. sativa* is a diploid, $2n = 12$. *N. sativa* is a variable species but can be distinguished from other species in the genus by its blue, petaloid sepals, triangular seeds and tuberculate capsular fruits with carpels united to the apex. A division into varieties or subspecies has been proposed on the basis of seed colour, type of branching, degree of hairiness, or division between wild and cultivated nigella. None is considered botanically acceptable, but classification of the numerous regional cultivars would be valuable. Ornamental cultivars have been given names by breeders to preserve exclusivity.

Nigella is an erect annual herb with a well-developed yellow-brown taproot, producing many secondary and tertiary roots. The stem, up to 70 cm, is profusely branched, subterete, ribbed, often becoming hollow with age, puberulous and light to dark green. The leaves are alternate and extipulate, the petiole, 1–6 cm, present only on basal leaves. The blade is 7 × 5 cm, pinnately dissected into thin sublinear lobes usually described as feathery, and normally green but may become red/brown with age.

The flowers are pale green when young, light blue when mature, solitary amd terminal, initially pale green becoming pale blue or white; the pedicel is 4–8 cm, inserted on a 2 mm diameter yellow/brownish, depressed receptacle. There are five ovate sepals up to 17 × 12 mm, tapering at the base into a claw, 2–3 mm. There are normally eight petals, each with a short glabrous claw; the stamens, in eight groups of three to seven, are initially upright becoming horizontal with age. The ovary is compound, 4–9 mm, with a free stigma 5–9 mm. Flowering is protandrous and pollination mainly by insects. In older flowers the stamens bend, as noted, and self-pollination may occur.

The fruit is a capsule, up to 16 × 12 mm in diameter, greyish initially, yellow/brown when mature, with persistent semi-erect stigma, opening at the base, containing numerous dark brown but generally black seeds. The seeds are obpyramidal, up to 3 × 2 mm in diameter, the testa pitted and wrinkled, the embryo minute and embedded in greyish fatty endosperm (Fig. 12.3). Nigella seed is incorrectly known in many languages

12.3. Nigella seeds.

as black cumin due to an apparent resemblance between the seeds; however, the taste is quite different. Nigella seed may also be confused with onion seed.

An analysis of European seed gave (approximately per 100 g) moisture 4 g, protein 22 g, fat 41 g, carbohydrates 17 g, fibre 8 g; ash 4.5 g consisting of Na 0.5 g, K 0.5 g, Ca 0.2 g, P 0.5 g, Fe 10 mg, thiamine 1.5 mg, pyridoxine 0.7 mg, tocopherol 34 mg and niacin 6 mg (Takruri and Damah, 1998). Analysis of Ethiopian seed gave (per 100 g) moisture 6.6 g, protein 13.8 g, fat 32.2 g, fibre 16.4 g, N 2.2 g; ash 7.5 g containing Ca 519 mg, P 594 mg and Fe 17 mg, plus thiamine 0.62 mg and niacin 9.5 mg (Agren and Gibson, 1968; Nergiz and Otles, 1993).

The fat component contains approximately stearic 2%, palmitic 12%, myristic 0.3%, oleic 45%, linoleic 36%, linolenic 0.3% and eicosenic 0.5% acids, but these ratios vary among cultivars. Other constituents include the glucoside melanthin at 1–1.5%, which hydrolyses to the toxic malantho-

genin, the alkaloids damascenine, nigellone, nigellimine, nigellimine-N-oxide and nigellicine, the sterols cholesterol, campestrol, stigmasterol, β-sitosterol and α-spinasterol, and tannins. Expressed fatty oil contains (approximately) eicosenoic 0.5%, linoleic about 60%, linolenic 0.3%, myristic 0.3%, oleic 25%, palmitic 12% and stearic 3% acids, and these ratios can also vary with the cultivar. Seeds contain about 0.5% volatile oil (see also the Products and specifications section).

Ecology

Nigella is grown under a wide range of environments, but flourishes in cooler regions with a temperature range of 5–25°C, with the optimum being 12–14°C. Plants are frost-sensitive at any growth stage and this limits its range in Europe and highland areas of the tropics. In the northern hemisphere, nigella is sown in late spring–early summer, but in regions with wet and dry seasons, just after the first rains. Regional cultivars can be grown from sea level to 2500 m, apparently with a reduction in yield with increasing altitude, although this is not supported by data. A rainfall of 400–500 mm will produce good crops, but above this the soil must be free-draining, as plants are susceptible to water-logging at any growth stage. Cultivars able to withstand considerable moisture stress have developed in North Africa and West Asia, while wild plants can be found growing in almost arid conditions. However, for cultivated crops, a rainfall below the optimum generally results in lower seed yield.

Soils and fertilizers

Nigella produces the highest yield on a sandy loam with a neutral reaction, but will grow well on soils with pH 5–8. As a smallholder crop it is sown successfully on almost any land available. Nigella responds favourably to fertilizers, but little is applied to smallholder crops. A compound fertilizer in India applied at the equivalent of 80 kg N, 17.5 kg P and 33 kg K ha^{-1} produced a seed yield of 1 t ha^{-1}. Seed and seed oil composition were affected by the level of applied N and P in Saudi Arabia (El-Sayed *et al.*, 2000).

Cultivation

The cultivations required for umbelliferous crops are suitable for nigella (see introduction to Chapter 10). Nigella is propagated by seed, or can be sown in seedbeds and transplanted, but as this is difficult and mortality high, seed should be sown where it is to grow. Fresh seed should be used whenever possible, but if kept dry and dark, seed remains viable for 2–3 years, although emergence can be erratic. Germination is epigeal, the hypercotyl about 1 mm, the cotyledons opposite and light green; the first non-seed leaf normally has a three-lobed blade. Germination is usually within 14 days and accelerated by a warm seedbed. A seed rate of 20–30 kg ha^{-1} in 25–40 cm rows is common, and as with umbelliferous crops, seed may be mixed with an inert filler to ensure a spacing of about 15 cm between plants; thinning may be necessary. Smallholders frequently broadcast seed at 20–25 kg ha^{-1}, and thinnings are eaten as a vegetable or fed to livestock.

Plants begin flowering 80–100 days after sowing depending on the average temperature; higher temperature accelerates flowering, but if prolonged generally reduces the number of flowers, or flowers drop after pollination. Seed reaches physiological maturity in 130–150 days, again depending on temperature, and should be harvested before it is ripe to avoid shattering. Nigella is not usually irrigated, as a crop is commercially unprofitable if it cannot be grown on available rainfall. Time of application and amount of water can affect the rate of plant growth, maturity, and essential oil yield and constituents. For example, the important thymoquinone content was highest at 58% when watering was delayed for 12 days in Iran (Mozaffari *et al.*, 2000). Weeding is normally manual, although small mechanical weeders are available. Nigella is often intercropped with barley and wheat in Ethiopia, and strip-cropped in North Africa and elsewhere. A rotation including nigella is unusual, and successive crops may be grown for several years.

Pests and diseases

Little information is available on the pests of nigella, except as an ornamental when it is damaged by a number of polyphagous garden insects. In the field, larvae of *S. litura* may cause a 40% loss in a particular season or crop. Empty carpels often occur on plants probably due to the larvae of borer beetles. Most records of diseases also come from ornamentals. A leaf-spot caused by *Cercospora nigellae* was recorded as causing minor damage in Ethiopia.

Harvesting

Plants are cut either by hand or reaper/binder when the majority of pods are mature, turning yellow but not fully ripe, as pods shatter easily. Plants are stooked to dry, then threshed; smallholder crops are threshed manually, generally by beating with canes. A yield of 1 t ha^{-1} has been recorded from experimental trials, but smallholder yield is normally low, often only a few hundred kilogrammes ha^{-1}, and reflects the generally poor standards of husbandry.

Processing and storage

Seed must be thoroughly cleaned by hand or machine winnowing, and damaged seed should be removed prior to bagging. However, as most seed is used domestically or sold at local markets little attention is paid to these requirements. Good-quality, sound seed should preferable be stored in clean sacks under dry conditions to maintain its flavour and value; sacks containing damaged seed rapidly deteriorate due to lipase action.

Products and specifications

The only commercially important product is the dried fruit, normally know as nigella seed, and an essential oil of minor importance.

Seed. The dried fruit is the spice nigella, and although much is sold loose in local markets, packeted seed is widely available. Seed for retail sale must be sound and of uniform dark brown to black colour. When crushed it has a strong, pungent odour reminiscent of carrot, and the taste is aromatic, oily, pungent, peppery and nutty. For domestic use, seed should be fresh, and can be used whole but preferably freshly ground. Nigella is used to flavour meat and vegetables dishes, pickles, is a minor ingredient in curries, fruit pies and other confections and in sauces, vinegar and alcoholic beverages. Crushed seed is commonly mixed with dough or sprinkled on bread before baking, which produces dark grey to almost black loaves, and is an essential constituent of the Middle East *choereg* rolls. In the USA, the regulatory status generally recognized as safe has been accorded to black cumin (nigella), and black cumin (nigella) oil, GRAS 2342 and GRAS 2237, respectively.

Nigella seed is an important constituent of both Eastern and Western herbal remedies and is reported to have anthelmintic and carminative properties. In Asia and South-east Asia, particularly Malaysia, the seed or extracts are regarded as a virtual panacea! A decoction of seeds is traditionally used to treat headache, rheumatic pains, asthma, coughs, nausea, and in northern India to induce abortion. Crushed seeds in vinegar are applied in skin disorders such as ringworm, eczema and baldness. In Egypt, a tea made from powdered nigella seeds, fenugreek, garden cress, *Commiphora* spp. and dried leaves of *Cleome* spp., *Ambrosia maritima* L. and *Centaurium pulchellum* (Sw.) Druce is used to treat diabetes. Similar to many such herbal remedies, most are of no value. Sachets of partially crushed seeds were previously used to protect woollen goods and linen against insects, but have been replaced by more potent products.

The effectiveness of nigella seeds in veterinary medicine is better documented. The lipid fraction of ether-extracted seeds exhibits galactagogue action in buffalo and goat. Aqueous extracts have a broncho-dilatory effect attributed to thymoquinone, but other compounds may also be involved. Powdered seeds are effective against the tapeworm *Moniezia expansa* in sheep.

Essential oil. An essential oil can be obtained by distilling whole or cracked seeds to yield about 0.5%. The oil is a mobile yellow liquid, the odour unpleasant and the taste likened to that of juniper. The oil is placed in the dry

woody group by Arctander (1960), and oil of *N. damascena* in the quite different winy, sweet floral group. The seven main constituents are (approximately) para-cymene 31%, thymoquinone 25%, ethyl linoleate 9%, α-pinene 9%, ethyl hexadecanoate 3%, ethyl oleate 3%, and β-pinene 2%. A high ratio of eicosadienoic acid to eicosamonoenoic acid combined with a high level of C_{20} fatty acids, is characteristic of nigella seed oils and could be used to identify genuine oil. The oil is of little commercial importance and normally produced to order. The oil has microbial activity (Minakshi and Banerjee, 1999), and the activity of minor oil constituents has been investigated (Abdel-Fattah *et al.*, 2000; El-Dakhakhny *et al.*, 2000; Salem and Hossain, 2000; Enomoto *et al.*, 2001).

Saffron

The *Iridaceae*, iris family, comprises some 1000 species in 70 genera including the genus *Crocus* L., which contains about 85 species distributed mainly in the Mediterranean region through West Asia to Iran. Many are ornamentals, but one is a spice producer, the saffron crocus, *Crocus sativus* L., whose exact origin is uncertain, but was probably Asia Minor, where it has a long history of cultivation and is thus widely dispersed, and ultimately to China and Japan. The spice was probably known to the ancient Middle Eastern civilizations of Babylon and Assyria, as the name *krokos* predates Greek. A very early depiction of saffron is a painted fresco in the palace of Minos on Crete, dated to about 1650 BC, while another at Akrotini on the neighbouring island of Santorini is dated to 1500 BC.

Saffron was originally made from the dried stigmas of wild plants, probably *Crocus cartwrightianus*, known and traded by the Phoenicians, and valued by the Persians, Greeks and Romans, who used saffron to colour and spice foods, saffron water to perfume their baths, houses and temples, and an extract of saffron as a medicinal narcotic. Saffron was named in early Hebrew as *carcom*, in the Sanskrit medical glossary *Bhavaprakasa* as *kunkuma*, and mentioned many times in the *Ain-i-Akbari* of AD 1590 compiled by Abdul Fazi. Iran has been a major producer since the early Persian empires, and exported saffron to China's Yuen dynasty (AD 1280–1368), where it was known as *sa-fa-lang*.

Expansion of the Arabs along the Mahgreb in the 9th and 10th centuries, and on into the Iberian peninsular, carried the plant and later its cultivation into Spain and Portugal, which became the major European producers, and saffron was known as the Alicante or Valencia crocus. Saffron was widely known in Europe but the plant is frost-sensitive and this limited its range. It was introduced to England early in the 14th century during the reign of King Edward III by a returning pilgrim who hid the corms in his staff, according to the traveller and historian, Richard Hakluyt (1552–1616). Saffron was cultivated at Saffron Walden, Essex (hence the name) by growers known as 'crokers', and later in Cambridgeshire until the end of the 18th century. So widespread was its use that any spice trader was commonly known as a 'saffron grocer'. The household accounts for the year AD 1418–1419 of Dame Alice de Bryene stated that she used 'three-quarters of a pound of saffron bought from Stourbridge Fair'; at that time one pound of saffron was selling at the same price as a horse or a cow!

Records show that saffron was cultivated on a commercial scale in Spain in the 9th century AD, in France and Germany in the 12th, and so severe were the penalties for adulteration that in Nurnberg in the mid-15th century, persons convicted of adulterating saffron were either burned or buried alive! Today it is cultivated mainly in those areas with surplus labour and few cash crops, including Spain, especially in the provinces of Valencia and Alicante, North Africa, Turkey, the southern Russian republics, in Iran, Kashmir until the civil war, and to a lesser extent in China. In 1990, a Tasmanian farmer began growing saffron, and by 1996 had established the first local commercial saffron farm. Saffron has been used to flavour food from antiquity to the present, and it remains a common, sometimes essential ingredient in many dishes, especially those based on rice or fish. Its use in traditional medicines and later herbal remedies was limited by its high cost and restricted availability. Saffron, similarly to safflower (Weiss, 2000), is also source of a water-soluble, deep yellow dye made fast by a mordant, used by the Greeks and Chinese to colour the robes of their rulers, and to dye the hair of ladies at the court of Henry VIII. Today it is evident in the robes of Buddhist monks and the round cast mark on the foreheads of Hindus.

Botany

C. sativus L. is known in English as saffron, in French and German as *safran*, in Italian as *zafferano*, in Spanish as *azafran*, in Portuguese as *acafrao*, in Arabic and Persian as *zafaran* and in Indian generally as *kesar* or *kunkuma*. The name saffron is considered to have been derived, together with other European names, from the Arabic *zafaran*. Saffron is known only under cultivation, with its putative ancestor considered to be the diploid *C. cartwrightianus* ($2n = 16$) (Mathew, 1999). Saffron is a triploid, $3n = 24$, and infertile. *In vitro* corm production from apical buds is possible, and of stigmas from stigma explants, but the techniques are currently of laboratory interest only (Ebrahimzadah *et al.*, 2000).

C. sativus is a small bulbous plant typical of the genus. The corms are generally 3–5 cm in diameter, the neck distinct, the base flattened, producing basal grey-green linear leaves, up to 2.5 mm wide, the length variable up to 12 cm, sometimes longer, and the margins hairy, surrounded at the corm apex by cylindrical sheaths (Fig. 12.4). Leaves appear before flowering and remain for some weeks afterwards, the period depending mainly on the cultivar and weather conditions.

The flowers are funnel-shaped, and there are one to six, with six petals. They are fragrant, and the colour is variable according to region but usually with the throat whitish, segments generally reddish–purple to dark lilac/blue, occasionally white (Fig. 12.5). The thin corolla resembling a normal flower stalk is divided into two distinct whorls of three perianth lobes each, with

Fig. 12.4. Saffron crocus corm.

three stamens opposite the outer lobes; the anthers are yellow, longer than the filaments, and have a central red style with three bright red stigma branches, each 25–35 mm. The main flowering period is usually September–October in the northern hemisphere. The seeds are small, dark brown, globose and papillose. The stigma colour is mainly due to crocin, and the taste to picrocrocin. An analysis of Greek saffron samples gave the following: moisture (103°C) 8.5–9.5%, total ash 5%, essential oil content 1.0–1.2%, colouring power (440 nm) 120–150. Composition is also discussed in the Products and specifications section.

Plants exude nectar from glands in the ovary wall, which is below ground at flowering. Since the corolla is very narrow, the nectar quickly accumulates and rises up the tube to become accessible to insects. Dissection of the reproductive parts of flowers showed no structural barrier to growth of the pollen tube, and other factors may be involved (Caiola, 1999). Saffron is a triploid, as noted, and unable to produce normal pollen grains or a fertile ovary. It is thus propagated by separating corms, and this is the reason why isolated populations are very homogenous.

Ecology

Saffron flourishes in a Mediterranean-type climate with warm sunny days and an annual rainfall of 350–500 mm, but up to 700 mm in some districts. An annual average summer temperature of 20–25°C is normal in the Mediterranean region but up to 30°C elsewhere. A winter temperature down to 2°C is tolerated. Rain or very low temperature just before or at flowering is especially damaging, and thus the most productive areas are those where no rain falls during flowering. Saffron is normally grown at low altitudes although wild plants, usually from abandoned fields, can be found on hillsides in North Africa and Iran. Low windbreaks may be necessary in areas where strong dry winds are common, either to prevent flowers being desiccated or wind erosion on sandy soils.

Soils and fertilizers

Saffron is usually grown on neutral to slightly acid soils with sandy loams pre-

Fig. 12.5. Saffron flowers – Iran.

ferred, but smallholders grow crops on any free-draining land available. In Spain, saffron is also grown on soils with a substantial gravel or small stone content. Standing water invariably kills the bulbs at any growth stage, and thus in higher rainfall regions saffron is planted on hillsides. Saline soils and irrigation water are not tolerated. Few saffron growers apply fertilizers, which they consider detrimental to flower production (E.A. Weiss, personal communcation). However, as most growers are virtually subsistence farmers, a shortage of cash is probably a contributory reason. Overall mulching when plants cease flowering helps to conserve moisture and provide some nutrients, but suitable material is often unavailable in the generally dry regions. Plant or animal residues are sometimes ploughed in before an exhausted plot is replanted. Where saffron is grown in rotation with cereals, as on the sandy loams of Azerbaijan, organic manures are applied to the preceding crop since direct application to saffron causes corm decay.

Cultivation

The cultivations for umbelliferous crops in the introduction to Chapter 10 are suitable for saffron, but in smallholder crops the work is normally carried out manually. Occasionally the land is tractor-ploughed when an exhausted crop is replanted, but in general plots are too small. Corms should be sound, with a distinct neck, and planted in late summer, 5–10 cm deep, depending on the soil type. Depth of planting is important as too deep a planting reduces the number of daughter corms produced, and in general planting depth should be as shallow as local conditions allow. Row widths vary according to local preference, commonly 10–15 cm, and the plantation is hand-weeded as necessary.

Partial mechanization of saffron growing is successful in Italy, but not harvesting (Galigani and Pegna, 1999). Lands are ploughed, harrowed and corms planted as close as possible in 20–30 cm rows using modified horticultural equipment. Following harvest, corms are mechanically lifted, cleaned and stored for the next season. Between 2000 and 3000 kg corms ha^{-1} are

required, depending on row width. In Greece and Spain, mechanization is limited to preplanting operations, and sometimes to lift harvested corms.

Weed control when necessary is normally by hand, although herbicides are occasionally used in smallholder crops where saffron is grown adjacent to other crops, as in the Nile Delta of Egypt, where chemical residues were found in spice samples collected from local markets (Abou-Arab and Donia, 2001). Simazine and atrazine reportedly gave the best control of local weeds in Greece.

Few crops are irrigated and Middle Eastern and Iranian growers state that a relatively dry soil is more productive (E.A. Weiss, personal observation). However, many plots are divided into smaller areas, levelled and surrounded by ridges to impound rainfall. One grower in Iran planted his crop on wide ridges, to conserve the irregular rainfall. However, in the very dry regions of Morocco fields are irrigated weekly from September to November, and approximately 350–500 m^3 ha^{-1} are applied (Ait-Oubahou and El-Otami, 1999).

Saffron is seldom intercropped, except occasionally with a low-growing legume harvested before the bulbs flower, or sown after plants have died back. Plots are often sown with an alternative crop, usually a cereal, for a minimum of 1 but up to 3 years before replanting, or left fallow in drier areas to conserve subsoil moisture. In the higher rainfall regions of Italy and Greece, saffron is planted on raised beds or ridges to assist drainage. Underplanting saffron on small terraced plots in existing olive or almond groves is common in Morocco.

Pests and diseases

Little information is available on the pests and diseases of saffron, but those locally recorded from other types of crocus are probably relevant. Corms are occasionally infested by unidentified eelworms, but the damage is usually slight. A common disease of saffron in Europe is due to infection by *Fusarium* spp., which causes corm rot; in Italy *F. oxysporum* f.sp. *gladioli* was responsible (Primo and Cappelli, 2000), and another corm rot may also be due to infection by *Penicillium* spp. Corm decay due to *Rhizoctonia crocorum* is considered the most important fungal disease in Greece. Rats and moles, which eat corms, may require control in Europe.

Harvesting

Plants flower in autumn and are hand-picked (Fig. 12.6). Flowers are collected daily in the early morning just before they open, a picker taking the flower between thumb and index finger and severing it with the fingernail. Flowers are then taken to the homestead where stigmas are removed as soon as possible. One flower provides about 6–7 mg dried stigmas and yield varies from 5 to 15 kg ha^{-1} depending mainly on the season. Between 100,000 and 200,000 flowers or 5 kg fresh stigmas and styles may be required to produce 1 kg of dried saffron, mainly depending on how and when flowers are picked. Reported average yield of dried saffron ha^{-1} in Italy is 7–10 kg but up to 16 kg in special districts, 6–15 kg in Spain but up to 20 kg, 6–10 kg but up to 18 kg in Turkey, 5–10 kg in Iran, 4–8 kg in Greece, 2–5 kg in India and Sri Lanka, and 2–3 kg in Morocco. Iranian growers stated that a poor season occurred every 3–4 years, but that yield was

not cyclical. The yields quoted are from commercial crops of saffron, but trials at experimental stations with selected local strains and optimum growing, harvesting and drying methods have yielded up to 30 kg dried saffron ha^{-1}. This high standard of production is unlikely to be commercially viable, but local yields can usually be increased by planting only sound corms, timely weeding, and improved post-harvest procedures.

Since no effective method of mechanized harvesting is available, this operation is carried out manually, hence the high labour requirement. It has been estimated that picking requires 45–55 min per 1000 flowers, removing stigmas and drying another 100–130 min, and 370–470 h are necessary to produce 1 kg dried spice. Commercial saffron growing is thus limited to smallholders with few alternative methods of earning an income.

The commercial life of a North African or Iranian plantation is about 5–7 years, although plants may continue flowering at a steadily diminishing rate for up to 15 years. In general, the second to fourth years are most productive. Many European growers treat saffron as an annual crop, but in France and Spain a plantation may also be left for 3–4 years. A major factor affecting the life of a crop is the difficulty of controlling perennial weeds.

Fig. 12.6. Harvesting saffron flowers – Kerman, Iran.

Processing

Stigmas are removed from flowers as soon as possible after harvesting, and often air-dried in the shade; in India, Sri Lanka and also some parts of Spain they are dried over a low fire (toasting), and in Europe generally special ovens or heated drying rooms are used to ensure a high-quality product. In China, layers of stigmas are sometimes placed between weighted boards in ovens and produce saffron 'cakes' for local use. The various drying methods produce the three main types of spice: hay, powder and cake. Drying reduces stigma weight by about 80%, and drying by any method requires considerable skill to ensure a high-quality spice. The effect of temperature and other drying components on spice colour and taste remain to be fully determined. In Greece and Italy, between 20 and 30°C is considered the optimum, and the drying period must be accurately controlled to preserve stigma colour.

Dried stigmas must not be allowed to become moist after drying, and to prevent bleaching were packed in tightly closed tins, more commonly nowadays in sealed, black plastic bags. Smallholders sell to traders who visit farms on a regular basis, or the dried saffron is taken to a local market. In Iran in 1998, the price ex-farm for 5 g dried stigmas was the equivalent of $2.5–3, one-fifth the bazaar price in the nearby town of Kerman. When sold retail in developed countries as a powder or in phials containing 0.5 g, prices are calculated to be $8000–10,000 kg^{-1}!

Products and specifications

Saffron. The only important product of *C. sativus* is the spice saffron, the dried flower stigmas (Fig. 12.7). Saffron contains (approximately) 2% each of crocin-1 and picrocin, small amounts of crocin-2, crocin-3 and crocin-4, α-, β- and γ-crocetin, *cis*-dimethyl-crocetin, and other carotenoids, plus starch at 13%, vitamins B1 and B2, fixed oil at 8–13% and volatile oil at 0.5–1.5%, described later. A detailed review of the chemical composition of saffron is contained in Winterhalter and Straubingar (2000). The activity of various components of saffron

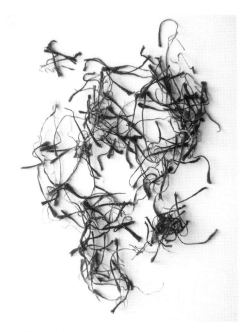

Fig. 12.7. Saffron spice – threads.

have been investigated (Jagadeeswaran *et al.*, 2000). The bitter principle is due to picrocin, a colourless derivative of a monoterpene aldehyde, matairesinol monoglucoside. The colour is mainly due to crocin, a water-soluble, bright yellow digentiobiose ester, reported as saffron yellow 30% and saffron red about 0.5%. A pinch of saffron placed in water expands immediately and the yellow colour diffuses. This dye is not persistent unless treated with a mordant, as noted.

Good-quality saffron should be less than a year old, brilliant orange not yellow, with a strong scent and a pungent bitter-sweet taste reminiscent of honey. Genuine saffron is expensive and is frequently adulterated, especially the powder sold in packets. Cheap saffron is suspect and may be the florets of safflower; to test, add dilute sulphuric acid, which turns saffron blue but not safflower. Methods of detecting adulteration have been reviewed (Winterhalter and Straubingar, 2000). However, saffron from different geographical regions or produced by varying drying methods differs substantially in quality, colour and aroma, as noted in the literature. Thus, comparative data is essential to establish the degree to which a particular sample has been adulterated. A method using molecular genetic analysis developed in China can identify genuine saffron from substitutes or adulterants, very important as saffron is an ingredient of many local traditional medicines (Ma *et al.*, 2001). Microbial contamination of the spice can occur if post-harvesting conditions or storage are poor, and gamma-irradiation used for decontamination. This may, however, result in loss of vital aroma and colour components (Zareena *et al.*, 2001).

Saffron, usually referred to as threads in recipes, is used mainly as a domestic spice to colour and flavour a range of foods including rice, cheese and fish dishes, especially bouillabaisse, and is a common ingredient in Mediterranean and Arabian foods. However, saffron is disliked by some who consider the taste unpleasant. When used domestically, the threads should preferably be slightly crushed and soaked in a small amount of hot water before adding to dishes, which aids uniform dispersal. Vinegar flavoured with saffron, garlic and thyme gives a special flavour to marinades and salads. Saffron is not a common ingredient of curries, where the yellow colour is due to turmeric, which, incidentally, is not a substitute for saffron. The use of saffron by food processors is limited by its high price, and other yellow colourants including carthamin from safflower are (sometimes illegally) substituted; a major commercial use is to colour certain cheeses and sausages. Saffron is also added to non-alcoholic drinks, and is an ingredient of vermouths and bitters. An extract (tincture) is used in fragrances and perfumes to give an oriental note.

In medicine, saffron is considered an anodyne, antispasmodic, appetizer, emmenagogue, expectorant, sedative, and incorrectly an aphrodisiac; it is recommended for the treatment of coughs, whooping cough, stomach gas, gastrointestinal colic and insomnia, and in China for the treatment of depression, shock and to ease the pain of childbirth. An Australian patent has been granted for use of an aqueous extract in the treatment of baldness. However, these varied uses were exceeded by Herr J.F. Hertodt of Germany in his *Crocologia* of 1670, which claimed saffron was able to cure anything from a cold to the plague! Currently it is little used for any medical purpose, but remains an ingredient of herbal drinks. Saffron must apparently be used with caution as a medicine, because the bitter component may occasionally produce an allergenic reaction. No risk is associated with food or therapeutic use of the spice up to 1.5 g; toxic reaction may occur at 5 g, and is reportedly fatal at 10–12 g. In the USA average maximum use level in food is 1% in baked products. Regulatory status in the USA is generally regarded as safe, GRAS 182.10 and 182.20, and approved as a food colourant exempt from certification, GRAS 73.50.

Essential oil. An essential oil can be obtained by distillation or solvent-extraction, but is seldom produced and has no commercial importance. The oil is placed in the delicate, sweet, leafy-floral group by Arctander (1960), and is a direct substitute for the spice in most applications. About 40 compounds have been identified in the oil, many in trace amounts only; the main constituents are safrol (safranal and oxysafranal), pinene, cineole, napthalene, 2-butenoic acid, lactone and 2-phenylethanol. If safrol undergoes demethoxylation it produces the well-known narcotic methylenodioxy amphetamine (MDA) (Emboden, 1979). Natural safrole oil, usually traded as sassefrass oil, is not derived from saffron (Weiss, 1997).

Unrelated species

A number of plants have the prefix saffron or are used as substitutes, including meadow saffron, *Crocus autumnalis* L., widely distributed in Europe, which is very poisonous in all its parts and must never be used as an alternative to saffron; *Crocus napolitanus* flowers are used as an antiseptic in Italy. The yellow or orange flowers of safflower, *C. tinctorius* L., often known as the saffron thistle, are a common colouring substitute for saffron, while fresh petals of the garden marigold, *Calendula officinalis*, will suffice in an emergency!

References

___

Abdel-Barry, J.A. and Al-Hakiem, M.H.H. (2000) Acute interperitoneal and oral toxicity of the leaf gluco-cidic extract of *T. foenum-graecum* in mice. *Journal of Ethnopharmacology* 70(1), 65–68.

Abdel-Fattah, A.F. *et al.* (2000) Antiococeptive effects of *N. sativa* oil and its major component thymo-quinone in mice. *European Journal of Pharmacology* 400(1), 89–97.

Abou-Arab, A.A.K. and Donia, M.A.A. (2001) Pesticide residues in some Egyptian spices and medicinal plants as affected by processing. *Food Chemistry* 72(4), 439–445.

Adamson, A.D. and Robbins, S.R.J. (1975) *The Market for Cloves and Clove Products in the United Kingdom.* Tropical Products Institute, London.

Agarwal, S.K. *et al.* (1999) Chemotaxonomical study of Indian *Illicium griffithii* and *I. verum* fruits. *Journal of Medicinal and Aromatic Plant Sciences* 21(4), 945–946.

Agnihotrei, P. *et al.* (1997) Variability, correlation and path analysis in fennel. *Journal of Spices and Aromatic Crops* 6(1), 51–54.

Agren, G. and Gibson, R. (1968) *Food composition tables for use in Ethiopia.* CNU-ENI Report No. 16. Uppsala, Sweden.

Ahlert, B. *et al.* (1998) Improvement of quality of black pepper at farm level. *Zeitschrift für Arznei-Gewurz pfalzen* 3(1), 15–20.

Ai, X.Z. *et al.* (1997) Effect of fertilizers on growth and yield of ginger. *China Vegetables* 1, 18–21.

Ai, X.Z. *et al.* (1998) Effect of temperature on photosynthesis characteristics of ginger leaf. *China Vegetables* 3, 1–3.

Ait-Oubabou, A. and El-Otmani, M. (1999) Saffron cultivation in Morocco. In: Negbi, N. (ed.) *Saffron.* Harwood Academic Publishers, Amsterdam, The Netherlands, pp. 87–94.

Akgul, A. (1986) Studies on the essential oils from Turkish fennel seeds. In: *Progress in Essential Oil Research.* W. de Gruyter and Co., New York.

Akgul, A. and Bayrak, A. (1989) The essential oil content and composition of myrtle (*M. communis*). *Doga Turk Tarum ve Ormancilic D* 13(2), 143–147.

Akhil, B. and Nath, S.C. (1997) Foliar epidermal characters in 12 species of cinnamon from North India. *Phytomorphology* 47(2), 127–134.

Akhil, B., Nath, S.C. and Boissya, C.L. (2000) Systematics and diversities of *Cinnamomum* species used as 'tejpat' spice in North-East India. *Journal of Economic and Taxanomic Botany* 24(2), 361–374.

Akhtar, M. and Mahmood, I. (1996) Effect of plant based products and some plant oils on nematodes. *Nematologia Mediterranea* 24(1), 3–5.

Al-Habori, M. and Raman, A. (1998) Antidiabetic and hypocholesterolaemic effect of fenugreek. *Phytological Research* 12(4), 233–242.

Ali, S.A. *et al.* (1999) Efficiency of fungicides in controlling powdery mildew of coriander. *Journal of Soil Crops* 9(2), 266–267.

Aliyu, L. and Lagoke, S.T.O. (2001) Profitability of chemical weed control in ginger (*Zingiber officinale*) production in northern Nigeria. *Crop Protection* 20(3), 237–240.

Amvam Zollo, P.H. *et al.* (1998) Aromatic plants of tropical Africa. Part XXXII. *Flavour and Fragrance Journal* 13(2), 107–114.

Anac, O. (1986) Essential oil contents and chemical composition of Turkish laurel leaves. *Perfumer and Flavourist* 11(5), 73–75.

Anandaraj, M. and Sarma, Y.R. (1994) Biological control of black pepper diseases. *Indian Cocoa, Arecanut and Spices Journal* 18(1), 22–23.

Anitescu, G., Doneanu, C. and Radulescu, V. (1997) Isolation of coriander oil: comparison between steam distillation and supercritical carbon dioxide extraction. *Flavour and Fragrance Journal* 12, 173–176.

Anon. (1976) *Major Economic Trees of Southern China*. Agricultural Printer, Peking, China.

Anon. (1984) *Note sur le Secteur des Huiles Essentielles*. Centre National D'Etudes Industrielles, Ministry of Economics, Tunis, Tunisia.

Anzidei, M. *et al.* (1996) Preliminary tests for *in vitro* selection of *F. vulgare* plants resistant to *Phomopsis foeniculi*. *Instituto Sperimentale per l'Assestamento Forestale e per l'Alpicoltara* 1996, 527–529.

Arasaratnam, S. (1978) *Francois Valentijn's Description of Ceylon*. The Hakluyt Society, British Museum, London.

Araujo, A.C., Sacramento, C.K. and Silva, P.H. (1989) *Economic Evaluation of Chemical Harvesting of Cloves in South-East Bahia*. Boletin Technico No. 164-CEPLAC, Brazil.

Arctander, S. (1960) *Perfume and Flavour Materials of Natural Origin*. Mrs G. Arctander, 6665 Valley View Boulevard, Las Vegas, Nevada.

Arganosa, G.C., Sosulski, F.W. and Slinkard, A.E. (1998a) Seed yields and essential oils of biennial caraway grown in western Canada. *Journal of Herbs, Spices and Medicinal Plants* 1(9), 9–17.

Arganosa, G.C., Sosulski, F.W. and Slikard, A.E. (1998b) Seed yield and essential oil of northern-grown coriander. *Journal of Herbs, Spices and Medicinal Plants* 6(2), 23–32.

Arimura, C.T., Finger, F.L. and Casali V.W. (2000) A fast method for *in vitro* propagation of ginger. *Tropical Science* 40(2), 86–91.

Armstrong, J.E. and Drummond, B.A. (1986) Floral biology of *Myristica fragrans*. *Biotropica* 18(1), 32–38.

Arnone, J.A. (1997) Temporal response of community fine root populations to long term elevated atmospheric carbon dioxide and soil nutrient patches in model tropical ecosystems. *Acta Oecologica* 18(3), 367–376.

Arnott, H.J. (1999) Scanning microscopy of black pepper (*P. nigrum*). *Scanning* 21(2), 113.

Arthamwar, D.N. *et al.* (1995) Correlation and regression studies in mustard. *Journal of Maharashtra Agricultural University* 20(2), 237–239.

Ashraf, M. and Bhatty, M.K. (1975) Studies on the Pakistani species of the familly *Umbelliferae*: fennel seed oil. *Pakistan Journal of Science and Industrial Research* 18, 236–240.

Ashurst, P.R. *et al.* (1972) The development of oils of pimento. *Journal of the Science and Research Council, Jamaica.* 3(1), 50. (See also *Proceedings of the Conference on Spices, 1973*. Trop. Prod. Institute, London, pp. 209–214).

Asllani, U. (2000) Chemical composition of Albanian myrtle oil (*M. communis*). *Journal of Essential Oil Research* 12(2), 140–142.

Babu, K.N. *et al.* (1993) Biotechnology in spices. *Indian Horticulture* 38(3), 46–50.

Badoc, A. and Lamarti, A. (1991) A chemotaxonomic evaluation of *Anetheum graveolens* of various origins. *Journal of Essential Oil Research* 3, 269–278.

Bagaturiya, N.S. *et al.* (1985) Changes in essential oil composition of laurel leaves during its accumulation. *Subtropical Kul'tury* 5, 103–106.

Balasubrahmanyam, V.R. *et al.* (1994) *Betelvine-Piper betle* L. National Botanical Research Institute, Lucknow, India.

Bandoni, A.I., Mizrahi, I. and Juarez, M.J. (1998) Composition and quality of the essential oil of coriander from Argentina. *Journal of Essential Oil Research* 10, 581–584.

Baratta, M.T. *et al.* (1998) Chemical composition, antimicrobial and antioxidative activity of laurel, sage, rosemary, oregano and coriander essential oils. *Journal of Essential Oil Research* 10(6), 618–627.

Barazani, O. *et al.* (1999) Chemical variation among indigenous populations of *F. vulgare* in Israel. *Planta Medica* 65(5), 486–489.

Barnard, D.R. (1999) Repellency of essential oils to mosquitoes. *Journal of Medical Entomology* 36(5), 625–629.

Bartley, J.P. (1995) A new method for the determination of pungent compounds in ginger. *Journal of the Science of Food and Agriculture* 68, 215–222.

Bartley, J.P. and Jacobs, A.L. (2000) Effects of drying on flavour compounds in Australian-grown ginger. *Journal of the Science of Food and Agriculture* 80(2), 209–215.

Bartschat, D., Beck, T. and Mosandi, A. (1997) Stereoisometric flavour compounds 79: celery, celeriac and fennel. *Journal of Agricultural and Food Chemistry* 45(12), 4554–4557.

Basavaraja, P.K. *et al.* (1998) Effect of copper ore tailings on chilli yield and quality. In: *Proceedings of the National Seminar on Water and Nutrient Management for Sustainable Production of Quality Spices*, Madikeri, Karnataka, India, pp. 99–102.

Baser, K.H.C. (1997) Tibbi ve Aromatik: Bitkilerin Ylac ve Alkollu Ycki Sanayilerinde Kullanimi. *Ystabul Ticaret Odasi* 30, 113.

Baser, K.H.C., Kurkcuoglu, M. and Ozek, T. (1992) Composition of Turkish cumin seed oil. *Journal of Essential Oil Research* 4, 133–138.

Bassen, B. and Brunn, W. (1999) Ochratoxin A in paprika powder. *Deutsche Lebensmittel-Rundschau* 95(4), 142–144.

Bavappa, K.V.A., Gurusingha, P.A. and Anadaswamy, B. (1981) *Pepper Technical Bulletin* No. 4. Department of Minor Export Crops, Colombo, Sri Lanka.

Beis, S.H. *et al.* (2000) Production of essential oil from cumin seeds. *Chemistry of Natural Compounds* 36(3), 265–268.

Belavadi, V.V., Venkateshal and Vivak, H.R. (1997) Significance of style in corolla tubes for honey bee pollinators. *Current Science* 73(3), 287–290.

Bellardi, M.G. *et al.* (1998) The deterioration of *A. graveolens*. *Informatars Agrario* 54(28), 70–72.

Bellomaria, B., Valentini, G. and Arnold, N. (1999) The essential oil of *F. vulgare*. *Rivista Italiano EPPOS* 27, 43–48.

Benezet, H. *et al.* (2000) Vanilla viruses in Reunion. *Phytoma* 526, 40–42.

Bernard, T. *et al.* (1989) Extraction of essential oils by refining of plant materials. *Flavour and Fragrance Journal* 4(2), 85–90.

Bernath, J. *et al.* (1996) Production-biological investigation of fennel populations of different genotypes. *Instituto Sperimentale per l'Assestamento Forestale e per l'Alpicoltara* 1996, 287–292.

Bhai, R.S. and Thomas, J. (2000) Phytophthora rot – a new disease of vanilla (*Vanilla planifolia*) in India. *Journal of Spices and Aromatic Crops* 9(1), 73–75.

Bharat Rai *et al.* (1997) Efficacy of fungicides and antibiotics on seed mycoflora and seed germination of some spices. *Indian Phytopathology* 50(2), 261–265.

Bisht, J.K. *et al.* (2000) Performance of ginger and turmeric with fodder tree based silvi-horti system. *Indian Journal of Agricultural Sciences* 70(7), 431–433.

Boelens, M.H. and Sindreu, R.J. (1986) Chemical composition of laurel leaf oil obtained by steam distillation and hydrodiffusion. In: *Progress in Essential Oil Research*. W. de Gruyter, Berlin, Germany, pp. 99–110.

Bomme, U. (1997) Choice of varieties for dill shoot production. *Gemuse (Munchen)* 33(3), 189–190.

Bones, A.M. and Rossiter, J.T. (1996) The myrosinase–glucosinolate system, its organization and biochemistry. *Physiologia Plantarum* 97(1), 194–208.

Bosland, P.W. and Votava, E.J. (2000) *Peppers: Vegetable and Spice Capsicums*. CAB International, Wallingford, UK.

Bottero, J. (1985) The cuisine of ancient Mesopotamia. *Biblical Archeologist* 3, 36–47.

Bouriquet, G. (1954) *Le Vanillier et la Vanille dans la Monde*. Editions Paul Lechavalier, Paris, France.

Bourrel, C. *et al.* (1995) Etude des proprietes bacteriostatiques et fongistatiques en milieu solide de 24 huiles essentielles prealablement analysees. *Rivista Italia EPPOS* 16, 3–12.

Bouwmeester, H.J. *et al.* (1998) Biosynthesis of the monoterpenes limonene and carvone in the fruit of caraway. *Plant Physiology* 117(3), 901–912.

Bradu, B.L. and Sobti, S.N. (1988) *C. tamala* in N.W. Himalayas. *Indian Perfumer* 32(4), 334–340.

de Brandao, A.L., Tafani, R.R. and Lima, L.P. (1978) Economic viability of black pepper cultivation in the cacao region of Bahia. *Boletin Technico Cent. de Pesquisas do cacau*. No.64. Bahia, Brazil.

Brandjes, P.J. and Lauckner, F.B. (1997) On-farm assessment of two limimg materials in cabbage and hot pepper cultivation in Guyana. *Experimental Agriculture* 33(2), 225–235.

Brenec, P. and Sauvaire, Y. (1996) Chemotaxonomic value of sterols and steriod sapogenins in the genus *Trigonella*. *Biochemical Systematics and Ecology* 24(2), 157–164.

Bretschneider, E. (1870) *On the Study and Value of Chinese Botanical Works*. Foochow, China.

Bretschneider, E. (1895) *Botanica Sinicum. Notes on Chinese Botany from Native and Western Sources*, Part 3. Kraus Reprint 1967, Nendeln, Lichtenstein.

Brown, E.G. (1955) Cinnamon and cassia: sources, production and trade. *Colonial Plant and Animal Products* 5, 257–280.

Brown, E.G. (1956) Cinnamon and cassia: sources, production and trade. *Colonial Plant and Animal Products* 6, 96–116.

Buckenhuskes, H.J. (1999) Current requirements on paprika powder for the food industry. *Zeitschrift fur Arznei-Gewurz pfalzen* 4(3), 111–118.

Burkill, I.H. (1966) *The Economic Products of the Malay Peninsula* (2 vols). Government Printer, Kuala Lumpur, Malaysia.

Burton, R. (1860) *The Lake Regions of Central Africa.* London.

Caiola, M.G. (1999) Reproductive biology of saffron and its allies. In: Negbi, M. (ed.) *Saffron.* Harwood Academic Publishers, Amsterdam, The Netherlands, pp. 31–44.

Carpana, E. *et al.* (1996) Prophylaxis and control of American foul brood in Italy. *Apicoltore Moderno* 87(1), 11–16.

Cavero, J., Zaragoza, C. and Gil-Ortega, R. (1996) Tolerance of direct seeded pepper under plastic mulch to herbicides. *Weed Technology* 10(4), 900–906.

Cecilio, F. *et al.* (2000) Turmeric: plant with medicinal and spice properties and other potential uses. *Ciencia Rural* 30(1), 171–177.

Chand, K., Jain, M.P. and Jain, S.C. (1999) Seed-borne nature of *A. alternata* in cumin, its detection and location in seed. *Journal of Mycology and Plant Pathology* 29(1), 137–138.

Chand, S., Sahrawat, A.K. and Prakash, D.V. (1997) *In vitro* culture of *P. anisum. Journal of Plant Biochemistry and Biotechnology* 6(2), 91–95.

Chandra, R., Sheo, G. and Desai, A.R. (1999) Growth, yield and quality performance of turmeric genetypes in mid-altitudes of Meghalaya. *Journal of Applied Horticulture* 1(2), 142–144.

Chang, K.C. (1976) *Early Chinese Civilization: Anthropological Perspectives.* Harvard University Press, Cambridge, Massachusetts.

Chattopadhyay, S.B. and Maiti, S. (1990) *Diseases of Betelvine and Spices,* 2nd edn. Indian Council of Agricultural Research, New Delhi, India.

Chaudhary, O.P. and Rakesh Kumar (2000) Studies on honeybee foraging and pollination in cardamom (*Ellataria cardamomum*). *Journal of Spices and Aromatic Crops* 9(1), 37–42.

Chaudhury, R. and Chandel, K.P.S. (1994) Germination studies and cryopreservation of seeds of black pepper (*P. nigrum*). *Cryo-Letters* 15(3), 145–150.

Chezhiyan, N. and Shanmugasundaram, K.A. (2000) BSR.2 – a promising turmeric variety from Tamil Nadu. *Indian Journal of Arecanut, Spices and Medicinal Plants* 2(1), 24–26.

Chhabra, M.L., Yadava, T.P. and Kiran, K. (1996) Translocation pattern of ^{14}C-assimilates from the siliquae. *Cruciferae News* 18, 48–49.

Chhaya, G. and Mishra, S.H. (1999) Antihepatotoxic activity of p-methoxy benzoic acid from *C. spinosa. Journal of Ethnopharmacology* 66(2), 187–192.

Chiu, A.M. and Zacharisen, M.C. (2000) Anaphylaxis to dill. *Annals of Allergy, Asthma and Immunology* 84(5), 559–560.

Chkhaidze, D.K. and Kechakmadze, V. (1991) Labour efficiency in bay laurel management; harvesting of dry leaves in relation to harvest methods. *Subtropical Kul'tury* 4, 9–14.

Chkhaidze, D.K. and Vadachkoriya, T.T. (1985) Effect of methods and frequency of harvesting on laurel dry leaf yield. *Subtropical Kul'tury* 3, 123–127.

Chkhaidze, D.K. and Vadachkoriya, T.T. (1989) Effect of methods and frequency of harvesting on laurel dry leaf yield. *Subtropical Kul'tury* 4, 101–108.

Chkhaidze, D.K. and Vadachkoriya, T.T. (1989) The effect of planting density on the yield of laurel leaves. *Subtropical Kul'tury* 5, 101–108.

Choo, L.-C. *et al.* (1999) Essential oil of nutmeg pericarp. *Journal of the Science of Food and Agriculture* 79(13), 1954–1957.

Chu, E.Y. *et al.* (1997) Evaluation of arbuscular mycorrhizal fungiinnoculation on the incidence of *Fusarium* root rot of black pepper. *Fitopathologia Brasiliera* 22(5), 205–209.

Clark, L. (1998) Dermal contact repellants for starlings: foot exposure to natural plant products. *Journal of Wilderness Management* 61(4), 1352–1358.

Correcher, V., Muniz, J.L. and Gomez-Ros, J.M. (1998) Dose dependence and fading effect of the thermoluminescence signals in gamma irrradiated peppers. *Journal of the Science of Food and Agriculture* 76(2), 149–155.

de Costa, W.A. and Perera, M.K. (1998) Effects of bean population and row arrangement on productivity of chillie/dwarf bean intercropping in Sri Lanka. *Journal of Agronomy and Crop Science* 180(1), 53–58.

Cu, J.Q., Perineau, F. and Goepfert, G. (1990) GC/MS analysis of star anise oil. *Journal of Essential Oil Research* 2, 91–92.

Cui, Z.-F. *et al.* (2000) Yield effect of ginger in plastic house. *China Vegetables* 3, 14–16.

Dabek, A.J. and Martin, P.J. (1987) Tetracycline therapy of clove trees affected by sudden death disease in Zanzibar. *Journal of Phytopathology* 119(1), 75–80.

Damodaran, A. (1998) Coffee as a mixed crop in pepper gardens. In: *Proceedings of the National Seminar on Water and Nutrient Management for Sustainable Production of Quality Spices*. Madikeri, Karnataka, India, pp. 164–169.

Darlington, C.D. and Wylie, A.P. (1955) *Chromosome Atlas of Flowering Plants*. Allen and Unwin, London.

Das, A.B. *et al.* (2000) Variation of 4C DNA content and caryotype in nine cultivars of fenugreek (*T. foenum-graecum*). *Journal of Herbs, Spices and Medicinal Plants* 7(1), 25–32.

Das, N. (1997) Reaction of coriander varieties to root knot nematode. *Annals of Agricultural Research (Orissa)* 19(1), 94–95.

Das, N. (1999) Effect of organic mulching on root-knot nematode population, rhizome rot incidence and yield of ginger. *Annals of Plant Protection Science* 7(1), 112–114.

Das, P., Rai, S. and Das, A.B. (1999) Cytomorphological barriers in seed set of cultivated ginger. *Iranian Journal of Botany* 8(1), 119–129.

Daswir, Z.H. (1986) Effects of fertilizer application and cultural practice on the growth of young clove. *P.P.T. Industri. Bogor.* 12(1/2), 17–21.

Deeksha, D. and Srivastava, N.K. (2000) Distribution of photosynthetically fixed 14CO$_2$ into curcumin and essential oil in relation to primary metabolites in developing turmeric leaves. *Plant Science (Limerick)* 152(2), 165–171.

Deli, J. *et al.* (2001) Carotenoid composition in the fruits of red paprika (*Capsicum annuum* var. *lycopersiciforme rubrum*) during ripening; biosynthesis of carotenoids in red paprika. *Journal of Agricultural and Food Chemistry* 49(3), 1517–1523.

Della Porta, G. *et al.* (1998) Isolation of clove bud and star arise essential oil by supercritical CO$_2$ extraction. *Lebensmittel-Wissenschaft und Technologia* 31(5), 454–460.

Devi, S. (1999) Micropropagation of ginger. *PRAP Report* 1999(7), 11–12.

Dhamayanthi, K.P. and Krishnamoorthy, B. (1999) Somatic chromosome number in nutmeg (*M. fragrans*). *Journal of Spices and Aromatic Crops* 8(2), 205–206.

Dhananiya, G. *et al.* (1999) Modulation of some gluconeogenic enzyme activities in diabetic rat liver and kidney. *Indian Journal of Experimental Biology* 37(2), 196–199.

Dorman, H.J.D. and Deans, S.G. (2000) Antimicrobial agents from plants; antibacterial activity of plant volatile oils. *Journal of Applied Microbiology* 88(2), 308–316.

Dorman, H.J.D *et al.* (2000a) *In vitro* evaluation of antioxidant activity of essential oils and their components. *Flavour and Fragrance Journal* 15(1), 12–16.

Dorman, H.J.D., Surai, P. and Deans, S.G. (2000b) *In vitro* antioxidant activity of a number of plant essential oils and phytoconstituents. *Journal of Essential Oil Research* 12(2), 241–248.

Dragland, S. and Aslaksen, T.H. (1996) Trial cultivation of caraway (*C. carvi*). Norway.

Dufournet, R. and Rodriguez, H. (1972) Travaux pour l'amelioration de la production du giroflier sur la cote oriental de Madagascar. *Agronomie Tropicale* 27, 633–638.

Dyal, S.K. *et al.* (1996) An agri-silvi-horticutural system to optimize production and cash returns from Shivalik foothills. *Indian Journal of Soil Conservation* 24(2), 150–155.

Ebanoidze, N.E. (1997) No-waste technology and equipment for laurel production. *Traktory i Sel'skokhozyaistvennye Mashiny* 1, 30–32.

Ebanoidze, N.E. (1998) Apparatus for heavy trimming of bushes of *L. nobilis*. *Traktory i Sel'skokhozyaistvennye Mashiny* 1, 27–29.

Ebrahimzadah, H., Radiabian, T. and Karamian, R. (2000) *In vitro* production of floral buds and stigma-like structures on floral organs of *Crocus sativus* L. *Pakistan Journal of Botany* 32(1), 141–150.

Ehlers, D. and Pfister, M. (1997) Compound of vanillons (*V. pompona*). *Journal of Essential Oil Research* 9(4), 427–431.

Ehlers, D. *et al.* (1998) High performance liquid analysis of nutmeg and mace oils by supercritical CO$_2$ extraction. *International Journal of Food Science and Technology* 33, 215–223.

Ehlers, D. *et al.* (1999) Vanillin content and several ratios of components – influence of different extraction parameters on reference values. *Deutsche Lebensmittel-Rundschau* 95(4), 123–129.

Ehlers, D. *et al.* (2000) Investigating fennel oils: carbon dioxide extraction and steam distillation. *Deutsche Lebensmittel-Rundschau* 96(9), 330–335.

Eikani, M.H., Goodarzani, I. and Mirza, M. (1999) Supercritical carbon dioxide extraction of cumin seeds. *Flavour and Fragrance Journal* 14(1), 29–31.

El-Dakhakhny, M. *et al.* (2000) Effect of *N. sativa* oil on gastric secretion and ethanol induced ulcer in rats. *Journal of Ethnopharmacology* 72(1/2), 299–304.

El-Sayed, K.A. *et al.* (2000) Effect of different fertilizers on the amino acid, fatty acid, and essential oil composition of *Nigella sativa* seeds. *Saudi Pharmaceutical Journal* 8(4), 175–182.

Emboden, W. (1979) *Narcotic Plants*. Macmillan, New York.

Enomoto, S. *et al.* (2001) Hematological studies in black cumin oil from seeds of *Nigella sativa*. *Biological Pharmaceutical Bulletin* 24(3), 307–310.

Ernst, E. and Pittler, M.H. (2000) Efficacy of ginger for nausea and vomiting trials. *British Journal of Anaesthesia* 84(3), 367–371.

Evenhuis, A., Verdam, B. and Wander, J.G. (1999) Crop management and anthracnose development in caraway. *Netherlands Journal of Agricultural Sciences* 47(1), 24–49.

Faber, B., Bangert, K. and Mosandi, A. (1997) GC–IRMS and enantioselective analysis in biochemical studies in dill. *Flavour and Fragrance Journal* 12(5), 305–314.

Fait, A. *et al.* (2000) Ecological variability in the germination response to temperature in indigenous populations of *F. vulgare* in Israel. *Acta Horticulturae* 517, 467–477.

Fehrmann, A., Schulz, H. and Pank, F. (1996) Non-destructive NIRS-measurement in caraway and fennel fruits. *Bundenanstalt für Zuchtungforschund und Kulturpflazen* 2(1), 418–421.

Fiorini, C. *et al.* (1997) Composition of flower, leaf and stem essential oils from *L. nobilis*. *Flavour and Fragrance Journal* 12(2), 91–93.

la Fleur, J.D. (2000) *Pieter van den Broecke's Journal of Voyages to Cape Verde, Guinea and Angola 1605–1612*. Hakluyt Society, British Museum, London.

Fock-Heng, P.A. (1965) Cinnamon of the Seychelles. *Economic Botany* 19, 257–261.

Formacek, K. and Kubeczka, K.H. (1982) The chemical composition of star anise oil. In: *Essential Oils Analysis by Capillary Chromatography and Carbon C-13 NMR Spectroscopy*. John Wiley & Sons, New York.

Foster, W. (1949) *The Red Sea at the Close of the 17th Century*. Hakluyt Society, British Museum, London.

Fournier, G., Leboeuf, M. and Cave, A. (1999) *Annonaceae* essential oils: a review. *Journal of Essential Oil Research* 11, 131–142.

Frank, J., Pank, F. and Ahne, R. (1996) Thousand seed weight determination on annual caraway by image analysis. *Bundenanstalt für Zuchtungforschund und Kulturpflazen* 2(1), 226–227.

Frau, A. *et al.* (2001) Micropropagation of five clones of Sardinian myrtle. *Informatore Agrario* 57(17), 65–67.

Freire, A.S. *et al.* (1985) Effect of herbicides on *S. aromaticum* seedlings. *CEPLAC* 362, 364–365.

Fujita, Y., Fujita, S. and Yoshikana, H. (1974) Biogenesis of essential oils in camphor tree. *Nippon Nogeikagaku Kaishi* 48(11), 633–636.

Gabuniya, N.A. and Ebanoidze, N.E. (1998) Machinery for harvesting laurel, *L. nobilis*. *Traktory i Sel'skokhozyaistvennye Mashiny* 6, 22–23.

Galigani, P.F. and Pegna, F.G. (1999) Mechanized saffron cultivation. In: Negbi, M. (ed.) *Saffron*. Harwood Academic Publishers, Amsterdam, The Netherlands, pp. 115–126.

Ganesh, D.S., Sreenath, H.L. and Jayashree, G. (1996) Micropropagation of vanilla through node culture. *Journal of Plantation Crops* 24(1), 16–22.

Gao, S., Bian, Y.-Y. and Chein, B.-J. (1999) Tissue culture of ginger to control virus diseases and rapid high-yielding cultivation. *China Vegetables* 3, 40–41.

Gare, B.N., Khade, K.K. and Bhoi, P.G. (1996) Performance of mustard cultivars at different dates of sowing. *Journal of Maharashtra Agricultural University* 21(1), 147–148.

Garg, S.C. and Rajshree, J. (1996) Volatile constituents of *Piper betle*. *Indian Journal of Chemistry, Section B* 35(8), 874–875.

Garg, S.N. *et al.* (1999) Variation in the rhizome essential oil and curcumin contents and oil quality in land races of *C. longa* of the North Indian plains. *Flavour and Fragrance Journal* 15(5), 315–318.

Gaydou, E.M. and Randriamiharisoa, R. (1987) Multidimensional analysis of gas chromatographic data to differentiation of clove bud and stem essential oils from Madagascar. *Perfum. Flav.* 12(6), 45–51.

Geetha, C.K., Aravindakshan, M. and Wahid, P.A. (1993) Influence of method of fertilizer application on nutrient absorption of black pepper vines. *Southern India Horticulture* 41(2), 95–100.

Geetha, S. and Shetty, S.A. (2000) *In vitro* propagation of *Vanilla planifolia*. *Current Science (India)* 79(6), 886–889.

George, P.S. and Ravishankar, G.A. (1997) *In vitro* multiplication of *Vanilla planifolia* using axillary bud explants. *Plant Cell Reports* 16(7), 490–494.

de Geus, J.G. (1973) *Fertilizer Guide for the Tropics and Subtropics*, 2nd edn. Centre d'Etude Azote, Zurich, Switzerland.

Gil, A. *et al.* (1999) Coriander. Yield response to plant populations. *Journal of Herbs, Spices and Medicinal Plants* 6(3), 63–73.

Gildmeister, E. and Hoffman, F. (1959) *Die Aetherischen Oele.* Akademie Verlag, Berlin, Germany.

Gill, B.S. *et al.* (1999) Response of turmeric to nitrogen in relation to application of farmyard manure and straw mulch. *Journal of Spices and Aromatic Crops* 8(2), 211–214.

Giridharan, M.P. and Balnakrishnan, S. (1991) Effect of gamma-irradiation on yield and quality of ginger. *Indian Cocoa, Arecanut, and Spices Journal* 14(3), 100–103.

Gomez, R., Pardo, J.E. and Varon, R. (1998) Effect of different storage conditions on paprika grown in the open and greenhouse. *Rivista della Society Italiana di Scienza dell Alimentzione* 27(1), 23–28.

Gopalakrishnan, M., Menon, N. and Mathew, A.G. (1982) Changes in the composition of clove oil during maturation. *Journal of Food Science and Technology* 19(6), 190–192.

Gopalakrishnan, M., Narayanan, C.S. and Mathew, A.G. (1988) Chemical composition of Indian clove bud, stem, and leaf oils. *Indian Perfumari* 32(3), 229–235.

Gopalakrishnan, N. (1994) Studies on the storage quality of carbon dioxide extracted cardamom and clove bud oils (analysis of fresh essential oil). *Journal of Agricultural and Food Chemistry* 42, 796–798.

Gopalam, A. and Zacharia, T.J. (1989) Lipophilic profits of extractable fat of nutmeg. *Journal of Plantation Crops* 16 (suppl.), 107–112.

Gopalan, B. *et al.* (2000) Supercritical carbon dioxide extraction of turmeric. *Journal of Agricultural and Food Chemistry* 48(6), 2189–2192.

Gopalkrishnan, M. *et al.* (1993) Analysis and odour profiles of four new Indian genotypes of *P. nigrum*. *Journal of Essential Oil Research* 5(3), 247–253.

Gouveia, J.P.G. *et al.* (1999) Study of drying kinetics of ginger in fixed bed dryer. *Revista Brasileira de Armazenamento* 24(2), 20–27.

Gowda, K.K., Melanta, K.R. and Prasad, T.R. (1999) Influence of NPK on the yield of ginger. *Journal of Plantation Crops* 27(1), 67–69.

Grant, K.L. and Schneider, C.D. (2000) Alternative therapies. *American Journal of Health-System Pharmacy* 57(12), 1121–1122.

Green, C. and Espinosa, F. (1989) Jamaican and central American pimento (allspice): characterisation of flavour differences and other distinguishing features. *Journal of Food Science* 18, 3–20.

Guaday, J.M. *et al.* (1997) Extraction, separation and identification of volatile organic compounds from paprika oleoresin (Spanish type). *Journal of Agricultural and Food Chemistry* 45(5), 1868–1872.

Guenther, E. (1950a) *The Essential Oils*, Vol. 4. D. van Nostrand Co., New York, pp. 241–256.

Guenther, E. (1950b) *The Essential Oils*, Vol. 4. D. van Nostrand, New York, pp. 634–645.

Guenther, E. (1952) In: *The Essential Oils*, Vol. 5. D. van Nostrand, New York, pp. 148–157.

Gulati, B.C. (1982) Essential oils of *Cinnamomum* species. In: *Cultivation and Utilization of Aromatic Plants.* CSIR, Jammu Tawi, India, pp. 607–619.

Gunasekaran, M. and Krishnaswamy, V. (1999) Influence of seed size and placement on seed quality characteristics of clove. *Indian Journal of Arecanut, Spices and Medicinal Plants* 1(3), 94–97.

Gupta, A.P. *et al.* (1999) Simultaneous determination of curcuminoides in *Curcuma* samples using high performance thin layer chromatography. *Journal of Liquid Chromotography and Related Techniques* 222(10), 1561–1569.

Gupta, C.R. and Sengar, S.S. (1998) Effect of varying levels of nitrogen and potassium fertilization on growth and yield of turmeric under rainfed conditions. In: *Proceedings of the National Seminar on Water and Nutrient Management for Sustainable Production of Quality Spices*, Madikeri, Karnataka, India, pp. 47–51.

Gupta, S.K. (1996) Effect of freezing on total lipids and fatty acid composition of mustard seeds (*B. juncae*). *Cruciferae News* 18, 58–59.

de Guzman, C.C. and Siemonsma, J.S. (eds) (1999) *Plant Resources of South East Asia*. No 13. *Spices.* Backhuys Publishers, Leiden, The Netherlands, p. 217.

Hafizoglu, H. and Reuanen, M. (1993) Studies on the composition of *L. nobilis* from Turkey with special reference to laurel berry fat. *Fat Science and Technology* 95, 304–308.

Haldankar, P.M. *et al.* (1999) Factors affecting epicotyl grafting in nutmeg. *Journal of Medicinal and Aromatic Plant Sciences* 21(4), 940–944.

Halesh, D.P. *et al.* (1998) Response of fenugreek (*T. foenum-graecum*) to nitrogen and phosphorus. In: *Proceedings of the National Seminar on Water and Nutrient Management for Sustainable Production of Quality Spices*. Madikeri, Karnataka, India, pp. 76–79.

Hallstrom, H. and Thuvander, A. (1997) Toxicological evaluation of myristicin. *Natural Toxins* 5(5), 186–192.

Halva, S. *et al.* (1986) Yield and glucosalinate of mustard seeds and volatile oils of caraway seeds and coriander fruit. *Journal of Agricultural Sciences, Finland* 58, 163–167.

Halva, S., Huopalahti, R. and Makinen, S. (1987) Studies on the fertilization of dill and basil. *Journal of Agricultural Sciences, Finland* 59, 19–24.

Hammer, K. *et al.* (2000) *Pimpinelle anisoides* Brig. *Genetic Research on Crop Evolution* 47(2), 223–225.

Harberd, D.J. (1972) A contribution to the cytotaxonomy of *Brassica* and its allies. *Botanical Journal of the Linnean Society* 65, 1–23.

van Harten, A.M. (1970) Malegueta Pepper. *Economic Botany* 24(2), 208–215.

Harvey, J. (1975) Gardening books and plant lists of Moorish Spain. *Garden Histology* 3(2), 10–21.

Hasanah, K. *et al.* (1984) Effects of growth inhibitors and storage temperature on viability of clove seed. *P.P.T. Industri. Bogor.* 8(49), 47–54.

Hashem, M.Y. (2000) Suggested procedures for applying carbon dioxide to control stored medicinal plant products from insect pests. *Zeitschrift für Pflanzenkrankheiten und Pflanzenschutz* 107(2), 212–217.

Havkin-Frenkel, D., Dorn, R. and Leuster, T. (1997) Plant tissue culture for production of secondary metabolites. *Food Technology (Chicago)* 51(11), 56–58.

Hazra, P., Roy, A. and Bandopadhyay, A. (2000) Growth characters as rhizome yield components of turmeric. *Crop Research (Hisar)* 19(2), 235–240.

Hiskias, Y., Leseman, D.E. and Vetten, H.J. (1999) Occurrence, distribution and relative importance of viruses infecting hot pepper and tomato in the major growing areas of Ethiopia. *Journal of Phyotopathology* 147(1), 5–11.

Hoppe, B. (1996) Status of German medicinal and spice plant production. *Gemuse (Munchen)* 32(4), 283–284.

Howard, L.R. *et al.* (2000) Changes in phyrochemical and antioxidant activity of selected pepper species as influenced by maturity. *Journal of Agricultural and Food Chemistry* 48(5), 1713–1720.

Hu, S.-L., Ao, P. and Fu, G.-F. (1995) Pharmacognostical studies on the root of *P. Nigrum* – determination of the essential oil and piperine. *Acta Horticulturae* 426, 179–182.

Hu, S.-L., Ao, P. and Fu, G.-F. (1996) Pharmacognostical studies on the roots of *P. nigrum. Acta Horticulturae* 426, 171–174.

Huberman, M. *et al.* (1997) Role of ethylene biosynthesis and auxin content and transport in high temperature induced abscission of pepper reproductive organs. *Journal of Plant Growth Regulators* 16(3), 129–135.

Huntingford, G.W.B. (1976) *The Periplus of the Erythrean Sea.* The Hakluyt Society, British Museum Library, London.

Husain, A. *et al.* (1988) *Major Essential Oil Bearing Plants Of India.* Central Institute Medicinal and Aromatic Plants, Lucknow, India.

Hussain, M.S. (1995) Response of growth, yield and essential oil of coriander and dill to different nitrogen sources. *Egyptian Journal of Horticulture* 22(1), 1–10.

Hussein, M.A. and Amla, B. (1998) *In vitro* embryogenesis of cumin hypocotyl segments. *Advances in Plant Science* 11(1), 125–127.

Ibrahim, J. and Goh, S.H. (1992) Essential oils of *Cinnamomum* species from Peninsular Malaysia. *Journal of Essential Oil Research* 4(2), 161–171.

Indian Spices Board (1998) *Spice Statistics.* Ministry of Commerce, Government of India, Cochin, India.

Ipe, C.V. and Varghese, C.A. (1990) Economics of nutmeg cultivation in Kerala. *Journal of Plantation Crops* 18(1), 29–33.

Islam, A.K., Edwards, D.G. and Asher, C.J. (1980) pH optima for crop growth. *Plant and Soil* 54, 339–357.

Jacobs, M. (1965) The genus *Capparis* from the Indus to the Pacific. *Blumea* 12, 104–112.

Jacobson, R.L. and Schlein, Y. (1999) Lectins and toxins in the plant diet of *Phlebotomus papatasi* can kill *Leishmania major* promastigotes in the sandfly. *Annals Tropical Medicine and Parasitology* 93(4), 351–356.

Jagadeeswaran, R. *et al.* (2000) *In vitro* studies on the selective cytotoxic effect of crocetin and quercetin. *Fitoterapia* 71(4), 395–399.

Jagella, T. and Grosch, W. (1999) Flavour and off-flavour compounds of black and white pepper (*P. nigrum*). I, II and III. *European Food Research and Technology* 209(1), 16–21, 22–26, 27–31.

Jansen, P.C.M. (1981) *Spices, Condiments and Medicinal Plants in Ethiopia, their Taxonomy and Agricultural Significance.* Centre for Agricultural Publishing and Documentation, Wageningen, The Netherlands.

Jaren-Galen, M., Nienaber, U. and Schwartz, S.J. (1999) Paprika oleoresin extraction with super-critical carbon dioxide. *Journal of Agricultural and Food Chemistry* 47(9), 3558–3564.

Jarvie, J.K. *et al.* (1986) Screening for compatibility between clove and related myrtaceous species. *P.P.T. Industri Bogor* 11(3/4), 56–59.

Jasvir S., Sudharshan, M.R. and Selvan, M.T. (1999) Seasonal population of cardamom thrips on three cultivars of cardamom. *Journal of Spices and Aromatic Crops* 8(1), 19–22.

Jayaprakasha, G.L., Rao, L.J. and Sakariah, K.K. (1997) Chemical composition of the volatile oil from fruits of *C. zeylanicum. Flavour and Fragrance Journal* 12(5), 331–333.

Jayatilaka, A. *et al.* (1995) Simultaneous microstream distillation/solvent extraction for the isolation of semivolatile flavor compounds from cinnamon by series coupled-column GC. *Analytica Chimica Acta* 302, 147–162.

Jeliazkova, E.A. *et al.* (1997) Irradiation of seeds and productivity of coriander. *Journal of Herbs, Spices and Medicinal Plants* 5(2), 73–79.

Ji, X.-D. *et al.* (1991) Essential oil of the leaf, bark and branch of *Cinnamomum burmannii* Blume. *Journal of Essential Oil Research* 3, 373–375.

Jirovetz, L., Buchhbauer, G. and Eberhardt, R. (2000) Analysis and quality control of essential cinnamon oils of different origin using GC, GC-MS, and olfactometry – determination of cumarin and safrole content. *Ernahrung* 24(9), 366–369.

Jones, V. (1978) *Sail The Indian Sea.* Gordon and Cremonesi, London.

Jose, J. and Sharma, A.K. (1984) Chromosome studies in the genus *Piper* L. *Journal of the Indian Botanical Society* 63(3), 313–319.

Joseph, D. *et al.* (1999) Levels of trace elements of a few Indian spices by energy dispersive X-ray fluorescence. *Journal of Food Science and Technology* 36(3), 264–265.

Joshi, G.D. *et al.* (1996) Preliminary studies on utilization of nutmeg fruit spice rind. *Indian Cocoa, Arecanut and Spices Journal* 20(1), 5–9.

Joy, C.M., Pittappillil, G.P. and Jose, K.P. (2000) Quality improvement of nutmeg using solar tunnel dryer. *Journal of Plantation Crops* 28(2), 138–143.

Jubina, P.A. and Girija, V.K. (1998) Antagonistic rhizobacteria for management of *Phytophthora capsici,* the incitant footrot of black pepper. *Journal of Mycology and Plant Pathology* 28(2), 147–153.

Kalra, A. *et al.* (2000) Effects of planting date and dinocarp applications on the control of powdery mildew and yields of seed and seed oil in coriander. *Journal of Agricultural Sciences* 135(2), 193–197.

Kamruddin, A.T. and Wenur, F. (1994) Drying of black pepper by solar energy. In: *Proceedings of the 9th Internatioal Drying Symposium,* Vol B. Gold Coast, Australia, pp. 887–895.

Kanakamany, M.T. *et al.* (1985) Key for identification of different cultivars of pepper. *Indian Cocoa, Arecanut and Spices Journal* 9(1), 6–11.

Kandiannan, K. *et al.* (1994) Growth regulators in black pepper production. *Indian Cocoa, Arecanut and Spices Journal* 18(4), 119–123.

Karaca, A. and Kevseroglu, K. (1999) The research of some important agricultural characteristics of coriander (*C. sativum*) and fennel (*F. vulgare*) varieties of Turkish origin. *Onodokuzmayis Universitesi Fakultesi Dergisi* 14(2), 65–77.

Kataeva, N.V. and Popowich, E.A. (1993) Maturation and rejuvenation of *C. sativum* shoot clones during micropropagation. *Plant Cell, Tissue and Organ Culture* 34(2), 141–148.

Kato, O.R. and Albuquerque F.C. (1980) Relationship between support size and black pepper productivity. *Pesquisa em Andamento. Alt.* 4, 3–6.

Kaunzinger, A., Juchelka, D. and Mosandi, A. (1997) Progress in the authenticity assessment of vanilla. *Journal of Agricultural and Food Chemistry* 45(1), 1752–1757.

Kaya, N., Yilmaz, G. and Telci, I. (2000) Agronomic and technological properties of coriander populations planted on different dates. *Turkish Journal of Agriculture and Forestry* 24(3), 355–364.

Kekelidze, N.A. (1987) Essential oils of *L. nobilis* bark and wood. *Khimiya Prirodnykh Soedinenii* 3, 458–459.

Khan, M.M., Azam, Z.M. and Samiullah (1999) Changes in the essential oil constituents of fennel as influenced by soil and foliar levels of N and P. *Canadian Journal of Plant Sciences* 79(4), 587–591.

Khan, N.A. (1996a) Effect of gibberelllic acid on carbonic anhydrase, photosynthesis, growth and yield of mustard. *Biologia Plantarum* 38(1), 145–147.

Khan, N.A. (1996b) *Biologia Plantarum* 38(4), 601–603.

Khan, N.A. (1996c) Response of mustard to Ethrel spray and basal and foliar application of nitrogen. *Journal of Agronomy and Crop Science* 176(5), 331–334.

Khana, R.K. *et al.* (1988) Essential oil from fruit rind of *C. cecidodaphne. Indian Perfumari* 32(4), 295–300.

Khanna, N.M. (1999) Turmeric – natures precious gift. *Current Science* 76(10), 1352–1356.

Kieniewicz, J. (1986) Pepper gardens and market in precolonial Malabar. *Moyen Oriente et Ocean Indien* 111, 1–36.

Kikuzaki, H. *et al.* (1999) Antioxidative phenylpropanoides from berries of *P. dioica*. *Phytochemistry* 52(7), 1307–1312.

Kim, Y.-H. *et al.* (1996) Screening for antagonistic natural materials against *Alternaria alternata*. *Korean Journal of Plant Pathology* 12(1), 66–71.

Kishor, C. and Jain, M.P. (1999) Effect of fungal spore load on cumin seed infection. *Annals Agri Bio Research* 4(1), 103–105.

Klein, J.D., Cohen, S. and Hebbe, Y. (2000) Seasonal variation in rooting ability of myrtle. *Scientia Horticulturae* 83(1), 71–76.

Klimes, I. and Lamparsky, D. (1976) Vanilla volatiles. A comprehensive analysis. *International Flavourings and Food Additives* 7, 272–273, 291.

Koli, N.R., Dashora, S.L. and Sastry, E.V.D. (2000) Genetic variability induced by gamma-irradiation in M1 and M2 generations of cumin (*C. cyminum*). *Indian Journal of Agricultural Sciences* 70(6), 418–419.

Korikanthimath, V.S. (1994) Multistorey cropping systems with coffee, clove and pepper. *Indian Coffee* 58(10), 3–5.

Korikanthimath, V.S. (1999) Rapid clonal multiplication of elite cardamom selections for generating planting material, yield upgrading and its economics. *Journal of Plantation Crops* 27(1), 45–53.

Korikanthimath, V.S. (2000) Performance and economics of replanted cardamom (*Elattaria cardamomum*) plantation. *Journal of Spices and Aromatic Crops* 9(1), 31–36.

Korikanthimath, V.S. and Ravindra, M. (1998) Assessment of elite cardamom lines for dry matter distribution and harvest index. *Journal of Medicinal and Aromatic Plants Sciences* 20(1), 28–31.

Korikanthimath, V.S. *et al.* (1997) Coffee, cardamom, black pepper and mandarin mixed cropping system – a case study. *Journal of Spices and Aromatic Crops* 6(1), 1–7.

Korikanthimath, V.S. *et al.* (1998a) Influence of major nutrients on the yield and yield parameters of cardamom grown under controlled shade. *Journal of Medicinal and Aromatic Plants Sciences* 20(3), 700–702.

Korikanthimath, V.S. *et al.* (1998b) Mixed cropping of arabica coffee and cardamom. *Journal of Medicinal and Aromatic Plants Sciences* 20(2), 394–396.

Korikanthimath, V.S., Hiremath, G.M. and Hosmani, M. (1999) Generation of employment potential in coffee–cardamom cropping system. *Journal of Medicinal and Aromatic Plants Sciences* 21(3), 29–34.

Korikanthimath, V.S. *et al.* (1999) Labour utilization pattern in relation to other input requirements in a cardamom plantation. *Spices India* 2(12), 5–11.

Korla, B.N., Tiwari, S.S. and Goyal, R.K. (1999) Evaluation of ginger clones for quality attributes under rainfed and irrigated conditions. *Horticultural Journal* 12(1), 39–44.

Kotikal, Y.K. and Kulkarni, K.A. (1999) Management of rhizome fly *Mimegralla coeruleifrons*, a serious pest of turmeric. *Pest Management of Horticultural Ecosystems* 5(1), 62–66.

Krajaklang, M., Kliber, A. and Dry, P.R. (2000) Colour at harvest and post harvest behaviour influence paprika and chilli spice quality. *Postharvest Biology and Technology* 20(3), 269–278.

Krishnakumar, V. *et al.* (1997) Estimation of leaf area in vanilla. *Crop Research Hissar* (India) 13(1), 239–240.

Krishnakumar, V., George, C.K. and Potty, S.N. (1999) Organic farming in pepper. *Planters Chronicles* 95(8), 357–361.

Krishnamoorthy, B. *et al.* (1988) Quality parameters of cinnamon. *Indian Cocoa, Arecanut and Spices Journal* 12(2), 38.

Krishnamoorthy, B. *et al.* (1997) Genetic resources of tree spices and their conservation in India. *Plant Genetic Research News* 111, 53–58.

Krishnamoorthy, B. *et al.* (2000) Evaluation of selected Chinese cassia (*C. cassia*) acessions for chemical quality. *Journal of Spices and Aromatic Crops* 8(2), 193–195.

Krishnamurthy, S.J., Anke Gowda, S.J. and George, J.K. (1998) Impact of water stress on some physiological parameters in black pepper. In: *Proceedings of the National Seminar on Water and Nutrient Management for Sustainable Production of Quality Spices*. Madikeri, Karnataka, India.

Krishnamurthy, K.S., Ankegowda, S.J. and Sail, K.V. (2000) Water stress effects on membrane damage and activities of catalase, peroxidase and duperoxide dismutase enzymes in black pepper. *Journal of Plant Biology* 27(1), 39–42.

Kruger, H. and Hammer, K. (1999) Chemotypes of fennel. *Journal of Essential Oil Research* 11(1), 79–82.

Kumar, P.R., Kailappan, R. and Viswanathan, R. (1998) Pepper thresher. *Planters Chronicles* 93(5), 207–209.

Kumar, T.P. *et al.* (1999) Studies on yielding behaviour of black pepper. *Indian Journal of Arecanut, Spices and Medicinal Plants* 1(3), 88–90.

Kusterer, A. and Gabler, J. (2000) Diseases of dill; what is the importance of fungi, bacteria, viruses. *Gemuse (Munchen)* 36(12), 31–32.

Lakra, B.S. (1999) Assessment of losses due to stem gall in coriander. *Plant Disease Research* 14(1), 85–87.

Lakra, B.S. (2000) Management of stem gall of coriander incited by *P. macrosporus*. *Indian Journal of Agricultural Sciences* 70(5), 338–340.

Lal, H., Rathore, S.V. and Kumar, P. (1996) Influence of irrigation and mixtalol spray on growth, flowering and yield of coriander. *Indian Journal of Soil Conservation* 24(2), 141–146.

Langer, E., Greifenberg, S. and Gruenwald, J. (1998) Ginger: history and use. *Advances in Natural Therapy* 15(1), 25–44.

Laufer, B. (1919) *Sino-Iranica*. Field Museum of Natural History, Chicago.

Lawrence, B.M. (1969) Determination of the botanical origin of cinnamoms of commerce by T L chromatography. *Canadian Institute of Food Technology Journal* 2, 178–180.

Lawrence, B.M. (1977) Chemical evaluation of various bay oils. In: *Proceedings of the 7th International Conference on Essential Oils.* Kyoto, Japan, pp. 172–179.

Lawrence, B. M. (1987) Fenugreek extract. *Perfumer and Flavourist* 12, 60.

Lawrence, B.M. (1993a) Myrtle oil. *Perfumer and Flavourist* 18(2), 52–55.

Lawrence, B.M. (1993b) Progress in essential oils. *Perfumer and Flavourist* 18(3), 65–68.

Lawrence, B.M. (1994a) Progress in essential oils. *Perfumer and Flavourist* 19(4), 33–35.

Lawrence, B.M. (1994b) Progress in essential oils. *Perfumer and Flavourist* 19(6), 60–62.

Lawrence, B.M. (1995) Progress in essential oils. *Perfumer and Flavourist* 19(5), 92, 94.

Lawrence, B.M. (1996a) Progress in essential oils. *Perfumer and Flavourist* 21(2), 25–28.

Lawrence, B.M. (1996b) Progress in essential oils. *Perfumer and Flavourist* 21(3), 55–68.

Lawrence, B.M. (1996c) Progress in essential oils. *Perfumer and Flavourist* 21(4), 57–60.

Lawrence, B.M. (1997a) Progress in essential oils. *Perfumer and Flavourist* 22(1), 49–56.

Lawrence, B.M. (1997b) Progress in essential oils. *Perfumer and Flavourist* 22(3), 57–66.

Lawrence, B.M. (1997c) Progress in essential oils. *Perfumer and Flavourist* 22(5), 71–83.

Lawrence, B.M. (1998a) Progress in essential oils. *Perfumer and Flavourist* 23(1), 39–50.

Lawrence, B.M. (1998b) Progress in essential oils. *Perfumer and Flavourist* 23(2), 47–57.

Lawrence, B.M. (1998c) Progress in essential oils. *Perfumer and Flavourist* 23(4), 37–50.

Lawrence, B.M. (1999a) Progress in essential oils. *Perfumer and Flavourist* 24(2), 35–47.

Lawrence, B.M. (1999b) Progress in essential oils. *Perfumer and Flavourist* 24(5), 45–63.

Lawrence, B.M. (2000a) Progress in essential oils. *Perfumer and Flavourist* 25(2), 46–57.

Lawrence, B.M. (2000b) Progress in essential oils. *Perfumer and Flavourist* 25(3), 54–69.

Lawrence, B.M. (2000c) Progress in essential oils. *Perfumer and Flavourist* 25(5), 52–71.

Lee, M.-J. *et al.* (2000) Effects of selenium on growth, storage life and internal qualities of coriander (*C. sativum*). *Journal of the Korean Society of Horticultural Sciences* 41(5), 490–494.

Lee, T.M. and Asher, C.J. (1981) Nitrogen nutrition of ginger. *Plant and Soil* 62, 23–24.

Lenardis, A. *et al.* (2001) Response of coriander (*Coriandrum sativum*) to nitrogen availability. *Journal of Herbs, Spices and Medicinal Plants* 7(4), 47–58.

Leto, C., Carrubba, A. and Trapani, P. (1996) Effect of sowing date on seed fennel in a semi-arid Sicilian environment. *Instituto Sperimentale per l'Assestamento Forestale e per l'Alpicoltara* 1996, 541–522.

Lima, J.S. *et al.* (1999) Heavy metal transfer from domestic waste compost to plants. In: *Proceedings of the 12th International IFOAM Conference.* Tholey-Theley Publishers, Germany, pp. 154–159.

Lin, Z.K. and Hua, Y.F. (1987) Chemical constituents of 14 essential oils from *Lauraceae* growing in Sichuan. *Chemistry and Industry of Forest Products* 7(1), 46–64.

Lionnet, J.F.G. (1958) Seychelles vanilla. *World Crops* 10, 441–444, and 11, 15–17.

Liu, Y.-H. *et al.* (1999) Study on mechanical drying of star anise. *Journal of Guangxi Agricultural and Biological Sciences* 18(2), 132–135.

Locatelli, D.P., Galli, P. and Moroni, E. (2000) Solid impurities of animal origin found on dried officinal herbs. *Industrie Alimantari* 39(3), 829–834.

Lock, J.M., Hall, J.B. and Abbiw, D.K. (1977) The cultivation of Malegueta pepper in Ghana. *Economic Botany* 31(3), 321–330.

Lozano, M. and de Espinosa, V.M. (1999) Paprika of La Vera: drying process and evaluation of some chemical characteristics of the paprika. *Alimentaria* 300, 91–96.

Lucchesini, M. and Mensuali-Sodi, A. (2000) Effects of vessel permeability to gas exchanges on the *in vitro* plantlets of *M. communis*. *Agricultura Mediterranea* 130(1), 78–84.

Ly-Tio-Fane, M. (1958) *Mauritius and the Spice Trade*. Esclapan Ltd, Port Louis, Mauritius.

Lynrah, P.G., Barua, P.K. and Chakrabaty, B.K. (1998) Pattern of genetic variability in collection of turmeric genotypes. *Indian Journal of Genetics and Plant Breeding* 58(2), 210–217.

Ma, X.Q. *et al.* (2001) Authentic identification of Stigma Croci from its adulterants by molecular genetic analysis. *Planta Medica* 67(2), 183–186.

Madia, M., Getan, S. and Reyna, S. (1999) Wilt and crown rot of coriander caused by a complex of *Fusarium* spp. in Argentina. *Fitopatologia* 34(3), 155–159.

Malijan, H.V. (1982) *Paminta in our Forests*. Agroforestry Research Centre, the Philippines.

Mallavarapu, G.R. and Ramesh, S. (1998) Composition of essential oils of nutmeg and mace. *Journal of Medicinal and Aromatic Plant Sciences* 20(3), 746–748.

Mandavia, M.K. *et al.* (2000) Inhibitory effects of phenolic compounds on fungal metabolism of host–pathogen interaction in *Fusarium* wilt of cumin. *Allopathy Journal* 7(1), 85–92.

Manirakiza, P., Covaci, A. and Schepens P. (1999) Solid phase extraction and gas chromatography with mass spectrometric determination of capsaicin and some analogues from chilli peppers. *Journal of AOAC International* 82(6), 1399–1405.

Manirakiza, P., Covaci, A. and Schepens P. (2000) Single step clean-up and GC-MS quantification of organochlorine pesticide residues in spice powder. *Chromatographia* 52(11/12), 787–790.

Marcelis, L.F.M. and Hoffman-Eijer, L.R. (1997) Effects of seed number on competition and dominance among fruits in *C. annuum*. *Annals of Botany* 79, 687–693.

Markus, F. *et al.* (1999) Change in the carotenoid and antioxidant content of spice red pepper (paprika) as a function of ripening. *Journal of Agricultural and Food Chemistry* 47(1), 100–107.

Maron, R.R. and Fahn, A. (1979) Ultrastructure and development of oil cells in *L. nobilis* leaves. *Botanical Journal of the Linnean Society* 78, 31–40.

Martin, P.J. *et al.* (1988) Causes of irregular clove production in the islands of Zanzibar and Pemba. *Experimental Agriculture* 24, 105–114.

Martini, A. and Garbati Pegna, F. (1996) Mechanical harvesting trials on *F. vulgare*. *Instituto Sperimentale per l'Assestamento Forestale e per l'Alpicoltara* 1996, 537–540.

Martins, A.P. *et al.* (1998) Essential oils from four *Piper* species. *Phytochemistry* 49(7), 2019–2023.

Mary, S. *et al.* (1999) *In vitro* seed culture of vanilla (*V. planifolia*). *Journal of Plantation Crops* 27(1), 13–21.

Mastura, M. *et al.* (1999) Anticandidal and antidermatophytic activity of *Cinnamomum* species essential oils. *Cytobios* 98(38), 17–23.

Mather, T. *et al.* (1999) Effect of washing on persistance of residues of mancozeb in cardamom. *Journal of Spices and Aromatic Crops* 8(1), 73–76.

Matheson, J.K. and Bovill, E.W. (1950) *East African Agriculture*. Oxford University Press, London.

Mathew, A.G. (1990) Oil and oleoresin from Indian spices. In: *Proceedings of the 11th International Congress on Essential Oils and Flavours*. Aspect Publishers, London, pp. 189–195.

Mathew, A.G. (1993) Green and white pepper. *International Pepper News Bulletin* 17(3), 10–13.

Mathew, B. (1999) The saffron plant (*Crocus sativus* L.) and its allies. In: Negbi, M. (ed.) *Saffron*. Harwood Academic Publishers, Amsterdam, The Netherlands, pp. 19–30.

Mathew, F.M. (1974) Karyomorphological studies in *P. nigrum*. *Journal of Plantation Crops* 1, 15–18.

Mathew, K.M. *et al.* (1999) *In vitro* culture sysems in vanilla. In: *Proceedings of a Symposium on Plant Tissue Culture and Biotechnology*. Universities Press India, New Delhi, India, pp. 171–179.

Mathew, P.A., Rema, J. and Krishnamoorthy, B. (2000) Softwood grafting in clove (*S. aromatica*) and related species. *Journal of Spices and Aromatic Crops* 8(2), 215.

Mathew, P.J., Mathew, P.M. and Vijyaraghava, K. (2001) Graph clustering of *Piper nigrum* L. (black pepper). *Euphytica* 118(3), 257–264.

Mathur, R. *et al.* (2000) The hottest chilli variety in India. *Current Science* 79(3), 287–288.

Maurya, K.R. *et al.* (1984) Prediction of maturity in ginger on calcareous soils. *Indian Perfumer* 28(1), 4–7.

Mazza, G. (1983) Gas chromatographic–mass spectrometric investigation of the volatile components of myrtle berries. *Journal of Chromatography* 264, 304–311.

McHale, D. *et al.* (1989) Transformation of the pungent principles in extracts of ginger. *Flavour and Fragrance Journal* 4, 9–15.

McReynolds, R.B. and Abraham, G. (1999) Screening vegetables for tolerance to pre-emergence and post-emergence herbicides. *Western Society of Weed Science* 1999, 39–41.

Meijer, T.M. and Schmid, H. (1948) Uber die Konstitution des Eugenins. *Helvetia Chemical Acta* 31, 1603–1607.

Mekhtieva, N.P. (1997) Essential oils of *Pimpinella squamosa*. *Chemistry of Natural Compounds* 33(5), 595–596.

Mendes, M.M., Gazarini, L.C. and Rodrigues, M.L. (2001) Acclimation of *Myrtus communis* to contrasting Mediterranean light environments – effect on structure and chemical composition of foliage and plant water relations. *Environmental and Experimental Botany* 45(2), 165–178.

Menon, A.N. *et al.* (2000) Essential oil composition of four popular Indian cultivars of black pepper. *Journal of Essential Oil Research* 12(4), 431–434.

Merkli, A., Christen, P. and Kapetandis, I. (1997) Production of diosgenin by hairy root culture of *T. foenum-graecum. Plant Cell Reports* 16(9), 632–636.

van der Mheen, H.J. (2000) Oil and carvone production of winter caraway harvested from umbels and seeds; a comparison. *PAV-Bulletin Akker.* 4, 31–33.

Milanez, D. *et al.* (1987) *Culture of Black Pepper.* Bulletin no. 33, EMCAPA, Campinas, Brazil.

Miller, J.I. (1969) *The Spice Trade of the Roman Empire.* Clarendon Press, Oxford.

Minakashi, D.C. *et al.* (1999) Antimicrobial screening of some Indian spices. *Phytotherapy Research* 13(7), 616–618.

Minguez-Mosquera, M.I. and Hornero-Mendez, D. (1997) Changes in provitamin A during paprika processing. *Journal of Food Protection* 60(7), 853–857.

Minguez-Mosquera, M.L. and Perez-Galvez, A. (1998) Colour quality in paprika oleoresins. *Journal of Agricultural and Food Chemistry* 46(12), 5124–5127.

Miraldi, E. (1999) Comparison of the essential oils from 10 *F. vulgare* fruit samples of different origins. *Flavour and Fragrance Journal* 14(6), 379–382.

Mitra, J. and Raghu, K. (1998) Detrimental effects of soil residues of DDT on chillies (*C. annuum*). *Fresenius Environmental Bulletin* 7(1/2), 8–13.

Monsalve, R.I. *et al.* (1993) Characterization of a new oriental-mustard (*B. juncea*) allergen, Bra j IE. *Biochemical Journal (London)* 293(3), 625–632.

Mozaffari, F.S. *et al.* (2000) Effect of water stress on the seed oil of *N. sativa. Journal of Essential Oil Research* 12(1), 36–38.

Mozzetti, C. *et al.* (1995) Variation in enzyme activities in leaves and cell suspensions as markers of incompatibility in different *Phytophthora*/pepper interactions. *Physiological and Molecular Plant Pathology* 46, 95–107.

Muchlis, N.A. and Rusli, S. (1978) A review of several problems of essential oil quality. *P.P.T. Industri Bogor* 28(1), 51–58.

Muckensturm, B. *et al.* (1996) Phytochemical and chemotaxonomic studies of *F. vulgare. Biochemical Systematics and Ecology* 25(4), 353–358.

Mukhopadhyay, D. and Sen, S.P. (1997) Augmentation of growth variables and yield components of plants yielding spices by foliar application of diazotropic bacteria. *Indian Journal of Agricultural Research* 31(1), 1–9.

Mulas, M., Perinu, B. and Francesconi, A.H.D. (2000) Regeneration and silvicultural management of myrtle. *Monti e Boschi* 51(1), 45–50.

Nair, K.P.P. *et al.* (1997) The importance of potassium buffer power in the growth and yield of cardamom. *Journal of Plant Nutrition* 20(7/8), 987–997.

Nair, P.K.U. and Sasikumaran, S. (1991) Effect of some fungicides on quick wilt of black pepper. *Indian Cocoa, Arecanut and Spices Journal* 14(3), 95–96.

Nakasone, Y. *et al.* (1999) Evaluation of pungency in the diploid and tetraploid types of ginger and antioxidative activity of their methanol extracts. *Japanese Journal of Tropical Agriculture* 43(2), 71–75.

Nanda, R., Bhargava, S.C. and Mamta, G. (1996) Effect of flowering time on rate of dry matter accumulation, duration of filling period, total oil content of *Brassica* seeds. *Indian Journal of Plant Pathology* 1(2), 88–92.

Nayak, A.K., Rao, G.G. and Chinchmalatpure, A.R. (2000) Conjunctive use of surface water for dill (*A. graveolens*) cultivation in salt affected black soils with saline ground water. *Indian Journal of Agricultural Sciences* 70(12), 863–865.

Nazeem, P.A., Wahid, P.A. and Nair, P.C. (1993) Nutrient deficiency symptoms in clove: visual symptoms. *South India Horticulture* 41(6), 360–365.

Nazeem, P.A. *et al.* (1996) Tissue culture system for *in vitro* pollination and regeneration of plantlets from *in vitro* raised seeds of ginger. *Acta Horticulturae* 426, 467–472.

Neeta Singh (1996) Effect of rate of cooling to liquid nitrogen temperatures on seed viability. *Seed Research* 22(2), 104–107.

Nehra, S.C. *et al.* (1998) Seed yield and quality of coriander as influenced by varieties, spacing, and fertility levels. In: *Proceedings of the National Seminar on Water and Nutrient Management for Sustainable Production of Quality Spices*. Madikeri, Karnatak, India, pp. 73–75.

Nemeth, E. (ed.) (1998) Caraway – the genus *Carum*. In: *Medicinal and Aromatic Plants – Industrial Profiles*. Harwood Academic Publishers, Amersterdam, The Netherlands.

Nemeth, E. and Szekely, G. (2000) Floral biology of medicinal plants. I. *Apiacea* species. *International Journal of Horticultural Science* 6(3), 133–136.

Nemeth, E., Bernath, J. and Petheo, F. (1999) Study on flowering dynamic and fertilization properties of caraway and fennel. *Acta Horticulturae* 502, 77–83.

Nergiz, C. and Otles, S. (1993) Chemical composition of *N. sativa* seeds. *Food Chemistry* 48, 259–261.

Nguyen, T.T. *et al.* (1998) Identification of components of *Illicium griffithii* essential oil from Vietnam. *Journal of Essential Oil Research* 10(4), 433–435.

Nicholls, C.S. (1971) *The Swahili Coast 1798–1856*. Allen & Unwin, London.

Nitta, A. (1993) Nutmeg found in Molucca Islands. *Journal of Japanese Botany* 68(1), 47–52.

Nolan, L.L., McClure, C.D. and Labbe, R.G. (1996) Effect of *Allium* spp. and herb extracts on food-borne pathogens. *Acta Horticulturae* 426, 277–285.

Nybe, E.V. and Nair, P.C.S. (1987) Nutrient deficiency symptoms in black pepper. *Agricultural Research Journal of Kerala* 25(1), 52–65, and 25(2), 132–150.

Nybe, E.V., Sivaraman, A. and Nair, P.C. (1979) Studies on the morphology of ginger types. *Indian Cocoa, Arecanut and Spices Journal* 3(1), 7–13.

Nyrdjanah, N., Hardjo, S. and Mirdna, M. (1991) Distillation method influences the yield and quality of clove leaf oil. *Industrial Crops Research Journal* 3(2), 18–26.

Odoux, E. (2000) Changes in vanillin and glucovanillin concentrations during various stages of the process traditionally used for curing *V. fragrans* beans in Reunion. *Fruits (Paris)* 55(2), 119–125.

Odoux, E. *et al.* (2000) The effect of the nature of the soil amendments on the main aromatic compounds in *Vanilla fragrans*. *Fruits (Paris)* 55(1), 63–71.

Ogbalu, O.K. (1999) The effects of different traditional nutrients on the infestation of pepper fruits by the pepper fly, *Atherigona orientalis* in Nigeria. *Journal of Agronomy and Crop Science* 182(1), 65–71.

Omer, E.A. *et al.* (1997) Effect of some growth regulators on the growth parameters and oil content of cumin with the disease incidence under two types of soil. *Journal of Horticulture* 24(1), 29–41.

Onyenekwe, P.C. (2000) Assessment of oleoresin and gingerol contents in gamma-irradiated ginger rhizomes. *Nahrung* 44(2), 130–132.

Onyenekwe, P.C. and Ogbadu, J.C. (1993) Volatile constituents of the essential oil of *Monodora myristica*. *Journal of the Science of Food and Agriculture* 61, 379–381.

Onyenekwe, P.C. and Hashimoto, S. (1999) The composition of the essential oil of dried Nigerian ginger. *European Food Research Technology* 209(6), 407–410.

Orjala, J. *et al.* (1993) Derivatives with antimicrobial and molluscicidal activity from *P. aduncum* leaves. *Planta Medica* 59(6), 546–551.

da Orta, G. (1563) *Colloquies on the Simples and Drugs of India*. New edition, 1895. Lisbon, Portugal.

Ortuno, A. *et al.* (1998) Distribution and changes of diosgenin during development of *T. foenum-graecum* plants. *Food Chemistry* 63(1), 51–54.

Ortuno, A. *et al.* (1999) Regulation of diosgenin expression in *T. foenum-graecum* plants by different growth regulators. *Food Chemistry* 65(2), 227–232.

Overdieck, R. (1998) Vanilla – a contribution to knowledge and analysis of vanilla (*V. planifolia*). *Deutsche Lebensmittel-Rundschau* 94(2), 53–59.

Owadally, A.L., Ramtohul, M. and Heerasingh, J. (1981) Ginger production and research in cultivation. *Revista Agriculture et Sucre de l'Ile Maurice* 60(3/4), 131–148.

Owusu, E.O., Nkansah, G.O. and Dennis, E.A. (2000) Effect of sources of nitrogen on growth and yield of hot pepper (*C. frutecens*). *Tropical Science* 40(2), 58–62.

Ozek, T., Bozan, B. and Baser, K.H.C. (1998) Supercritical carbon dioxide extraction of volatile components from leaves of *Laurus nobilis*. *Chemistry of Natural Compounds* 6, 746–750.

Padmanabha, K.K. and Rangaswamy, J.R. (1996) Characterisation of interaction products of *alpha*-pinene and linalool with phosphine. *Indian Journal of Chemistry* 35(6), 611–614.

Padmini, K., Venugopal, M.N. and Sasikumar, B. (2000) Performance of hybrids, open pollinated progenies and inbreds of cardamom under nursery conditions. *Indian Journal of Agricultural Sciences* 70(8), 550–551.

PakHeng Young *et al.* (1996) Effect of different day and night temperature regimes on the growth of hot pepper seedlings. *Journal of the Korean Society for Horticultural Science* 37(5), 617–621.

Palau, E. and Torregrosa, A. (1997) Mechanical harvesting of paprika pepper in Spain. *Journal of Agricultural Engineering* 66(3), 195–201.

Pande, V.K., Sonune, A.V. and Philip, S.K. (2000) Solar drying of coriander and methi leaves. *Journal of Food Science and Technology* 37(6), 592–595.

Pandev, Y.R. *et al.* (1997) *In vitro* propagation of ginger. *Kasetsart Journal of Natural Science* 31(1), 10–19.

Pank, F., Neumann, M. and Kruger, H. (2000) Results of the selection of bitter fennel with small shaped fruits. *Zeitschrift für Arznei-Gewurz pfalzen* 5(1), 40–48.

Pappas, A.C. and Elena, K. (1997) Occurance of *F. oxysporum* f.sp. *cumini* in the island of Chios, Greece. *Journal of Phytopathology* 145(5/6), 271–272.

Parish, R.L. and Bracy, R.P. (1995) Rolling beds after precision seeding enhances bed integrity at harvest. *Journal of Vegetable Crop Production* 1(2), 41–50.

Parry, J.W. (1965) *Spices: Their Morphology, Histology and Chemistry.* Chemical Publishing Co., New York.

Patil, S.P. *et al.* (1997) Allergy to fenugreek. *Annals of Allergy, Asthma and Immunology* 78(3), 297–300.

Pearson, M.N. (1996) *Spices of the Indian Ocean*, Vol. 2. Variorum, Amsterdam.

Pearson, M.N., Kershaw, S. and Gupta, D. (1999) Effects of two naturally occuring mild polyviruses on the symptom expression of vanilla necrosid potyvirus in *V. fragrans. New Zealand Journal of Crop and Horticultural Science* 27(4), 325–330.

Perucka, I. and Oleszek, W. (2000) Extraction and determination of capsiacinoids in hot pepper *C. annuum. Food Chemistry* 71(2), 287–291.

Peterson, L.E., Clarke, R.J. and Menery, R.C. (1993) Umbel initiation and stem elongation in fennel (*F. vulare*) initiated by photoperiod. *Journal of Essential Oil Research* 5(1), 37–43.

Pett, B. and Kembu, A.B. (1999) Factors influencing vanilla mass propagation. *Pacific Regional Agricultural Programme* 7, 25–27.

Piccaglia, R. (1998) Aromatic plants: a world of flavouring compounds. *Agro Food Industries Hi-Tech* 9(3), 12–15.

Piccaglia, R. and Marotti, M. (2001) Characterization of some Italian types of wild fennel (*Foeniculum vulgare*). *Journal of Agricultural and Food Chemistry* 49(1), 239–244.

Pico, B. and Neuz, F. (2000) Minor crops of Mesoamerica in early sources (II); herbs as condiments. *Genetic Resources of Crop Evolution* 47(5), 541–552.

Pillai, P.K.T., Vigaykumar, G. and Namhair, B. (1978) Flowering behaviour, cytology and pollen germination in ginger. *Journal of Plantation Crops* 6(1), 12–13.

Pilone, N. (1990) Variation in the potential for natural rhizogenisis in capers. *Informatare Agrario* 46(13), 69–70.

Pino, J.A. (1990) Chemical and sensory properties of black pepper oil (*P. nigrum*). *Nahrung* 34(6), 555–560.

Pino, J.A. (1999a) Volatile components of spices: dill. *Alimentaria* 301, 102–103.

Pino, J.A. (1999b) Volatile components of spices: fennel. *Alimentaria* 301, 95–98.

Pino, J.A. (1999c) Volatile components of spices: clove. *Alimentaria* 301, 71–74.

Pino, J.A. and Borges, P. (1999a) Volatile components of spices: bay laurel. *Alimentaria* 301, 67–70.

Pino, J.A. and Borges, P. (1999b) Volatile components of spices: coriander. *Alimentaria* 301, 75–78.

Pino, J.A. and Borges, P. (1999c) Volatile components of spices: black pepper. *Alimentaria* 301, 47–53.

Pino, J.A. and Borges, P. (1999d) Volatile components of spices: cumin. *Alimentaria* 301, 63–66.

Pino, J.A. and Borges, P. (1999e) Volatile components of spices: nutmeg. *Alimentaria* 301, 55–61.

Pino, J. and Rosado, A. (1996) Chemical compositiion of the leaf oil of *Pimenta dioica* L. from Cuba. *Journal of Essential Oil Research* 8, 331–332.

Plucknett, D.L. (1978) Cassia – a tropical oil crop. *Hawaiian Agric. Expt. Sta. Journal,* Series 2345, Hawaii.

Pohlan, J. and Heynoldt, J. (1986) Investigations on cultivation of pepper in Vietnam. *Beit. Trop. Land. Veter.* 24(1), 13–24.

Poornima, K. and Sivagami, V. (1999) Occurance and seasonal population behaviour of phytonematodes in turmeric. *Pest Management of Horticultural Ecosystems* 5(1), 42–45.

Popoola, A.V. *et al.* (1994) Extraction, spectroscopic and colouring potential studies of the dye in ginger. *Pakistan Journal of Scientific and Industrial Research* 37(5), 217–220.

Potter, T.L. (1996) Essential composition of cilantro. *Journal of Agricultural and Food Chemistry* 44(7), 1824–1826.

Potter, T.L. and Ferguson, I.S. (1990) Composition of coriander leaf volatiles. *Journal of Agricultural and Food Chemistry* 38, 2054–2056.

Potter, T.L., Ferguson, I.S. and Craker, L.E. (1993) Composition of Vietnamese coriander leaf oil. *Acta Horticulturae* 344, 305–311.

Prabha, P. and Bohra, A. (1999) Seed mycoflora associated with some important spice seeds. *Advances in Plant Science* 12(1), 195–198.

Prabhakaran, P.V. (1997) Quantitative determination of loss of yield in black pepper (*P. nigrum*) in Kannuu District, Kerala, India. *Journal of Spices and Aromatic Crops* 6(1), 31–36.

Prakashchandra, K.S. and Chandrasekharappa, G. (1984) Lipid profile and fatty acid composition of fat extracted from arils (mace) of *M. fragrans* and *A. hirsutus*. *Journal of Food Science and Technology (India)* 21(1), 40–42.

Prasad, B.K. *et al.* (2000) Decay of chilli fruits in India during storage. *Indian Pathology* 53(1), 42–44.

Pratibha, Y. *et al.* (1999) Antidermatophytic activity of essential oil of *Cinnamomum tamala*. *Journal of Medical and Aromatic Plant Science* 21(2), 347–351.

Premanand, D., Samuel, R. and Das, A.B. (1999) Genetic advances, heritability and path analysis in ginger. *Journal of Plantation Crops* 27(1), 27–30.

Primo, P. and Cappelli, C. (2000) Preliminary characterization of *Fusarium oxysporum* f.sp *gladioli* causing corm rot of saffron in Italy. *Plant Disease* 84(7), 806.

Prithi, J.S. (1998) *Major Spices Of India*. Indian Council of Agricultural Research, Pusa, New Delhi, India, pp. 159–175.

Pruthy, J.S. (1993) Technological innovations in pepper technology; super-critical fluid extraction. *International Pepper News Bulletin* 17(3), 14–24.

Purseglove, J.W. (1981a) *Spices*, Vol. 1, pp. 101–125; (1981b) Vol. 2, pp. 174–228; (1981c) Vol. 2, pp. 447–531; (1981d) Vol. 2, pp. 532–643. Longmans, Harlow, UK.

Putievsky. E. (1983) Effects of daylength and temperature on growth and yield components of three seed spices. *Journal of Horticultural Science* 58(2), 271–275.

Putievsky, E. *et al.* (1984) The essential oils from cultivated bay laurel. *Israel Journal of Botany* 33(1), 47–52.

Putnam, G.G. (1924) *Salem Vessels and Their Voyages*. The Essex Institute, Salem, Massachusetts.

Quick, C.R. (1961) How long can seeds remain alive? In: *Seeds: Yearbook of Agriculture 1961*. US Department of Agriculture, Washington, DC, pp. 94–96.

Quilitzsch, R. and Pank, F. (1996) Estimation of fruit ripeness in fennel and caraway by colour measurements. *Bundenanstalt für Zuchtungforschund und Kulturpflazen* 2(1), 404–407.

Rai, B. (1995) Development of hybrid cultivars in oilseed brassicas; status and strategies. *Journal of Oilseed Research* 12(2), 239–244.

Rai, S. and Hossain, M. (1998) Comparative studies of three traditional methods of seed rhizome storage of ginger practiced in Shikkim and Darjeeling hills. *Environmental Ecology* 16(1), 34–36.

Rai, S., Das, A.B. and Das, P. (1997) Estimation of 4C DNA and karyotype analysis in ginger. *Cytologia* 62(2), 133–141.

Rai, S., Das, A.B. and Das, P. (1999) Variations in chlorophylls, carotenoids, protein and secondary metabolites amongst ginger cultivars and their association with rhizome yield. *New Zealand Journal of Crop and Horticultural Science* 27(1), 79–82.

Ramana, K.V. and Eapen, S.J. (1999) Nematode pests of spices and their management. *Indian Journal of Arecanut Spices and Medicinal Plants* 1(4), 146–153.

Ramarosov-Raonizafinimanana, B., Gaydoux, E.M. and Bombarda, I. (1998) 4-Demethylsterols and triterpene alcohols from two vanilla species (*V. fragrans* and *V. tahitensis*). *Journal of the American Oil Chemists' Society* 75(1), 51–55.

Ramarosov-Raonizafinimanana, B., Gaydoux, E.M. and Bombarda, I. (2000) Long-chain aliphatic beta-diketones from epicuticular wax of Vanilla bean species. *Journal of Agricultural and Food Chemistry* 48(10), 4739–4743.

Rana, K.S. (1997) Development of rhizome rot of ginger during storage. *Journal of Mycology and Plant Pathology* 27(2), 221–222.

Randhawa, G.S. *et al.* (1996a) Agronomic technology for production of fenugreek (*T. foenum-graecum*) seeds. *Journal of Herbs, Spices and Medicinal Plants* 4(3), 43–49.

Randhawa, G.S. *et al.* (1996b) Effect of plant spacings and nitrogen levels on the seed yield of dillseed. *Acta Horticulturae* 426, 623–628.

Rangappa, M. *et al.* (1997) Cilantro response to nitrogen fertilizer rates. *Journal of Herbs, Spices and Medicinal Plants* 5(1), 63–68.

Rao, D.V.R. (1999) Effect of gamma irradiation on growth, yield and quality of turmeric. *Advances in Horticulture and Forestry* 6, 107–110.

Rapisarda, A., Iauk, L. and Ragus, S. (1998) Biometric analysis of the fruits of subspecies and varieties of *F. vulgare*. *Rivista Italia EPPOS* 24, 19–22.

Ravender, S. and Kundu, D.K. (2000) Soil salinity effect on germination of wheat, castor, safflower, and dill seed in Gujarat. *Indian Journal of Agricultural Science* 70(7), 459–460.

Ravindran, P.N. and Babu, K.N. (1994) Chemotaxonomy of south Indian pepper. *Journal of Spices and Aromatic Crops* 3(1), 6–13.

Ravindran, P.N. and Peter, K.V. (1995) Biodiversity of major spices and their conservation in India. In: *Proceedings of the South-East Asian National Co-ordinators Meeting on Plant Genetic Resources*. National Research Centre for Spices, Marikunnnu, Calicut, Kerala, India, pp. 123–134.

Reddy, B.N., Sivaraman, K. and Savandan, A.K. (1991) High plant density approach to boost black pepper production. *Indian Cocoa, Arecanut and Spices Journal* 15(4), 35–36.

Reddy, K. and Rolston, M.P. (1999) Coriander seed production: nitrogen, row spacing, sowing rate and time of sowing. *Journal of Applied Seed Production* 17, 49–53.

Remashree, A.B. *et al.* (1997) Histological studies on ginger rhizome. *Phytomorphology* 47(1), 67–75.

Remashree, A.B. *et al.* (1998) Developmental anatomy of ginger rhizomes: ontogeny of buds, roots, and phloem. *Phytomorphology* 48(2), 155–166.

Remashree, A.B. *et al.* (1999) Development of oils cells and ducts in ginger. *Journal of Spices and Aromatic Crops* 8(2), 163–170.

Remaud, G.S. *et al.* (1997) Detection of sophisticated adulterations of natural vanilla flavours and extracts. *Journal of Agricultural and Food Chemistry* 45(3), 859–866.

Rethinam, P. and Sandandan, K. (1994) Nutrition and management of seed spices. *Advances in Horticulture* 9(1), 499–513.

Rhizopoulous, S. (1990) Physiological responses of C. *spinosa* to drought. *Journal of Plant Physiology* 136(3), 341–348.

Rodrigo, M. *et al.* (1992) Composition of capers (C. *spinosa*); influence of cultivar, size and harvest date. *Journal of Food Science* 57(5), 1152–1154.

Rohricht, C., Manicke, S. and Kohler, A. (1999) Effect of soil and fertilizer on coriander. *Zeitschrift für Arznei-Gewurz pfalzen* 4(2), 75–78.

Rosengarten, F. (1969) *The Book of Spices*. Livingstone Publishers, Wynewood, USA.

Rovio, S. *et al.* (1999) Extraction of clove using pressurized hot water. *Flavour and Fragrance Journal* 14(6), 399–404.

Rumi, K. *et al.* (1999) Studies on curcumin and essential content of different cultivars of turmeric grown in Manipur. *Indian Journal of Arecanut, Spices and Medicinal Plants* 1(3), 91–92.

Rumpel, J. and Kaniszewski, S. (1998) Outdoor soilless culture of vegetables; status and prospects. *Journal of Vegetable Crop Production* 4(1), 3–10.

Rushenberger, W.S.W. (1838) *A Voyage Round the World*. Carey, Lee & Blanchard, Philadelphia, USA.

Rusli, S. and Nurdjannah, N. (1993) Handling and processing of pepper. *Indonesian Agricultural Research Development Journal* 15(2), 38–43.

Rusli, S., Laksmanahardja, P. and Simarmata, J.P. (1979) Distillation methods affecting yield and quality of clove leaf oil. *P.P.T. Industri. Bogor* 34(7–9), 1–12.

Sairam, C.V. *et al.* (1997) Capital requirements for adoption of coconut based inter cropping systems in Kerala. *Indian Coconut Journal (Cochin)* 27(10), 2–4.

Salem, M.L. and Hossain, M.S. (2000) Protective effect of black seed oil from N. *sativa* against murine cytomegalvirus infection. *International Journal of Immunopharmacology* 22(9), 729–740.

Sallal, A.K. and Alkofahi, A. (1996) Inhibition of the haemolytic activities of snake and scorpion venoms in *in vitro* plant extracts. *Biomedical Letters* 53(212), 211–215.

Samuel, M.R. and Bavappa, R.V.A. (1981) Chromosome numbers in the genus *Piper*. *Current Science (India)* 50(4), 197–198.

Sandanandan, A.K. and Hamza, S. (1998) Effect of organic farming on nutrient uptake, yield and quality of ginger. In: *Proceedings of the National Seminar on Water and Nutrient Management for Sustainable Production of Quality Spices*. Madikeri, Karnataka, India, pp. 89–94.

Sandanandan, A.K. *et al.* (1993) Effect of coconut–pepper mixed cropping on soil fertility and crop productivity. In: *Advances in Coconut Research and Development*. Oxford University Press, London.

Sandanandan, A.K., Kandiannan, K. and Hamza, S. (1997) Soil nutrients and water management for sustainable spice production. In: *Proceedings of the National Seminar on Water and Nutrient Management for Sustainable Production of Quality Spices*. Madakiri, Karnataka, India, pp. 12–20.

Sandanandan, A.K., Kandiannan, K. and Hamza, S. (1998) Soil nutrients and water management for sustainable spice production. In: *Proceedings of the National Seminar on Water and Nutrient Management for Sustainable Production of Quality Spices*. Madikeri, Karnataka, India, pp. 12–20.

Sandford, K.J. and Heinz, D.E. (1971) Effects of storage on the volatile composition of nutmegs. *Phytochemistry* 10, 1245–1250.

Sanjay, M., Kumar, A. and Kishore, V.V. (1999) A study of large-cardamom curing chambers in Sikkim. *Biometrics and Bioenergy* 16(6), 463–473.

Sanjeev, A. and Sharma, R.K. (1999) Effect of storage of seeds of fennel, coriander and coriander powder on quality. *Indian Journal of Arecanut Spices and Medicinal Plants* 1(2), 63–64.

Sarath-Kumara, S.J., Jansz, E.R. and Dharmadasa, H.M. (1985) Some physical and chemical characteristics of Sri Lankan nutmeg oil. *Journal of the Science of Food and Agriculture* 36, 93–100.

Saravanakumar, M., Panicker, S. and Usha, R. (1998) Partial characterisation and early detection of a virus associated with necrosis of small cardamom in south India. *Indian Journal of Virology* 14(1), 59–63.

Sarma, Y.R. and Anandaraj, M. (2000) Diseases of spice crops and their management. *Journal of Arecanut Spices and Medicinal Plants* 2(1), 6–20.

Sato, K. *et al.* (1999) Direct connection of supercritical fluid extraction and supercritical fluid chromatography as a rapid quantitative method for capsaicinoids in placentas for *Capsicum. Journal of Agricultural and Food Chemistry* 47(11), 4665–4668.

Satyendra, G., Arun, S. and Thomas, P. (1998) Improved bacterial turbidimetric method for detection of irradiated spices. *Journal of Agricultural and Food Chemistry* 46(12), 5110–5112.

Schiff, S., Anzidei, M. and Bennici, A. (1996) Cell culture, calogenesis and morphogenesis of *F. vulgare in vitro. Instituto Sperimentale per l'Assestamento Forestale e per l'Alpicoltara* 1996, 523–525.

Schmid, R. (1972) A resolution of the *Eugenia–Syzygium* controversy (*Myrtaceae*). *American Journal of Botany* 59, 423–436.

Schreiber, L. and Reiderer, M. (1996a) Determination of diffusion of octadecanoic acid in isolated cuticular waxes and their relationship to cuticular water permability. *Plant Cell and Environment* 19(9), 1075–1082.

Schreiber, L. and Reiderer, M. (1996b) Ecophysiology of cuticular transpiration. *Oecologia* 107(4), 426–432.

Schulz, H. *et al.* (2000) Estimation of minor components in caraway, fennel and carrots by NIRS. *International Agrophysics* 14(2), 249–253.

Sedlakova, J., Kocourkova, B. and Kuban, V. (2001) Determination of essential oils content and composition in caraway (*Carum carvi*). *Czech Journal of Food Science* 19(1), 31–36.

Senanayake, U.M. and Wijesekera, R.O.B. (1990) The volatiles of the *Cinnamomum* species. In: *Proceedings of the 11th Congress on Essential Oils, Fragrances and Flavours,* Vol. 4. Aspect Publishing, London, pp. 103–120.

Seneviratne, K.G.S. *et al.* (1985) Influence of shade on rooting and growth of black pepper. *Journal of Plantation Crops* 13(1), 41–43.

Setia, R.C. *et al.* (1996) Influence of paclobutrazol on growth and development of fruit. *Plant Growth Regulation* 20(3), 307–316.

Sewell, A.C., Mosandi, A. and Bohles, H. (1999) False diagnosis of maple syrup urine disease owing to ingestion of herbal tea. *New England Journal of Medicine* 341(10), 769.

Shah, J.J. and Raju, E.C. (1975) Ontogeny of the shoot apex of ginger. *Norwegian Journal of Botany* 22(3), 227–236.

Shah, N.C. (1997) Traditional uses of turmeric. *Journal of Medicinal and Aromatic Plant Sciences* 19(4), 948–954.

Shanthaveerabhadraiah, S.M. *et al.* (1997) Effect of growth regulators and spacing on multiplication of planting units in cardamom. *Journal of Spices and Aromatic Crops* 6(1), 49–50.

Shashidhara, G.B. *et al.* (1998) Effect of organic and inorganic fertilizers on growth and yield of chilli. In: *Proceedings of the National Seminar on Water and Nutrient Management for Sustainable Production of Quality Spices.* Madikeri, Karnataka, India, pp. 56–91.

Sheffield, F.M.L. (1950) The clove tree of the Seychelles. *East African Agricultural and Forestry Journal* 16, 3–8.

Shekhawat, J.S. (1999) Flower and fruit development in *C. decidua. Arid Zone Research Association of India* 1999, 383–386.

Sheoran, R.S., Pannu, R.K. and Sharma, H.C. (1999) Influence of sowing time and phosphorus fertilization on yield attributes of fenugreek (*T. foenum-graecum*). *Indian Journal of Arecanut Spices and Medicinal Plants* 1(1), 15–18.

Sherlija, K.K., Unnikrishnan, K. and Ravindran, P.N. (1998) Bud and root development of turmeric rhizome. *Journal of Spices and Aromatic Crops* 8(1), 49–55.

Shih, Y.C. *et al*. (1999) Effect of seedling age on dry matter and nitrogen content of transplanted chilli pepper. *Journal of the Chinese Society of Horticultural Science* 45(2), 263–272.

Shikiev, A.Sh., Abbosov, R.M. and Mamedova, Z.A. (1978) Composition of *Myrtus communis* essential oil. *Chemical Abstracts* 89, 176379.

Siddagangaiah *et al*. (1996) Standardisation of rooting media for propagation of vanilla (*V. planifolia*). *Journal of Spices and Aromatic Crops* 5(2), 131–133.

Simandi, B. *et al*. (1999) Supercritical carbon dioxide extraction and fractionation of fennel oil. *Journal of Agricultural and Food Chemistry* 47(4), 1635–1640.

Singh, A.K. *et al*. (2000) Reaction of ginger germplasm to *P. zingiberi* under field conditions. *Indian Phytopathology* 53(2), 210–212.

Singh, D.K. *et al*. (2000) Control of bacterial wilt of ginger with antibiotics. *Journal of Research of Birsa Agricultural University* 12(1), 41–43.

Singh, R.P. *et al*. (1995) Effect of different levels of nitrogen, phosphorus and potash on aphid infestation and yield of mustard. *Indian Journal of Entomology* 57(1), 18–21.

Singh, Y.N. and Blumenthal, M. (1997) Kava – an overview. *Herbal-Gram* 39, 33–55.

Sitepu, D. (1993) Disease management of pepper. *Indian Agricultural Research and Development Journal* 15(2), 31–37.

Sivaraman, K. and Palaniappan, S.P. (1996) Turmeric, maize and onion intercropping systems: PAR interception. *Journal of Spices and Aromatic Crops* 5(2), 139–142.

Small, E. (1996) Confusion of common names for toxic and edible star anise (*Illicium* spp.). *Economic Botany* 50(3), 337–339.

Smallfield, B.M. *et al*. (2001) Coriander spice oil – effect of fruit crushing and distillation time on yield and composition. *Journal of Agricultural and Food Chemistry* 49(1), 118–123.

Smith, M.K. and Hamill, S.D. (1996) Field evaluation of micropropagated and conventionally propagated ginger in subtropical Queensland. *Australian Journal of Experimental Agriculture* 36, 347–354.

Somogyi, N. (2000) The importance of international cooperation in the Hungarian spice paprika breeding programme. *Acta Horticulturae* 524, 251–254.

Somogyi, N., Pek, M. and Mihaly, A. (2000) Applied spice paprika (*C. annuum* var. *longum*) growing technologies and processing in Hungary. *Acta Horticulturae* 536, 389–396.

Sontakke, B.K. (2000) Occurance, damage and biological observations on rhizome fly *Mimegralla coeruleifrons* infesting ginger. *Indian Journal of Entomology* 62(2), 146–149.

Sozzi, G.O. and Chiesa, A. (1995) Improvement of caper (*C. spinosa*) seed germination by breaking seed-coat induced dormancy. *Scientia Horticulturae* 62(4), 255–261.

Sreasil, Z. (1997) Content of oil and individual fatty acids in some species of alternative oil-bearing crops. *Rostlinna Vyroba* 43(2), 59–64.

Sreenarayanan, V.V. and Viswanathan, R. (1989) An improved turmeric boiler. *Spices of India* 2(10), 15–16.

Srinivasan, K. *et al*. (1999) Mixed cropping systems in cardamom – an analysis. *Journal of Spices amd Aromatic Crops* 8(1), 63–66.

Sritharan, R., Jacob, V.J. and Balasubramaniam, S. (1994) Thin layer chromatographic analysis of essential oils from *Cinnamomum* species. *Journal of Herbs, Spices and Medicinal Plants* 2(2), 49–63.

Srivastava, L.S. *et al*. (1998) Micro-organisms associated with ginger in Sikkim. *Journal of Hill Research* 11(1), 120–122.

Sthapit, V.M. and Tuladhar, P.M. (1993) Sugandha kokila, *C. cecidodaphne* oil. *Journal of Herbs, Spices and Medicinal Plants* 1(4), 31–35.

Stobart, T. (1977) *Herbs, Spices and Flavourings*. Penguin Books, Harmondsworth, Middlesex, UK.

Stone, S.M.H. (2000) Economic, ecological and social implications of betel production, consumption and market system on Yap, Federated States of Micronesia. *TRI News* 19, 20–24.

Strasil, Z. (1997) Content of oil and individual fats in some species of alternative oil-bearing crops. *Rostlinna Vyroba* 43(1), 59–64.

Suchandra, C. *et al*. (2000) Effect of gamma-irradiation on the volatile oil constituents of turmeric. *Food Research International* 33(2), 103–106.

Suh M.-C. *et al*. (1999) Isoforms of acyl carrier protein in seed-specific fatty acid synthesis. *Plant Journal* 17(6), 679–688.

Sumathikutty, M.A. *et al*. (1989) Quality criteria of pepper grades from pure cultivars. *Indian Perfumer* 33(2), 147–150.

Sunitibala, H., Damayanti, M. and Sharma, G.J. (2001) *In vitro* propagation and rhizome formation in *Curcuma longa* Linn. *Cytobios* 105(409), 71–82.

Suryanto, S. and Lambert, K. (1994) Amelioration of tropical deep peat for lowland vegetable production. *Acta Horticulturae* 369, 455–468.

Sushi Kumar *et al.* (2001) Higher yields and profits from new crop rotations permitting integration of mediculture and agriculture in the Indo-Gangetic plains. *Current Science* 80(4), 563–566.

Sushi Sharma (1999) Effect of sowing dates on powdery mildew of fenugreek. *Journal of Mycology and Plant Pathology* 29(1), 144–145.

Svendsen, A.B. and Scweffer, J.J.C. (1985) *Essential Oils and Aromatic Plants.* Martinus Nijhoff/Dr W. Junk, Dordrecht, The Netherlands.

Takruri, H.R. and Damah, M.A. (1993) Study of the nutritional value of black cumin seeds (*N. sativa*). *Journal of the Science of Food and Agriculture* 76, 404–410.

Tansi, S. (1999) Propagation methods for caper. *Agricoltura Mediterranea* 129(1), 45–49.

Taranalli, A.D. and Kuppast, I.J. (1996) Study of wound healing activity of *T. foenum graecum* in rats. *Indian Journal of Pharmacological Science* 58(3), 117–119.

Tawfik, A.A. (1998) Plant regeneration in callus culture of cumin. *Acta Horticulturae* 457, 389–393.

Thakral, K.K., Singh, G.R. and Bhatia, A.K. (1997) Split application of nitrogen in coriander seed crop. In: *Proceedings of the National Seminar on Water and Nutrient Management for Sustainable Production of Quality Spices.* Madakeri, Karnataka, India, pp. 68–72.

Theilade, I. (1996) Revision of the genus *Zingerber* in Peninsular Malaysia. *The Gardeners Bulletin* 48(1/2), 207–236.

Theilade, I. *et al.* (1993) Pollen morphology and structure in ginger. *Grana* 32(6), 338–342.

Thomas, B.V., Schreiber, L. and Weisskopf A.A. (1998) Simple method for quantitation of capsiaciniodes in pepper using capillary gas chromatography. *Journal of Agricultural and Food Chemistry* 46(7), 2655–2663.

Tidbury, G.E. (1949) *The Clove Tree.* Crosby Lockwood, London.

Tikhonova, V.L. and Kruzhalina, T.N. (1997) The effect of deep freezing seeds on growth and development of some medicinal plants. *Phytochemistry* 46(8), 1313–1317.

Toben, H.M. and Rudolphi, K. (1996) *P. syringae* pv. *coriandricola* incitant of bacterial umbel blight and seed decay of coriander in Germany. *Journal of Phytopathology* 144(4), 169–178.

Todorova, V., Pevicharova, G. and Todorov, Y. (1999) Total pigment content in red pepper cultivars for grinding. *Capsicum and Eggplant Newsletter* 18, 25–27.

Tomar, S.S. and Gautam, D.S. (1995) Economic attributes of inputs on yield of mustard. *Gujarat Agricultural University Research Journal* 21(1), 157–160.

Tsang, M.M. and Shintaku, M. (1998) Hot air treatment for the control of bacterial wilt of ginger. *Applied Engineering in Agriculture* 14(2), 159–163.

Tsuda, M. (1989) The production and diseases of pepper in Brazil. *Agriculture and Horticulture* 64(10), 1159–1166.

Tucker, A.O. and Maciarello, M.J. (1999) Volatile oils of *I. floridanum* and *I. parviflora* of the south-eastern United States and their potential economic utilization. *Economic Botany* 53(4), 435–438.

Tucker, A.O. *et al.* (1991) Volatile leaf oils of Caribbean *Myrtaceae*: three varieties of *P. racemosa* of the Dominican Republic. *Journal of Essential Oil Research* 3(5), 323–329.

UNCTAD/GATT (1974) *Essential Oils and Oleoresins.* Geneva, Switzerland.

UNCTAD/GATT (1982) *Spices* (2 vols). Geneva, Switzerland.

UNCTAD/WTO (1998) *Spices*, Vols 1 and 2. Geneva, Switzerland.

Vadhera, I., Tiwari, S.P. and Dave, G.S. (1998) Plant parasitic nematodes associated with ginger (*Z. officinale*) in Madhya Pradesh. *Indian Journal of Agricultural Sciences* 68(7), 376–370.

Variyar, P.S., Gholap, A.S. and Thomas, P. (1997) Effect of gamma-irradiation on the volatile oil constituents of fresh ginger rhizome. *Food Research International* 30(1), 41–43.

Varma, O.P., Singh, P.V. and Karan, S. (1995) Effect of storage conditions on germination of Indian mustard seeds. *Seed Research* 21(2), 117–118.

Vasanthakumar, K. and Mohanakumaran, N. (1989) Nutrient status of cardamom at different stages of maturity. *Spices of India* 2(11), 17–23.

Vasanthakumar, K. and Sheels, V. (1990) Sprinkler irrigation in cardamom plantations. *Spices of India* 3(3), 3–9.

Ved, D.K. *et al.* (1998) Ecodistribution mapping of the priority medicinal plants of south eastern India. *Current Science* 75(3), 205–210.

Velayodhan, K.C., Amalraj, V.A. and Muralidharan, V. (1996) The conspectus of the genus *Curcuma* in India. *Journal of Economic and Taxonomic Botany* 26(2), 375–382.

Velisek, J. *et al.* (1995) Chemometric investigation of mustard seed. *Potravinarske Vedy* 13(1), 1–12.

Veloso, C.A.C. and Carvalho, E.J.M. (1999) Nutrient uptake and extraction by black pepper. *Scientia Agricola* 56(2), 443–447.

Venkatesha, J. *et al.* (1997) Effect of method of storage on the viability of seed rhizome in turmeric. *Current Research (Bangalore)* 26(6/7), 114–115.

Venkatesha, J., Khan, M.M. and Farooqi, A.A. (1998) Effect of major nutrients (NPK) on growth, yield and quality of turmeric cultivars. In: *Proceedings of the National Seminar on Water and Nutrient Management for Sustainable Production of Quality Spices*. Madikeri, Karnataka, India, pp. 52–58.

Venugopal, M.N. (1999) Natural disease escapes as sources of resistance to cardamom mosaic virus causing katte disease of cardamom. *Journal of Spices and Aromatic Crops* 8(2), 145–151.

Venugopal, M.N. and Padmini, K. (1999) Studies on the crossability between elite types of cardamom. *Journal of Spices and Aromatic Crops* 8(2), 185–187.

Verma, T.S. and Joshi, S. (2000) Status of paprika development in India. *Indian Journal of Arecanut Spices and Medicinal Plants* 2(2), 39–44.

Verman, O.P., Singh, P.V. and Karan Singh (1955) Effect of storage conditions on germination of Indian mustard seed (*B. junceae*). *Seed Research* 21(2), 117–118.

Vernin, C. *et al.* (1990) La canelle, Premiere partie. Analyse d'huile essentielle de canelle de Ceylan et de Chine. *Parfume, Cosmetics et Aromes* 93, 85–90.

Vernin, G. and Parkanyi, C. (1994) Ginger oil (*Zingiber officinale* Roscoe). In: Charalambous, G. (ed.) *Spices, Herbs and Edible Fungi*. Elsevier Science, Amsterdam, The Netherlands, pp. 579–594.

Vernin, G. *et al.* (1994) GC/MS analyses of clove essential oils. In: Charalambous, G. (ed.) *Spices, Herbs and Edible Fungi*. Elsevier, Amsterdam.

Vinay, S. and Bisen, R.K. (1999) Response of nitrogen and phosphorus on seed crop of coriander. *Environmental Ecology* 17(1), 238–239.

de Waard, P.W.F. (1964) Pepper cultivation in Sarawak. *World Crops* 16(3), 24–30.

Wahid, P. and Usman, K. (1984) Effect of organic and inorganic fertilizers on growth of young clove trees. *P.P.T. Industri. Bogor.* 9(50), 11–17.

Wahid, P. and Zaubin, R. (1993) Crop improvement and cultivation of black pepper. *Indonesian Agricultural Research and Development Journal* 15(2), 27–30.

Wahid, P.H. (1977) Fertilizer trials on mature clove trees. *P.P.T. Industri. Bogor.* 25, 27–37.

Wake, C.H.H. (1979) The changing pattern of Europe's spice imports ca. 1400–1700. *Journal of European Economic History* V111, 361–403.

Wander, J.G.N. (1997) Improving the harvest reliability and quality of plant-producing carvone. *PAV Bulletin Akkerbou*, 11–14 February.

Wander, J.G. and Boumeester, H.J. (1997) Effects of nitrogen fertilization on dill seed and carvone production. *Industrial Crops and Products* 7(2/3), 211–216.

Watt, G. (1908) *The Commercial Products of India*. John Murray, London.

Weiss, E.A. (1966) Sunflower trials in Western Kenya. *East African Agricultural and Forestry Journal* 31(4), 405–408.

Weiss, E.A. (1997) *Essential Oil Crops*. CAB International, Wallingford, UK.

Weiss, E.A. (2000) *Oilseeds*, 2nd edn. Blackwell Science, Oxford, UK.

Werkhoff, P. and Guntert, M. (1997) Identification of some ester compounds in bourbon vanilla beans. *Lebensmittel-Wissenschaft und Technologie* 30(4), 429–431.

Whiley, A.W. (1981) Effect of plant density on time to first harvest maturity etc. of ginger grown in S.E. Queensland. *Tropical Agriculture (Trin)* 58(3), 245–251.

Wijesekera, R.O.B. *et al.* (1976) Cinnamon. *Natural Products Monograph* No.1. ISIR, Colombo, Sri Lanka.

Willems, W.B. (1971) *Report on Clove Production, Marketing and Processing in Indonesia*. London.

Winterhalter, P. and Straubinger, M. (2000) Saffron – renewed interest in an ancient spice. *Food Reviews International* 16(1), 39–59.

Woolf, A. (1999) Essential oil poisoning. *Clinical Toxicology* 37(6), 721–727.

Wu, J.J. and Yang, J.S. (1994) Effects of irradiation on the volatile compounds of ginger rhizomes. *Journal of Agricultural and Food Chemistry* 2(11), 2574–2577.

Xu, K. (1999) The influence of mulching with straw on the field microclimate and ginger growth. *China Vegetables* 2, 15–17.

Xu, K. and Kang L.-M. (1999) Effect of growth period on yield of ginger. *China Vegetables* 4, 30–31.

Xu, K. and Li, M.-G. (2000) Effects of mulching with straw on the photosynthetic characteristics of ginger. *China Vegetables* 2, 18–20.

Xu, K. and Xu, F. (1999) Effect of nitrogen on the growth and yield of ginger. *China Vegetables* 6, 12–14.

Yadav, A.C., Batra, B.R. and Malik, Y.S. (1997) Irrigation requirement of fennel. In: *Proceedings of the National Seminar on Water and Nutrient Management for Sustainable Production of Quality Spices.* Madikeri, Karnataka, India, pp. 136–140.

Yadav, A.C., Lal, S. and Mangal, J.L. (2000) Salinity effects on growth, flowering and yield of coriander. *Crop Research (Hisar)* 59(3), 169–174.

Yadav, H.D. *et al.* (1996) Response of winter spices to sodic water irrigation in light textured sodic soil. *Harayana University Journal of Research* 26(1), 51–55.

Yadav, J.P. and Blyth, J.B. (1996) Combined productivity of *Prosopis cinararia*–mustard agro silvicultural practice. *International Tree Crops Journal* 9(1), 47–58.

Yadav, R.K. (1999a) Genetic variability in ginger. *Journal of Spices and Aromatic Crops* 8(1), 81–83.

Yadav, R.K. (1999b) Variability in a collection of fenugreek (*T. foenum-graecum*) germplasm. *Journal of Spices and Aromatic Crops* 8(2), 217.

Yodpetch, C. (2000) Study on the optimum fertilizer rate on yield and quality of three long cayenne peppers. *Kesetsart Journal of Natural Science* 32(5 suppl.), pp. 37–45.

Yoshikawa, M. *et al.* (1997) Medicinal foodstuffs; fenugreek seed. *Chemical and Pharmacological Bulletin* 45(1), 81–87.

Yothasiri, A. *et al.* (1997) Effect of types and sizes of seed rhizomes on growth and yield of turmeric. *Kasetsart Journal of Natural Sciences* 31(1), 10–19.

Yu, K., Zhao, D.W. and Jiang, X.M. (1993) Studies on the nitrogen absorption rate of ginger using isotopes. *Acta Horticulturae Sinica* 20(2), 161–165.

Yu, X.-C. *et al.* (1996) Studies on the relationship between population, canopy photosynthesis and yield formation in ginger. *Journal of Shandong Agricultural University* 27(1), 83–86.

Yusof, N. (1990) Sprout inhibition by gamma-irradiation in fresh ginger. *Journal of Food Processing Preservation* 14(2), 113–122.

Zang, Z.-Q. *et al.* (1999) Selection of extrainers in extraction of paprika with supercritical carbon dioxide. *Transactions of the Chinese Society of Agricultural Engineering* 15(2), 208–212.

Zanzibar, (var). *Annual Reports of the Department of Agriculture.* Government Printer, Dar-es-Salaam, Tanzania.

Zareena, A.V. *et al.* (2001) Chemical investigation of gamma-irradiated saffron (*Crocus sativus*). *Journal of Agricultural and Food Chemistry* 49(2), 687–691.

Zawirska-Woitasiak, R. and Wasowicz, E. (2000) Enantiometric composition of limonene and carvone in seeds of dill and caraway. *Polish Journal of Food and Nutritional Science* 9(3), 9–13.

Zawirska-Woitasiak, R. *et al.* (1998) Aroma characteristics of dill seed varieties grown in Poland. *Polish Journal of Food and Nutritional Science* 7(2), 181–192.

Zewdie, Y. and Bosland, P.W. (2000) Capsaicinoid inheritance in an interspecific hybrid of *C. annuum* X *C. chinense. Journal of the American Society for Horticultural Science* 125(4), 448–453.

Zhao, G.-X. and Wei, Z.-X. (1999) An anatomical study on the stems of two species of vanilla. *Botanica Yunnanica* 21(1), 65–67.

Zhao, J. and Cranston, P.M. (1995) Microbial decontamination of black pepper by ozone and the effect of the treatment on volatile oil constituents of the spice. *Journal of the Science of Food and Agriculture* 68, 11–18.

Zheljazkov, V. and Jekov, D. (1996) Heavy metal content in some essential oils and plant extracts. *Acta Horticulturae* 426, 426, 427–433.

Zhu, L., Ding, D. and Lawrence, B.M. (1994) The *Cinnamomum* species in China. *Perfumer and Flavourist* 19(4), 17–22.

Zupancic, A. *et al.* (2001) The impact of fertilizing on fenugreek yield (*Trigonella foenumgraecum* L.) and diosgenin content in the plant drug. *Rostlinna Vyroba* 47(5), 218–224.

Zwaving, J.H. and Bos, R. (1992) Analysis of the essential oils of five *Curcuma* species. *Flavour and Fragrance Journal* 7, 19–22.

Glossary

Absolute. Highly concentrated, alcohol-soluble, usually liquid; normally obtained by solvent-extracting plant material, oils or **concretes**. Absolutes are considered most accurately to reflect the taste and odour of the original material.

Acetyl value. Number of milligrams of potassium hydroxide required to neutralize acetic acid liberated by hydrolysis when 1 g of the acetylated fat or oil is saponified. It measures the free hydroxyl groups.

Acid number (value). Number of milligrams of potassium hydroxide required to neutralize the free acid from 1 g of the substance.

Adulteration. Is normally used to designate materials added for fraudulent purposes. Legitimate adulteration is usually defined, i.e. fixative or preservative added.

Alcohols. Organic compounds containing one or more hydroxyl groups attached directly to carbon atoms. Primary alcohols give aldehydes on oxidation, secondary alcohols give ketones.

Alkaloids. Organic substances existing in combination with organic acids in great variety in plants. Alkaloids are usually derivatives of nitrogen ring compounds, and are frequently colourless, crystalline solids with a bitter taste. Often toxic to humans, although widely used in medicines.

Anthelmintic. A compound or drug used to eliminate intestinal worms.

Antioxidant. A substance that inhibits reactions, usually detrimental, due to oxygen or peroxides; commonly included in many retail products.

Antiseptic. A substance inhibiting the growth and development of microorganisms; usually applied to human pathogens.

Antispasmodic. Relieves spasms.

Aphrodisiac. Many spices or their derivatives have, at some time, been promoted as stimulating sexual activity, usually erroneously.

Aril. An expansion of the funicle arising from the placenta and enveloping the seed. Mace is the aril of nutmeg.

Aroma. An intangible concept described by Arctander (1960) as odour plus flavour.

Aromatherapy. Use of essential oils or creams to treat muscular and similar ills in humans; of little value other than to promote a sense of well-being.

Aromatic chemicals. Any chemical with an aroma or flavour, and usually synthetic. Not to be confused with the chemical definition of compounds containing a benzene ring structure.

Artefacts. Compounds produced during distillation and other processes, especially during storage of essential oils.

Artificial. Similar to imitation. A flavouring that may contain all natural ingredients but have no natural counterpart.

ASEAN. Trading group of South-east Asian countries: Indonesia, Malaysia, Myanmar, the Philippines, Singapore, Thailand and Vietnam.

Bark. The tissue external to the vascular cambium: the secondary phloem, cortex and periderm.

Biotechnology. Currently includes gene transfers and commonly known as genetic engineering.

Blender. Material added to another, or a combination, which generally enhances the odour or flavour.

Boiling point. Components of essential oils have varying boiling points: low-boiling-point compounds will distill-over first, those with higher boiling points progressively later. Control of distillate content is thus possible by varying distillation temperature or duration.

Brix number. Percentage of total soluble solids using a Brix hydrometer calibrated at 20°C.

Carminative. A compound which relieves flatulence.

Chemotaxonomy. Use of chemical analysis of plants or plant parts, especially their essential oils, to assist in classification. Is particularly useful where there are no (or minor) visual botanic differences.

Chemotype. Plants of the same species, which are botanically similar but whose seed composition, or essential oil, differs.

Cohobation. Return of all or part of the distillate to charge for further distilling. Usually occurs automatically.

Colour. From comparison with standard colour charts, or calorimeters; commonly obtained spectrophotometrically.

Compound. Frequently used to describe a group of plant constituents. Is ill-defined.

Concrete. Concentrated solvent-extracted solid or waxy material; its main use is as raw material for production of absolutes.

Conspecific. Belonging to the same species; sometimes used to denote doubt as to its correct identity.

Constituent. Specific component of a plant, seed or its oil, accurately defined.

Contamination. Previously a major problem affecting traditionally dried and processed spice seeds; usually microbial. The degree of permissable contamination is controlled by national legislation in producing and importing countries.

Cracking. Breaking whole seed into pieces to facilitate dehulling and flaking.

Cultivar. A further definition of a plant following the generic and species name and often preceded by the abbreviation cv. Usually applied to those specially bred or selected and not occurring in the wild; e.g. *Carum carvi* cv. *Karzo*. *See also* **Variety**.

DBH. Diameter at breast height. A standard measurement in forestry.

Decorticate. To remove shell or hard covering from seeds.

Diaphoretic. Promoting or inducing profuse sweating.

Diuretic. Increases the flow of urine.

Dispersal agent. Inert substance used as carrier for commercial oleoresins to facilitate handling and storage.

Distillation. A method of obtaining the essential oil of plants, usually by heating the material in water or by steam, and cooling the distillate.

Escape. Applied to introduced plants that escape from cultivation and become established locally.

Emmenagogue. A substance that induces menstruation.

Essence. A general and ambiguous term for a concentrated flavouring material.

Essential oil. The commonly used term for a volatile oil normally secreted in special glands in various plant parts and not found in living cells. Obtained by distillation, extraction or expression. Most essential oils are terpenoids, some benzene derivatives. Not to be confused with fixed or vegetable oils.

Esters. Organic compounds formed by union of an acid and an alcohol with elimination of water. Oil quality is often defined by ester content expressed as a percentage of the desired major component(s), linalool, linalyl acetate, etc., or as total esters.

Evapotranspiration. The total loss of water from the soil by evaporation and by transpiration from plants.

Extracts. Any concentrated material normally obtained by solvent-extracting natural raw materials. The term is widely misused.

Feral. Applied to plants that escape from cultivation and become locally established; *see* **Escape**. The term is more appropriately applied to animals.

Fixed oils. Non-volatile, fatty, generally vegetable, frequently contained in seeds. Usually obtained by pressure extraction.

Flow-point. On heating, a solid or semi-solid (usually an absolute) becomes soft and flows downward through a small orifice. The temperature at which the sample forms a hemi-spherical protuberance at the orifice is the flow-point. The temperature at which the first drop falls is the drop-point.

Folded oil. A concentrated oil produced by removing unwanted compounds from whole oil; commonly twofold, tenfold, etc. and considered to have that many times the strength of the desired component or odour.

FYM. Farmyard manure; usually containing animal wastes and straw, more or less decomposed.

Gas–liquid chromatography (GLC). A technique for the separation of the constituents of liquid or gaseous mixtures; in combination with mass spectroscopy it is a powerful tool for the qualitative and quantitative analysis of complex mixtures of chemical compounds including essential oils.

Glucosinolates. Sulphur compounds present in some seeds, which render the meal toxic to certain animals.

Headspace. The space in a container between the contents and the closure; in perfumery the volatile compounds evaporated by flowers and representing their true odour.

Herb. Any non-woody vascular plant. Also the green parts of plants used as food flavouring. See also definition in the Introduction.

Imitation. Concentrate, generally flavouring, containing all or some portion of non-natural materials. In many countries, such additives must be defined in consumer products, i.e. beverages, ice cream and sweets.

In vitro. Outside the living plant and in an artificial environment.

Inner bark. The secondary phloem, living tissue outside the cambium. Cinnamon quills are the inner bark.

Isolates. Obtained by fractionating oils. Isolates have specific uses and an essential oil may be used solely as a source of isolates.

Malesia. The biogeographical region including Malaysia, Indonesia, the Philippines, Singapore, Brunei and Papua New Guinea.

MERCOSUR. The South American Free Trade Area. Includes Brazil, Paraguay, Uruguay, with Bolivia as an associate.

Mucilage. A gelatinous substance similar to gum but which swells in water without dissolving and forms a slimy mass.

Mycorrhiza. A symbiotic association of roots with a fungal mycelium, which may form a layer outside the root (ectotrophic) or within the outer root tissue (endotrophic).

NAFTA. North American Free Trade Area. Includes Canada, Mexico and the USA.

Narcotic. A substance producing narcosis or stupor; in excess may cause stupor, coma, convulsions and death. Commonly applied to addictive drugs.

Native. Naturally occuring in a region or country.

Nutraceuticals. Foods or drinks containing ingredients, often spices and herbs, considered to promote human health or well-being. Also known as functional foods.

Naturalized. Introduced to a region or country and becoming established as a wild plant; *see also* **Feral** *or* **Escape**.

Odour. There is no accurate description for this term. It commonly attempts to explain what an individual perceives via their nose as the scent of a material. A highly esoteric language is used to describe odour by technicians and perfumers (see odour grouping as suggested by Arctander (1960), pp. 679–694).

Oleoresin. Can be a natural exudate (gum or resin) or solvent-extracted from plant materials. Prepared oleoresins are normally dark, very viscous to gummy or plastic semi-solids, and considered to be the most concentrated liquid form of the raw material.

Optical rotation. Angle through which the plane of polarization of light is rotated when polarized light passes through a layer of liquid. Unless otherwise specified, measurement is by sodium light in a 1 mm layer at 20°C. Oils may be dextrorotatory (+) or laevorotatory (−) according to whether the plane of polarization is rotated to the left or right.

Organic. Generally applied to manures composed of plant or animal residues or by-products. Also means to be grown without the use of agrochemicals or chemical fertilizers.

Organoleptic. The total sensory impression in the mouth, nose and throat of a spice, including aroma, flavour, pungency, bitterness, etc.

Outer bark. The non-living layer of fibrous or corky tissue outside the cambium in woody plants, which may be shed or retained; the periderm or rhytidome.

Oxidization. A general term indicating the changes in composition or odour when exposed to air; most are undesirable. Thus, powdered spices or the essential oil should be stored in full airtight containers.

Melting point. A common method of determining melting point is to place a small portion of the solid (usually an absolute) in a 1 mm glass capillary tube closed at the base and suspended in a water bath with a thermometer. The temperature at which the sample becomes completely clear is the melting point (cf. **Flow-point**).

Photosensitive. Reacts to light. Many spice powders and essential oils are affected by light, resulting in decomposition or undesirable reactions among constituents.

Pungency. The hot sensation experienced when eating certain spices, including chillies and ginger. Measured in **Scoville heat units**.

Rectification. A second distillation of an essential oil to remove unwanted components or solids.

Refractive index. Ratio of the velocity of light in a vacuum to its velocity in a substance. It varies with the wavelength of light used.

Resinoid. Prepared by solvent-extracting exudates, highly lignified plant material, or animal substances. Commonly but incorrectly used when describing the physical condition of absolutes.

Rhizome. The underground rootstock of plants that produces roots, stems and leaves. Dried rhizomes provide the familiar spices ginger and turmeric.

Scoville heat units. A method of measuring pungency of spices. Hot is about 30,000 units; the hottest chilli pepper currently discovered reportedly rates a fiery 855,000 units!

Sesquiterpenes. *See* **Terpenoids**.

Solubility. Maximum quantity of an oil that can be dissolved in a stated volume (v/v) or weight (w/w) of alcohol of given concentration, e.g. 70%, 90%, etc., normally at 20°C, or as specified.

Solvent. Hexane, petroleum ether, acetone or methanol used to extract essential oil or other derivatives from plant materials or their extracts. There are legal requirements in many countries regarding their use and residues in products for human consumption.

Specific gravity. Ratio of the weight of a volume of material to the weight of an equal volume of water at 4°C.

Spice. See definition in the Introduction.

Stockfeed. Specially formulated foods for consumption by domestic animals, birds and fish, usually kept under intensive conditions.

Supercritical fluid extraction. Method of extraction using a liquid gas, frequently carbon dioxide.

Subspecies. Subdivision of a species, in rank between a variety and a species; usually written subsp.

Synthetics. Mixtures or formulations, usually of synthetic chemicals but sometimes also containing natural products, which simulate to some degree the original product; most are only partially successful.

Terpenoids. Volatile aromatic hydrocarbons, including monoterpenes, sesquiterpenes, diterpenes and higher polymers. A wider use of the term would include terpene hydrocarbons, alcohols, ketones and camphors. Terpenes are of great importance in essential oil composition and their presence or absence has a marked effect on odour and usage. Terpenoid distribution is now considered of major value in plant classification.

Tocopherols. Methylated derivatives of tocol, which are widely distributed in vegetable lipids. They have vitamin E activity and can protect unsaturated fats against oxidation.

Tonne (t). Equal to 1000 kg (2205 lb).

Ultrasonic extraction. Method of obtaining plant aromatic materials using very-high-frequency vibrations; is specialized and uncommon.

Unsaponifiable matter. All constituents not saponified by alcoholic caustic potash, but soluble in petroleum ether or ethyl ether.

Vacuum distillation. Distillation in equipment from which most air has been removed; this reduces the pressure acting on the material to be distilled, with the result that it will boil and distil at a lower temperature. It is used to distil liquids containing compounds that decompose at high temperature.

Variety. A botanical variety, which is a subdivision of a species; an agricultural or horticultural variety is referred to as a **Cultivar.**

Viscosity. Rate of flow of a liquid. Measured in cgs units. The absolute unit of viscosity is the poise. Absolute viscosity of water at 20.2°C is 1000 centipoise (one hundredth of a poise).

Volatile oil. *See* **Essential oil.**

Wax. A mixture of esters of higher alcohols and higher fatty acids. Plant waxes are removed from **concretes** to produce **absolutes.**

Weight per millilitre. Weight in grams of 1 ml of a liquid weighed in air at the specified temperature.

Whole oil. Specifically an oil free of all impurities, water and additives and true to the original material. The term is often misused.

WONF. With other natural flavours. An extract or flavour too weak to be used alone, and reinforced with other natural flavouring material containing similar or complementary components. This identification is a legal requirement in certain countries including the USA.

Appendix A

European Spice Association Specifications of Quality Minima for Herbs and Spices

Subject	Specification
Extraneous matter	Herbs 2%, spices 1%.
Sampling	(For routine sampling.) Square root of units/lots to a maximum of 10 samples. (For arbitration purposes.) Square root of all containers. e.g. 1 lot of pepper may = 400 bags, therefore square root = 20 samples.
Foreign matter	Maximum 2%.
Ash	Refer to Annex.*
Acid-insoluble ash	Refer to Annex.
H_2O	Refer to Annex.
Packaging	Should be agreed between buyer and seller. If made of jute and sisal, they should conform to the standards set by CAOBISCO Ref. C502-51 -sj of 20-02-95 (see Annex II). However, these materials are not favoured by the industry, as they are a source of product contamination, with loose fibres from the sacking entering the product.
Heavy metals	Shall comply with national/EU legislation.
Pesticides	Shall be utilized in accordance with manufacturers' recommendations and good agricultural practice and comply with existing national and/or EU legislation.
Treatments	Use of any EU-approved fumigants in accordance with manufacturers' instructions, to be indicated on accompanying documents. (Irradiation should not be used unless agreed between buyer and seller.)
Microbiology	*Salmonella* absent in (at least) 25 g. Yeast and moulds 10^5 g^{-1} target, 10^6 g^{-1} absolute maximum. *E. coli.* 10^2 g^{-1} target, 10^3 g^{-1} absolute maximum. Other requirements to be agreed between buyer and seller.
Off odours	Shall be free from off odour or taste.
Infestation	Should be free in practical terms from live and/or dead insects, insect fragments and rodent contamination visible to the naked eye (corrected if necessary for abnormal vision).
Aflatoxins	Should be grown, harvested, handled and stored in such a manner as to prevent the occurrence of aflatoxins or minimize the risk of occurrence. If found, levels should comply with existing national and/or EU legislation.
Volatile oil	Refer to Glossary.
Adulteraflon	Shall be free from.
Bulk density	To be agreed between buyer and seller.
Species	To be agreed between buyer and seller.
Documents	Should provide: details of any treatments the product has undergone; name of product; weight; country of origin; lot identification/batch number; year of harvest.

*Annex not included

Appendix B

World Spice Harvest Calendar

Spice	Origin	\[Months (January–December)\]											
		1	2	3	4	5	6	7	8	9	10	11	12
Allspice	Jamaica								×	×			
(Pimento)	Guatemala							×	×				
	Mexico										×	×	×
Aniseed	China									×			
	Egypt						×	×					
	India		×	×	×								
	Morocco								×	×	×		
	Spain								×	×	×		
	Syria				×	×							
Basil	Europe					×	×						
Cardamom	Guatemala									×	×		
	India	×							×	×	×	×	×
	Tanzania	×	×					×	×	×	×	×	×
Caraway	The Netherlands							×	×				
Cassia	China						×	×					
	India						×	×				×	×
	Indonesia						×	×	×				
	Vietnam						×	×					
Celery seed	France					×	×						
	India						×	×					
Chillis	India	×	×	×	×	×	×	×	×	×	×	×	×
	Morocco	×	×	×	×	×	×					×	×
	Sri Lanka							×	×				
	Tanzania							×	×				
Cinnamon	Madagascar	×	×	×	×	×	×	×	×	×	×	×	×
	Seychelles	×	×	×	×	×	×	×	×	×	×	×	×
	Sri Lanka						×	×					
Cloves	Indonesia					×	×	×	×				
	Madagascar										×	×	×
	Sri Lanka	×	×										×
	Tanzania			×	×						×	×	

Spice	Origin	Months (January–December)											
		1	2	3	4	5	6	7	8	9	10	11	12
Laurel (bay leaves)	Greece								×	×			
	Morocco							×	×				
	Portugal							×	×				
	Turkey								×	×			
	Yugoslavia								×	×			
Pepper (black)	Brazil								×	×			
	India	×	×	×									
	Indonesia								×	×	×		
	Madagascar						×	×	×	×	×		
	Malaysia						×	×	×				
	Sri Lanka										×	×	×
Pepper (white)	Brazil								×	×			
	Indonesia									×	×	×	
	Malaysia						×	×	×				
Rosemary	Europe					×	×						
Saffron	Spain											×	×
Sage	Europe									×	×		
Savoury	Europe						×						
Tarragon	Europe							×	×				
Thyme	France					×	×						
	Morocco						×	×					
Turmeric	India		×	×	×	×							
Vanilla	Madagascar					×	×	×	×				

Index